Lecture Notes in Computer Science 6111

Commenced Publication in 1973
Founding and Former Series Editors:
Gerhard Goos, Juris Hartmanis, and Jan van Leeuwen

Aurélio Campilho Mohamed Kamel (Eds.)

Image Analysis and Recognition

7th International Conference, ICIAR 2010
Póvoa de Varzim, Portugal, June 21-23, 2010
Proceedings, Part I

Volume Editors

Aurélio Campilho
University of Porto, Faculty of Engineering
Institute of Biomedical Engineering
4200-465 Porto, Portugal
E-mail: campilho@fe.up.pt

Mohamed Kamel
University of Waterloo, Department of Electrical
and Computer Engineering
Waterloo, Ontario, N2L 3G1, Canada
E-mail: mkamel@uwaterloo.ca

Library of Congress Control Number: 2010928206

CR Subject Classification (1998): I.4, I.5, I.2.10, I.2, I.3.5, F.2.2

LNCS Sublibrary: SL 6 – Image Processing, Computer Vision, Pattern Recognition, and Graphics

ISSN 0302-9743
ISBN-10 3-642-13771-7 Springer Berlin Heidelberg New York
ISBN-13 978-3-642-13771-6 Springer Berlin Heidelberg New York

springer.com

Printed in Germany

Typesetting: Camera-ready by author, data conversion by Scientific Publishing Services, Chennai, India
Printed on acid-free paper 06/3180

Preface

ICIAR 2010, the International Conference on Image Analysis and Recognition, held in Póvoa do Varzim, Portugal, June 21-23, was seventh in the ICIAR series of annual conferences alternating between Europe and North America. The idea of organizing these conferences was to foster the collaboration and exchange between researchers and scientists in the broad fields of image analysis and pattern recognition, addressing recent advances in theory, methodology and applications. During the years the conferences have become a forum with a strong participation from many countries. This year, ICIAR was organized along with AIS 2010, the International Conference on Autonomous and Intelligent Systems. Both conferences were organized by AIMI—Association for Image and Machine Intelligence.

For ICIAR 2010, we received a total of 164 full papers from 37 countries. The review process was carried out by members of the Program Committee and other reviewers; all are experts in various image analysis and pattern recognition areas. Each paper was reviewed by at least two reviewers, and checked by the Conference Chairs. A total of 89 papers were finally accepted and appear in the two volumes of these proceedings. The high quality of the papers is attributed first to the authors, and second to the quality of the reviews provided by the experts. We would like to sincerely thank the authors for responding to our call, and to thank the reviewers for their careful evaluation and feedback provided to the authors. It is this collective effort that resulted in the strong conference program and high-quality proceedings.

This year included a competition on "Fingerprint Singular Points Detection" and a challenge on "Arabidopsis Thaliana Root Cell Segmentation Challenge," which attracted the attention of ICIAR participants.

We were very pleased to be able to include in the conference program keynote talks by three well-known experts: Alberto Sanfeliu, Universitat Politècnica de Catalunya, Spain; Edwin Hancock University of York, UK and José Santos-Victor, Institute for Systems and Robotics, Instituto Superior Técnico, Portugal. We would like to express our sincere gratitude to the keynote speakers for accepting our invitation to share their vision and recent advances in their specialized areas.

We would like to thank Khaled Hammouda, the webmaster of the conference, for maintaining the Website, interacting with the authors and preparing the proceedings. Special thanks are also due to the members of the local Organizing Committee for their advice and help. We are also grateful to Springer's editorial staff, for supporting this publication in the LNCS series. We would like to acknowledge the professional service of Viagens Abreu in taking care of the registration process and the special events of the conference.

Finally, we were very pleased to welcome all the participants to ICIAR 2010. For those who did not attend, we hope this publication provides a good view into the research presented at the conference, and we look forward to meeting you at the next ICIAR conference.

June 2010

Aurélio Campilho
Mohamed Kamel

ICIAR 2010 – International Conference on Image Analysis and Recognition

General Chair

Aurélio Campilho
University of Porto, Portugal
campilho@fe.up.pt

General Co-chair

Mohamed Kamel
University of Waterloo, Canada
mkamel@uwaterloo.ca

Local Organizing Committee

Ana Maria Mendonça
University of Porto
Portugal
amendon@fe.up.pt

Jorge Alves Silva
University of Porto
Portugal
jsilva@fe.up.pt

António Pimenta Monteiro
University of Porto
Portugal
apm@fe.up.pt

Pedro Quelhas
Biomedical Engineering Institute
Portugal

Gabriela Afonso
Biomedical Engineering Institute
Portugal
iciar10@fe.up.pt

Conference Secretariat

Viagens Abreu SA
Porto, Portugal
congresses.porto@viagensabreu.pt

Webmaster

Khaled Hammouda
Waterloo, Ontario, Canada
hammouda@pami.uwaterloo.ca

Advisory Committee

M. Ahmadi	University of Windsor, Canada
P. Bhattacharya	Concordia University, Canada
T.D. Bui	Concordia University, Canada
M. Cheriet	University of Quebec, Canada
E. Dubois	University of Ottawa, Canada
Z. Duric	George Mason University, USA
G. Granlund	Linköping University, Sweden
L. Guan	Ryerson University, Canada
M. Haindl	Institute of Information Theory and Automation, Czech Republic
E. Hancock	The University of York, UK
J. Kovacevic	Carnegie Mellon University, USA
M. Kunt	Swiss Federal Institute of Technology (EPFL), Switzerland
J. Padilha	University of Porto, Portugal
K.N. Plataniotis	University of Toronto, Canada
A. Sanfeliu	Technical University of Catalonia, Spain
M. Shah	University of Central Florida, USA
M. Sid-Ahmed	University of Windsor, Canada
C.Y. Suen	Concordia University, Canada
A.N. Venetsanopoulos	University of Toronto, Canada
M. Viergever	University of Utrecht, The Netherlands
B. Vijayakumar	Carnegie Mellon University, USA
J. Villanueva	Autonomous University of Barcelona, Spain
R. Ward	University of British Columbia, Canada
D. Zhang	The Hong Kong Polytechnic University, Hong Kong

Program Committee

A. Abate	University of Salerno, Italy
P. Aguiar	Institute for Systems and Robotics, Portugal
M. Ahmed	Wilfrid Laurier University, Canada
N. Alajlan	King Saud University, Saudi Arabia
J. Alirezaie	Ryerson University, Canada
H. Araújo	University of Coimbra, Portugal
N. Arica	Turkish Naval Academy, Turkey
J. Barbosa	University of Porto, Portugal
J. Barron	University of Western Ontario, Canada
J. Batista	University of Coimbra, Portugal
C. Bauckhage	York University, Canada
A. Bernardino	Technical University of Lisbon, Portugal
G. Bilodeau	École Polytechnique de Montréal, Canada
J. Bioucas	Technical University of Lisbon, Portugal

B. Boufama	University of Windsor, Canada
T.D. Bui	Concordia University, Canada
J. Cardoso	University of Porto, Portugal
E. Cernadas	University of Vigo, Spain
F. Cheriet	École Polytechnique de Montréal, Canada
M. Cheriet	University of Quebec, Canada
M. Coimbra	University of Porto, Portugal
M. Correia	University of Porto, Portugal
L. Corte-Real	University of Porto, Portugal
J. Costeira	Technical University of Lisbon, Portugal
A. Dawoud	University of South Alabama, USA
M. De Gregorio	Istituto di Cibernetica "E. Caianiello" - CNR, Italy
Z. Duric	George Mason University, USA
N. El Gayar	Nile University, Egypt
M. El-Sakka	University of Western Ontario, Canada
P. Fieguth	University of Waterloo, Canada
M. Figueiredo	Technical University of Lisbon, Portugal
G. Freeman	University of Waterloo, Canada
V. Grau	University of Oxford, UK
M. Greenspan	Queen's University, Canada
L. Guan	Ryerson University, Canada
F. Guibault	École Polytechnique de Montréal, Canada
M. Haindl	Institute of Information Theory and Automation, Czech Republic
E. Hancock	University of York, UK
C. Hong	Hong Kong Polytechnic, Hong Kong
K. Huang	Chinese Academy of Sciences, China
J. Jiang	University of Bradford, UK
B. Kamel	University of Sidi Bel Abbès, Algeria
G. Khan	Ryerson University, Canada
M. Khan	Saudi Arabia
Y. Kita	National Institute AIST, Japan
A. Kong	Nanyang Technological University, Singapore
M. Kyan	Ryerson University, Canada
J. Laaksonen	Helsinki University of Technology, Finland
Q. Li	Western Kentucky University, USA
X. Li	University of London, UK
R. Lins	Universidade Federal de Pernambuco, Brazil
J. Lorenzo-Ginori	Universidad Central "Marta Abreu" de Las Villas, Cuba
G. Lu	Harbin Institute, China
R. Lukac	University of Toronto, Canada
A. Mansouri	Université de Bourgogne, France
A. Marçal	University of Porto, Portugal
J. Marques	Technical University of Lisbon, Portugal

M. Melkemi	Univeriste de Haute Alsace, France
A. Mendonça	University of Porto, Portugal
J. Meunier	University of Montreal, Canada
M. Mignotte	University of Montreal, Canada
A. Monteiro	University of Porto, Portugal
M. Nappi	University of Salerno, Italy
A. Padilha	University of Porto, Portugal
F. Perales	University of the Balearic Islands, Spain
F. Pereira	Technical University of Lisbon, Portugal
E. Petrakis	Technical University of Crete, Greece
P. Pina	Technical University of Lisbon, Portugal
A. Pinho	University of Aveiro, Portugal
J. Pinto	Technical University of Lisbon, Portugal
F. Pla	Universitat Jaume I, Spain
P. Quelhas	Biomedical Engineering Institute, Portugal
M. Queluz	Technical University of Lisbon, Portugal
P. Radeva	Autonomous University of Barcelona, Spain
B. Raducanu	Autonomous University of Barcelona, Spain
S. Rahnamayan	University of Ontario Institute of Technology (UOIT), Canada
E. Ribeiro	Florida Institute of Technology, USA
J. Sanches	Technical University of Lisbon, Portugal
J. Sánchez	University of Las Palmas de Gran Canaria, Spain
B. Santos	University of Aveiro, Portugal
A. Sappa	Computer Vision Center, Spain
G. Schaefer	Nottingham Trent University, UK
P. Scheunders	University of Antwerp, Belgium
J. Sequeira	Ecole Supérieure d'Ingénieurs de Luminy, France
J. Shen	Singapore Management University, Singapore
J. Silva	University of Porto, Portugal
B. Smolka	Silesian University of Technology, Poland
M. Song	Hong Kong Polytechnical University, Hong Kong
J. Sousa	Technical University of Lisbon, Portugal
H. Suesse	Friedrich Schiller University Jena, Germany
S. Sural	Indian Institute of Technology, India
S. Suthaharan	USA
A. Taboada-Crispí	Universidad Central "Marta Abreu" de las Villas, Cuba
M. Vento	University of Salerno, Italy
J. Vitria	Computer Vision Center, Spain
Y. Voisin	Université de Bourgogne, France
E. Vrscay	University of Waterloo, Canada
L. Wang	University of Melbourne, Australia

Z. Wang	University of Waterloo, Canada
M. Wirth	University of Guelph, Canada
J. Wu	University of Windsor, Canada
F. Yarman-Vural	Middle East Technical University, Turkey
J. Zelek	University of Waterloo, Canada
L. Zhang	The Hong Kong Polytechnic University, Hong Kong
L. Zhang	Wuhan University, China
G. Zheng	University of Bern, Switzerland
H. Zhou	Queen Mary College, UK
D. Ziou	University of Sherbrooke, Canada

Reviewers

A. Abdel-Dayem	Laurentian University, Canada
D. Frejlichowski	West Pomeranian University of Technology, Poland
A. Mohebi	University of Waterloo, Canada
Y. Ou	University of Pennsylvania, USA
R. Rocha	Biomedical Engineering Institute, Portugal
F. Sahba	University of Toronto, Canada

Supported by

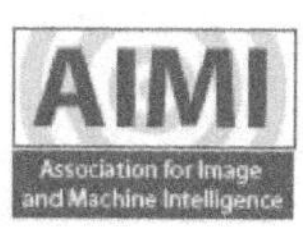

AIMI – Association for Image and Machine Intelligence

Department of Electrical and Computer Engineering
Faculty of Engineering
University of Porto
Portugal

INEB – Instituto de Engenharia Biomédica
Portugal

PAMI – Pattern Analysis and Machine Intelligence Group
University of Waterloo
Canada

Table of Contents – Part I

Image Morphology, Enhancement and Restoration

Image Segmentation

Feature Extraction and Pattern Recognition

Computer Vision

Shape, Texture and Motion Analysis

Coding, Indexing and Retrieval

Face Detection and Recognition

Table of Contents – Part II

Biomedical Image Analysis

Biometrics

Applications

PageRank Image Denoising

Panganai Gomo*

EECE, University of Birmingham, B15 2TT, United Kingdom
panganai.gomo@gmail.com

Abstract. We present a novel probabilistic algorithm for image noise removal. The algorithm is inspired by the Google PageRank algorithm for ranking hypertextual world wide web documents and based upon considering the topological structure of the photometric similarity between image pixels. We provide computationally efficient strategies for obtaining a solution using the conjugate gradient algorithm. Comparisons with other state-of-art denoising filters, namely the total variation minimising filter and the bilateral filter, are made.

Keywords: Image Denoising, PageRank, Markov Processes, Diffusion Maps, Spectral Graph Theory, Total Variation Minimisation, Bilateral Filter.

1 Introduction

The problem of image denoising is well established and studied within the literature. In this paper we propose a new image denoising algorithm. Inspired by the Google PageRank algorithm [6,23] which assigns rank to web pages based upon a Markov process encoding the link structure of the world wide web we propose an iterative algorithm that assigns image intensities according to a Markov process encoding the topological structure of an image. Due to the distribution of the eigenvalues of the Markov process we propose an efficient solution using the conjugate gradient algorithm of Hestenes and Stiefel [17]. We compare the performance of this denoising algorithm to the closely related state-of-art methods; total variation minimisation and bilateral filters.

1.1 Previous Work

A typical formulation is to assume the noise is generated by a zero mean Gaussian processes allowing the decomposition

$$f = u + \eta, \tag{1}$$

where f is the observed signal, u is the original signal and η is a zero mean Gaussian process with finite variance. A powerful regularisation framework was proposed in [25]. This framework is known as the nonlinear total variation noise

* I would like to thank Dr. Mike Spann for suggestions and comments.

A. Campilho and M. Kamel (Eds.): ICIAR 2010, Part I, LNCS 6111, pp. 1–10, 2010.

removal algorithm. The minimisation of their regularisation problem leads to a nonlinear anisotropic partial differential equation diffusion filter. This algorithm smooths over subsets of pixels considered similar whilst avoiding blurring across edges hence achieving the desired effect of a structural image denoising algorithm. This algorithm belongs to a class of nonlinear partial differential equation diffusion filters. Much literature has been devoted to the study of such filters and can be found in references such as [24,30,9,5]. Related filters have been developed that seek to average image intensities by means of a nonlinear combination of nearby image values. The measure of similarity and weighting of pixels is based upon geometric closeness and photometric similarity. Such filters are exemplified by the bilateral filter [27] and the non-local means filter [7,8].

More recently image denoising algorithms are being developed that diffuse image intensities based upon models that capture the topological structure of images. The topological structure of the image is modeled by photometric and geometric similarities within the image. Algorithms based upon graph formulations have appeared in the form of normalized cuts [26] for image segmentation, heat kernel smoothing for image denoising [31] and random walks and Markov processes for image denoising [2,15]. A common idea these methods share is that a priori beliefs about the correlations or similarities between pixels are captured in a kernel matrix. This matrix reflects the graph adjacency structure and can be studied by forming the graph Laplacian [12].

Graph like data structures are ubiquitous in science; one obvious area being the hyperlink structure of the world wide web. This is a graph of documents related to one another by hyperlinks. A successful algorithm, known as PageRank, for exploiting the hyperlink structure of the world wide web in order to rank web page documents, giving a measure of authority or influence of web pages, was proposed in [6,23]. Much literature has been devoted to the study of this algorithm with a survey of the literature to be found in [19].

2 Diffusion Models

The model input space is the discrete image vector, that is the image ordered as a one dimensional vector $\mathbf{f}$, so an $n \times n$ matrix becomes an $N \times 1$ vector where $N = n \times n$. As we will be dealing with stochastic matrices we normalise the vectors in L_1 norm such that

$$\mathbf{f} \leftarrow \frac{\mathbf{f}}{\|\mathbf{f}\|_1} \tag{2}$$

where $\|\mathbf{f}\|_1 = \sum_{i=1}^{N} |f|$. We choose $\mathbf{u}_0$ such that $\{u_i = \frac{1}{N} : \text{ for } i = 1, \ldots, N\}$ and $\|\mathbf{u}_0\|_1 = 1$. The construction of the graph is motivated by the considerations that pixels within a neighbourhood are likely to have been generated by the same process. We want the weights used to model the topological structure of the image to capture the photometric similarities. The typical Gaussian kernel is used to model the photometric similarity within a neighbourhood. We only make comparisons in an 8−connected neighbourhood for computational speed and pixels outside this region are considered geometrically far. We form the weight matrix W with the entries given by the following

$$W = \begin{cases} W_{ij} = w(x_i, x_j) = e^{\frac{d(f(x_i), f(x_j))^2}{\sigma^2}} & i, j \text{ in the 8-connected region} \\ W_{ij} = 0 & \text{otherwise} \end{cases} \quad (3)$$

This gaussian similarity kernel is standard in image processing [26,7,13,14]. This measures is positive preserving $w(x_i, x_j) \geq 0$ and symmetric $w(x_i, x_j) = w(x_j, x_i)$. The measure $d(f(x_i), f(x_j))$[1] models the photometric similarities between image intensities. In the experiments in this paper we only consider the Euclidean distance between the intensities. The resulting matrix W is symmetric and positive definite. The degree matrix can be defined as

$$D = \begin{cases} D_{ij} = d_i = \sum_j W_{ij} & \text{for } i = j \\ 0 & \text{otherwise} \end{cases} \quad (4)$$

In graph theory normalisation of the weight matrix can be carried out in a number of ways [12]. We recall the normalised weight matrix M_n is the matrix whose entries are determined by

$$M_{n,ij} = \frac{W_{ij}}{\sqrt{d_i}\sqrt{d_j}} \quad (5)$$

and for the random walker kernel matrix M_{rw}

$$M_{rw,ij} = \frac{W_{ij}}{d_i}. \quad (6)$$

From the kernel[2] matrices we can define various graph Laplacian operators as in table (1). In image processing the normalised Laplacian has been applied to problems in image segmentation [26] and image denoising [31]. The matrix M_{rw} defines a random walk on a graph [20]. It exhibits the Markov property [21]. We can study this matrix in order to derive algorithms for diffusion models on images. We note this matrix contains the geometric information contained in the signal **f**. The weightings in this kernel matrix directly model the local

Table 1. Definition of Graph Laplacian Operators

Laplacian	Matrix
Unormalised Laplacian	$L\text{un} = D - W$
Normalised Laplacian	$L\text{n} = I - D^{\frac{-1}{2}} W D^{\frac{-1}{2}} = I - M\text{n}$
Random Walker	$L\text{rw} = I - D^{-1} W = I - M\text{rw}$

[1] One should not confuse this notation $d(\cdot, \cdot)$ with that for the degree of a node d_i.

[2] It is worth noting that the construction of this model is related to the Markov random field formulations where a Gibbs prior is used to model the prior probability distribution of the image [3,16].

geometry defined by the immediate neighbours of each node. Indeed M_{rw} defines the Markov matrix where its entries are transition probabilities. The entries of the matrix

$$M_{\mathrm{rw},ij} = p(\mathbf{x}(t + \Delta t) = \mathbf{x}_j | \mathbf{x}(t) = \mathbf{x}_i).$$

represent the probability of transition in one time step from node x_i to x_j. As in [21] we take $\Delta t = \sigma$. The parameter σ has two interpretations. First it is the radius of the domain of influence used to determine the graph structure. Second it is the time stepping parameter at which a random walker jumps from state to state. We also introduce a parameter $\alpha = 1 - \Delta t$ and hence Δt is in the bound $[0, 1]$. We call the parameter α the mixing probability.

To motivate an algorithm to denoise a signal we may consider running Markov matrix forward in time on the initial data. That is we generate

$$\mathbf{u}^{n+1} = M_{rw}\mathbf{u}^n.$$

Equivalently we could apply powers of the Matrix M_{rw} on the initial data

$$\mathbf{u} = M^k \mathbf{u}^0.$$

The powers of M_{rw} allow us to integrate the local geometry of $\mathbf{f}$ and hence reveal the relevant structure in the signal. We diffuse the original data through the structure of M_{rw}. From a random walk point of view the notion of a cluster can be viewed as a region of the graph for which the probability of escaping is low. So we can recognise the powers of M_{rw} are of prime interest. In fact it is the powers of Markov matrices that make Monte Carlo methods such as Metropolis-Hastings algorithm and simulated annealing successful [11,1]. Indeed such techniques have been applied to image denoising [15]. Unfortunately Markov Chain Monte Carlo techniques can be slow and may require many iterations until convergence. A classical approach to view the action of the powers of Markov matrix operators on a vector is through spectral analysis. We refer the reader to [28] for an introduction to this subject. From this theory of Markov chains as $k \to \infty$ we obtain a stationary distribution

$$\mathbf{p} = M^{\infty}\mathbf{u}^0 \tag{7}$$

where

$$\mathbf{p} = M\mathbf{p}. \tag{8}$$

It is not difficult to show that the stationary distribution is given by

$$p_i = \frac{D_{ii}}{\sum_j D_{jj}}. \tag{9}$$

which is the transpose of the left eigenvector of M_{rw}. This eigenvector therefore contains no useful information for denoising an image as the image would be equalised as a uniform distribution. We may do better by seeking an eigenvectors associated with other eigenvalues. In conclusion of prime interest in developing our denoising algorithms are the action powers of the Matrix M_{rw} on the initial data and it's eigendecomposition.

2.1 PageRank Denoising

Consider the dynamics

$$\mathbf{u}^n = [\alpha U + (1-\alpha)M_{rw}]^n \mathbf{u}^0 \tag{10}$$

where $A = [\alpha U + (1-\alpha)M_{rw}]$ is known as the PageRank matrix. U is a uniform $N \times N$matrix with entries given by

$$U = \begin{bmatrix} \frac{1}{N} & \frac{1}{N} & \cdots & \frac{1}{N} \\ \frac{1}{N} & \frac{1}{N} & \cdots & \frac{1}{N} \\ \vdots & \vdots & \ddots & \vdots \\ \frac{1}{N} & \frac{1}{N} & \cdots & \frac{1}{N} \end{bmatrix} \tag{11}$$

and $U\mathbf{u}^n = \mathbf{e}$ where $\mathbf{e}$ is the uniform $N \times 1$ vector with entries given by $\mathbf{e} = [\frac{1}{N}, \frac{1}{N}, \ldots, \frac{1}{N}]^T$. Applying this iteration to our data produces a weighted combination of a uniform signal $\alpha\mathbf{e}$ and the local variation or detail in the image $(1-\alpha)M_{rw}\mathbf{u}^n$. The parameter α is the probability of mixing the image intensities at each iteration. Varying this parameter produces different quality in the denoising of the image. In the theory of web page ranking the matrix M_{rw} would model and capture the link structure of the world wide web. One then imagines a random web surfer who at each time step is at a web page deciding to follow a web page at the next time step according to the decision: with probability α they rest by jumping to a web page uniformly and at random with probability $(1-\alpha)$ jumps to one of the hyperlinks on the web page [22]. In this setting α is termed the teleportation probability and chosen within the range 0.1 to 0.2 with a typical choice of 0.15. Convergence of the algorithm results in a stationary distribution which corresponds to the ranking of the web pages.

2.2 Implementation

As the matrix A is a Markov matrix there is strong theory to suggest the existence and convergence to the stationary distribution [18]. By simple algebra the PageRank dynamics of equation (10) can be rewritten as

$$\mathbf{u}^{n+1} = \alpha\mathbf{f} + (1-\alpha)M_{rw}\mathbf{u}^n. \tag{12}$$

We prefer this formulation as we do not explicitly form and store the dense matrix U. Secondly this iteration corresponds to what is known as personalised PageRank [19]. From Markov theory if a stationary distribution exists then

$$\mathbf{u}^* = \alpha\mathbf{f} + (1-\alpha)M_{rw}\mathbf{u}^*. \tag{13}$$

where $\mathbf{u}^*$ is satisfied. Using simple algebra and recalling our definition $\alpha = 1 - \Delta t$ we can show that the stationary distribution can be found by solving the linear system

$$\mathbf{u}^* = (1-\Delta t)(I - \Delta t M_{rw})^{-1}\mathbf{f}. \tag{14}$$

3 Evaluation and Comparison

Here we evaluate computational performance, investigate denoising performance with respect to the mixing probability α and make comparisons with related state-of-the-art algorithms; namely the total variation minimisation filter (TV) and the bilateral filter (BF). To facilitate this study we use the structural similarity index (SSIM) [29]. This index uses a measure of similarity based upon the luminance, contrast and structure of the signal. A score is generated in the bound $[0, 1]$, where one indicates absolute fidelity between the signal and target. The SSIM index works on the hypothesis that the human visual system is highly adapted to extract structural information. As our algorithm is based upon smoothing whilst preserving topological structure this measure is better suited, than the mean square error (MSE) and peak signal-to-noise ratio (PSNR), to evaluate the performance of the algorithm as it provides an objective measure of structural integrity. Our algorithm can be computed in two ways. Firstly as the stationary distribution of the Markov chain corresponding to the matrix A. This leads to the power method for solution of the linear system. The convergence rate of the power method is given by the ration $\frac{|\lambda_2|}{|\lambda_1|}$ [17] and we know that $|\lambda_2| \leq (1-\alpha)$ [18] therefore rate of convergence is contingent on this parameter. Secondly we can compute the solution of the linear system, As this system is symmetric positive definite the ideal candidate for the linear solver is the conjugate gradient (CG) method [17]. Moreover the eigenvalues of the matrix $I - \Delta t M_{rw}$ are well clustered around 1 implying convergence is guaranteed [10]. We compare the convergence of the two methods and the rates of convergence with respect to the mixing probability. Table (2) shows the computational requirements of these algorithms.

Table 2. Computational requirements. Operations per iteration: IP (inner product) counts, SAXPY counts AXPY operations per iteration, SpMV (sparse matrix vector) multipliers per iterations, and storage counts the number of matrices and vectors required for the method. We use BLAS parlance [4].

Method	IP	SAXPY	SpMV	Storage
Power		1	1	Sparse Matrix + $2N$
CG	2	3	1	Sparse Matrix + $6N$

3.1 Experiments and Results

The first set of experiments investigate the relationship between the noise variance, parameter α and the number of iterations required till algorithm convergence. We compute the optimal[3] α for denoising an image corrupted with zero mean Gaussian noise with standard deviaiton in the range $[0.005 \ldots 0.1]$

[3] The optimality is the parameter value required to maximise the SSIM measure when comparing the denoised image to the original uncorrupted image.

allowing us to investigate these relationships. These experiments reveal that α decreases with increasing noise variance and and the conjugate gradient algorithm outperforms the power method. The power method is terminated when $\|\mathbf{u}^{n+1} - \mathbf{u}^n\|_2 \leq 1 \times 10^{-8}$ and the conjugate gradient method is terminated when $\|(I - \Delta t M_{rw})\mathbf{u} - (I - \Delta t)\mathbf{f}\|_2 \leq 1 \times 10^{-8}$ The results are presented in figure (1). The second set of experiments make comparisons of the denoising performance of the PageRank (PR) denoising algorithm against the total variation minimisation filter and the bilateral filter. We compare the performance for an image corrupted with Gaussian white noise with zero mean. The signal variance is varied and performance is measured by generating the SSIM score for optimal algorithm parameters. The PageRank denoising algorithm significantly outperforms the other algorithms in low noise environments with competitive performance in higher noise environments. The figures show that the PageRank

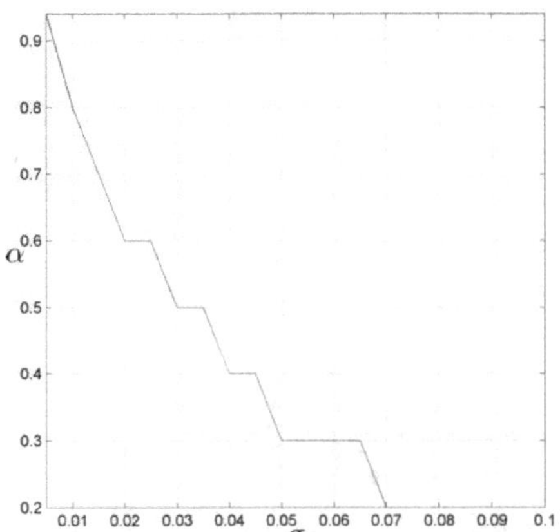

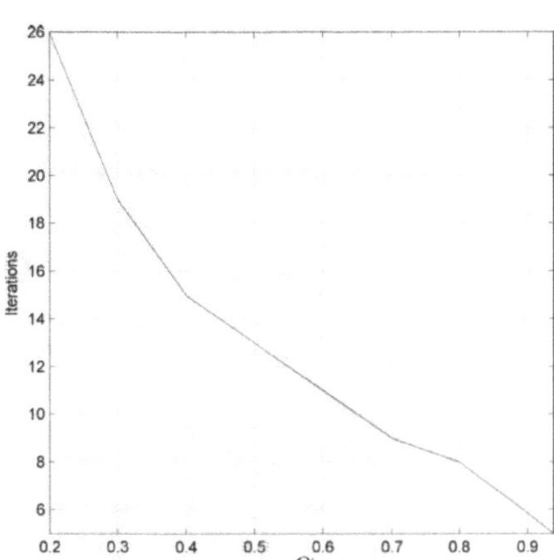

Fig. 1. The left graph plots the optimal α versus noise standard deviation σ. The right graph investigates number of iterations to convergence versus α. The y−axis is the ratio of number of iterations of the conjugate gradient method to the number of iterations using the power method.

Table 3. Comparison of image denoising algorithms using the SSIM measure. The column **f** shows the score for comparison of the original image with the unfiltered image. The columns BF, TV and PR are comparisons with the respective filters.

σ	**f**	BF	TV	PR
0.005	0.995	0.864	0.861	0.995
0.010	0.980	0.861	0.864	0.982
0.015	0.957	0.863	0.864	0.965
0.020	0.928	0.864	0.865	0.948
0.040	0.780	0.855	0.860	0.877
0.060	0.637	0.805	0.831	0.819

Fig. 2. The top left Albert image is corrupted with zero mean Gaussian noise and $\sigma = 0.060$. The top right is after the application of the TV filter, bottom left is after the application of the BF filter and the bottom right after the application of the PR filter. Note how the related anisotropic filters (TV and BF) maintain structural detail but fail to render fine detail such as Albert's hair. Note how well the PageRank filter removes Gaussian noise but renders fine detail, such as Albert's hair, well.

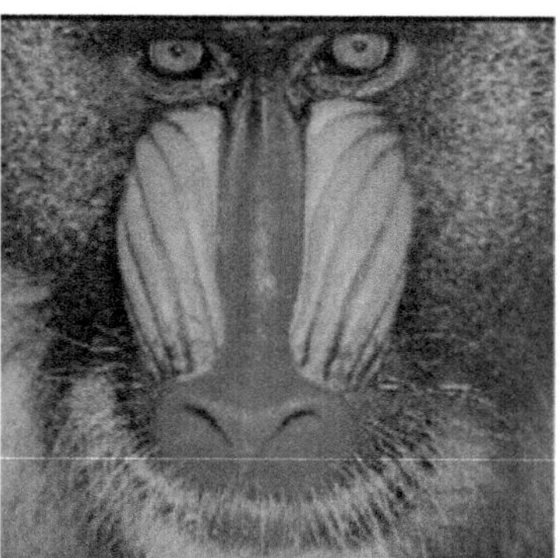
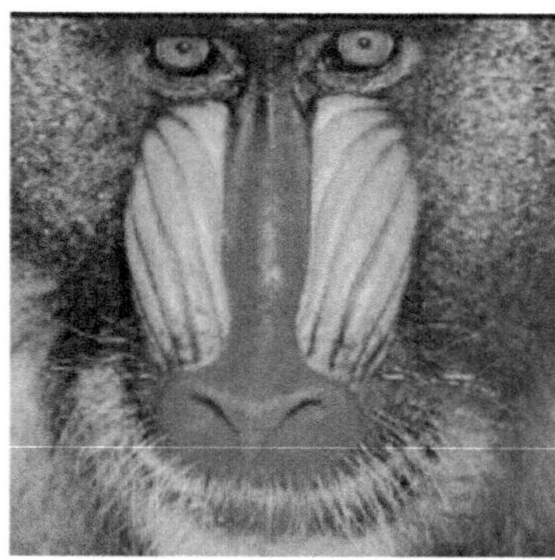

Fig. 3. The left Mandrill image is corrupted with zero mean Gaussian noise with $\sigma = 0.015$ for each colour channel. The right image is after application of the PageRank denoising algorithm with optimal parameter choice.

denoising algorithm renders fine detail very well whilst removing Gaussian noise. The results are in table (3) and figure (2). This is particularly evident in the mandrill image, figure (3), which is an application of the PageRank denoising algorithm to a colour image.

4 Conclusion

We presented a novel image denoising algorithm inspired by the Google PageRank algorithm for ranking hypertextual web documents. A method for efficient computation of the denoised image was discussed and benchmarked. Finally we objectively compared the performance of the algorithm versus related state-of-art denoising filters. The comparisons revealed the new algorithm significantly outperforms the other algorithms in low noise environments with competitive performance in higher noise environments. The highlight of the novel algorithm being that it renders fine detail well whilst removing gaussian noise.

References

1. Andrieu, C., de Freitas, N., Doucet, A., Jordan, M.I.: An introduction to mcmc for machine learning. Machine Learning 50, 5–43 (2003)
2. Azzabou, N., Paragios, N., Guichard, F.: Random walks, constrained multiple hypothesis testing and image enhancement. In: Leonardis, A., Bischof, H., Pinz, A. (eds.) ECCV 2006. LNCS, vol. 3951, pp. 379–390. Springer, Heidelberg (2006)
3. Barbu, A.: Learning real-time mrf inference for image denoising. In: IEEE International Conference on Computer Vision and Pattern Recognition (2009)
4. Barrett, B., Berry, M., Chan, T.F., Demmel, J., Donato, J.M., Dongarra, J., Eijkhout, V., Pozo, R., Romine, C., Van der Vorst, H.: Templates for the Solution of Linear Systems: Building Blocks for Iterative Methods. SIAM, Philadelphia (1994)
5. Bresson, X., Chan, T.F.: Fast dual minimization of the vectorial total variation norm and applications to color image processing. Inverse Problems and Imaging 2, 455–484 (2008)
6. Brin, S., Page, L.: The anatomy of a large-scale hypertextual web search engin. Computer Networks and ISDN Systems 30, 107–117 (1998)
7. Buades, A., Coll, B., Morel, J.-M.: A non-local algorithm for image denoising. In: IEEE Computer Society Conference on Computer Vision and Pattern Recognition (2005)
8. Buades, A., Coll, B., Morel, J.M.: A review of image denoising algorithms, with a new one. Multiscale Modeling & Simulation 4, 490–530 (2005)
9. Chambolle, A.: An algorithm for total variation minimization and applications. Journal of Mathematical Imaging and Vision 20, 89–97 (2004)
10. Chan, R.H., Jin, X.Q.: An Introduction to Iterative Toeplitz Solvers. SIAM, Philadelphia (2007)
11. Chib, S., Greenberg, E.: Understanding the metropolis-hastings algorithm. The American Statistician 4, 327–335 (1995)
12. Chung, F.: Spectral Graph Theory. American Mathematical Society, Providence (1997)
13. Coifman, R.R., Lafon, S.: Diffusion maps. Applied and Computational Harmonic Analysis 21, 5–30 (2006)
14. Duchenne, O., Audibert, J.-Y., Keriven, R., Ponce, J., Segonne, F.: Segmentation by transduction. In: IEEE Conference on Computer Vision and Pattern Recognition (2008)
15. Estrada, D., Fleet, F., Jepson, A.: Stochastic image denoising. In: British Machine Vision Conference (2009)

16. Geman, S., Geman, D.: Stochastic relaxation, gibbs distributions and the bayesian restoration of images. IEEE Transactions on Pattern Analysis and Machine Intelligence 6, 6 (1984)
17. Golub, G.H., Van Loan, C.F.: Matrix computations. In: Johns Hopkins Studies in the Mathematical Sciences (1996)
18. Haveliwala, T., Kamvar, S.: The second eigenvalue of the google matrix. Technical report, Stanford (2003)
19. Langville, A.N., Meyer, C.D.: Deeper inside pagerank. Internet Mathematics 1, 335–380 (2003)
20. Meila, M., Shi, J.: A random walks view of spectral segmentation. In: AI and Statistics (2001)
21. Nadler, B., Lafon, S., Coifman, R.R., Kevrekidis, I.G.: Diffusion maps, spectral clustering and eigenfunctions of fokker-planck operators. In: Advances in Neural Information Processing Systems (2005)
22. Ng, A.Y., Zheng, A.X., Jordan, M.I.: Link analysis, eigenvectors and stability. In: 17th International Joint Conference on Artificial Intelligence (2001)
23. Page, L., Brin, S., Motwani, R., Winograd, T.: The pagerank citation ranking: Bringing order to the web. Technical report, Stanford University (1999)
24. Perona, P., Malik, J.: Scale-space and edge detection using anisotropic diffusion. IEEE Transactions on Pattern Analysis and Machine Intelligence 12, 629–639 (1990)
25. Rudin, L.I., Osher, S., Fatemi, E.: Nonlinear total variation based noise removal algorithms. Physica D 60, 259–268 (1992)
26. Shi, J., Malik, J.: Normalized cuts and image segmentation. IEEE Transactions on Pattern Analysis and Machine Intelligence 22, 888–905 (2000)
27. Tomasi, C., Manduchi, R.: Bilateral filtering for gray and color images. In: Sixth International Conference on Computer Vision (1998)
28. Trefethen, L.N., Embree, M.: Spectra and Pseudospectra: The Behavior of Nonnormal Matrices and Operators. Princeton University Press, Princeton (2005)
29. Wang, Z., Bovik, A.C., Sheikh, H.R., Simoncell, E.P.: Image quality assessment: From error visibility to structural similarity. IEEE Transactions on Image Processing 13, 600–612 (2004)
30. Weickert, J., Ter Haar Romeny, B.M., Viergever, M.A.: Efficient and reliable schemes for nonlinear diffusion filtering. IEEE Transactions on Image Processing 7, 398–410 (1998)
31. Zhang, F., Hancock, E.R.: Graph spectral image smoothing using the heat kernel. Pattern Recognition 41, 3328–3342 (2006)

Structural Similarity-Based Approximation of Signals and Images Using Orthogonal Bases

Dominique Brunet[1], Edward R. Vrscay[1], and Zhou Wang[2]

[1] Department of Applied Mathematics, Faculty of Mathematics, University of Waterloo, Waterloo, Ontario, Canada N2L 3G1
[2] Department of Electrical and Computer Engineering, Faculty of Engineering, University of Waterloo, Waterloo, Ontario, Canada N2L 3G1
dbrunet@uwaterloo.ca, ervrscay@uwaterloo.ca, zhouwang@ieee.org

Abstract. The structural similarity (SSIM) index has been shown to be an useful tool in a wide variety of applications that involve the assessment of image quality and similarity. However, in-depth studies are still lacking on how to incorporate it for signal representation and approximation problems, where minimal mean squared error is still the dominant optimization criterion. Here we examine the problem of best approximation of signals and images by maximizing the SSIM between them. In the case of a decomposition of a signal in terms of an orthonormal basis, the optimal SSIM-based coefficients are determined with a surprisingly simple approach, namely, a scaling of the optimal L^2 coefficients. We then examine a very simple algorithm to maximize SSIM with a constrained number of basis functions. The algorithm is applied to the DCT approximation of images.

1 Introduction

The structural similarity (SSIM) index [9] was proposed as a measure to predict visual distortions between two images. If one of the images being compared is assumed to have perfect quality, the SSIM value can also be interpreted as a perceptual quality measure of the second image. When tested with large-scale independent subject-rated image quality databases [6,4], SSIM has demonstrated superior performance in comparison with traditional image distortion measures such as the mean square error (MSE), which is the most widely employed metric in the image processing literature [8]. In the past few years, SSIM has found a wide range of applications, ranging from image compression, restoration, fusion, and watermarking, to video streaming, digital camera design, biometrics, remote sensing and target recognition [8]. In most of the existing works, however, SSIM has been used for quality evaluation and algorithm comparison purposes only. Much less has been done on using SSIM as an optimization criterion in the design and tuning of image processing algorithms and systems [10,3,1,2,5].

A fundamental issue that has to be resolved before effectively employing SSIM-based optimization in various image processing applications is how to decompose a signal or image as a linear combination of basis functions optimally *in the SSIM sense* (as opposed to the usual L^2 sense). This is a nontrivial

A. Campilho and M. Kamel (Eds.): ICIAR 2010, Part I, LNCS 6111, pp. 11–22, 2010.

problem, given the non-convex property of SSIM. It has been addressed in the particular contexts of image compression [3,5] and image restoration [1,2]. In this paper, however, we "step back" in an effort to understand the problem mathematically. We analyze the simpler case where the SSIM function is defined over nonoverlapping blocks (as opposed to "local SSIM," which involves overlapping patches) and an orthonormal basis is employed. The unique global maximum of the SSIM function over a block may be found by examining its partial derivatives with respect to the expansion coefficients. In this way, an optimal SSIM-based approximation is obtained, as opposed to the well-known L^2-based result, i.e., mean-squared-error (MSE). We obtain the remarkable result that the optimal SSIM-based approximation may, in fact, be determined from the optimal L^2-based approximation: The zeroth-order coefficients are the same, and the higher order SSIM coefficients are obtained from their Fourier counterparts by scaling.

In closing this section, we mention that partial derivatives of the SSIM have been used before. In fact, formulas for the more complicated "local SSIM" case appear in [11]. In that paper, they were used in a numerical gradient ascent/descent algorithm for finding the maxima and minima of the SSIM function over spheres of constant MSE with respect to a reference image. In this case, however, the formulas for the derivatives are very complicated and stationary points cannot, in general, be determined analytically. The study presented below is intended to be a first step toward a deeper understanding of the relationship between SSIM- and L^2-based approximations.

2 SSIM-Based Approximations of Signals/Images

Very briefly, the "local SSIM," that is, the SSIM computed between two local image patches, say **a** and **b**, measures the similarities of three elements of these patches: (i) the local patch luminances or brightness values, (ii) the local patch contrasts and (iii) local patch structures. These three components are then multiplied to form a local SSIM index between **a** and **b**. The "closer" that **a** and **b** are to each other, the closer the value of the SSIM to 1. It is possible to express this SSIM function as a product of only two components. It is this form of the SSIM that is employed in this paper.

In what follows, we let $\mathbf{x}, \mathbf{y} \in \mathbf{R}^N$ denote two N-dimensional signal/image blocks, e.g., $\mathbf{x} = (x_1, x_2, \cdots, x_N)$. We consider a variation of the SSIM function described above, a global SSIM between $\mathbf{x}$ and $\mathbf{y}$ defined as follows,

$$S(\mathbf{x}, \mathbf{y}) = S_1(\mathbf{x}, \mathbf{y}) S_2(\mathbf{x}, \mathbf{y}) = \left[\frac{2\bar{\mathbf{x}}\bar{\mathbf{y}} + \epsilon_1}{\bar{\mathbf{x}}^2 + \bar{\mathbf{y}}^2 + \epsilon_1}\right] \left[\frac{2 s_{\mathbf{xy}} + \epsilon_2}{s_{\mathbf{x}}^2 + s_{\mathbf{y}}^2 + \epsilon_2}\right], \tag{1}$$

where

$$\bar{\mathbf{x}} = \frac{1}{N}\sum_{i=1}^{N} x_i, \quad s_{\mathbf{x}}^2 = \frac{1}{N-1}\sum_{i=1}^{N}(x_i - \bar{\mathbf{x}})^2, \quad s_{\mathbf{xy}} = \frac{1}{N-1}\sum_{i=1}^{N}(x_i - \bar{\mathbf{x}})(y_i - \bar{\mathbf{y}}). \tag{2}$$

The small positive constants $\epsilon_1, \epsilon_2 \ll 1$ are added for numerical stability and can be adjusted to accomodate the perception of the human visual system. It will be convenient to denote the special case where these parameters are zero as follows,

$$S_0(\mathbf{x}, \mathbf{y}) = \frac{4\bar{\mathbf{x}}\bar{\mathbf{y}} s_{\mathbf{xy}}}{(\bar{\mathbf{x}}^2 + \bar{\mathbf{y}}^2)(s_{\mathbf{x}}^2 + s_{\mathbf{y}}^2)}. \tag{3}$$

The functional form of the component S_1 in Eq. (1) was originally chosen in an effort to accomodate Weber's law of perception [9]; that of S_2 follows the idea of divisive normalization [7].

Note that $-1 \leq S(\mathbf{x}, \mathbf{y}) \leq 1$, and $S(\mathbf{x}, \mathbf{y}) = 1$ if and only if $\mathbf{x} = \mathbf{y}$. The component $S_1(\mathbf{x}, \mathbf{y})$ measures the similarity between the means of $\mathbf{x}$ and $\mathbf{y}$: If $\bar{\mathbf{x}} = \bar{\mathbf{y}}$, then $S_1(\mathbf{x}, \mathbf{y}) = 1$, its maximum possible value. This will be important in the discussion below.

Unless otherwise specified, we consider $\mathbf{x}$ to be a given signal and $\mathbf{y}$ to be an approximation to $\mathbf{x}$. We shall generally consider $\mathbf{y}$ to be an element of a particular subset $A \subset \mathbf{R}^N$ – details to be given below – and look for solutions to the problem

$$\mathbf{y}_A = \arg \max_{\mathbf{y} \in A} S(\mathbf{x}, \mathbf{y}). \tag{4}$$

In the case $A = \mathbf{R}^N$, $\mathbf{y} = \mathbf{x}$ and $S(\mathbf{x}, \mathbf{y}) = S(\mathbf{x}, \mathbf{x}) = 1$.

We start with a set of (complete) orthonormal basis functions $\mathbf{R}^N$, to be denoted as $\{\psi_0, \psi_1, \cdots, \psi_{N-1}\}$. We assume that only the first element has nonzero mean: $\overline{\psi}_k = 0$ for $1 \leq k \leq N-1$. We also assume that ψ_0 is "flat", i.e., constant: $\psi_0 = N^{-1/2}(1, 1, \cdots, 1)$, which accomodates the discrete cosine transform (DCT) as well as Haar multiresolution system on $\mathbf{R}^N$.

The L^2-based expansion of $\mathbf{x}$ in this basis is, of course,

$$\mathbf{x} = \sum_{k=0}^{N-1} a_k \psi_k, \qquad a_k = \langle \mathbf{x}, \psi_k \rangle, \ \ 0 \leq k \leq N-1. \tag{5}$$

It follows that

$$\bar{\mathbf{x}} = a_0 N^{-1/2}. \tag{6}$$

The expansions of the approximation $\mathbf{y}$ will be denoted as follows,

$$\mathbf{y} = \mathbf{y}(\mathbf{c}) = \sum_{k=0}^{N-1} c_k \psi_k, \tag{7}$$

where the notation $\mathbf{y}(\mathbf{c})$ acknowledges the dependence of the approximation on the coefficients c_k. It also follows that

$$\bar{\mathbf{y}} = c_0 N^{-1/2}. \tag{8}$$

In this study, the approximation spaces A in (4) will be the spans of subsets of the set of basis functions $\{\psi_k\}_{k=0}^{N-1}$ which include ψ_0: From (6) and (8), the inclusion of ψ_0 automatically maximizes component $S_1(\mathbf{x}, \mathbf{y})$ of the SSIM function. At this point, we do not specify exactly which other ψ_k basis functions will be used, but consider all possible subsets of $M < N$ basis functions:

$$A = \text{span}\{\psi_0, \psi_{\gamma(1)}, \cdots, \psi_{\gamma(M-1)}\}, \tag{9}$$

where $\gamma(k) \in \{1, 2, \cdots, N-1\}$ and $c_{\gamma(M)} = \cdots = c_{\gamma(N-1)} = 0$. Of course, we are interested in finding the *optimal* M-dimensional subset, in the SSIM sense.

Before studying optimal SSIM approximations, however, it is most helpful to review the well-known L^2-based case.

Proposition 1. *For a given* $\mathbf{x} \in \mathbf{R}^N$*, the* M *coefficients* c_k *of the optimal* L^2*-based approximation* $\mathbf{y} \in A$ *to* $\mathbf{x}$ *are given by* $c_0 = \langle \mathbf{x}, \psi_0 \rangle$ *and the* $M-1$ *remaining Fourier coefficients* $a_k = \langle \mathbf{x}, \psi_k \rangle$ *of greatest magnitude, i.e.,*

$$c_k = a_{\gamma(k)} = \langle \mathbf{x}, \psi_{\gamma(k)} \rangle, \quad 1 \le k \le M-1, \tag{10}$$

where $|a_{\gamma(1)}| \ge |a_{\gamma(2)}| \ge \ldots \ge |a_{\gamma(M-1)}| \ge |a_l|$ *with* $l \in \{1, 2, \cdots, N-1\} \setminus \{\gamma(1), \cdots, \gamma(M-1)\}$.

Proof. For an arbitrary $\mathbf{c} \in \mathbf{R}^N$, let $\Delta(\mathbf{x}, \mathbf{y}(\mathbf{c})) = \|\mathbf{x} - \mathbf{y}(\mathbf{c})\|_2$, the L^2 error of approximation of $\mathbf{x}$ by $\mathbf{y}(\mathbf{c})$. For any $p \in \{0, 1, 2, \cdots, N-1\}$, consider the change in this error produced by altering the coefficient c_p by ϵ, i.e., $\mathbf{c} \to \mathbf{c} + \epsilon \hat{\mathbf{e}}_p$. Because the squared L^2-error is a quadratic form in the c_k, its Taylor series in ϵ,

$$\Delta^2(\mathbf{x}, \mathbf{y}(\mathbf{c} + \epsilon \hat{\mathbf{e}}_p)) = \Delta^2(\mathbf{x}, \mathbf{y}(\mathbf{c})) + \epsilon \frac{\partial}{\partial c_p} \Delta^2(\mathbf{x}, \mathbf{y}(\mathbf{c})) + \cdots, \tag{11}$$

is finite – in fact, a quadratic polynomial:

$$\Delta^2(\mathbf{x}, \mathbf{y}(\mathbf{c} + \epsilon \hat{\mathbf{e}}_p)) = \|\mathbf{x} - \mathbf{y}(\mathbf{c})\|_2^2 + 2\epsilon(c_p - \langle \mathbf{x}, \psi_p \rangle) + \epsilon^2. \tag{12}$$

We see that the only stationary point occurs when $c_p = a_p = \langle \mathbf{x}, \psi_p \rangle$ and that it is, in fact, a global minimum. Since ψ_0 is a basis element of A, we set $c_0 = a_0 = \langle \mathbf{x}, \psi_0 \rangle$. Now, for some choice of distinct $M-1$ indices $\gamma(k) \in \{1, 2, \cdots, N-1\}$, set $c_k = a_{\gamma(k)}$, $1 \le k \le M-1$, and $c_l = 0$ otherwise. Then

$$\begin{aligned} \|\mathbf{x} - \mathbf{y}(\mathbf{c})\|_2^2 &= \|\mathbf{x}\|_2^2 - 2\sum_{k=0}^{N-1} c_k \langle \mathbf{x}, \psi_k \rangle + \sum_{k=0}^{N-1} c_k^2 \\ &= \|\mathbf{x}\|_2^2 - a_0^2 - \sum_{k=1}^{M-1} a_{\gamma(k)}^2. \end{aligned} \tag{13}$$

Clearly, the smallest L^2 approximation error is produced if the $M-1$ coefficients $a_{\gamma(k)}$ in Eq. (13) are those with the largest magnitudes. □

We now consider the optimal SSIM-based approximation of an element $\mathbf{x} \in \mathbf{R}^N$ in the M-dimensional subspace A defined in Eq. (9).

Proposition 2. *The coefficients of the optimal SSIM-based approximation of* $\mathbf{x}$ *in the* M*-dimensional subspace* A *defined in Eq. (9) are given by*

$$c_0 = a_0, \quad c_{\gamma(k)} = \alpha a_{\gamma(k)} \quad \text{for } 1 \leq k \leq M-1, \tag{14}$$

where the $a_{\gamma(k)} = \langle x, \psi_{\gamma(k)} \rangle$ *are the* $M-1$ *optimal* L^2*-based coefficients from Proposition 1 and the scaling coefficient* α *is given by*

$$\alpha = \frac{-\epsilon_2 + \sqrt{\epsilon_2^2 + (\frac{4}{N-1}\sum_{k=1}^{M-1} a_{\gamma(k)}^2)(s_{\mathbf{x}}^2 + \epsilon_2)}}{\frac{2}{N-1}\sum_{k=1}^{M-1} a_{\gamma(k)}^2}. \tag{15}$$

Proof. Without loss of generality, we assume that the last $N-M$ coefficients c_k are zeros. After some simple algebra, we find that

$$s_{\mathbf{xy}} = \frac{1}{N-1}\sum_{k=1}^{N-1} a_k c_k \quad \text{and} \quad s_{\mathbf{y}}^2 = \frac{1}{N-1}\sum_{k=1}^{N-1} c_k^2. \tag{16}$$

The dependence of the SSIM function in Eq. (1) on the c_k is as follows: (i) the first term in Eq. (1) depends only on the coefficient c_0 and (ii) the second term in Eq. (1) is independent of c_0. The choice $c_0 = a_0$ maximizes the first term in Eq. (1), giving it the value of 1. In terms of the remaining c_k,

$$S(\mathbf{x}, \mathbf{y}(\mathbf{c})) = \frac{\frac{2}{N-1}\sum_{k=1}^{M-1} a_k c_k + \epsilon_2}{s_{\mathbf{x}}^2 + \frac{1}{N-1}\sum_{k=1}^{M-1} c_k^2 + \epsilon_2}. \tag{17}$$

We now look for stationary points which will be candidates for solutions to the approximation problem in (4). Logarithmic differentiation yields the following partial derivatives with respect to c_k for $1 \leq k \leq M-1$:

$$\frac{\partial S}{\partial c_k} = S\left[\frac{2\bar{\mathbf{x}}}{2\bar{\mathbf{x}}\bar{\mathbf{y}} + \epsilon_1}\frac{\partial \bar{\mathbf{y}}}{\partial c_k} + \frac{2}{2s_{\mathbf{xy}} + \epsilon_2}\frac{\partial s_{\mathbf{xy}}}{\partial c_k} - \frac{2\bar{\mathbf{y}}}{\bar{\mathbf{x}}^2 + \bar{\mathbf{y}}^2 + \epsilon_1}\frac{\partial \bar{\mathbf{y}}}{\partial c_k} - \frac{1}{s_{\mathbf{x}}^2 + s_{\mathbf{y}}^2 + \epsilon_2}\frac{\partial s_{\mathbf{y}}^2}{\partial c_k}\right]. \tag{18}$$

After some additional (yet simple) algebra, we obtain the following conditions for a stationary point,

$$\frac{\partial S}{\partial c_k} = \frac{S}{N-1}\left[\frac{2a_k}{2s_{\mathbf{xy}} + \epsilon_2} - \frac{2c_k}{s_{\mathbf{x}}^2 + s_{\mathbf{y}}^2 + \epsilon_2}\right] = 0, \quad 1 \leq k \leq M-1. \tag{19}$$

If $a_p = 0$ for any $1 \leq p \leq M-1$, then $c_p = 0$. Otherwise, we have that

$$\frac{a_k}{c_k} = \frac{2s_{\mathbf{xy}} + \epsilon_2}{s_{\mathbf{x}}^2 + s_{\mathbf{y}}^2 + \epsilon_2}, \quad 1 \leq k \leq M-1. \tag{20}$$

Note that the RHS of each equation is independent of k, implying that

$$\frac{a_1}{c_1} = \frac{a_2}{c_2} = \cdots = \frac{a_{M-1}}{c_{M-1}} = C \quad \text{(constant)}. \tag{21}$$

Hence

$$c_k = \alpha a_k \quad \text{for } 1 \leq k \leq M-1, \tag{22}$$

where $\alpha = 1/C$. We now rewrite Eq. (20) as follows,

$$(2s_{\mathbf{xy}}(\mathbf{c}) + \epsilon_2)c_k = (s_{\mathbf{x}}^2 + s_{\mathbf{y}}^2(\mathbf{c}) + \epsilon_2)a_k, \tag{23}$$

and employ (22) and the expansions in Eq. (16) to arrive at the following quadratic equation in α:

$$\frac{2\alpha^2}{N-1}\sum_{k=1}^{M-1} a_k^2 + \alpha\epsilon_2 = s_{\mathbf{x}}^2 + \frac{\alpha^2}{N-1}\sum_{k=1}^{M-1} a_k^2 + \epsilon_2. \tag{24}$$

The roots of this equation are

$$\alpha_{1,2} = \frac{-\epsilon_2 \pm \sqrt{\epsilon_2^2 + (\frac{4}{N-1}\sum_{k=1}^{M-1} a_k^2)(s_{\mathbf{x}}^2 + \epsilon_2)}}{\frac{2}{N-1}\sum_{k=1}^{M-1} a_k^2}. \tag{25}$$

Notice that $\alpha_1 \geq 1$ and $\alpha_2 \leq -1$. Substituting $c_k = \alpha_{1,2}a_k$ into Eq. (17) we observe that α_1 and α_2 correspond to the scaling factors for, respectively, a local maximum and a local minimum.

Now that the natures of the critical points have determined, we examine the behaviour of $S(\mathbf{x}, \mathbf{y})$ "on the boundaries," i.e., as $|c_k| \to \infty$. In this case, $|S(\mathbf{x}, \mathbf{y})| \to 0$, which allows us to conclude that $c_k = \alpha_1 a_k$ at the global maximum, thus proving Eq. (15). For the remainder of the proof, we let $\alpha = \alpha_1$.

For a given $M < N$, it now remains to determine which subset of $M-1$ coefficients c_k should be chosen in order to maximize the structural similarity $S(\mathbf{x}, \mathbf{y}(\mathbf{c}))$. The global maximum value of the structural similarity, which we denote as $S_{\max}$, is found by substituting $c_k = \alpha a_k$, $1 \leq k \leq M-1$, into Eq. (17):

$$S_{\max}(\mathbf{x}, \mathbf{y}(\mathbf{c})) = \frac{\frac{2\alpha}{N-1}\sum_{k=1}^{M-1} a_k^2 + \epsilon_2}{s_{\mathbf{x}}^2 + \frac{\alpha^2}{N-1}\sum_{k=1}^{M-1} a_k^2 + \epsilon_2}. \tag{26}$$

From Eq. (24), we have the interesting result that

$$S_{\max}(\mathbf{x}, \mathbf{y}(\mathbf{c})) = \frac{1}{\alpha}. \tag{27}$$

Substitution of this result into Eq. (26) yields a quadratic equation in $S_{\max}$. Only the positive root of this equation is admissible and it is given by

$$S_{\max}(\mathbf{x}, \mathbf{y}(\mathbf{c})) = \frac{\epsilon_2 + \sqrt{\epsilon_2^2 + (\frac{4}{N-1}\sum_{k=1}^{M-1} a_k^2)(s_{\mathbf{x}}^2 + \epsilon_2)}}{2(s_{\mathbf{x}}^2 + \epsilon_2)}. \tag{28}$$

From this expression, it is clear that the maximum possible value of $S_{\max}$ is achieved if the $M-1$ Fourier coefficients a_k with the largest magnitudes are employed in the summation. □

A few remarks regarding these results are in order.

1. An important consequence of Proposition 2 is that the optimal SSIM-based approximation $\mathbf{y}(\mathbf{c})$ of $\mathbf{x}$ may be obtained by first computing the best L^2 approximation of $\mathbf{x}$ that includes ψ_0 (which is almost always the case), then setting $c_0 = a_0$ and finally scaling the remaining Fourier coefficients according to Eqs. (14) and (15).
2. Since $c_0 = a_0$, it follows that $\bar{\mathbf{x}} = \bar{\mathbf{y}}(\mathbf{c})$. Regarding the other coefficients, the fact that the scaling factor $\alpha > 1$ for $M < N$ implies that the SSIM-based approximation $\mathbf{y}$ represents a contrast-enhanced version of $\mathbf{x}$.
3. In the special case that $\epsilon_2 = 0$, the optimal scaling factor α in Eq. (15) has the simple form
$$\alpha = \left[\sum_{k=1}^{N-1} a_k^2\right]^{1/2} \left[\sum_{k=1}^{M-1} a_k^2\right]^{-1/2}. \tag{29}$$
4. In the special case $M = N$, we have $\alpha = 1$ and $S_{\max} = 1$, as expected since there is no approximation, i.e., $\mathbf{y} = \mathbf{x}$.
5. The existence of such a simple analytic solution to this problem is made possible by the simplicity of the approach – we have been considering "global" SSIM, i.e., the entire signal/image (or block), as opposed to "local SSIM" where overlapping patches/neighbourhoods are employed. In the latter case, the derivatives of the SSIM function with respect to the c_k coefficients are not as straightforward. The above approach applies directly to (nonoverlapping) block-based coding, which includes DCT and Haar wavelet coding.
6. The assumption that the first function ψ_0 is "flat" may be relaxed, in which case Eqs. (6) and (8) would have to be modified. This, however, will not change the condition that $c_0 = a_0$ in the SSIM-optimality condition.

We finally remark that SSIM-based approximation (with no stability constants) may be viewed as a kind of (inverse) variance-weighted L^2 approximation of signals after their means have been subtracted out, as shown by Richter and Kim [5]. To see this result, let $\mathbf{x}, \mathbf{y} \in \mathbf{R}^N$ and define $\mathbf{x}_0 = \mathbf{x} - \bar{\mathbf{x}}$ and $\mathbf{y}_0 = \mathbf{y} - \bar{\mathbf{y}}$ so that $\bar{\mathbf{x}}_0 = \bar{\mathbf{y}}_0 = 0$. Then

$$\begin{aligned} \|\mathbf{x}_0 - \mathbf{y}_0\|^2 &= \sum_{k=1}^{N} (x_{0,k} - y_{0,k})^2 \\ &= (N-1)[s_{\mathbf{x}_0}^2 + s_{\mathbf{y}_0}^2 - 2s_{\mathbf{x}_0\mathbf{y}_0}]. \end{aligned} \tag{30}$$

From this, the definition of the SSIM function $S_0(\mathbf{x}, \mathbf{y})$ in Eq. (1) and a little algebra, we find that

$$1 - S_0(\mathbf{x}_0, \mathbf{y}_0) = \frac{1}{N-1} \frac{\|\mathbf{x}_0 - \mathbf{y}_0\|^2}{s_{\mathbf{x}_0}^2 + s_{\mathbf{y}_0}^2}. \tag{31}$$

This, along with Proposition 2, gives an idea of the link between L^2- and SSIM-based approximations.

3 SSIM-Based Image Reconstruction from a Constrained Number of DCT Coefficients

Proposition 2 provides a very simple procedure to optimize L^2-based expansions from the SSIM point of view. The procedure, however, has very limited applicability. When the the same number M of coefficients are employed in the L^2- and SSIM-based optimizations, we find that the latter generally yields very little, if any, noticeable improvement in perceptual quality. Indeed, the greatest increase in the SSIM value is usually found for small values of M, in which case both the L^2- and SSIM-based optimizations yield poor approximations.

On the other hand, SSIM-based optimization may yield significant improvements in perceptual quality when it is employed to decide the allocation of a prescribed number of coefficients/bits. We illustrate below with an application to the block-based discrete cosine transform (DCT).

Our simple algorithm starts with the set of zeroth-order coefficients for all blocks. The goal is to add K of the remaining higher-order DCT coefficients to this set. At each step of the selection process, we estimate the gain in structural similarity (with respect to the original image, using the original DCT coefficients) produced by adding a DCT coefficient that has not yet been employed. The unused coefficients from all blocks of the image are examined. The DCT coefficient yielding the greatest increase in the SSIM is then added to the set. For comparison, we perform a similar algorithm in which the decrease in the L^2 error of approximation is used as the criterion for selection at each step.

We define the following quantities:

1. BSSIM: the average value of the structural similarities, (Eq. 1), of all non-overlapping blocks of the image,
2. MSSIM: the average of the *weighted* SSIM, computed with a Gaussian sliding window with the parameters as in [9]. Our method will give us a BSSIM-optimal reconstructed image, but not necessarily a MSSIM-optimal image,
3. $\mathbf{x}^i$: the i-th block of the image $\mathbf{x}$ being approximated,
4. $V^i_{K_i} := \frac{1}{N-1}\sum_{k=1}^{K_i}(a^i_{\gamma(k)})^2$: the variance of the DCT approximation of the $\mathbf{x}^i$ using K_i non-zero higher-order DCT coefficients.
5. $\mathbf{c}^i_{K_i}$: the set of K_i non-zero higher-order coefficients which, along with the zeroth-order coefficient, define the SSIM-based approximation $\mathbf{y}^i$ to block $\mathbf{x}^i$.

For a given block $\mathbf{x}^i$, the gain in structural similarity produced by adding the first non-zero coefficient, c^i_1, is, from Eq. (28),

$$S(\mathbf{x}^i,\mathbf{y}^i(\mathbf{c}^i_1)) - S(\mathbf{x}^i,\mathbf{y}^i(\mathbf{c}^i_0)) = \frac{\sqrt{\epsilon_2^2 + 4V^i_1(s^2_{\mathbf{x}^i}+\epsilon_2)} - \epsilon_2}{2(s^2_{\mathbf{x}^i}+\epsilon_2)}. \tag{32}$$

For a given block $\mathbf{x}^i$, the gain in SSIM produced by adding the coefficient $c^i_{K_i+1}$ ($K_i > 0$) to the existing set of SSIM-optimized DCT coefficients is given by

$$S(\mathbf{x}^i, \mathbf{y}^i(\mathbf{c}^i_{K_i+1})) - S(\mathbf{x}^i, \mathbf{y}^i(\mathbf{c}^i_{K_i})) = \frac{\sqrt{\epsilon_2^2 + 4V^i_{K_i+1}(s^2_{\mathbf{x}^i} + \epsilon_2)} - \sqrt{\epsilon_2^2 + 4V^i_{K_i}(s^2_{\mathbf{x}^i} + \epsilon_2)}}{2(s^2_{\mathbf{x}^i} + \epsilon_2)}. \quad (33)$$

After examining all blocks, the coefficient $c^{i^*}_{K_{i^*}+1}$ yielding the highest gain in SSIM, according to either Eq. (32) ($K_{i^*} = 0$) or Eq. (33) ($K_{i^*} > 0$) is then added to the set. The algorithm is terminated when K coefficients have been added.

As an example, the 512×512 8bpp *Lena* test image was decomposed with the discrete cosine transform (DCT) over 8×8 nonoverlapping pixel blocks. In Fig. 1 are shown the results for a "total budget" of $K = 10,000$ non-zero higher-order coefficients. (4096 zeroth-order c^i_0 coefficients are also employed in the expansion, but not counted, in this simple scheme.) In the first row of this figure, we show (a) the original *Lena* image, (b) the "BSSIM map" of the optimal BSSIM DCT approximation and (c) the "BSSIM map" of the optimal L^2 DCT approximation that employs the same budget. In (b) and (c), the greyscale assigned to each block is proportional to its SSIM value, with black representing 0, and white representing 1.

Note that the BSSIM map in (b) is, for the most part, "lighter" than that in (c) which, of course, is expected, since the former corresponds to SSIM optimization. However, there are some blocks, most notably those containing edges, in which the SSIM values of (b) are lower than those of (c). This is revealed in the lower portion of Fig. 1, where three representative patches – (i) a part of the hat, (ii) the face and (iii) the shoulder – of the *Lena* image are presented. In the first column are shown the three patches from the original image. The second column of the array presents the patches from the SSIM-optimized image. The third column shows the patches from the L^2-optimized image. We see that the BSSIM-optimized approximation demonstrates less blocking effects on smooth regions than its L^2-optimized counterpart. It also preserves more details such as the fine textures of the hat – in fact, the improvement afforded by the SSIM method over the L^2 method is quite remarkable here.

On other hand, the BSSIM-optimized approximation performs less efficiently on edges. Here, the contrasts – as determined by the scaling factor α for each block – are too large, giving rise to noticeable blocking effects. This could be explained in part by the fact that the BSSIM-optimal image enhances the contrast of the image locally on each block without taking neighbouring blocks into account. An MSSIM-optimized approximation could correct such blocking problems around edges.

The excessively large contrasts exhibited in these blocks may also be due to a small number of coefficients being assigned to them in the BSSIM procedure, with its "total bit budget" for the entire image. From Proposition 2, the scaling function α in Eq. (15) is greater than one, and approaches one as the number of coefficients M approaches N, the size of the block. In the case $M = 2$

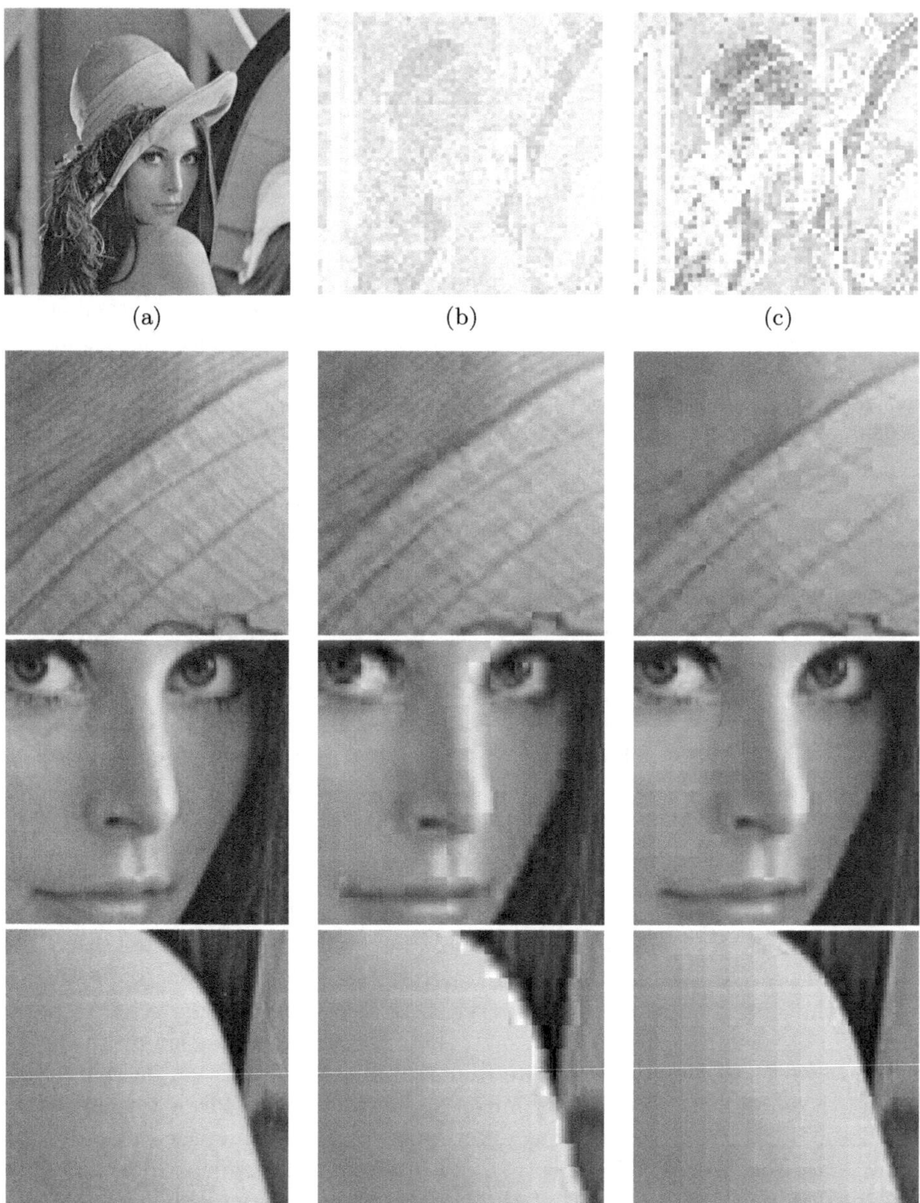

Fig. 1. Top row. (a): Original 512×512 8bpp *Lena* image. (b): BSSIM map of BSSIM-optimized block-DCT approximation of *Lena* with 10,000 non-zero higher-order DCT coefficients. (c): BSSIM map of L^2-optimized block DCT approximation of *Lena* with 10,000 non-zero higher-order DCT coefficients. **Bottom rows.** First column: Patches of the original image. Second column: Patches of the BSSIM-optimized image. Third column: Patches of the L^2-optimized image. Second row: Hat, cropped from (81, 169) to (176, 264). Third row: Face, cropped from (257, 257) to (362, 362). Fourth row: Shoulder, cropped from (401, 289) to (496, 384).

(corresponding to only one higher-order coefficient), it is not too hard to show that, for small ϵ_2 and nonzero $s_{\mathbf{x}}$, $\alpha \approx s_{\mathbf{x}}$. This implies that blocks with large variance – typically those with edges – can have large α values when a low number of coefficients is being used. This feature is demonstrated in the figure, particularly along the shoulder.

In Fig. 2 are presented plots of the BSSIM values vs. the number of non-zero coefficients for the SSIM- and L^2-based optimization methods. For this particular example, we see that the greatest increase in BSSIM from the L^2-based method occurs when the number K of non-zero higher-order coefficients is between 2000 and 3000. As K increases toward 10,000, the difference between the two methods decreases.

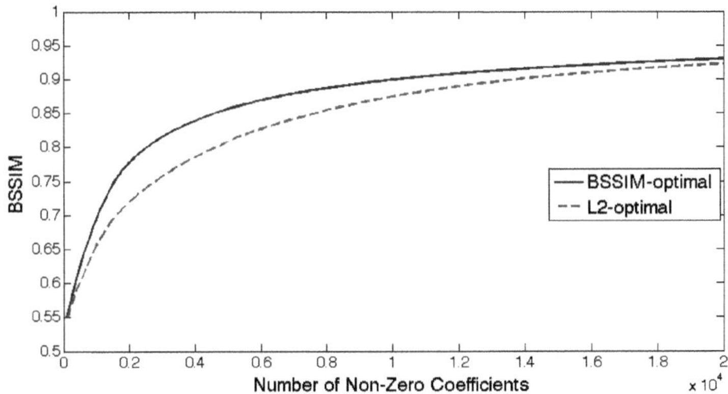

Fig. 2. Graphs of the BSSIM values of the BSSIM-optimized and L^2-optimized 8×8 block-DCT approximations of *Lena* image, as a function of the number of non-zero higher-order DCT coefficients employed

4 Concluding Remarks

We have mathematically examined the problem of SSIM-based approximations of signal/image blocks in terms of an orthogonal basis. In general, the non-convex nature of the SSIM function complicates optimization problems. In this case, however, a closed-form solution can be found since the partial derivatives of the SSIM function with respect to the expansion coefficients are rather straightforward and stationary points can be determined. The optimal SSIM approximation is found to be related to the well-known optimal L^2 approximation in a quite simple manner: The zeroth-order coefficients of both expansions are the same, and higher-order SSIM coefficients are obtained from their Fourier counterparts by simple scaling. On its own, this algorithm does not yield any significant improvement in the usual L^2-optimized expansions. But we have shown its potential use in the construction of approximations that are subject to fixed "bit budgets." Only a simple example was studied here – there are many possibilities for further exploration, including coding/compression and enhancement.

The results of Section 2 have inspired an investigation of more general cases which will be reported elsewhere. These cases include a generalization of the basis set to the nonorthonormal and overcomplete cases. Other avenues of future research include the combination of SSIM-optimization with prior image models in a Bayesian framework.

Acknowledgements. This research was supported in part by the Natural Sciences and Engineering Research Council of Canada (NSERC) in the form of a Postgraduate Scholarship (DB), Discovery Grants (ERV, ZW) and a Strategic Grant (ZW). It was also supported in part by the Ontario Ministry of Research & Innovation in the form of an Early Researcher Award (ZW).

References

1. Channappayya, S.S., Bovik, A.C., Caramanis, C., Heath, R.W.: Design of linear equalizers optimized for the structural similarity index. IEEE Trans. Image Processing 17(6), 1624–1639 (2008)
2. Channappayya, S.S., Bovik, A.C., Caramanis, C., Heath, R.W.: SSIM-optimal linear image restoration. In: Proc. IEEE Int. Conf. Acoust., Speech, and Signal Processing, Las Vegas, Nevada (March 2008)
3. Channappayya, S.S., Bovik, A.C., Heath, R.W.: Rate bounds on ssim index of quantized images. IEEE Trans. Image Processing 17(6), 857–872 (2008)
4. Ponomarenko, N., Egiazarian, K.: Tampere Image Database, TID 2008 (2008), http://www.ponomarenko.info/tid2008.htm
5. Richter, T., Kim, K.J.: A MS-SSIM optimal JPEG 2000 encoder. In: Data Compression Conference, Snowbird, Utah, March 2009, pp. 401–410 (2009)
6. Sheikh, H.R., Seshadrinathan, K., Moorthy, A.K., Wang, Z., Bovik, A.C., Cormack, L.K.: Image and video quality assessment research at LIVE, http://live.ece.utexas.edu/research/quality/
7. Wainwright, M.J., Schwartz, O., Simoncelli, E.P.: Natural image statistics and divisive normalization: Modeling nonlinearity and adaptation in cortical neurons. In: Rao, R., Olshausen, B., Lewicki, M. (eds.) Probabilistic Models of the Brain: Perception and Neural Function, pp. 203–222. MIT Press, Cambridge (2002)
8. Wang, Z., Bovik, A.C.: Mean squared error: Love it or leave it? A new look at signal fidelity measures. IEEE Signal Processing Magazine 26(1), 98–117 (2009)
9. Wang, Z., Bovik, A.C., Sheikh, H.R., Simoncelli, E.P.: Image quality assessment: From error visibility to structural similarity. IEEE Trans. Image Processing 13(4), 600–612 (2004)
10. Wang, Z., Li, Q., Shang, X.: Perceptual image coding based on a maximum of minimal structural similarity criterion. In: Proc. IEEE Int. Conf. Image Proc., San Antonio, TX (September 2007)
11. Wang, Z., Simoncelli, E.P.: Maximum differentiation (MAD) competition: A methodology for comparing computational models of perceptual quantities. Journal of Vision 8(12), 1–13 (2008)

A Neighborhood Dependent Nonlinear Technique for Color Image Enhancement

Rupal Patel[1] and Vijayan K. Asari[2]

[1] Old Dominion University, 231 Kaufman Hall, Norfolk, VA, USA
rpate017@odu.edu
[2] University of Dayton, 300 College Park, Dayton, OH, USA
vijayan.asari@notes.udayton.edu

Abstract. An image enhancement algorithm based on a neighborhood dependent nonlinear model is presented to improve visual quality of digital images captured under extremely non-uniform lighting conditions. The paper presents techniques for adaptive and simultaneous intensity enhancement of extremely dark and bright images, contrast enhancement, and color restoration. The core idea of the algorithm is a nonlinear sine transfer function with an image dependent parameter. Adaptive computation of the control parameter increases flexibility in enhancing the dark regions and compressing overexposed regions in an image. A neighborhood dependent approach is employed for contrast enhancement. A linear color restoration process is used to obtain color image from the enhanced intensity image by utilizing the chromatic information of the original image. It is observed that the proposed algorithm yields visually optimal results on images captured under extreme lighting conditions, and also on images with bright and dark regions.

Keywords: image enhancement, dynamic range compression, intensity transformation, sine nonlinearity, contrast enhancement.

1 Introduction

The fact that camera does not see exactly the way human eyes do, introduces limitations in the formation and display of an image of a real world scene. In the nature, the scene luminance ranges in the span of two to six orders of magnitude, thereby producing a very high dynamic range radiance map. The dynamic range represents the amount of contrast that a given device can record. Currently available standard electronic cameras can measure light between 8 to 10 stops (2 to 4 orders of magnitude). A high end camera with wider dynamic range can measure light up to 14-16 stops, which is still inferior to human eye that can see details in a scene containing a contrast range of nearly 24 stops (more than six orders of magnitude). In addition, the dynamic range of a camera is limited by noise levels, meaning that details captured in dark shadow or bright areas may exhibit excessive noise and rendered as black or white [1]. Human eye is capable of handling the wide dynamic range radiance map due to its complex structure and adaptive mechanism. On the other hand, the camera aperture is fixed and sets global exposure when capturing an image. Furthermore, image

A. Campilho and M. Kamel (Eds.): ICIAR 2010, Part I, LNCS 6111, pp. 23–34, 2010.

display devices, like monitors and printers, also demonstrate limited dynamic range. Consequently, images captured under extremely bright or ill lighting conditions suffer from saturation and underexposure respectively. When displayed on LDR devices, important features and fine details are not visible [1].

In order to improve visual quality of images while dealing with the technical limitations of recording and display devices, compressing the dynamic range (mapping of the natural range of luminance to a smaller range [4]) is important. Several image processing techniques exist that can perform dynamic range compression such as, logarithmic compression, gamma correction, histogram equalization, etc. However, these techniques are not sophisticated enough to preserve all the features and fine details. Also, they may not be able to enhance all the regions proportionately. For example, in logarithmic enhancement, the low intensity pixel values can be enhanced at the loss of high intensity values [5]. In these techniques, regions of the scene where the slope of the mapping operator is low can become difficult to see [6].

To address the contrast issues, several advanced image processing techniques have been developed to compress the dynamic range along with local contrast enhancement, such as adaptive histogram equalization [7], Retinex [8], multi-scale retinex (MSR) [10,13], AINDANE [14], and LTSNE [15]. Among them, Histogram Equalization is a fairly simple and fast algorithm but works well only on the images possessing uni-modal or weakly bi-modal histograms [14]. Many variations have been made to the original HE technique to improve contrast and details. The drawback of the advanced HE algorithms is that it makes the image look unnatural while bringing out the object details. Successful efforts have been made to imitate human visual system based on Retinex theory derived by E. Land [9]. The basic concept of the Retinex theory is to separate illumination and reflectance components of an image. Retinex based algorithms accomplish two of the requirements of a lightness-color constancy algorithm for machine vision [8-10]. These are: (i) dynamic range compression and (ii) color independence from the spectral distribution of the scene illuminant. Multi Scale Retinex (MSR) [11] theory was developed based on a center/surround method in which the best results were obtained by averaging three images resulting from three different surround sizes. Later, a color restoration step was added to overcome a graying out effect caused by the method. However, the biggest problem with both MSR and standard Retinex is the separate nonlinear processing of three color bands. It not only produces strong "halo" effect and incorrect color artifacts, but also makes the algorithm computationally intensive.

In recent years, a more promising technique called AINDANE (Adaptive Integrated Neighborhood Dependent Approach for Nonlinear Enhancement) [14] has been developed. It involves itself in adaptive luminance enhancement and adaptive contrast enhancement. The enhanced image can be obtained by a linear color restoration process based on the chromatic information in the original image. This method handles enhancement of dark or ill-illuminated images very well, however, it does not provide solution for overexposed images. In order to obtain fine details and balance between over and underexposed regions in images, an innovative technique named LTSNE has been developed [15], which also forms the basis for the proposed algorithm. In LTSNE, the major contribution has been the simultaneous enhancement and compression of dark and bright pixels using a nonlinear sine squared function with image dependent parameters. As an improvement over LTSNE, in the proposed algorithm NDNE (Neighborhood Dependent Nonlinear Enhancement of Color Images), computation of the image dependent parameters has been simplified to reduce processing time and yield improved visual quality. It provides means for preserving dark as well as light regions in any high contrast scene. In the following section, the proposed algorithm is discussed in detail. Experimental results of our algorithm are discussed in section 3 followed by conclusions in section 4.

2 Algorithm

The proposed algorithm is implemented in three steps: adaptive intensity enhancement, contrast enhancement, and color restoration. The goal of this algorithm is to enhance the visual quality of images captured under extremely non-uniform lighting conditions. Hence the primary step is the adaptive intensity enhancement of dark and bright pixels. After intensity enhancement, the contrast is degraded in the intensity-enhanced image, hence contrast enhancement process is applied to restore contrast and in turn, preserve or enhance important visual details. Finally, after the contrast enhancement, the enhanced color image is obtained by performing a linear color restoration process on the enhanced intensity image using the chromatic information in the input image. The key contribution of this algorithm is the computation of control parameter involved in adaptive intensity enhancement and is discussed in detail in the following section.

2.1 Adaptive Intensity Enhancement

In this algorithm, the color image is first converted to gray scale image using NTSC standard [16] as follows:

$$I(x,y) = \frac{76.245 I_R(x,y) + 149.685 I_G(x,y) + 29.07 I_B(x,y)}{255} \quad (1)$$

where $I_R(x,y)$, $I_G(x,y)$ and $I_B(x,y)$ are red, green and blue color values respectively of a pixel located at (x, y) position in the image. The intensity image is further normalized by:

$$I_n(x,y) = \frac{I(x,y)}{255} . \quad (2)$$

Compressing the dynamic range of the intensity image is an efficient method of image enhancement. Thus, an enhancement and compression process is performed on the normalized intensity image using a nonlinear sine transfer function. The particularly designed sine transfer function increases the luminance of the dark pixels and reduces the luminance of overexposed pixels simultaneously and is defined as:

$$I_E\,(x,y) = Sin^2\,(I_n\,(x,y)^q * \pi/2)\,. \quad (3)$$

The key contribution of this paper is the computation of the image dependent parameter q used in (3) corresponding to the mean intensity value of the pixel and is defined as:

$$q = \frac{I_{M_n}(x,y)}{c1*(1- I_{M_n}(x,y)+ \varepsilon)} + c2 \tag{4}$$

where I_{M_n} is the normalized mean intensity value of the pixel at location (x, y), c1 and c2 are constants determined empirically, and ε = 0.01 is a numerical stability factor introduced to avoid division by zero when I_{M_n}= 1. The role of the control parameter q in intensity transfer function can be described well in the following manner. The transfer function is a decreasing function of the q parameter. Therefore, to boost the intensity, value of q parameter should be kept small and to lower the intensity, q should be large. Now q is directly proportional to the ratio of $I_{M_n}(x, y)/(1 - I_{M_n}(x, y))$. Hence, when the mean intensity of the pixel $I_{M_n}(x, y)$ is very high, it generates larger q and vice versa. This can be clearly seen in figure 1 which shows the plot of parameter q for normalized mean intensity values ranging from 0 to 1.

For the mean intensity values close to 0, there is a strong possibility of noise being enhanced in the extreme dark regions. Therefore, c2 is added in (4) to counteract the noise enhancement. The range of c2 in this experiment is empirically determined to be .13 to .4. Note that addition of c2 has almost negligible effect on the pixels with intensity values close to 1 for which equation 1 produces much larger q compared to that of dark pixels. For the exceedingly bright pixels, the transfer function may produce very low or almost black intensity values (when intensity value is close to 1, ratio of $I_{M_n}(x, y)$ / (1-$I_{M_n}(x, y)$) produces much larger q). To avoid this phenomenon, the denominator is multiplied with c1, thus, q is inversely proportional to c1. It reflects the fact that for overexposed images, c1 in the range of 2 to 4 gives good results. If the image is very dark, then c1 value in the range of 5 to 8 helps sufficiently to boost the luminance.

In this method, using the sine square transfer function, the luminance of dark pixels is greatly pulled up, and the intensity of bright pixels is pulled down and the well illuminated pixels are left unaltered. A set of curves of the sine squared transfer function is provided in figure 2 for various mean intensity values ranging from 0.01 to 0.99. In addition, the transfer function compresses dynamic range while preserving fine details and provides good enhancement results. In this method, the mean image is computed using a Gaussian smoothing operator. The Gaussian mask is defined as follows:

$$G(x,y) = K \cdot e^{\left(\frac{-(x^2+y^2)}{c^2}\right)} \tag{5}$$

where c is the Gaussian surround space constant and K is the constant to ensure that the area under the Gaussian is 1. K is determined by evaluating the following integral across the Gaussian kernel:

$$\iint K \cdot e^{\left(\frac{-(x^2+y^2)}{c^2}\right)} \cdot dxdy = 1 \tag{6}$$

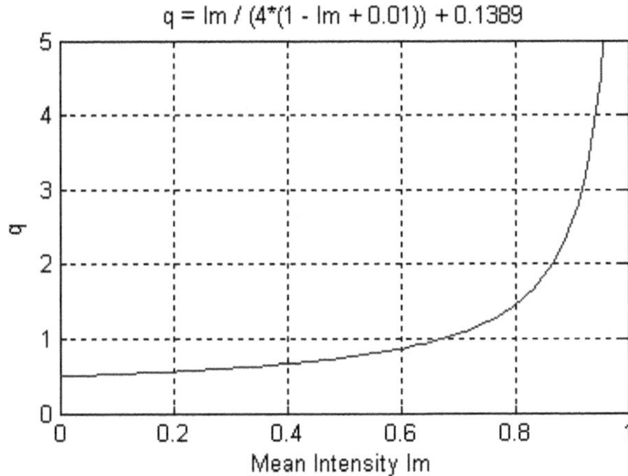

Fig. 1. Plot of q values for I_{M_n} *(mean intensity Im)* for c1 =4 and c2 = .1389

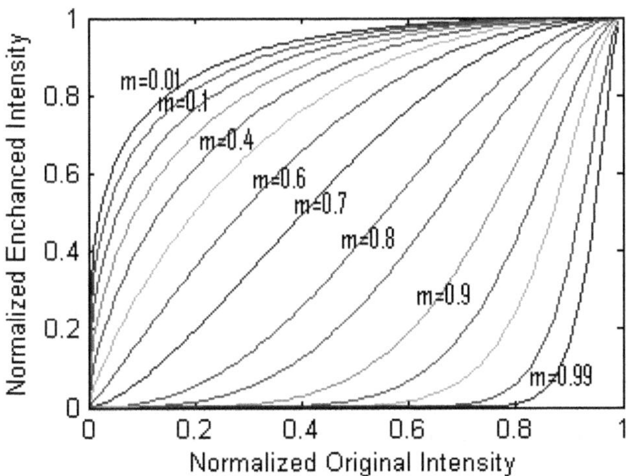

Fig. 2. Curves of the transfer function corresponding to mean values ranging from 0.01 to 0.99 for c1 = 4 and c2 = 0

Choosing the right scale is very important as it determines the size of the neighborhood for 2D discrete spatial convolution with a Gaussian kernel. Convolution using a small scale uses few neighboring pixels, thus luminance information of the nearest neighboring pixels is available. On the other hand, large scale convolution provides information of global luminance distribution. In other words, Gaussian smoothing with small scale preserves details, where as large scale convolution provides global tonality, which helps produce more natural looking enhanced images [14]. In this method, a multi scale Gaussian is used to produce the mean intensity image I_{M_n} as multiple convolutions yield more complete information about the overall luminance distribution. The neighborhood averaging with multi scale Gaussian can be described as follows:

$$I_M(x,y) = \sum_{m=0}^{M-1}\sum_{n=0}^{N-1} I(m,n)\, G_{i(m+x,n+y)} \tag{7}$$

where G_i indicates the weighted sum of i Gaussian functions with different scales. In this method, we used a combination of small scale (1-5% of the image size), medium scale (10-15 % of the image size), and a large scale (25-45% of the image size) Gaussians to obtain optimal results.

2.2 Contrast Enhancement

Contrast enhancement is performed in a similar way as in AINDANE [14]. As illustrated in [14], during the dynamic range compression, the mid-tone and low frequency components responsible for fine details and local contrast are degraded. To compensate this degradation and restore the contrast of the luminance-enhanced image, a center-surround contrast enhancement method is performed as follows:

$$S(x,y) = 255 \cdot I_E(x,y)^{E(x,y)} \tag{8}$$

where the exponent is defined by:

$$E(x,y) = R(x,y)^P = \left(\frac{I_{conv}(x,y)}{I(x,y)}\right)^P . \tag{9}$$

$S(x,y)$ is the contrast-enhanced pixel intensity, $R(x,y)$ is the intensity ratio between low-pass filtered $I_{conv}(x,y)$ and original intensity image $I(x,y)$. P is an image dependent parameter determined by the global standard deviation of the input intensity image $I(x,y)$. As shown in (9), if the center pixel is brighter than surrounding pixels then the ratio $R(x,y)$ is smaller than 1 and hence, the intensity of this pixel is pulled up. Likewise, if the center pixel is darker than the neighboring pixels then the ratio $R(x,y)$ is grater than 1 and the intensity of the pixel is lowered. By performing this method, contrast and fine details of the compressed luminance image can be sufficiently improved while maintaining the image quality.

2.3 Color Restoration

Recall from Section 2.1 that the color image was first converted to a grayscale image using the NTSC standard. We now convert our enhanced gray scale image back to color image using the chromatic information of the original image as follows:

$$S_j(x,y) = S(x,y)\frac{I_j(x,y)}{I(x,y)} \cdot \lambda_j \qquad j \in \{R,G,B\} \tag{10}$$

where j = R, G, B represents the red, green, and blue spectral bands respectively, $I_R(x,y)$, $I_G(x,y)$ and $I_B(x,y)$ are R, G, and B color values in the original color image, $I(x, y)$ is intensity image computed using (1), $S(x, y)$ is the enhanced intensity image computed using (8), and S_R, S_G and S_B are the RGB values obtained to form the enhanced color image. The parameter λ adjusts the color hue of the enhanced color images. It takes different values in different spectral bands. Normally its value is close to 1. However, when all λs are equal to 1, according to eqn. (10) the chromatic information of the input color image is preserved for minimal color shifts [14].

3 Results and Discussion

The NDNE algorithm has been applied to a large number of digital images captured under varying lighting conditions for performance evaluation and comparison with other state of the art techniques. Results as well as detailed discussion about specific characteristics of our algorithm are presented in this section.

The sample images in figure 3 taken under varying lighting conditions are used for evaluating the effectiveness of the proposed algorithm. Figure 3(a) shows the image captured under extremely dark lighting condition. After processing it with NDNE, the visual quality is highly improved as can be seen in figure 3(b). In figure 3(c) the entire sample image is overexposed. By applying NDNE, the details in the overexposed regions are made visible which can be seen from figure 3(d). For example, we can see the letters indicating the name of the river which is not visible in the original image. Image in figure 3(e) was captured under medium scale lighting conditions where some objects are well lit while others are not. The algorithm produces well balanced image in which, the regions that are already sufficiently illuminated are left unaltered which can be verified from figure 3(f). By decreasing the intensity around the lamp, the details are enhanced while the intensity around the window as well as the mirror frame is increased. The image in figure 3(g) contains both extremely dark and bright regions. Again, as can be seen in figure 3(h), the resulting image shows much greater details.

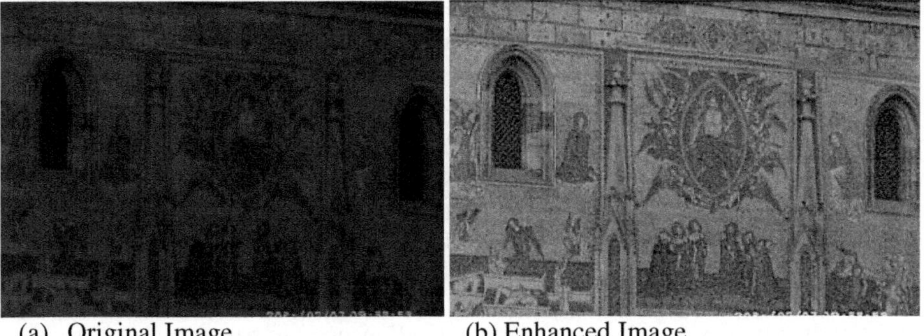

(a) Original Image (b) Enhanced Image

Fig. 3. Image enhancement results by NDNE

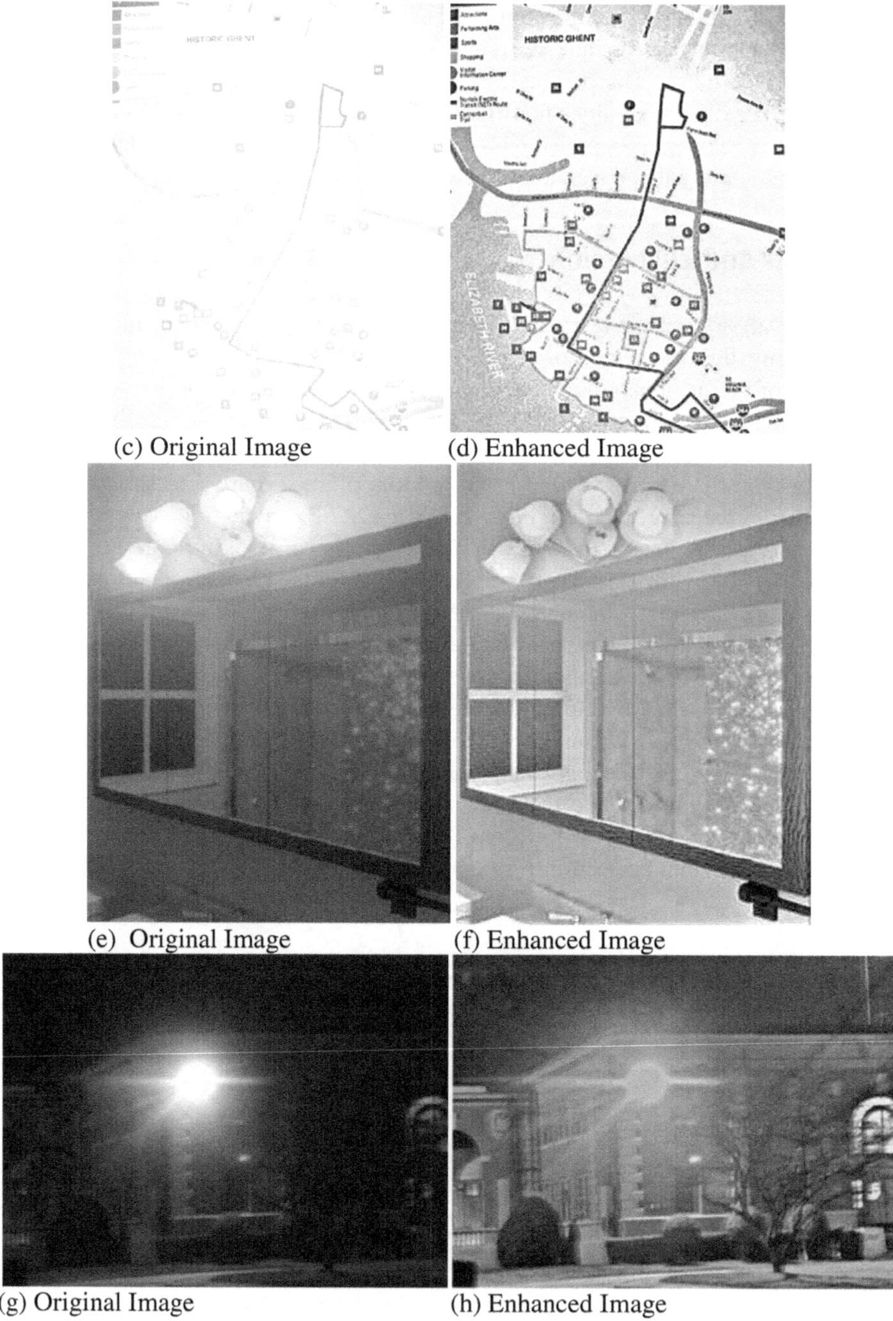

(c) Original Image (d) Enhanced Image

(e) Original Image (f) Enhanced Image

(g) Original Image (h) Enhanced Image

Fig. 3. (*Continued*)

3.1 Comparison with AINDANE and LTSNE

In figure 4 the sample image is provided for comparison with the performance of AINDANE, LTSNE and NDNE. It can be observed that the images produced by NDNE possess more details with high visual quality in both the underexposed and over exposed regions than those processed by above mentioned techniques. As we can see in figure 4(b), AINDANE over enhances the bright areas. In the image processed by LTSNE (figure 4(c)), the bright areas are compressed well, however, it creates dark halo around the bright areas. The resultant image (figure 4(d)) of the proposed algorithm shows better balance in enhanced and compressed regions yielding more details.

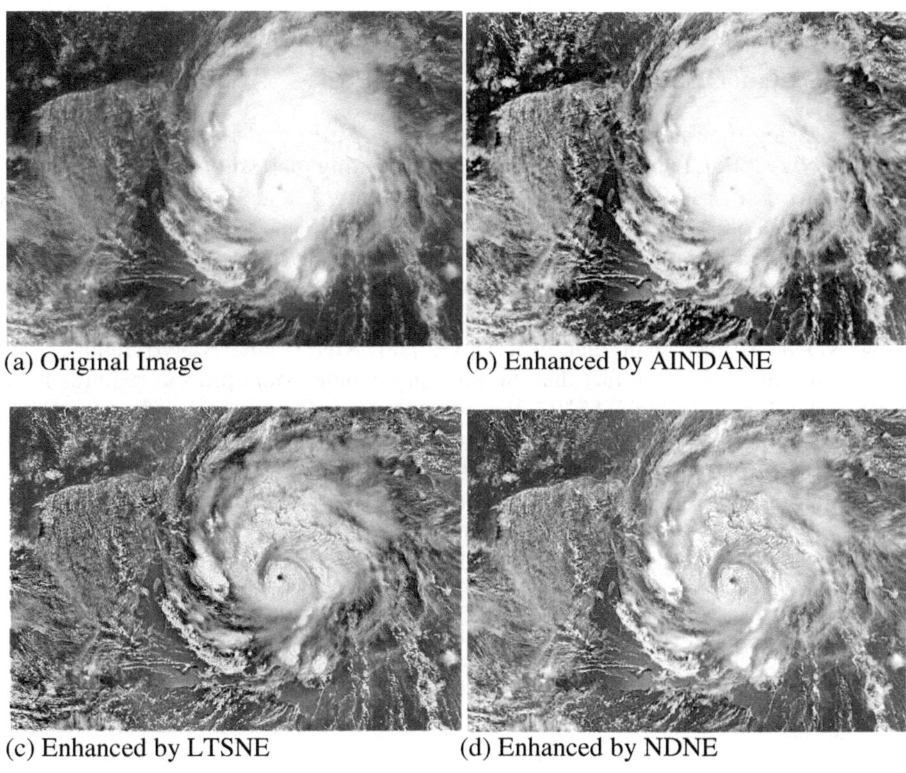

(a) Original Image (b) Enhanced by AINDANE

(c) Enhanced by LTSNE (d) Enhanced by NDNE

Fig. 4. Performance comparison with AINDANE, LTSNE, and NDNE

3.2 Statistical Evaluation

A statistical method proposed in [17] is used to evaluate the visual quality of the enhanced images in order to compare the performance of different enhancement techniques. In this method, visual quality of an image is measured in terms of brightness and contrast using the statistical properties such as image mean and mean of the block standard deviation. The mean and mean of standard deviations of the original images and the enhanced images are plotted in figure 5. The diamonds, circles, and squares represent the images enhanced using the AINDANE, LTSNE, and NDNE algorithms respectively. The numbers inside the shapes indicate the enhanced image corresponding to the original image number. Image 1(figure 3(g)) was captured under very dark lighting condition. All three techniques increase the luminance as well as contrast. Image 2(figure 4(a)) is captured under mixed lighting condition and contains both dark and bright regions. All three methods enhance the image sufficiently and the results fall in the visually optimal region. However, LTSNE and NDNE posses higher contrast as they perform compression as well as enhancement. Image 3(figure 3(c)) is very bright for which AINDANE enhances the extremely brighter pixels whereas LTSNE and NDNE compresses them. As can be seen in figure 5, resultant image of NDNE is closer to the optimally visual region as it compresses the overexposed regions with more details.

3.3 Computational Speed

The processing time needed for enhancing images of different sizes is compared between AINDANE, LTSNE, and NDNE. The computing platform is an Intel Pentium 4 system, processor running at 3.06 GHz, 1GB memory, and Windows XP® Professional Edition operating system. AINDANE, LTSNE, and NDNE implemented in C++ are applied to process the same set of images. The processing time needed to enhance images of various sizes is provided in Table 1 for comparison between AINDANE, LTSNE, and NDNE. Table 1 show that the time required to process an image using NDNE is less than that of LTSNE and AINDANE. NDNE requires less processing time due to the fact that the intensity enhancement process requires fewer and simpler functions. In LTSNE, the computation of image dependent parameter involves computationally expensive logarithm and tangent functions. Whereas in NDNE, these functions are replaced with a division operation in order to reduce processing time.

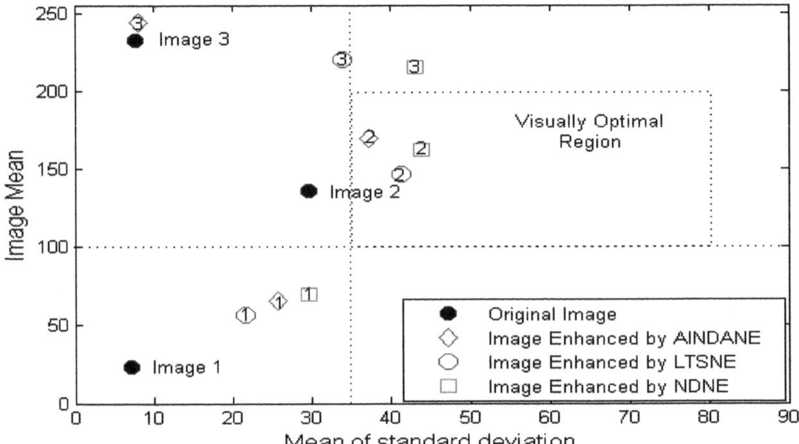

Fig. 5. Comparison of visual quality of images enhanced by AINDANE, LTSNE, and NDNE

Table 1. Comparison of processing time of AINDANE, LTSNE and NDNE

Image size (pixels)	Processing time by AINDANE (seconds)	Processing time by LTSNE (seconds)	Processing time by NDNE (seconds)
360 × 240	0.25	0.19	0.173
640 × 480	1.4	0.687	0.527
1024 × 768	2.8	1.716	1.28
2000 × 1312	6.7	4.572	3.438

4 Conclusion

In this paper, we presented a new nonlinear image enhancement algorithm NDNE to improve the visual quality of digital images captured under complex lighting conditions. The method performs adaptive luminance enhancement, contrast enhancement, and color restoration steps. Dividing the enhancement process in three steps increased the flexibility and provided more control for fine tuning. This method allows for corrections of non uniform illumination, shadows, and other traditionally difficult lighting issues. Effectiveness of the algorithm depending on the statistical information (mean and mean of standard deviation) of the original and enhanced images has been evaluated based on its capability to automatically refine the image quality. The algorithm has been tested on a large datasets and the performance has been verified. The images enhanced by NDNE possess improved visual quality compared with those produced by other techniques in terms of better contrast and luminance enhancement of images containing underexposed and overexposed regions. Moreover, the processing speed of NDNE is faster than LTSNE, and AINDANE. NDNE algorithm would be a more efficient and fast image enhancement technique that could be useful in many applications. We envision a number of improvements that would set all the parameters automatically to make the algorithm fully adaptive.

References

1. Digital Photography Color Management Basics. International Color Consortium (April 2005), `http://www.color.org/ICC_white_paper_20_Digital_photography_color_management_basics.pdf`
2. Bianco, C.: How Vision Works (April 1, 2000), http://health.howstuffworks.com/eye.htm (January 13, 2010)
3. Kolb, H.: How the Retina Works. American Scientist 91(1) (2003)
4. Dynamic Range Compression, `https://ccrma.stanford.edu/~jos/fp/Nonlinear_Filter_Example_Dynamic.html`
5. Gonzalez, R., Woods, R.: Digital Image Processing, pp. 167–168. Addison Wesley Publishing Company, Reading (1992)

6. Bonnier, N., Simoncelli, E.P.: Locally Adaptive Multiscale Contrast Optimization. In: Proc. 12th IEEE International Conference on Image Processing, vol. 1, pp. 949–952 (2005)
7. Pizer, S.M., Amburn, E.P.: Adaptive histogram equalization and its variations. In: SPIE Proceedings, vol. 1092, pp. 290–300 (1989)
8. Funt, B., Ciurea, F., McCann, J.: Retinex in Matlab. In: Proc. CIC'8 8th Color Imaging Conference, Scottsdale, Arizona, pp. 112–121 (2000)
9. Land, E., McCann, J.: Lightness and Retinex Theory. Journal of the Optical Society of America 61(1) (1971)
10. Rahman, Z., Jobson, D., Woodell, G.: Multiscale Retinex for Color Image Enhancement. In: Proceedings of the IEEE International Conference on Image Processing. IEEE, Los Alamitos (1996)
11. Rahman, Z., Jobson, D., Woodell, G.: Multiscale Retinex for Color Rendition and Dynamic Range Compression. In: Tescher, A.G. (ed.) Proceedings of the SPIE Applications of Digital Image Processing XIX, vol. 2847 (1996)
12. Rahman, Z., Jobson, D., Woodell, G.: Multiscale Retinex for Bridging the Gap between Color Images and Human Observations of the Scene. IEEE Transactions on Image Processing: Special Issue on Color Processing 6, 965–976 (1997)
13. Rahman, Z., Woodell, G., Jobson, D.: A Comparison of the Multiscale Retinex with Other Enhancement Techniques. In: Proceeding of the IS&T 50th Anniversary Conference, IS&T, pp. 426–431 (1997)
14. Tao, L., Asari, K.V.: An Adaptive and Integrated Neighborhood Dependent Approach for Nonlinear Enhancement of Color Images. SPIE Journal of Electronic Imaging 14(4), 1.4 – 1.14 (2005)
15. Arigela, S., Asari, K.V.: A Locally Tuned Nonlinear Technique for Color Image Enhancement. WSEAS Transactions on Signal Processing 4, 514–519 (2008)
16. Valensi, G.: Color Television System. US Patent 3, 534, 153 (October 1970)
17. Jobson, D.J., Rahman, Z., Woodell, G.: Statistics of Visual Representation. In: SPIE Proceedings, vol. 4736, pp. 25–35 (2002)

Queue and Priority Queue Based Algorithms for Computing the Quasi-distance Transform

Raffi Enficiaud

INRIA Paris-Rocquencourt, Domaine de Voluceau, BP 105,
78153 Le Chesnay Cedex, France
raffi.enficiaud@{inria.fr,mines-paris.org}
http://cmm.ensmp.fr/~enficiau

Abstract. The quasi-distance transform introduced by Beucher shows interesting properties for various tasks in image processing such as segmentation, filtering and images simplification. Despite its simple formulation, a naive and direct implementation of the transform leads to poor results in terms of computational time. This article proposes a new algorithm for computing the quasi-distance, based on a front propagating approach by means of queues and hierarchical queues. Quantitative analysis of the running time are provided, and show an exponential downscale of the complexity compared to the original algorithm.

1 Introduction

Morphological distance transforms on binary sets are one of the major tools widely used in the context of Mathematical Morphology. It provides for instance an elegant way for counting the number convex overlapping grains in binary images. In this context, each grain is identified by a local maxima in the transformed images [5]. The morphological distance transform may also be used for grain segmentation, where the inverse of the transformed binary image is used as a topographical map on which a watershed is applied (see Fig. 1 for illustrations). More precisely, a morphological distance transform is defined as the result of a sequence of pairs of erosions, each of them combined with a set subtracting function. It is an instance of a wider class of transforms, namely the *residual transforms* that we will be discussing below, which will serve as an introduction for the extension of the morphological distances to grey level images.

1.1 Residual Transforms

We now give an a formal definition of the residual transforms in the binary functions over sets, which will introduce the framework for numerical functions detailed in the next section. A residual transform Θ on a set X is defined by the input of a family of pairs of functions $\left\{(\psi_i, \zeta_i)_{(i)}, i \in \mathbb{N}^* | \forall i, \psi_i \geq \zeta_i\right\}$, together with a family of functions $(r_i)_{(i)}$ of two arguments, and such as:

$$\Theta : \begin{cases} \mathcal{P}(E) \to \mathcal{P}(E) \\ \mathbf{X} \quad \mapsto \bigcup_i r_i\left(\zeta_i(\mathbf{X}), \psi_i(\mathbf{X})\right) \end{cases} \tag{1}$$

A. Campilho and M. Kamel (Eds.): ICIAR 2010, Part I, LNCS 6111, pp. 35–44, 2010.

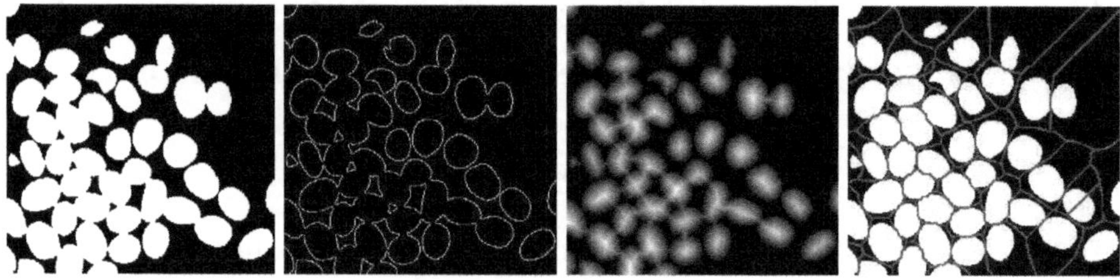

Fig. 1. An example of the distance transform for binary functions and its possible use for convex set counting and segmentation. From left to right: original image; $\epsilon^{(0)} \setminus \epsilon^{(1)}$; morphological distance transform; watershed on the distance map (grains separation). The local maxima in the distance map also provide the number of convex grains in the image.

The $(r_i)_{(i)}$ are called the *residuals* while $(\psi_i, \zeta_i)_{(i)}$ are the *primitives.* The definition above is quite general, and we will illustrate it with first the distance transform, and second the ultimate erosion transform. In the case of a distance transform in a binary set framework, it is easy to see from the sequential behaviour of the distance transform itself that it can be achieved with residuals being the set difference $(\cdot, *) \mapsto \cdot \setminus *$ for all i, and the primitives family being $\zeta_i = \epsilon^{(i)}, \psi_i = \epsilon^{(i+1)}$ for all i, where $\epsilon^{(i)}$ is the unitary erosion iterated i times and taking $\epsilon^{(0)} = Id_{\mathcal{P}(E)}$. Such an example is in fact interesting in order to introduce another notion in the residual transforms. For any sets A, B, C such that $C \subset B \subset A$, we have $(A \setminus B) \cup (B \setminus C) = A \cup \overline{C}$: the intermediate B has no effect. By noticing that the family of erosions $\{\epsilon^{(i)}\}$ is a decreasing map of i and taking as a convention that $\epsilon^{(+\infty)} = \varnothing$, an inference on the previous subset rule easily leads to $\Theta = Id_{\mathcal{P}(E)}$. In fact what is interesting in the binary distance transform does not lie in the results of residuals themselves, but rather in the occurring step of each of these residuals. This is the reason why the notion of an indicator function, denoted q in the sequel, should also be introduced. In the special case of the binary distance transform, there is no ambiguity in the choice of the occurrence indication since the residual family r_i produces a non-overlapping family of sets (see Fig. 1). As indicator function q, we can safely take $q : x \mapsto 1 + \arg_i(x \in r_i)$. As a second example, by keeping the residuals as being the set subtraction and taking as primitives $\psi_i = \epsilon^{(i)}$ and $\zeta_i : \mathbf{X} \mapsto \gamma_X^{rec} \circ \epsilon^{(i-1)}(\mathbf{X})$, where $\gamma_{\mathbf{X}}^{rec}$ is the opening by reconstruction under the set $\mathbf{X}$, we define the Θ transform as being the ultimate erosion (as convention, we use $A \setminus \varnothing = A$ for any possibly empty set A, and $A \subset B \Rightarrow A \setminus B = \varnothing$). Again, the indicating function q here is unambiguously defined since for each of connected component, the *activity* of the associated family of residuals is non empty at the step just before the complete removal of the connected component. From these two examples, it becomes clear that the residual transforms are a powerful unifying framework for setting up morphological transforms. However the discussion remained in the binary case, and the next section discusses about the possible extensions onto valued sets.

1.2 The Quasi-distance: A Morphological Distance for Grey Level Images

We consider in this section an image function $\mathcal{I} : E \rightarrow \mathbb{R}$ mapping each point of the universe E to a value in $\mathbb{R}$. The extension of the previous definitions to valued sets is not a direct transposition of the binary case. This is mainly due to the fact that two properties were implicitly used in the binary case. The first one is that the residuals in the binary case form a pairwise disjoint family of sets. This property does not hold any more for grey level images: the set difference cannot be used and a more appropriate choice should be made concerning the residuals to keep. The grey level difference $\forall i, r_i : (\cdot, *) \mapsto |\cdot - *|$ seems to capture the desired properties [1].

The second property, which in fact is a consequence of the first one, is that the indicator function for binary images is uniquely defined at each point. This remark led Beucher to consider the occurrence of the maximal residue only, which lead to the following definition of q:

$$q : x \mapsto \begin{cases} 1 + \bigvee \arg_i \bigvee \{r_i | r_i > 0\} & \\ 0 & \text{if } \forall i, r_i = 0 \end{cases} \tag{2}$$

From this setting and in case of the maximal residue at different steps ($\arg_i$ returning the set of the occurrences), only the greatest occurrence is kept. If all of the residues are null at x, it means that the activity of the operator is empty at x, which discard x from being of any interest.

The *quasi-distance transform* is the immediate application of the settings above, with the same primitives as in the binary case, $\zeta_i = \epsilon^{(i)}, \psi_i = \epsilon^{(i+1)}$. This transform is usually applied on the image and its inverse. The name "distance" comes from the analogy one may do with the classical distance transform on binary sets. The interesting result of the transform lays in the indicator function which acts almost similarly as in the binary case. However, the above definition does not lead to a proper distance function, since it does not meet the 1-Lipschitz property, as illustrated in Fig.2. In other terms, $\exists (x, y) \in \mathit{supp}(q)^2 \ / \ |q(x) - q(y)| > |x - y|$. The subset of points violating the 1-Lipschitz property are named in [2] the "hung up" points. In order to obtain a distance map meeting the 1-Lipschitz property, a regularizing procedure is required after the computation of the map q. To summarize, the quasi-distance transform should be processed in two steps: the computation of the distance (indicator) function itself through the family of successive erosions, and then the regularizing of the resulting distance map.

An example of the quasi-distance transform in the context of visual surveillance is provided on Fig.3. The residue function is such as the step of maximum contrast is kept: in this example the clothes have a high contrast with the surrounding scene. The residue is maximized at the first step on the boundary of the silhouette, which becomes the location from which a continuous distance outside the body is propagated. From this example, the quasi-distance acts as a contrast filter prior to a segmentation. This contrast property is used in [3] in order to keep the edges of maximum intensities on a grey level gradient image.

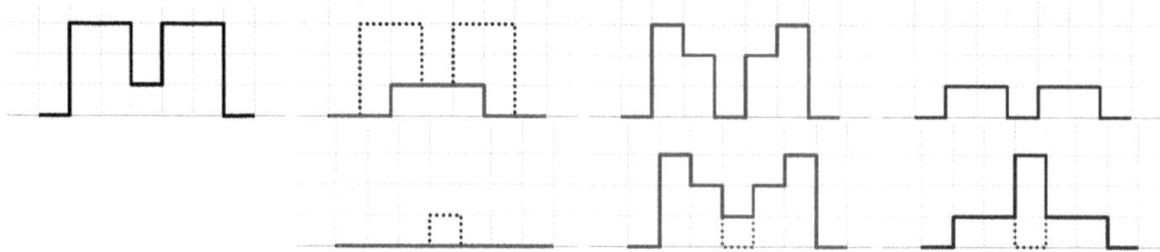

Fig. 2. The problem of "hung up" distance points. From left to right, top row: original set $f(\mathbf{X}) = f_\mathbf{X}$, $\epsilon^{(1)}(f_\mathbf{X})$, $r_1 = f_\mathbf{X} - \epsilon^{(1)}(f_\mathbf{X})$ and q_1. Bottom row: $\epsilon^{(3)}(\mathbf{X})$, $r_3 = \epsilon^{(2)}(f_\mathbf{X}) - \epsilon^{(3)}(f_\mathbf{X})$ and q_3. The central point of the map q_3 jumps from 0 to 2 and hence violates the 1-Lipschitz property.

Fig. 3. Example of quasi-distance transform. From left to right: original image, regularized distance on the image, and regularized distance on the inverse of the image.

The initial algorithms for the computation of the family of erosions and the regularizing are presented in Alg.1 and Alg.2 respectively. The first algorithm follows directly the definition of the erosion/residue process as previously described. By Q[E] we denote the subset E of points of Q. The iteration stops when no variation occurs between two successive erosions. The second algorithm is the regularization of the distance map. The subset of points that do not satisfy the 1-Lipschitz condition are looked for, and this property is enforced by adding a unit to the current distance step. Since this modification is able to create new non 1-Lipschitz points, the previous operations are iterated until the idempotence is reached.

From a computational point of view, the bottleneck of such an approach is the processing of the whole images at each step. In fact, the points that really need to be processed form a subset of the image which usually gets smaller while the iteration step increases. This assertion is asymptotically always true, in that extent that only the lowest points of the image keep eroding before reaching the idempotence. Focusing the computation on the points where the activity of the residual is not empty would avoid unnecessary operations. This remark is the main property we use in the design of our algorithms.

The sequel of this article is organized as follow: we propose two algorithms for the computation of the family of erosions and for its the regularization in section 2 and 3 respectively. Their performances in terms of computation time

Algorithm 1. Computation of q - Algorithm presented by S. Beucher [2]

```
Data: I
Result: Q, R
W1 ← 𝓘
Q, R, W2 ← 0
i ← 0
repeat
    i ← i + 1
    W2 ← ε(W1)
    residue ←W1−W2
    E ← (residue ≥ R and residue ≠ 0)
    Q[E] ← ∨{i,Q[E]}
    R ← ∨{residue,R}
    W1 ← W2
until residue = 0
```

Algorithm 2. Regularizing of q - Algorithm presented by S. Beucher [2]

```
Data: Q
Result: RQ
RQ ← Q
i ← 0
repeat
    W ← RQ−ε(RQ)
    E ← ¬(W ≤ 1)
    RQ[E] ← ε(RQ[E]) + 1
until E = ∅
```

are compared to the classical algorithms in section 4. Section 5 draws concluding remarks and further improvements.

2 Quasi-distance Transform Algorithm

In the following, we only deal with flat structuring elements the we consider as defining a neighbouring graph. We denote $\mathcal{N}_p(W)$ the neighbourhood of the point p in the image W or simply $\mathcal{N}_p$ if there is not ambiguity over the image. The algorithm for the quasi-distance transform is presented in Alg. 3. It falls into two distinct parts: an initialization followed by a propagation loop. The initialization looks for all the points of the input image $\mathcal{I}$ for which the activity of the residue is not empty or equivalently, the initial support of the residue r_1. Let us call E_0 this subset, it is given by $E_0 = \{x | \exists v \in \mathcal{N}_x, x < v\}$. All the points of E_0 are placed in the queue, but no further processing is performed at this stage.

The propagation loop falls into two parts. The first is the effective computation of the erosion of $\mathcal{I}$ and the update of the distance Q and the residue $R = \cup_i r_i$ maps accordingly (lines 11-13). Inside each neighbourhood and once the erosion is done, the residue r_i is computed and compared to the previous residues $\cup_{j<i} r_j$ stored in R. If the conditions of Equ.2 are met, the maps Q and R are updated. Note that the occurrence of the maximal residue is corrected regarding the definition of q. Since the element in the queue do not carry neighbouring information, two intermediate images $W1$ and $W2$ are needed. These two images

are involved for the computation of the successive erosions. This process being iterative, the changes of the eroded image $\epsilon^{(i+1)}$ should be reflected inside the reference image $\epsilon^{(i)}$ (line 15), which only occurs for the subset E_i.

The second part of the algorithm *propagates* the changes made on E_i to the neighborhood of E_i, because newly eroded points of E_i can erode their neighbours in turn. For each point p of E_i, the algorithm looks for the points above p inside its neighbourhood. These points form a new set E_{i+1}. This way, a point can be put in the propagation queue from several locations. The image C carries the states of the points and prevents them from being inserted more than once (line 17). The erosion and propagation steps of the algorithm are illustrated on figure 4.

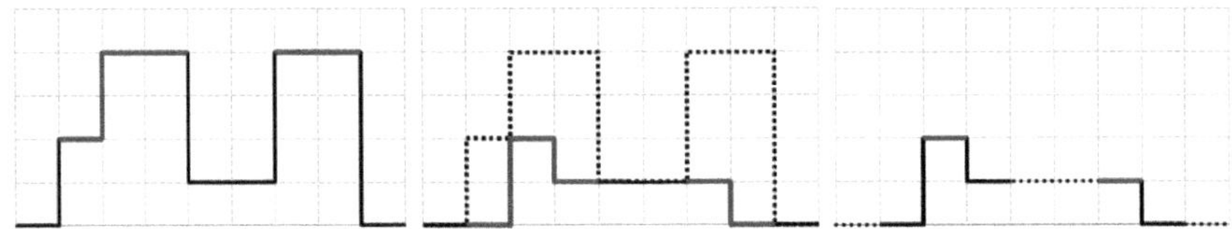

Fig. 4. Main steps of Alg.3. Left: the original profile and the subset E_0 (in red). Middle: result of the erosion of E_0. Right: new subset E_1 constructed from E_0.

The performance of this algorithm is directly related to the cardinality $\#E_i$ of the set E_i of points being eroded at step i: the fewer the processed points, the faster the step. It is worth noticing that first, this cardinality is bounded by the number of points of the image. Second, even if the sequence $(\#E_i)_i$ is not monotonically decreasing[1], the family of erosions always converges to the global minimum of the image, which means that the sequence $(\#E_i)_i$ has 0 as unique limit. This proves the convergence and justifies the stop criterion of the algorithm (line 8).

3 Distance Map Regularization Algorithm

The algorithm for the distance regularization is given in Alg. 4. Q stands for the non-regularized input distance map generated by the previous algorithm, while RQ stands for the regularized distance map. "*hq*" is a hierarchical queue, and it is assumed that it provides a method named "empty priority x", which removes all the elements at priority x from the queue.

The algorithm falls into three parts, also illustrated on Fig.5. The first part looks for the subset NL of points that do not verify the 1-Lipschitz constraint: $NL = \{x \mid \exists\, v \in \mathcal{N}_x, v < x, |x - v| > 1\}$. The aim of the second part is to find the points of NL which are also neighbours of $\overline{NL}$ or, in other words, to find the

[1] Consider the front around an isolated eroding point which describes a circle with a growing radius.

Algorithm 3. Computation of q

```
Data: I
Result: Q, R
C ← candidate, W1 ← I, W2 ← I, Q ← 0, R ← 0
f1, f2 ← ∅
forall the p ∈ W1 do
    forall the v ∈ {N(p, W1) \ p} do
        if (v < p) then f1 ← f1 + p, break
i ← 1
while f1 ≠ ∅ do
    forall the p ∈ f1 do
        C(p) ← candidate
        W2(p) ← ∧N_p(W1)
        residue ← W1(p) - W2(p)
        if residue ≥ R(p) then R(p) ← residue; Q(p) ← i
    forall the p ∈ f1 do
        W1(p) ← W2(p)
        forall the v ∈ N_p(W2) do
            if (v > p) and (C(v) ≠ in-queue) then f2 ← f2 + v; C(v) ← in-queue
    f1 ← f2; f2 ← ∅; i ← i + 1
```

interior frontier ∂NL of NL. The points of ∂NL are inserted into the priority queue at the corrected priority. Indeed, two neighbour points of NL are mutually dependant and regularizing one of them is likely to make the other violate the 1-Lipschitz property. To avoid any redundancy, the processing of the points of $NL \cap \overline{\partial NL}$ is postponed to a next propagation step. Since 0 and 1 valued points are left unchanged by the first regularizing step, all points neighbours to the points at height 1 should be processed before the others. From this first step, we propagate the regularization to points of higher value. The priority queue structure is perfectly suited for this kind of ordered sequential processing of heights (see Fig.5). Once the priority queue has been initialized, neighbour points of ∂NL that do not meet the 1-Lipschitz property are added. Since we know exactly the value these points should have, they are pushed at the corrected height into the priority queue. This is the aim of the third loop of the algorithm. Hence, processed points are either already in the queue from the initialization step, or added after all the heights below have been processed.

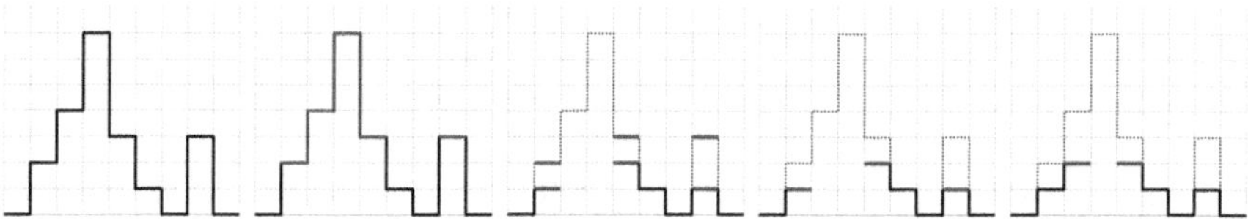

Fig. 5. Steps of the regularizing algorithm. From left to right: original profile, the subset NL of non 1-Lipschitz points (in red), the interior frontier ∂NL (in red) inserted in priority queue at the regularized priority (in green), regularization of the point of highest priority (in red) and propagation to their non 1-Lipschitz neighbours.

Algorithm 4. Regularization of the distance map

```
Data: I
Result: RQ
RQ ← Q
C ← none
hq ← ∅
forall the p ∈ RQ do
    forall the v ∈ N_p do
        if v > p + 1 then  C(v) ← in-queue
forall the p ∈ RQ do
    if C(p) = in-queue then
        forall the v ∈ N_p do
            least-neighbor ← I(v) + 1
            if (C(v) = none) and (p > least-neighbor) then
                RQ(p) ← least-neighbor
                hq ← hq + p at priority least-neighbor
while hq ≠ ∅ do
    pr ← highest priority of hq
    forall the p ∈ hq(pr) do
        forall the v ∈ N_p do
            if RQ(v) > pr + 1 then  hq ← hq + v at priority pr + 1, RQ(v) ← pr + 1
    hq ← empty priority pr
```

4 Results

We tested the proposed algorithms on 44 images of different dimension and content. We used a Pentium 4 PC at $2.8GHz$, with $512MB$ of DDR2 RAM and running Windows XP. For each image, we considered the average time of 10 realizations, which we then normalized by the number of pixels in the image. Processing time results are shown on Fig.6. Each figure is divided into two plots. The barplots show four bars per image: the two bright ones (first and third bars) represent the times associated with the proposed algorithms 3 and 4 respectively, while the two dark bars show the time per pixel for the classical algorithms 1 and 2. For each pairs of bars, one is for the original image and the other for the inverted image. The vertical unit is the $\log_{10}$ of the milliseconds per pixel. The line plots on the top show the difference between the new and classic algorithms. The dashed plot is for the inverted images while the plain one for the original images. Vertical unit is logarithmic (without dimension).

As we see on these figures, the proposed algorithms provide an improvement in terms of processing time. All time ratios are above 1, both for the quasi-distance algorithm and for its regularizing part. Since the points are processed only once in the regularization, the regularization algorithm performs very fast and offers a downscale $log_{10}ratio \in [1, 2]$ which is of an exponential magnitude. The proposed quasi-distance algorithm also reduces the overall complexity of the transform. The benefit is lower than for the regularization, for reasons that will be discussed. The log-ratio is however above 0.4 and often around 0.6, which means $10^{0.6} = 3.98$ less time spent on each pixels.

As mentioned, the performances are directly related to $\#E_i$ (see section 2). We see on Fig. 7 that this sequence quickly decreases during the firsts steps of the propagation. The queue approach suffers mainly from two major drawbacks.

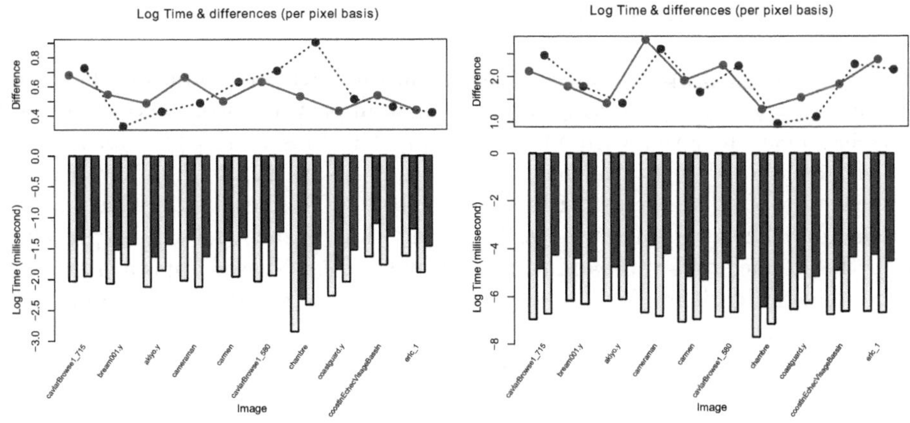

Fig. 6. Left: time per pixel for the Quasi-Distance algorithm 3 and 1. Right: time per pixel regularization algorithm 4 and 2. Please refer to the text for the legend.

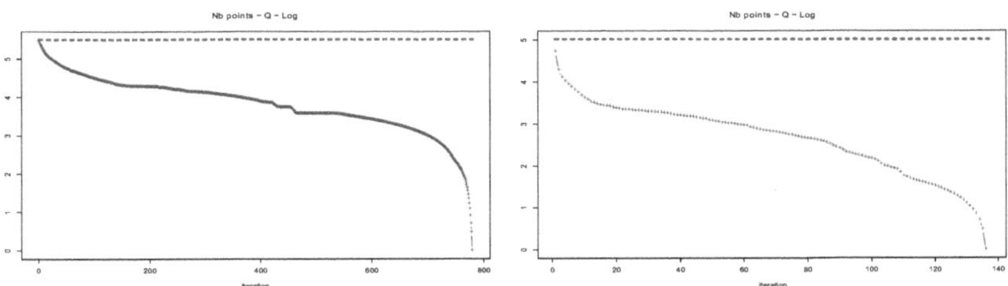

Fig. 7. Amount of points inside the queue of Alg.3 for two different images. The dashed line on the top indicates the number of points in the image. Units are logarithmic in number of pixels.

First, in order to avoid multiple insertion of a point and since the queue structure does not allow to do this, an additional work image implying an overhead is needed. Second, the implementations for computing the sequence of erosions can take advantage of the redundancy between two successive neighbourhoods (by eg. vectorized instructions [6]). As mentioned in introduction, there is no need to compute the erosion for all the pixels, and the latter kind of implementations necessarily suffer from a high computational costs at some step of the computation. The queue structure is well adapted when the subset of eroding points is sparse in the image but the natural order of the pixel being lost, not well adapted for taking into account the mutual information of its elements. From these observations, we also tested a combined approach which performs the vectorized processing at the first k steps, and then switches to the queued approach. The number of steps k depends naturally on the content of the image. We have determined $k \approx 14$ on the same set of images as providing the best results.

5 Conclusion and Future Work

In this article, we proposed a new algorithmic approach for computing the quasi-distance transform. Considering the results, an improvement of the computation time has been achieved without introducing any bias in the transform itself. On the conducted tests, the new approach is always faster than the original one: the new regularizing algorithm is often 10 to 100 times faster, which is very satisfactory. The quasi-distance itself is often 4 times faster, and we also proposed to further reduce the overhead introduced by the queue-based approach through an combined algorithm. We however think that these results can be improved.

As concluding remarks, we may say that the quasi-distance transform can also be subject to some modifications. An extension to colour images has been proposed in [7], which derives directly from the grey valued functions by means of lexicographical orders and colour distances. Besides, one may notice that from the initial setting of the quasi-distance, any global minimum has an influence over the entire image. However, it is often desirable this influence having a restricted support, for instance for taking into account an a priori knowledge concerning the size of objects of interest. By modifying the indicating function, one would benefit from the filtering properties of the residual transform while discarding some of its drawbacks. In such a case, the proposed algorithms can be easily extended to handle a new stop criterion, and the advantages in processing times would be even more important.

References

1. Beucher, S.: Numerical Residues. In: Proceedings of the 7th International Symposium on Mathematical Morphology, pp. 23–32 (2005)
2. Beucher, S.: Transformations résiduelles en morphologie numérique. Technical report of the Mathematical Morphology Center (2003)
3. Hanbury, A., Marcotegui, B.: Waterfall segmentation of complex scenes. In: 7th Asian Conference on Computer Vision (2006)
4. Beucher, S., Lantuéjoul, C.: Use of watersheds in contour detection. In: International Workshop on Image Processing, Real-Time Edge and Motion Detection/Estimation (1979)
5. Lantuéjoul, C., Beucher, S.: On the use of geodesic metric in image analysis. Journal of Microscopy 121, 39–49 (1981)
6. Brambor, J.: Algorithmes de la Morphologie Mathématique pour les architectures orientées flux. PhD Thesis, Mathematical Morphology Center (2006)
7. Enficiaud, R.: Algorithmes multidimensionels et multispectraux en morphologie mathématique: approche par méta-programmation. PhD Thesis, Mathematical Morphology Center (2007)

Denoising of Three Dimensional Data Cube Using Bivariate Wavelet Shrinking

Guangyi Chen[1], Tien D. Bui[2], and Adam Krzyzak[2]

[1] Department of Mathematics and Statistics, Concordia University, Montreal, Quebec, Canada H3G 1M8
guang_c@cse.concordia.ca
[2] Department of Computer Science and Software Engineering, Concordia University, Montreal, Quebec, Canada H3G 1M8
{bui,krzyzak}@cse.concordia.ca

Abstract. The denoising of a natural signal/image corrupted by Gaussian white noise is a classical problem in signal/image processing. However, it is still in its infancy to denoise high dimensional data. In this paper, we extended Sendur and Selesnick's bivariate wavelet thresholding from two-dimensional image denoising to three dimensional data denoising. Our study shows that bivariate wavelet thresholding is still valid for three dimensional data. Experimental results confirm its superiority.

Keywords: Denoising, high dimensional data, wavelet transforms, thresholding.

1 Introduction

Wavelet denoising for one-dimensional (1D) signal and two-dimensional (2D) images has been a popular research topic in the past two decades. The denoising problem to be solved can be defined as follows. Let $g(t)$ be a noise-free signal and $f(t)$ the signal corrupted with Gaussian white noise $z(t)$, i.e.,

$$f(t) = g(t) + \sigma_n z(t), \tag{1}$$

where $z(t)$ has a normal distribution with zero mean and unit variance, and σ_n is the noise variance. Our goal is to remove the Gaussian noise and recover the noise-free signal $g(t)$. The basic procedure of wavelet denoising [1] is to transform the noisy signal into the wavelet domain, threshold the wavelet coefficients, and then perform the inverse wavelet transform to obtain the denoised image. The thresholding may be undertaken on one wavelet coefficient alone or by considering the influence of other wavelet coefficients on the wavelet coefficient to be thresholded.

We briefly review the most popular wavelet denoising methods in the literature. Chen and Bui [2] extended the neighbouring wavelet thresholding idea to the multiwavelet case. They claimed that neighbour multiwavelet denoising outperforms neighbour single wavelet denoising and the term-by-term multiwavelet denoising [3] for some standard test signals and real-life signals. Chen et al. [4] proposed an

A. Campilho and M. Kamel (Eds.): ICIAR 2010, Part I, LNCS 6111, pp. 45–51, 2010.

image-denoising scheme by considering a square neighbourhood window in the wavelet domain. Chen et al. [5] also considered a square neighbourhood window, and tried to customize the wavelet filter and the threshold for image denoising. Experimental results show that both methods produce better denoising results. Chen and Kegl [6] have successfully applied the dual-tree complex wavelet to image denoising by using the complex ridgelet transform. Chen and Qian [7], [8] recently developed two methods for denoising hyperspectral data cube by using bivariate wavelet thresholding and wavelet packets, respectively. Sendur and Selesnick [9], [10] developed a bivariate shrinkage function for image denoising. Their results showed that the estimated wavelet coefficients depend on the parent coefficients. The smaller the parent coefficients are, the greater the shrinkage is.

In this paper, we extended the 2D bivariate wavelet thresholding proposed in [10], to 3D data cube denoising. We found that the bivariate thresholding formula is still valid for 3D datacube denoising. Experimental results show that the proposed method is better than denoising every spectral band separately. Our proposed method can denoise the whole data cube in one shot instead of performing a 2D wavelet thresholding on each spectral band of the data cube.

The paper is organized as follows. Section 2 extends the 2D bivariate wavelet denoising [10] to 3D data cube. Section 3 conducts experiments for denoising both a simulated 3D data cube and a real hyperspectral data cube. Our experimental results show that the proposed method outperforms band-by-band denoising. Finally, Section 4 draws the conclusions and gives future work to be done.

2 Bivariate Wavelet Thresholding for 3D Data Cube

In this section, we extended the 2D bivariate wavelet thresholding [10] to the 3D case. We found that the bivariate wavelet thresholding formula is still valid for denoising 3D data cube. However, the parameters in the thresholding formula have to be changed from the 2D case to the 3D case. For any given wavelet coefficient w_1, let w_2 be the parent of w_1, and define

$$y = w + n \tag{2}$$

where $w = (w_1, w_2)$ is the noise-free wavelet coefficients, $y = (y_1, y_2)$ the noisy coefficients, and $n = (n_1, n_2)$ the Gaussian white noise. The 2D bivariate thresholding formula [10] is given by

$$w_1 = y_1 \cdot \left(1 - \frac{\frac{\sqrt{3}}{\sigma}\sigma_n^2}{\sqrt{y_1^2 + y_2^2}}\right)_+. \tag{3}$$

where $(x)_+ = \max(x,0)$. The noise variance σ_n [1] can be approximated as

$$\sigma_n = \frac{median(|\, y_{1i} \,|)}{0.6745}, \quad y_{1i} \in \text{subband } HH_1. \tag{4}$$

and

$$\sigma = \sqrt{\left(\frac{1}{M}\sum_{y_{1i}\in S} y_{1i}^2 - \sigma_n^2\right)_+} \tag{5}$$

The HH_1 in (4) is the finest 2D wavelet coefficient subband, and M is the number of pixels in the 2D neighborhood window S.

The orthonormal basis of compactly supported wavelets of $L_2(R)$ is formed by the dilation and translation of a single function $\varphi(x)$

$$\varphi_{j,k}(x) = 2^{-j/2}\varphi((2^{-j} - k)$$

The function $\varphi(x)$ has a companion, the scaling function $\phi(x)$, and these functions satisfy the following relations

$$\begin{aligned} \phi(x) &= \sqrt{2}\sum_{k=0}^{L-1} h_k \phi(2x-k) \\ \varphi(x) &= \sqrt{2}\sum_{k=0}^{L-1} g_k \phi(2x-k) \end{aligned} \tag{6}$$

The h_k and g_k are the low-pass and high-pass filter coefficients, respectively. The wavelet transform of a signal is just the projection of the signal onto the wavelet bases. The above analysis is suitable for 1D signals. Let L denote the low pass output subband, and H the high pass output subband. The 3D wavelet transform can be applied recursively to the low-low-low (LLL_{i-1}) subband by performing a 1D wavelet transform along each of the three dimensions of the 3D data cube. The output subbands for the ith decomposition scale are the LLL_i, and seven high pass wavelet subbands HLL_i, HLH_i, HHL_i, HHH_i, LLH_i, LHL_i, and LHH_i. For denoising a 3D data cube, our study showed that the thresholding formula (3) is still valid.

Fig. 1. The Lena image used to build the simulated data cube

$$w_1 = y_1 \cdot (1 - \frac{\frac{\sqrt{3}}{\sigma}\sigma_n^2}{\sqrt{y_1^2 + y_2^2}})_+ . \quad (7)$$

The noise variance σ_n has to be approximated as

$$\sigma_n = \frac{median(|\, y_{1i} \,|)}{0.6745}, \ y_{1i} \in \text{subband } HHH_1 . \quad (8)$$

The subband HHH_1 is the finest high-high-high subband in the forward 3D wavelet transform. Also,

$$\sigma = \sqrt{(\frac{1}{M}\sum_{y_{1i}\in S} y_{1i}^2 - \sigma_n^2)_+} \quad (9)$$

where M is the number of pixels in the 3D neighbourhood cube S, and $(x)_+ = \max(0, x)$. In this paper, we have chosen S as a $7\times7\times7$ neighbourhood cube centered at y_1, the noisy 3D wavelet coefficient to be thresholded.

The proposed wavelet-denoising algorithm for 3D data cube can be summarized as follows.

1. Perform the forward 3D wavelet transform on the noisy data cube until a certain specified decomposition scale J.
2. Estimate the noise variance σ_n according to (8) and σ according to (9).
3. Threshold the 3D wavelet coefficients by using the bivariate thresholding formula (7).
4. Conduct the inverse 3D wavelet transform to obtain the denoised data cube.

It is worth explaining the parent-child relation in the 3D wavelet coefficients here. For any 3D wavelet coefficient y_1 and its parent y_2, both of them have to be in the same subands in two consecutive decomposition scales. For example, if y_1 is in subband LLH_i, then its parent has to be in subband LLH_{i-1}.

3 Experimental Results

In this section, we conducted some experiments to both simulated data cube and real life hyperspectral data cube. For the simulated data cube, we extracted a region of the Lena image with 128×128 pixels, and repeatedly pack this region for 128 times so that we can have a simulated data cube of size $128\times128\times128$. We also tested a real life hyperspectral data cube in this paper. This data cube was acquired by the Airborne Visible/Infrared Imaging Spectrometer (AVIRIS) in the Cuprite mining district, Nevada, by Jet Propulsion Laboratory (JPL) in 1997. The original scene with size of 614×512 pixels and 224 bands is available online at http : //aviris.jpl.nasa.gov/html/aviris.freedata.html.

The upper-right corner of the scene that consists of 350 x 350 pixels and 224 bands was extracted for the experiments in this paper. This scene is well understood mineralogically and it has been made a standard test site for validation and assessment of remote sensing methods. For our experiment in this paper, we extracted a small cube from this Cuprite data cube with size $128\times128\times128$. Fig. 2 shows the AVIRIS Cuprite scene at wavelength 827 nm (spectral band 50).

The noisy data cubes are obtained by adding Gaussian white noise to the original noise-free data cubes. The noise variance σ_n goes from 5 to 40 in the experiments conducted in this paper. The Daubechies-8 wavelet filter is used for the existing denoising methods. The neighbourhood window size is chosen as $7\times7\times7$ pixels and the wavelet transform is performed for three decomposition scales. Tables 1 and 2 tabulate the PSNR values of the denoised data cube resulting from the proposed method and the method that denoises each spectral band separately for the simulated and Cuprite hyperspectral data cubes at different levels of noise variance. The peak signal to noise ratio (PSNR) is defined as

$$PSNR = 10\log_{10}\left(\frac{N\times255^2}{\sum_{i,j,k}(B(i,j,k)-A(i,j,k))^2}\right) \tag{10}$$

where N is the number of pixels in the data cubes, and B and A are the denoised and noise-free data cubes. From the experiments conducted in this paper we found that the proposed method is better than denoising every spectral subband separately. Therefore, it is preferred in denoising real life noisy data cubes.

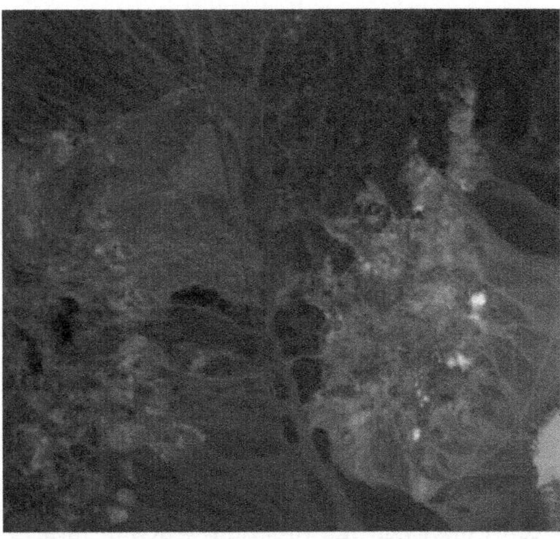

Fig. 2. The AVIRIS Cuprite scene at wavelength 827 nm (spectral band 50)

Table 1. PSNR values of the 2D bivariate thresholding and the proposed method for the simulated data cube

σ_n	Bivariate thresholding on each spectral band	Proposed Method
5	34.82	**38.86**
10	31.42	**34.55**
15	29.46	**32.48**
20	28.12	**31.12**
25	27.08	**30.11**
30	26.21	**29.31**
35	25.46	**28.64**
40	24.79	**28.06**

Table 2. PSNR values of the 2D bivariate thresholding and the proposed method for the cuprite data cube

σ_n	Bivariate thresholding on each spectral band	Proposed method
5	36.60	**39.53**
10	33.26	**35.85**
15	31.50	**34.10**
20	30.24	**32.94**
25	29.20	**32.06**
30	28.26	**31.33**
35	27.39	**30.70**
40	26.58	**30.14**

4 Conclusion

In this paper, we have extended the 2D bivariate wavelet thresholding proposed in [9], [10], to the 3D case. We found that the bivariate wavelet thresholding formula is still valid for 3D data denoising. Experimental results show that the proposed method is better than denoising every spectral band separately. Our proposed method denoises the whole data cube in one shot instead of performing a 2D wavelet thresholding on each spectral band of the datacube.

Further investigation will be carried out by exploiting both inter-scale and intra-scale relationships in the 3D wavelet coefficients. The parent-child relations in multiwavelet coefficients could also be investigated to achieve better denoising results. We could also extend the 2D bivariate wavelet thresholding technique to even higher dimensional data.

Acknowledgments. This work was supported by the Natural Sciences and Engineering Research Council of Canada (NSERC).

References

1. Donoho, D.L., Johnstone, I.M.: Ideal spatial adaptation by wavelet shrinkage. Biometrika 81(3), 425–455 (1994)
2. Chen, G.Y., Bui, T.D.: Multiwavelet denoising using neighbouring Coefficients. IEEE Signal Processing Letters 10(7), 211–214 (2003)
3. Bui, T.D., Chen, G.Y.: Translation-invariant denoising using multiwavelets. IEEE Transactions on Signal Processing 46(12), 3414–3420 (1998)
4. Chen, G.Y., Bui, T.D., Krzyzak, A.: Image denoising using neighbouring wavelet coefficients. Integrated Computer-Aided Engineering 12(1), 99–107 (2005)
5. Chen, G.Y., Bui, T.D., Krzyzak, A.: Image Denoising with Neighbour Dependency and Customized Wavelet and Threshold. Pattern Recognition 38(1), 115–124 (2005)
6. Chen, G.Y., Kegl, B.: Image denoising with complex ridgelets. Pattern Recognition 40(2), 578–585 (2007)
7. Chen, G.Y., Qian, S.E.: Simultaneous dimensionality reduction and denoising of hyperspectral imagery using bivariate wavelet shrinking and principal component analysis. Canadian Journal of Remote Sensing 34(5), 447–454 (2008)
8. Chen, G.Y., Qian, S.E.: Denoising and dimensionality reduction of hyperspectral imagery using wavelet packets, neighbour shrinking and principal component analysis. International Journal of Remote Sensing (to appear)
9. Sendur, L., Selesnick, I.W.: Bivariate shrinkage functions for wavelet-based denoising exploiting interscale dependency. IEEE Transactions on Signal Processing 50(11), 2744–2756 (2002)
10. Sendur, L., Selesnick, I.W.: Bivariate shrinkage with local variance estimation. IEEE Signal Processing Letters 9(12), 438–441 (2002)

Entropy of Gabor Filtering for Image Quality Assessment

Esteban Vazquez-Fernandez[1,2], Angel Dacal-Nieto[3], Fernando Martin[2], and Soledad Torres-Guijarro[4]

[1] GRADIANT - Galician R&D Center in Advanced Telecommunications, Spain
evazquez@gradiant.org
[2] Communications and Signal Theory Department, Universidade de Vigo
[3] Computer Science Department, Universidade de Vigo, Spain
[4] Laboratorio Oficial de Metroloxía de Galicia (LOMG), Spain

Abstract. A new algorithm for image quality assessment based on entropy of Gabor filtered images is proposed. A bank of Gabor filters is used to extract contours and directional textures. Then, the entropy of the images obtained after the Gabor filtering is calculated. Finally, a metric for the image quality is proposed. It is important to note that the quality of the image is image content-dependent, so our metric must be applied to variations of the same scene, like in image acquisition and image processing tasks. This process makes up an interesting tool to evaluate the quality of image acquisition systems or to adjust them to obtain the best possible images for further processing tasks. An image database has been created to test the algorithm with series of images degraded by four methods that simulate image acquisition usual problems. The presented results show that the proposed method accurately measures image quality, even with slight degradations.

1 Introduction

Image acquisition is a fundamental stage in every machine vision system. Obtaining the best quality images is critical to ensure a good performance. In this context, it is interesting to have a reliable way to measure the quality of the captured images or, from another point of view, to adjust the system to obtain the best possible images. Image quality assessment plays a fundamental role in this process, as well as in many image processing applications. It can be used to compare the performance of different methods (processing or acquisition) and to select the one which provides the best quality (or less image degradation); it can be used to measure the degradation itself after image processing operations; it also provides a metric to evaluate the performance of compression methods, like JPEG, or the quality of transmission channels (which is not covered in this work).

The most challenging problem in image quality assessment is the subjectivity inherent to perceived visual quality [1]. Several attempts to measure the quality of an image have been made, but it remains an open problem. Methods based on

A. Campilho and M. Kamel (Eds.): ICIAR 2010, Part I, LNCS 6111, pp. 52–61, 2010.

the measurement of Peak Signal to Noise Ratio (PSNR) or Mean Square Error (MSE) have been widely used due to their easy implementation, but the results show that they are not well suited to measure the human observer perceived quality [2]. Methods based on the use of previous knowledge of the Human Visual System (HVS) have shown a better performance in image quality assessment [3], [4]. HVS relays on the assumption that human observers pay more attention to details like structural information, which are more relevant to image quality measurement. Some previous contributions have pointed the use of entropy to measure image quality [5]. However, an entropy measure is unable to distinguish between noise and structural information. To solve this problem, a method based on image anisotropy has been proposed in [6].

Gabor filters have been extensively used in texture analysis and classification [7], [8], [9], but their use in image quality assessment remains little explored [10], [11]. The proposed method uses a bank of Gabor filters to model the linear filtering properties of single cells in visual cortex and to extract image contours and directional textures, which are directly related to HVS. Then, an estimation of the amount of visual information (randomness) perceived is calculated measuring the entropy of the outputs of the filter bank. The entropy value is directly related to the randomness of the image. Poorly defined transitions in the perceived image (Gabor response), which means less image quality, would produce a high entropy value. A metric is calculated by averaging the entropies obtained from the different Gabor filter bank outputs. This value can be used by itself as a reference, or can be normalized in relation to the original reference image, to show whether certain adjustment or process diminishes the image quality.

The paper is organized as follows. A theoretical background and the proposed algorithm are presented in Sect. 2. In Sect. 3 the developed test procedure to validate the method is shown. Results and discussion are presented in Sect. 4. Finally, some conclusions are given in Sect. 5.

2 Algorithm

2.1 Gabor Filters

Gabor filtering for image textural analysis has been introduced by Daugman [12]. The success of Gabor filters in this field is due to their aptitude to model the response of simple cortical cells in the visual system.

A 2D Gabor filter can be thought of as a complex plane wave modulated by a 2D Gaussian envelope and can be expressed in the spatial domain as:

$$\begin{aligned} G_{\theta,f,\sigma_1,\sigma_2}(x,y) &= \exp\left[\tfrac{-1}{2}\left(\tfrac{x'^2}{\sigma_1^2}+\tfrac{y'^2}{\sigma_2^2}\right)\right]\cos\left(2\pi f x'+\varphi\right) \\ x' &= x\sin\theta + y\cos\theta \\ y' &= x\cos\theta - y\sin\theta \end{aligned} \tag{1}$$

where f is the spatial frequency of the wave at an angle θ with the x axis, σ_1 and σ_2 are the standard deviations of the 2D Gaussian envelope, and φ is the phase.

Frequently in textural analysis applications, and also in this case, the Gaussian envelop is symmetric, so we have $\sigma = \sigma_1 = \sigma_2$.

A Gabor filter is suited to obtain local frequency information in a specific orientation (given by θ), which is directly related with image contours. A common practice in Gabor texture analysis is to use a bank of Gabor filters with different parameters tuned to capture several orientations and spatial frequencies. Attempts to systematize the design of the bank have been proposed [9], showing that increasing the number of frequencies and orientations has a little effect on the performance of the filter bank. However, the smoothing parameter, σ, is a significant factor to be carefully chosen in the bank design. Unfortunately, most of the times, it needs to be empirically chosen.

2.2 Image Entropy

The concept of entropy is associated with the amount of disorder in a physical system. Shannon redefined the entropy as a measure of the amount of information (uncertainty) in a source [13]. If the source is an image, it can be seen as a 2D array of information. The Shannon entropy is given by:

$$H(X) = -\sum_{i=1}^{n} p(x_i) \log_b p(x_i) \tag{2}$$

where $\Pr[X = x_i] = p(x_i)$ is the probability mass distribution of the source. This equation can be used to estimate the global entropy of an image characterized by its histogram:

$$H(I) = -\sum_{i=1}^{N} \text{hist}_{\text{norm}}(L_i) \log(\text{hist}_{\text{norm}}(L_i)) \tag{3}$$

where L_i represents the N intensity levels of the $m \times n$ image $I(x, y)$ and $\text{hist}_{\text{norm}}(L_i)$ is the histogram properly normalized to fit a probability distribution function:

$$\sum_{i=1}^{N} \text{hist}_{\text{norm}}(L_i) = 1 \tag{4}$$

The entropy of an image is an estimation of randomness, and is frequently used to measure its texture. As shown in Fig. 1, entropy can be thought as a measurement of the sharpness of the histogram peaks, which is directly related with a better defined structural information.

2.3 The Proposed Method

A flowchart of the proposed process is shown in Fig. 2. The input image is a grey level one; however the process can be easily applied to planes of a color space (like RGB). A bank of Gabor filters is used to extract contours and textural information. This stage converts the information to the HVS domain (cortex responses). The selected parameters for the filters are the following:

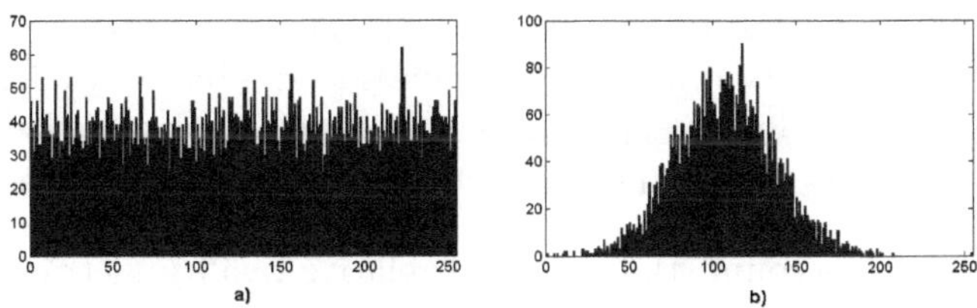

Fig. 1. Example of the entropy of different shape histograms. a) shows a higher entropy than b) ($Ha = 13.9627$; $Hb = 6.9216$).

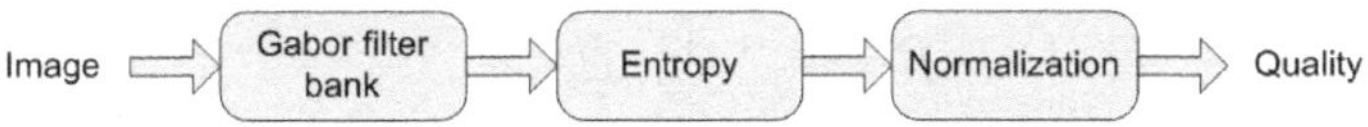

Fig. 2. Flowchart of the process

- Six different orientations are used. However, the empirical tests show that the number of filters and angles does not seem to be crucial:

$$\theta \in \left[0, \frac{\pi}{6}, \frac{\pi}{3}, \frac{\pi}{2}, \frac{2\pi}{3}, \frac{5\pi}{6}\right] \quad (5)$$

- Two phases are used, $\varphi_1 = 0$ for a symmetric filter (on the θ orientation) and $\varphi_2 = \frac{\pi}{2}$ for an anti-symmetric filter. This can be thought as real and imaginary parts of the same filter response.
- Two different spatial frequencies are used: $f_1 = \frac{1}{8}$ (spatial period λ of 8 pixels) and $f_2 = \frac{1}{4}$ (spatial period λ of 4 pixels).
- The standard deviation of the Gaussian envelope is empirically fixed to $\sigma = \frac{\lambda}{2}$ for all the filters.

24 filtered images are obtained, 12 with $\varphi_1 = 0$ and 12 with $\varphi_2 = \frac{\pi}{2}$. Each pair is combined to estimate the energy of the filtered images by:

$$E(x, y) = \sqrt{R_{\varphi_1}(x, y)^2 + R_{\varphi_2}(x, y)^2} \quad (6)$$

where $R_{\varphi_i(x,y)}$ is the Gabor response for the phase φ_i.

This process results in 12 energy images. The histogram of these energy images is computed, and their entropy estimated through Eq. (3). Entropy measures the amount of information or, in other words, the randomness of the image histogram.

This procedure combines the benefits of objective and subjective measurements. On the one hand, Gabor filtering provides features inherent to the visual perceived quality by modelling the behaviour of visual cells. On the other hand, this information is quantified by the use of entropy.

However, the amount of information in an image depends on its content as well as on its quality. E.g. there is less information in an image of a white sheet than in a written one. For this reason, the entropy of the Gabor filtered image is not an absolute quality measurement, unless compared to a reference image. This is not a problem for the applications proposed in Section 1, in which the interest lies in comparing the quality of images of the same scene, or the effect of certain processing.

Taking this into account, the proposed relative quality metric (Q_r) is computed averaging the entropy of the energies of the 12 Gabor filtered images. The result is inverted and multiplied by the entropy of the reference image, obtained the same way:

$$\begin{aligned} H_{ref} &= \tfrac{1}{12}\textstyle\sum_{i=1}^{12} H_{ref_i} \\ H &= \tfrac{1}{12}\textstyle\sum_{i=1}^{12} H_i \\ Q_r &= \tfrac{H_{ref}}{H} \end{aligned} \tag{7}$$

with H being the calculated entropy and H_{ref} the entropy of the reference image. As the entropy increases, the quality of the image decreases, so a $Q_r \in (0,1)$ means the quality of the image is lower than the reference (e.g. after the transmission through a noisy channel). If the resultant $Q_r > 1$, the quality of the image is higher than the reference (e.g. a noisy image which is enhanced by a median filtering, a blurred image which is enhanced by a fine tuning of the acquisition system, etc.).

3 Test Design

Two different test procedures have been developed to validate the performance of the proposed metric. The fist one is intended to model subtle variations in the image acquisition system. This is an objective quality test. The second one is intended to compare the proposed metric with the quality perceived by human observers. This is a subjective test.

3.1 Objective Test

For this purpose, an image database of natural scenes has been created. It is composed of 1100 images of 2136×1424 pixels. It was originated by 25 original images (see Figure 3) progressively degraded in 10 steps following 4 different procedures (see below), which becomes in $25 \times (10+1) \times 4$ quality tagged images.

The degradations introduced to the original images in this first database are:

- Blur: Gaussian blur has been applied by increasing the filter size in 10 steps (from 3×3 to 21×21 pixel blocks).
- Noise: Zero mean Gaussian noise has been added by increasing its standard deviation in 10 steps (from 5 to 25 in 8 bits per pixel grey scale images).
- Blur & Noise: Gaussian blur has been applied, followed by adding Gaussian noise (10 steps). It models the effect of sensor noise after an out of focus imaging.

Fig. 3. Original images of natural scenes used to create the image database

- Noise & Blur: Gaussian Noise has been added, followed by Gaussian blur (10 steps). It models the effect of software blurring operations after a noisy image acquisition (sensor noise).

The combination of noise and blur effects in different order, allows to simulate the effects of different acquisition systems, preprocessing operations, etc. [14]. Gaussian blur simulates the blur in an out of focus image. The Gaussian noise models the electronic noise which is produced in the camera sensor if the illumination, exposure time and gain parameters are not properly set. Figure 4 shows the effect of the 4 degradation procedures. This first database is intended to test the performance of the metric in the presence of subtle degradations.

3.2 Subjective Test

For the second test procedure, images from the LIVE Image Database [14] have been used. LIVE database contains images obtained by several distortion procedures, which have been subjectively classified and scored by human observers. The scores have been scaled and shifted to a 1 to 100 range, and a Difference Mean Opinion Score (DMOS) was computed. For our test, images distorted with white noise and Gaussian blur have been used. The database also contains images affected by JPEG compression, but it is not the aim of the proposed algorithm to test compression formats. The test is performed in a similar way to the first one. Images from the same scene have been sorted by their DMOS value (original and distorted ones). Then, the Q_r metric has been computed.

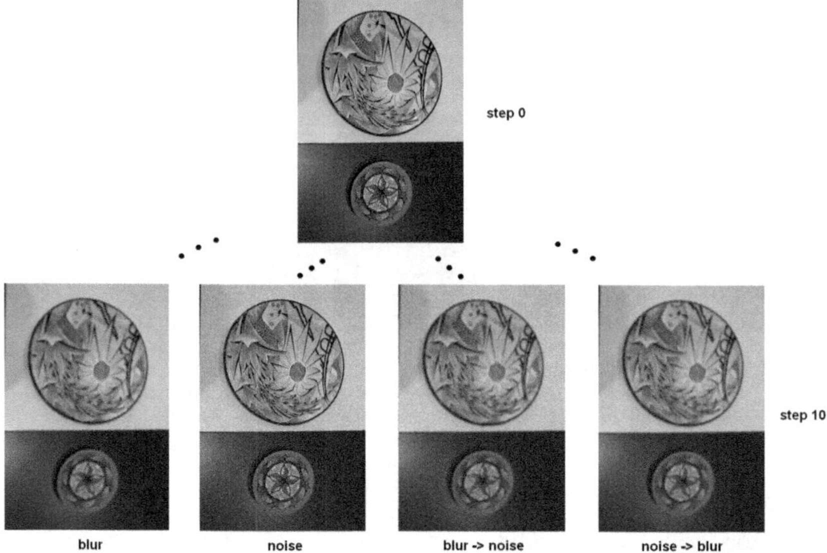

Fig. 4. Some steps for the blur, noise, blur & noise, and noise & blur degradation process of an example image

4 Results and Discussion

Figure 5 shows the results of the quality evaluation for the progressively degraded image shown in Figure 4. Similar results are obtained for the rest of the images used for the objective quality test.

It is interesting to note that most of the image quality assessment algorithms are tested using images that have a broad variation in quality. This is adequate when the objective is to model the quality perceived by an observer, e.g. to evaluate the performance of a compression algorithm or a transmission system. In these situations, the evaluation algorithm can be less precise (more tolerant), since variations in quality which are not perceived by the observer are not critical for the system.

However, if the objective is to select the best imaging system (or adjust it at its best) for a machine vision application, we have to be more strict in the performance of the method in a narrow error interval around the best possible image, which we call *Critical Peak*. In other words, we need to measure the quality of the image with a sufficiently high precision to obtain a strictly crescent/decreasing function.

As can be seen again in the example of Figure 4, the degradation applied is kindly subtle (low noise variance and small blurring mask) to test the Critical Peak performance. Figure 6 shows that all tested images have a strictly de-crescent function for their measured qualities in the test. The slope of the quality function varies significantly from image to image, because the degradation depends on the introduced distortion, as well as on the image content.

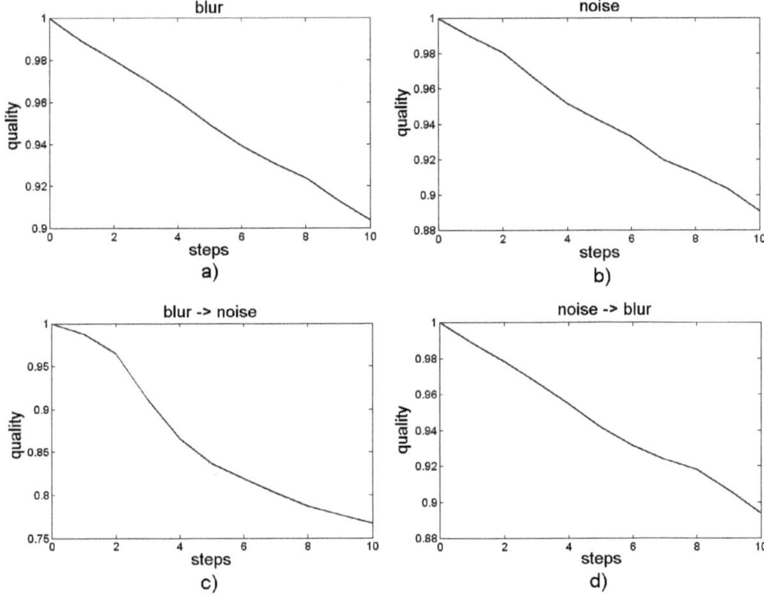

Fig. 5. Measured quality of the 10 step degraded images: a) blur; b) noise; c) noise after blur; d) blur after noise

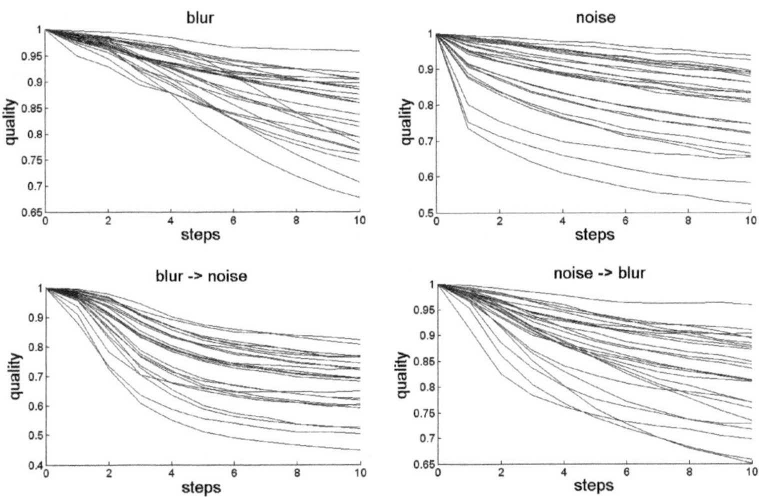

Fig. 6. Quality graphics of the test procedure. All images shown a strictly de-crescent function through the degradation procedures.

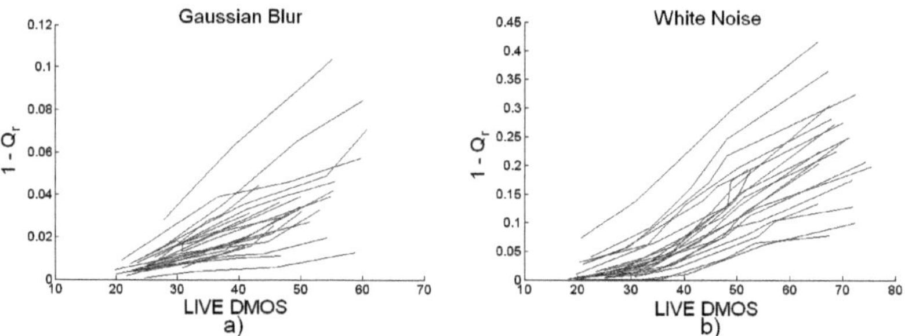

Fig. 7. LIVE images test. a) shows Gaussian blur distorted images; b) shows white noise distorted images.

For the second part of the test (LIVE images), crescent curves have been obtained, without taking into account the differences in scale (Q_r ranges from 0 to 1, where 1 is the higher quality value). The results are shown in Figure 7. As can be seen, the proposed metric also correlates with the quality perceived by human observers. Note that a higher DMOS value means less quality (larger difference to the reference image).

In this case, most of the curves are strictly crescent. However, there are some anomalies in some functions, which can be due to several factors (inherent to the image database): variance in the perceived qualities by different observers or from one day to another; the DMOS scaling system grades every image in a 1 to 5 discrete scale, which means that the minimum DMOS values for a distorted image is always higher than 20 (in a 1 to 100 re-scaled range); in addition, every single image is evaluated by comparison with the original one, but not with the whole sequence of distorted images.

There are also differences in the slopes, which produce dispersion in the curves of different images, due to the dependency of image content (besides its degradation). However, we obtain a good performance in quality evaluation for degraded sequences of the same image, which is the aim of the method.

5 Conclusions

Image quality assessment is an important tool that allows the user to perform a comparison between variations of an image. This can be useful when developing image processing algorithms and when designing imaging systems. A method based on the entropy of Gabor filtered images has been developed. It combines objective measures (entropy) with subjective ones (HVS).

An image database has been created to test our metric, by means of an original set of natural images and applying different degradation methods to this initial set. With these degradations, real world behaviours present in image acquisition and image processing systems are modeled. When tested with this database, the proposed metric works properly even in narrow ranges, which can be checked by its strictly de-crescent charts. A test using LIVE Image Database also confirms it is well suited to human observer perception of image quality.

To conclude, the combination of subjective characteristics, modelled by Gabor filtering, and objective features, like entropy, provide a useful and powerful starting point for further developments on Image Quality Assessment.

References

1. Wang, Z., Bovik, A.C., Lu, L.: Why Is Image Quality Assessment so Difficult? In: Proc. IEEE Int. Conf. on Acoustics, Speech and Signal Processing, vol. 4, pp. 3313–3316 (2002)
2. Wang, Z., Sheikh, H.R., Bovik, A.C.: No-reference perceptual quality assessment of JPEG compressed images. In: Proc. Int. Conf. on Image Processing, vol. 1, pp. 477–480 (2002)
3. Wang, Z., Bovik, A.C., Sheikh, H.R., Simoncelli, E.P.: Image quality assessment: from error visibility to structural similarity. IEEE Trans. Image Processing 13, 600–612 (2004)
4. Suresh, S., Babu, R.V., Kim, H.J.: No-reference image quality assessment using modified extreme learning machine classifier. Applied Soft Computing 9(2), 541–552 (2009)
5. Kirsanova, E.N., Sadovsky, M.G.: Entropy approach in the analysis of anisotropy of digital images. Open Syst. Inf. Dyn. 9, 239–250 (2004)
6. Gabarda, S., Cristóbal, G.: Blind image quality assessment through anisotropy. Journal of the Optical Society of America 24(12), 42–51 (2007)
7. Jain, A.K., Farrokhnia, F.: Unsupervised texture segmentation using Gabor filters. Pattern Recognition 24(12), 1167–1186 (1991)
8. Jain, A.K., Ratha, N.R., Lakhsmanan, S.: Object detection using Gabor filters. Pattern Recognition 30(2), 295–309 (1997)
9. Bianconi, F., Fernández, A.: Evaluation of the effects of Gabor filter parameters on texture classification. Pattern Recognition 40, 3325–3335 (2007)
10. Taylor, C.C., Pizlo, Z., Allebach, J.P., Bouman, C.A.: Image Quality Assessment with a Gabor pyramid model of the human visual system. In: Proc. SPIE Int. Symposium on Electronic Imaging Science and Technology, vol. 3016, pp. 58–69 (1997)
11. Zhai, G., Zhang, W., Yang, X., Yao, S., Xu, Y.: GES: A new image quality assessment metric based on energy features in Gabor Transform Domain. In: IEEE Proc. Int. Symposium on Circuit and Systems, pp. 1715–1718 (2006)
12. Daugman, J.G.: Uncertainty relation for resolution in space, spatial frequency, and orientation optimized by two-dimensional visual cortical filters. Journal of the Optical Society of America 2(7), 1160–1169 (1985)
13. Shannon, C.E.: The Mathematical Theory of Communication. The Bell System Technical Journal 27, 379–423, 623–656 (1948)
14. Sheikh, H.R., Wang, Z., Cormack, L., Bovik, A.C.: LIVE Image Quality Assessment Database Release 2, `http://live.ece.utexas.edu/research/quality`

Segmentation Based Noise Variance Estimation from Background MRI Data

Jeny Rajan, Dirk Poot, Jaber Juntu, and Jan Sijbers

Vision Lab
University of Antwerp
Antwerp 2610, Belgium
{jeny.rajan,dirk.poot,jaber.juntu,jan.sijbers}@ua.ac.be

Abstract. Accurate and precise estimation of the noise variance is often of key importance as an input parameter for posterior image processing tasks. In MR images, background data is well suited for noise estimation since (theoretically) it lacks contributions from object signal. However, background data not only suffers from small contributions of object signal but also from quantization of the intensity values. In this paper, we propose a noise variance estimation method that is insensitive to quantization errors and that is robust against low intensity variations such as low contrast tissues and ghost artifacts.

The proposed method starts with an automated background segmentation procedure, and proceeds then by correctly modeling the background's histogram. The model is based on the Rayleigh distribution of the background data and accounts for intensity quantization errors. The noise variance, which is one of the parameters of the model, is then estimated using maximum likelihood estimation. The proposed method is compared with the recently proposed noise estimation methods and is shown to be more accurate.

Keywords: Noise Estimation, MRI, Segmentation.

1 Introduction

Noise estimation in Magnetic Resonance Images (MRI) is important as it can play a key role in effective denoising of these images. It also finds application in quality assessment of MR images and various parameter estimation problems. Many techniques have been proposed in the literature to estimate the noise variance from MR images. A survey of these methods is given in [2]. The noise variance in MRI can be estimated from either complex or magnitude images. Usually, the estimation is done in magnitude MR image, since it is the usual output of the scanning process. Most of the methods in this category estimate the image noise variance from the background region of the image. Typical MR images usually include an empty region of air outside the tissue of interest. Especially in multi-slice or 3D images, there is an abundant number of background voxels available for noise estimation. Since the signal in these empty region of air is zero, the noise in these areas will be Rayleigh distributed.

A. Campilho and M. Kamel (Eds.): ICIAR 2010, Part I, LNCS 6111, pp. 62–70, 2010.

In previous work the noise variance was estimated by fitting a Rayleigh probability density function to the partial histogram of the MR image [1]. This approach proved to be highly effective as long as the noise variance is not too high. For high noise variance, however, the information in the signal region of the MR data may significantly contribute to the partial histogram and this may lead to a bias in the estimation of the variance. This is particularly applicable in the case of diffusion weighted MRI (DW-MRI) where the signal-to-noise ratio (SNR) is inherently low. The Rayleigh model of the background can also fail when ghosting artifacts are present. In this paper, we propose a method to reduce the influence of low signal-intensity areas (eg: scalp in DW-MRI of brain) and ghost effects in the noise estimation process. The improvement is achieved through background segmentation and there by estimating the noise variance by fitting the Rayleigh PDF to the histogram of the segmented background.

The paper is organised as follows. Section 2 discusses the segmentation algorithm proposed for extracting the image background from the MR image. In Section 3, the noise estimation procedure followed for the estimation of noise level from the segmented background is given. Comparative analysis of the proposed method with recently proposed approaches is shown in Section 4. Finally, conclusion and remarks are given in Section 5.

2 Background Segmentation

Segmentation of an image with low SNR is a challenging task. In this work, we combine the wavelet based bivariate shrinkage and morphological operations to achieve this goal. The image is first denoised using the wavelet based method to avoid the segmentation artifacts. Morphological operations are then applied to the denoised image for segmenting the image background. In the following subsection, we will discuss the approaches we followed for segmenting the signal and background from the noisy MR image.

2.1 Noise Reduction

Presence of noise can always lead to wrong segmentation. Considering high noise levels in MR images, especially in diffusion weighted images, an efficient denoising algorithm is a must. Since the algorithm which we are developing for noise estimation is fully automatic, the smoothing operation should be adaptive to varying noise levels. Another requirement for smoothing is anisotropic behavior of filter. i.e, the edges should be preserved while smoothing. Considering these requirements, we choose the bivariate shrinkage with the Dual Tree Complex Wavelet Transform (DTCWT)[9] for denoising the noisy MR image.

The DTCWT calculates the complex transform of a signal with two separate Discrete Wavelet Transform (DWT) decompositions. The two main disadvantages of DWT, the lack of shift invariance and poor directional selectivity, can be overcome by using DTCWT. The properties of DTCWT are discussed in detail in [9]. Along with DTCWT, we used the bivariate shrinkage method proposed in [3] to estimate the original wavelet coefficients from the noisy one. For

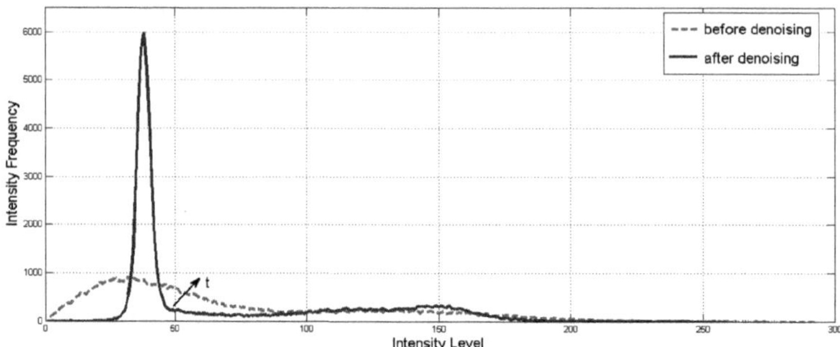

Fig. 1. Selection of threshold from the histogram of the denoised MR volume. It can be seen from the histograms that the accurate threshold selection for segmentation is not possible from noisy MR image without denosing.

the implementation of DTCWT, the wavelet software from [10] is used. In our work we applied a 4 level wavelet decomposition. The model proposed in [3] is a modification of Bayesian estimation problem where the statistical dependency between adjacent wavelet coefficients is also considered. This model can be written as

$$\hat{w}_1 = \frac{y_1}{\sqrt{y_1^2 + y_2^2}} \left(\sqrt{y_1^2 + y_2^2} - \frac{\sqrt{3}\sigma_n^2}{\sigma} \right)_+ \tag{1}$$

where $(g)_+$ is defined as

$$(g)_+ = \begin{cases} 0, & \text{if } g < 0 \\ g, & \text{otherwise} \end{cases} \tag{2}$$

and y_1 and y_2 are noisy observations of adjacent wavelet coefficients w_1 and w_2 (w_2 represents the parent of w_1). The noise variance $\hat{\sigma_n^2}$ is estimated from the finest scale wavelet coefficients [4].

2.2 Background Extraction

Once the image is denoised, a threshold t is to be estimated for creating a background mask. We computed this threshold from the histogram of the denoised image. The index of the first local minimum occurs after the maximum value (peak) in the histogram is considered as the threshold value t. An example for the selection of t is shown in Fig. 1. The MR volume is converted to a binary image based on this threshold value. Morphological closing is then applied on this binary image to fill the holes in the data area which is defined below

$$C = (A \oplus B) \ominus B \tag{3}$$

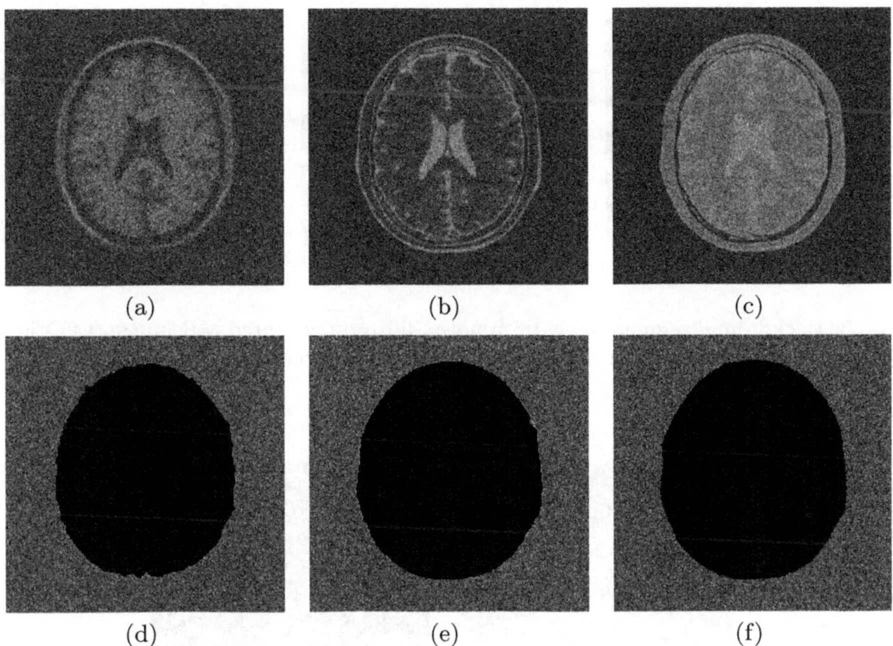

Fig. 2. Automatic background segmentation from noisy MR image (a),(b) and (c) Original simulated images : T1,T2 and PD respectively (d),(e) and (f) Image after background segmentation

where A is the binary image and B is the structuring element. $\oplus$ denotes the morphological dilation operation and $\ominus$ denotes the morphological erosion operation. We used a disk shape structuring element of size 5×5.

Morphological closing may still leave some holes inside the signal area and some noisy data in the background area. For an improved segmentation, we applied connected component analysis to select the largest connected component from the binary image C. The resulting mask (with all the holes filled) was then used to extract the background from the foreground areas of the MR image. The result of this operation is shown in Fig. 2. The algorithm was tested for T1, T2, Proton Density (PD), and Diffusion Weighted MR images with various noise levels.

One problem earlier reported with the background segmentation of Diffusion Weighted MR images of brain is the improper segmentation of the scalp [2,5]. Most of the conventional algorithms segment scalp as background and this may introduce a bias in the noise estimation. This wrong segmentation is mainly due to the high contrast difference between brain and scalp area. Contrary to conventional MR images, the intensity of scalp is very low in diffusion weighted images which makes the segmentation of scalp difficult here. As the noise level increases, it becomes more and more difficult to differentiate between scalp and the noisy background. Experiments with our proposed background segmentation

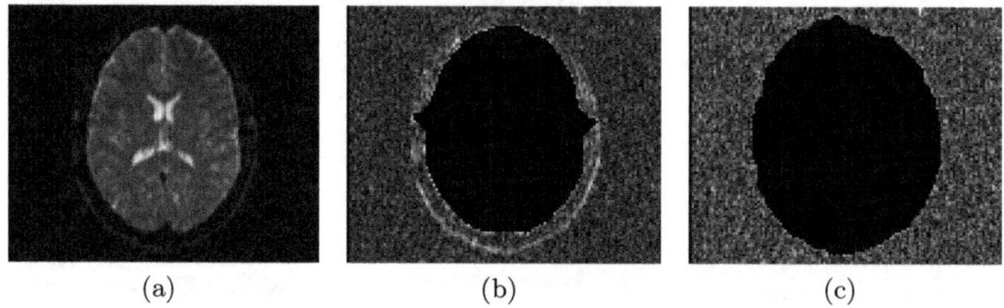

Fig. 3. Background segmentation from noisy diffusion weighted MR image (a) Original Image (b) Image segmented with the conventional approach (c) Image segmented with the proposed method

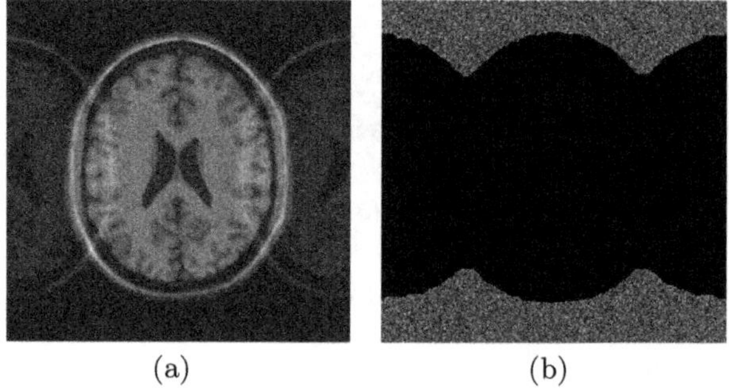

Fig. 4. Background segmentation from noisy MR image with ghost artifact (a) Original Image (b) Image segmented with the proposed method

algorithm show that the segmentation results are good for diffusion weighted MR images also. For diffusion weighted MR images, we used b_0 images for generating the mask and this mask is used to segment the background from all b value images. This will also help in reducing segmentation artifacts. Fig. 3 shows the segmentation of a DW-MR image with the proposed segmentation algorithm. Another issue reported with the method in [1] is the induction of error, if there is significant ghost artifacts. The proposed method will consider ghost data as part of MR signal, if there is significant ghost effect. The segmentation result of an MR image corrupted with ghost artifacts are shown in Fig. 4.

3 Estimation of the Noise Variance

The raw, complex MR data acquired in the Fourier domain is characterized by a zero mean Gaussian probability density function (PDF) [8]. After the inverse Fourier transform, the noise distribution in the real and imaginary components will be still Gaussian due to the linearity and orthogonality of the Fourier transform. But when it is transformed to a magnitude image, the noise distribution will be no longer Gaussian but Rician distributed [7]. If A is the original signal amplitude, then the PDF of the reconstructed magnitude image M will be

$$p_M(M_{i,j}|A_{i,j},\sigma_n) = \frac{M_{i,j}}{\sigma_n^2} e^{-\frac{M_{i,j}^2+A_{i,j}^2}{2\sigma_n^2}} I_0\left(\frac{A_{i,j}M_{i,j}}{\sigma_n^2}\right) u(M_{i,j}) \tag{4}$$

where $I_0(.)$ being the 0th order modified Bessel function of the first kind, $u(.)$ the Heaviside step function and σ_n^2 the variance of the noise. $M_{i,j}$ is the magnitude value of the pixel (i,j) and $A_{i,j}$, the original value of the pixel without noise.

In the image background, where the SNR is zero due to the lack of water-proton density in the air, the Rician PDF simplifies to a Rayleigh distribution with PDF[5]

$$p_M(M_{i,j}|\sigma_n) = \frac{M_{i,j}}{\sigma_n^2} e^{-\frac{M_{i,j}^2}{2\sigma_n^2}} u(M_{i,j}) \tag{5}$$

Once the segmentation algorithm is applied to the image, there will be only background, and the image variance reduces to the variance of a Rayleigh distribution. One straight forward approach to estimate the noise standard deviation from the segmented background is to use the equation which relates normal distribution and Rayleigh distribution [5] which is given below

$$\widehat{\sigma_n} = \sqrt{\sigma_M^2 \left(2 - \frac{\pi}{2}\right)^{-1}} \tag{6}$$

where σ_M^2 is the variance of the segmented background. One problem with this approach is the possibility of over-estimation of the noise level, if the segmented area also contains signal contributions. A maximum likelihood (ML) estimation of the noise standard deviation using the joint PDF of the histogram of the Rayleigh data was proposed in [1]

$$\widehat{\sigma_n} = \arg\min_{\sigma} \left[N_k \ln\left(e^{-\frac{l_0^2}{2\sigma^2}} - e^{-\frac{l_K^2}{2\sigma^2}}\right) - \sum_{i=1}^{K} n_i \ln\left(e^{-\frac{l_{i-1}^2}{2\sigma^2}} - e^{-\frac{l_i^2}{2\sigma^2}}\right)\right] \tag{7}$$

where l_i with $i = 0, ..., K$ denote the set of boundaries of the histogram bins, n_i represent the number of observations within the bin $[l_{i-1}, l_i]$ which are multinomially distributed and $N_k = \sum_{i=1}^{k} n_i$. A method to select the optimal number of bins is also described in [1]. In our experiments we used Eq. (7) for the estimation of the noise standard deviation from the segmented background. Usage of the background histogram, instead of the partial histogram, makes the proposed approach more robust to higher noise levels.

4 Results and Discussion

Experiments were conducted on both synthetic and real MR images (2D and 3D) to measure the improvement achieved with the proposed method over the existing ones.

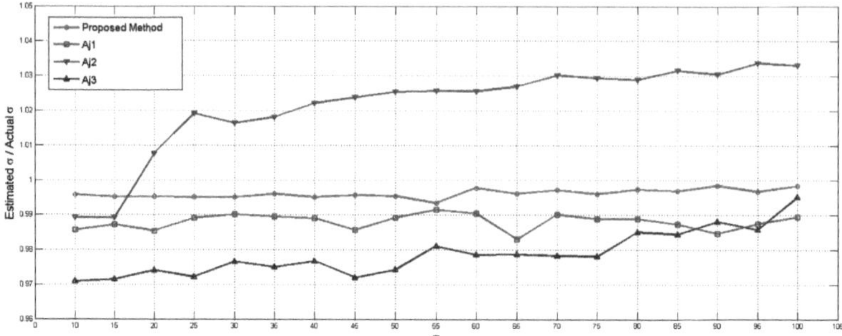

Fig. 5. Comparison of different noise estimators for Rician magnitude MR data : 100 experiments were considered for each σ value. The graph shows the mean of the estimated value divided by the actual value. The proposed method is compared with the estimators suggested in [5].

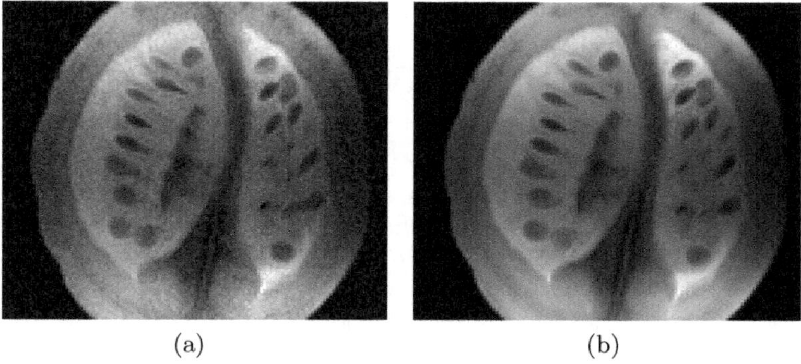

(a) (b)

Fig. 6. MR image of a cherry tomato (a) acquired with 1 average (b) acquired with 12 averages

Synthetic data: For simulations, we used the standard MR image phantom (T1, T2 and PD) with a voxel resolution of 1 mm^3 (8 bit quantization) from the Brainweb database [6]. Rician distributed data with varying σ were then generated from this noiseless image. The dimensions of the image were $181\times217\times181$. For computational simplicity (with wavelets), we resized it to $256\times256\times181$ by zero padding. The graph in Fig. 5 shows the mean of the estimated value ($\widehat{\sigma}$) divided by the actual value (σ). The value closer to 1 is the best estimator.

The result of the experiments on a 2D slice of the above mentioned MR data is shown in Fig. 5. In the experiment, for every σ, the $\widehat{\sigma}$ is estimated from the mean of 100 simulations. The proposed method is compared with the recently proposed estimators mentioned in [5]. In [2] a comparative analysis of these estimators with other popular methods can be seen. The labels in the graph, $Aj1$, $Aj2$ and $Aj3$ refer to the estimators based on local second order moment, local mean, and local variance in the MR data, respectively. These local moments

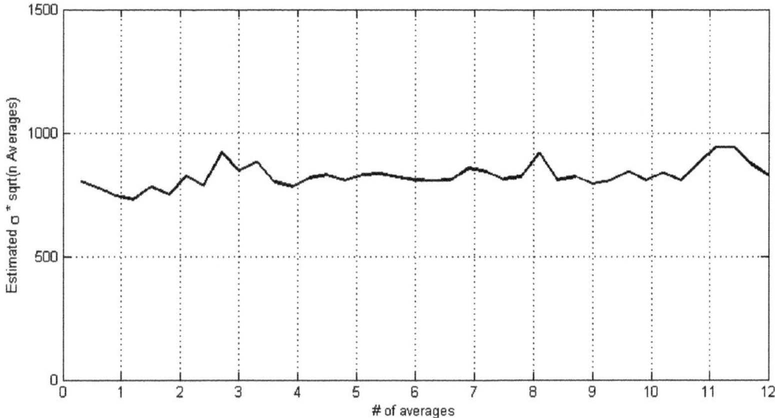

Fig. 7. Estimated σ as a function of the number of averages n used during the acquisition. MR image of a cherry tomato was used for this experiment.

were calculated using a 7×7 window. It can be seen from the graph that the $\widehat{\sigma}$ estimated with the proposed method is closer to the ground truth than other methods.

Real data: For the experiments on real data, we used the MR image of a cherry tomato. A set of MR images was reconstructed by averaging 1 to 12 acquired images. Fig. 6 shows the images reconstructed with 1 and 12 averages, respectively. Averaging was done in the complex k-space. The resulting noise variance as a function of the number of averages over n images was then estimated. The theoretical reduction of the noise standard deviation as a function of the number of images n over which the average was taken is known to be $\frac{1}{\sqrt{n}}$. Since the experimental setup for all the acquisitions were the same, except for averaging, $\widehat{\sigma} \times \sqrt{n}$ should be constant over n. It can be seen from Fig. 7 that the proposed method exhibits this property.

5 Conclusion

In this paper, we presented a method that needs prior segmentation of MR image background for the estimation of noise level. The proposed method minimizes the artifacts introduced in the segmentation process by conventional approaches. The reliability of the method is proved on both simulated and real MR images at various noise levels. Comparative analysis with the recently proposed methods shows that the proposed approach has an edge over the existing ones.

Acknowledgments. This work was financially supported by the IWT (Institute for Science and Technology, Belgium; SBO QUANTIVIAM) and the UIAP.

References

1. Sijbers, J., Poot, D., den Dekker, A.J., Pintjens, W.: Automatic estimation of the noise variance from the histogram of a magnetic resonance image. Physics in Medicine and Biology 52, 1335–1348 (2007)
2. Aja-Fernández, S., Antonio, T.V., Alberola-López, C.: Noise estimation in single and multiple magnetic resonance data based on statistical models. Magnetic Resonance Imaging 27, 1397–1409 (2009)
3. Şendur, L., Selesnick, I.W.: Bivariate shrinkage with local variance estimation. IEEE Signal Processing Letters 9, 438–441 (2002)
4. Donoho, D.L., Johnstone, I.M.: Ideal spatial adaptation by wavelet shrinkage. Biometrika 81, 425–455 (1994)
5. Aja-Fernández, S., Alberola-López, C., Westin, C.F.: Noise and signal estimation in magnitude MRI and rician distributed images: A LMMSE approach. IEEE Trans. Medical Imaging 17, 1383–1398 (2008)
6. Cocosco, C.A., Kollokian, V., Kwan, R.K.S., Evans, A.C.: BrainWeb: online interface to a 3D MRI simulated brain database. NeuroImage 5, S425 (1997)
7. Henkelman, R.M.: Measurement of signal intensities in the presence of noise in MR images. Medical Physics 12, 232–233 (1985)
8. Wang, Y., Lei, T.: Statistical analysis of MR imaging and its applications in image modelling. In: Proceedings of the IEEE International Conference on Image Processing and Neural Networks, vol. 1, pp. 866–870 (1994)
9. Kingsbury, N.: The dual tree complex wavelet transform: A new technique for shift invariance and directional filters. In: EUSIPCO, pp. 319–322 (1998)
10. Cai, S., Li, K., Selesnick, I.: Matlab implementation of wavelet transforms, `http://taco.poly.edu/WaveletSoftware/index.html`

Morphological Thick Line Center Detection

Miguel Alemán-Flores, Luis Alvarez, Pedro Henríquez, and Luis Mazorra

Departamento de Informática y Sistemas, Universidad de Las Palmas de Gran Canaria, 35017 Las Palmas de G.C., Spain

Abstract. Thick line center and width estimation are important problems in computer vision. In this paper, we analyze this issue in real situations where we have to deal with some additional difficulties, such as the thick line distortion produced by interlaced broadcast video cameras or large shaded areas in the scene. We propose a technique to properly extract the thick lines and their centers using mathematical morphological operators. In order to illustrate the performance of the method, we present some numerical experiments in real images.

1 Introduction

The detection of thick lines and their centers is an important issue in computer vision (see, for instance, [11]). In this paper, we analyze this problem in real application scenarios where we have to deal with some additional difficulties, such as the line distortion produced by interlaced broadcast video cameras or large shaded areas in the scene (see figures 1 and 2). The main tools we use are mathematical morphological operators, which are very convenient to extract geometric shape information. The main assumption we make is that, in the image, all lines of interest are brighter (or darker) than the background. As optional additional information for the applications we deal with, we can assume that the background has a rather uniform color. For instance, in figures 1 and 2, we can appreciate that the background is green (the color of the soccer field grass). This information is useful in order to avoid spurious line detection outside the region of interest, i.e. the field of play. In this application, we call thick lines all line markings in a soccer field, straight as well as curved, including touchlines, goal lines, the half-way line, goal and penalty areas, the center circle and penalty arcs.

This paper is structured as follows: In section 2, we present a short overview of some related previous works. Section 3 briefly explains the mathematical morphological operators we have used. In section 4, we introduce the simple deinterlacing method used to remove thick line noise. In section 5, we present the thick line detection method we propose in scenarios without large shaded areas. Section 6 presents the method proposed in scenarios with large shaded areas. Finally, in section 7 we give an account of our main conclusions.

A. Campilho and M. Kamel (Eds.): ICIAR 2010, Part I, LNCS 6111, pp. 71–80, 2010.

Fig. 1. Real HD video frame from a soccer match (image provided by Mediapro)

Fig. 2. Real image of a soccer stadium with a large shaded area (image provided by Mediapro)

2 Related Works

Line detection is an important task for the processing of video sequences from the broadcasting of sport events because it is useful for a wide variety of purposes, such as camera calibration, player tracking, ball tracking, mosaicing, detection of on-field advertisement billboards, change of the view point, detection of highlights (e.g. goals or fouls), automatic summarization or insertion of virtual objects. Most line detection methods applied to sport events assume that line and background colors are constant in the region of interest. Moreover, they also assume that lines are one-pixel wide. It is common to use a segmentation with dominant color detection using RGB space [3,10,7], or HSV space [9,2,4]. These methods use cumulative histograms [4], extract their peaks and, according to some criteria, decide which pixels belong to the lines and which ones belong to the background [3,10,7]. Some other methods to extract lines start with a segmentation with Gaussian mixture models [5,8]. These methods cannot be applied when dealing with interlaced images, HD definition images (where lines may be 8-pixels wide or even more), or scenarios with significant contrast variations between the background and the thick lines.

3 Mathematical Morphology

Mathematical morphology is a theory and a technique for the analysis and processing of geometric structures [6]. It can be stated in a continuous or discrete way and for binary or grayscale images. The basic morphological operators we will use in this paper to locate thick lines and their centers in an image are the following ones:

Disk morphological operators : Given a disk $D_s(x)$ of center x and radius s, we define:

Disk dilation: $I \oplus D_s(x) = \sup_{y \in D_s(x)} I(y)$
Disk erosion: $I \ominus D_s(x) = \inf_{y \in D_s(x)} I(y)$
Disk opening: $I \circ D_s(x) = (I \ominus D_s) \oplus D_s(x)$
Disk closing: $I \bullet D_s(x) = (I \oplus D_s) \ominus D_s(x)$

Disk morphological operators will be used to extract thick lines in the image. For instance, we can observe that, if the maximum line width in the image is s, and if the image lines are brighter than the background, then the morphological operation $I \circ D_s$ removes image lines.

Line morphological operators: Given a set of angle orientations Θ, $\theta \in \Theta$ and a segment $L_{s,\theta}(x)$ of center x, radius s and orientation θ,we define:

Line opening: $I \circ L_s(x) = \sup_{\theta \in \Theta}(\inf_{y \in L_{s,\theta}(x)} I(y))$
Line closing: $I \bullet L_s(x) = \inf_{\theta \in \Theta}(\sup_{y \in L_{s,\theta}(x)} I(y))$

Line morphological operators will be used to filter noise in the image and to clean line boundaries.

Morphological skeleton: Given a set X, the morphological skeleton is given by:

Morphological skeleton: $S = \cup_{s>0} \left(\cap_{\mu>0} \left(X \ominus D_s \backslash (X \ominus D_s) \circ D_\mu \right) \right)$
where D_s is a disk of radius s centered at 0

The skeleton represents, for a given shape X, the centers of the maximal disks included in X. Morphological skeleton will be used to find out the centers of thick lines.

4 A Simple Deinterlacing Procedure Using Morphological Line Filters

Interlaced video technology may introduce strong perturbations in thick lines, especially when the camera moves quickly and we work with HD video (1920×1080 frames). In figure 3, we illustrate this phenomenon for a real HD video

Fig. 3. Original interlaced image (left) and deinterlaced image using the proposed algorithm (right)

frame acquired in a sport event scene. In order to properly find out image thick lines, we need to preprocess the image to remove noise. Deinterlacing video is a major problem in computer vision and developing sophisticated deinterlacing techniques is beyond the scope of this paper. In fact, what we need is a simple and fast deinterlacing procedure which removes the image line noise. We propose the following simple deinterlacing procedure: we replace even lines by odd lines in the image and then we apply a line morphological operator to clean thick lines. This operation is performed independently in each one of the image RGB color channels.

5 Thick Line Detection in Scenarios without Shaded Regions

We start with the simplest case, where no thick line of interest is located in a shaded region. We use the morphological disk opening $I \circ D_s$ to find out the lines. We observe that if image lines are brighter than the background and the maximum line width is s, then, for each one of the RGB color channels, the opening operation removes the lines from the image. Therefore, a first approximation of the thick line region A, can be expressed as :

$$A = \left\{ x : \left\{ \begin{array}{l} (R(x) - R \circ D_s(x)) > t_R \\ (G(x) - G \circ D_s(x)) > t_G \\ (B(x) - B \circ D_s(x)) > t_B \end{array} \right. \right\} \quad (1)$$

where t_R, t_G, t_B are thresholds for each image channel. We observe that the proposed method is robust against illumination changes, as far as the contrast between the lines and the background remains high enough.

In case the background color does not change significantly in the image, as in the case of a soccer field, where the background is green, we can select the region of interest *a priori* in the image, according to the background color. To

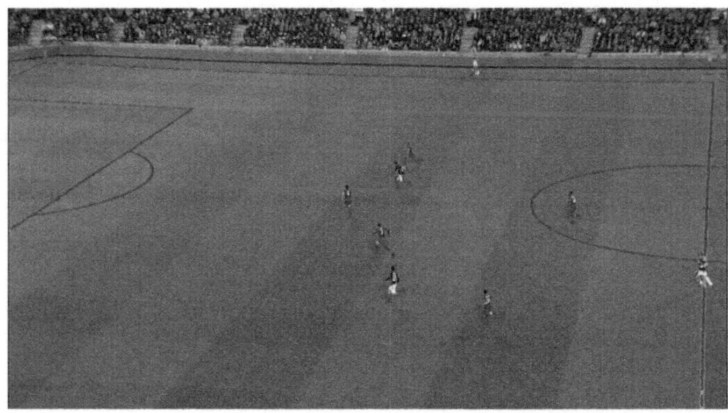

Fig. 4. Thick lines extracted with the proposed method for the image in figure 1

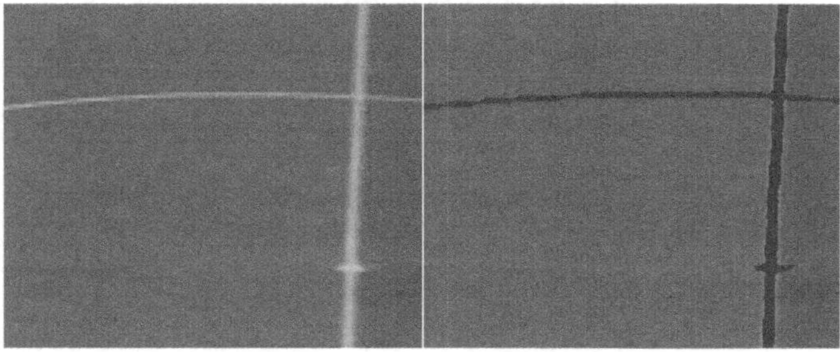

Fig. 5. Details of the thick lines extracted in image 1. We observe that lines of quite different widths are detected.

manage color information, it is more convenient to work in the HSV color space. Hue (H) is the main component concerning color information. Let us denote by $(H_s(x), S_s(x), V_s(x))$ the HSV channels of the image $(R \circ D_s, G \circ D_s, B \circ D_s)$. Then, the line background area C can be expressed as

$$C = \{x : \; t_{H_1} \leq H_s(x) \leq t_{H_2}\} \tag{2}$$

That is, we threshold the hue value H_s. We observe that, since H_s is computed after the opening process, the image thick lines we are interested in are included in C. In other words, the set $A \cap C$ represents the final set $\mathcal{B}$ of line points which corresponds to image thick lines located in the background region of interest. In the numerical experiments we present, the parameters are chosen in the following way: s, the maximum radius of line width, is set to 5 in order to be sure that all lines of interest in the image are included. t_{H_1} and t_{H_2} are chosen analyzing the peak of the histogram of H_s channel using standard histogram segmentation technique (see [1] for details). As the thick line area is very small with respect to the line background area, the parameters t_R, t_G and t_B are chosen in terms of a percentage $0 < p < 1$ with respect to the histogram of the corresponding image channel. For instance, t_R is chosen to satisfy

Fig. 6. Details of the thick lines extracted in image 1

$$p = \frac{|x \in C : (R(x) - R \circ D_s(x)) > t_R|}{|C|}$$

where |.| represents the cardinal (size) of the set. In the experiment we chose $p = 0.02$.

Figure 4 illustrates the thick lines extracted in a sample image without large shaded regions. Figures 5 and 6 show two zooms of the extracted lines. As observed, even very low contrast lines can be detected.

6 Thick Line Detection in Scenarios with Shaded Regions

In scenarios with large shaded regions, see figure 2, the method presented in the above section does not work properly because we cannot set a single threshold

Fig. 7. Thick lines extracted with the proposed method for the image in figure 2

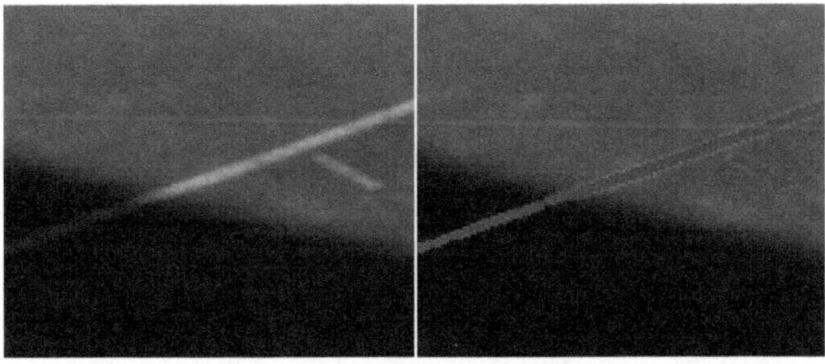

Fig. 8. Details of the thick lines extracted for the image in figure 2

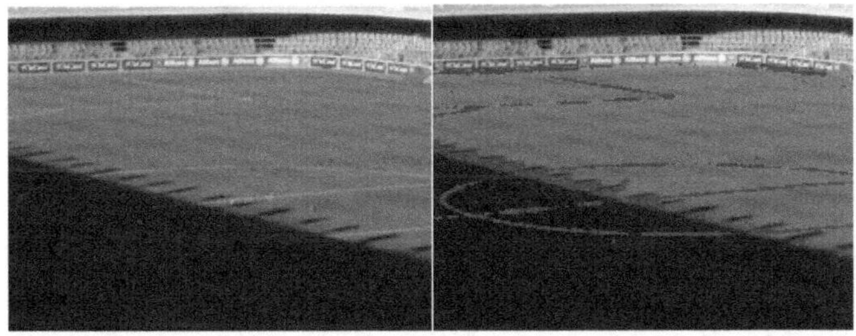

Fig. 9. Details of the thick lines extracted for the image in figure 2

configuration for t_R, t_G and t_B which works simultaneously in the shaded and lighted regions. Therefore, we first need to identify that there exist two regions of interest in the scene (the shaded and lighted areas where the thick lines of interest are located) and next set a different threshold configuration for each region.

Usually, the hue value of the background area does not change significantly from the lighted to the shaded areas (e.g., a soccer field has a similar hue value -green color- in the lighted and in the shaded areas). However the value channel $V_s(x)$ in the HSV space varies significantly from the lighted to the shaded area. In order to automatically identify whether we deal with large shaded area scenes we analyze the histogram of the value channel $V_s(x)$ but in the region of interest defined by the hue channel. Let $h(w)$ be the histogram of the $V_s(x)$ values in the region of interest, i.e. $H_s(x) \in [t_{H_1}, t_{H_2}]$. If we deal with just one region, then $h(w)$ has a profile with a single peak. If we deal with two regions, $h(w)$ has a profile with two peaks. Using a standard histogram segmentation technique (see [1] for details) we can automatically identify the number of significant peaks in $h(w)$ profile. Once we have separated the shaded and lighted regions, we apply the same procedure proposed in the above section to each region and we obtain the line region $\mathcal{B}$ for the whole image.

Figure 7 illustrates the thick lines extracted in a sample image with large shaded regions. Figure 8 shows that thick lines are properly extracted in the shaded and lighted areas. Figure 9 shows that the central circle is quite well extracted despite being in the shaded and lighted areas and with a very low contrast in the shaded area.

7 Thick Line Center Detection Using Morphological Skeleton

In the case of discrete lattices, the morphological skeleton can be stated in the following way: If we denote by D_n the disk of radius n centered in 0, then, the center points of the line of width n can be obtained as the set :

$$S_n = (\mathcal{B} \ominus D_n) \setminus ((\mathcal{B} \ominus D_n) \circ D_1))$$

where $\mathcal{B}$ is the detected thick line region. Therefore, the skeleton computation also provides, automatically, the line width.

Figures 10, 11, 12 and 13 show some examples of thick line center detection in different situations.

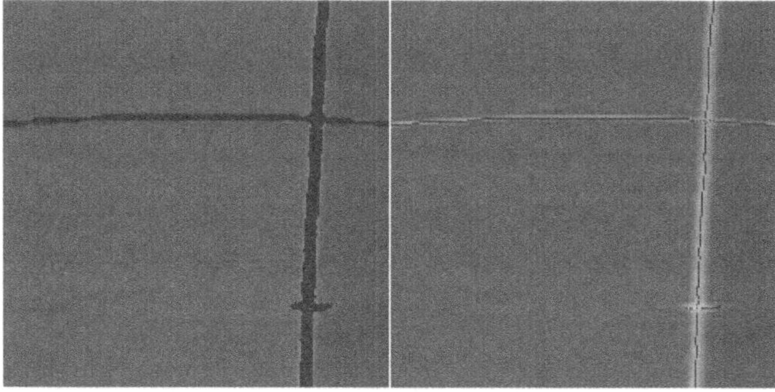

Fig. 10. Details of the thick line centers extracted for the image in figure 1

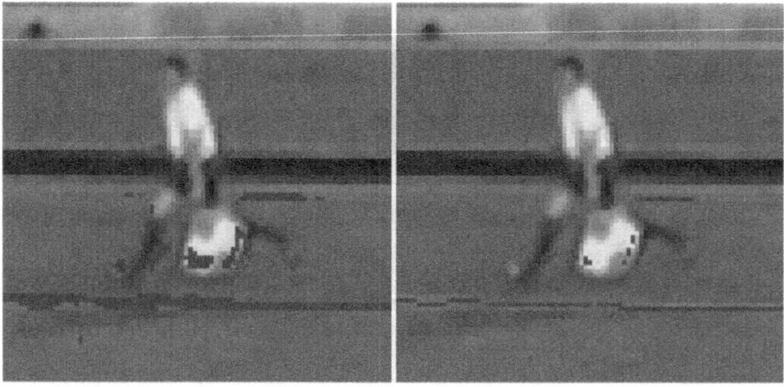

Fig. 11. Details of the thick line centers extracted for the image in figure 1

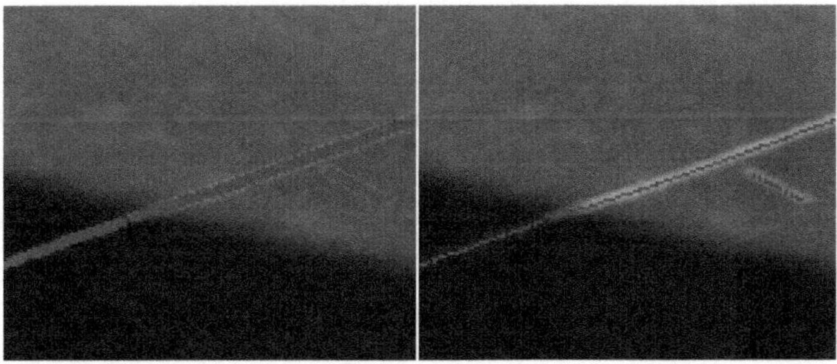

Fig. 12. Details of the thick line centers extracted for the image in figure 2

Fig. 13. Details of the thick line centers extracted for the image in figure 2

8 Conclusions

We have presented a new technique for image thick lines and thick line centers extraction based on morphological operators in real situations. The proposed method works properly even in complex scenarios where we have to deal with interlaced broadcast images of large shaded areas. The numerical experiments are very promising. Most of the significant thick line centers are extracted. The amount of spurious false thick lines detected is small and isolated. Moreover, these false detections could be easily removed in a postprocessing stage where we search for straight lines and ellipses in the image based on the extracted thick line centers.

Acknowledgement. We acknowledge Mediapro for providing us with the test images used in this paper. This work was partially funded by Mediapro through the Spanish project CENIT-2007-1012 i3media.

References

1. Alvarez, L., Esclarin, J.: Image quantization using reaction-diffusion equations. SIAM Journal on Applied Mathematics 57(1), 153–175 (1997)
2. Ekin, A., Tekalp, A.M., Mehrotra, R.: Automatic Soccer Video Analysis and Summarization. IEEE Transactions on Image Processing 12(7) (July 2003)

3. Khatoonabadi, S.H., Rahmati, M.: Automatic soccer players tracking in goal scenes by camera motion elimination. Image and Vision Computing 27, 469–479 (2009)
4. Liu, J., Tong, X., Li, W., Wang, T., Zhang, Y., Wang, H.: Automatic Player Detection, Labeling and Tracking in Broadcast Soccer Video. Pattern Recognition Letters 30, 103–113 (2009)
5. Liu, Y., Huang, Q., Ye, Q., Gao, W.: A new method to calculate the camera focusing area and player position on playfield in soccer video. In: Visual Communications and Image Processing 2005 (2005)
6. Serra, J.: Image Analysis and Mathematical Morphology. Academic Press, London (1982)
7. Wan, K., Yan, X., Yu, X., Xu, C.: Real-Time Goal-Mouth Detection In MPEG Soccer Video. In: MM 2003, November 2-8 (2003)
8. Wang, L., Zeng, B., Lin, S., Xu, G., Shum, H.: Automatic Extraction of Semantic Colors In Sports Video. In: ICASSP 2004 (2004)
9. Watve, A., Sural, S.: Soccer video processing for the detection of advertisement billboards. Pattern Recognition Letters 29, 994–1006 (2008)
10. Yoon, H., Bae, Y.J., Yang, Y.: A Soccer Image Sequence Mosaicking and Analysis Method Using Line and Advertisement Board Detection. ETRI Journal 24(6) (December 2002)
11. Zhang, Q., Couloigner, I.: Accurate Centerline Detection and Line Width Estimation of Thick Lines Using the Radon Transform. IEEE Transactions on Image Processing 16(2), 310–316 (2007)

Segmentation of Very High Resolution Remote Sensing Imagery of Urban Areas Using Particle Swarm Optimization Algorithm

Safaa M. Bedawi and Mohamed S. Kamel

Department of Electrical and Computer Engineering, University of Waterloo,
Waterloo, Ontario, Canada N2L 3G1
{mkamel,sbedawi}@pami.uwaterloo.ca

Abstract. As the improvement of the resolution of aerial and satellite remote sensed images, the semantic richness of the image increases which makes image analysis more difficult. Dense urban environment sensed by very high-resolution (VHR) optical sensors is even more challenging. Occlusions and shadows due to buildings and trees hide some objects of the scene. Fast and efficient segmentation of such noisy images (which is essential for their further analysis) has remained a challenging problem for years. It is difficult for traditional methods to deal with such noisy and large volume data. Clustering-based segmentation with swarm-based algorithms is emerging as an alternative to more conventional clustering methods, such as hierarchical clustering and k-means. In this paper, we introduce the use of Particle Swarm Optimization (PSO) clustering algorithm segmenting high resolution remote sensing images. Contrary to the localized searching of the K-means algorithm, the PSO clustering algorithm performs a globalized search in the entire solution space. We applied the PSO and K-means clustering algorithm on thirty images cropped from color aerial images. The results illustrate that PSO algorithm can generate more compact clustering results than the K-means algorithm.

Keywords: Swarm Intelligence, Particle Swarm Optimization (PSO), Remote Sensing, Aerial Images, and Clustering-Based Segmentation.

1 Introduction

With the rapid development of aerospace technologies and remote sensing sensor technologies, images of very high spatial resolution of the earth surface have been obtained more frequently and quickly than before (for example the recently launched satellites: GeoEye-1 and WorldView-2). Therefore, remote sensing images have significant applications in different areas such as urban planning, surveys and mapping, agricultural analysis, environmental monitoring and military intelligence, etc. Remote sensing image analysis, such as image segmentation, image classification and feature extraction, can be challenging because there are many uncertainties in remote sensing data and there is no definite mathematical model that truly captures the image data. Urban land cover information extraction is a hot topic within urban studies. Heterogeneous spectra of the VHR imagery caused by the inner complexity of dense urban

A. Campilho and M. Kamel (Eds.): ICIAR 2010, Part I, LNCS 6111, pp. 81–88, 2010.

areas and the occlusions and shadows caused by the variety of objects in urban area, for example buildings, roads, and trees - make it even more difficult and challenging, hindering exhaustive automatic or manual extraction.

In most cases, information needed for image analysis and understanding is not represented in pixels but in meaningful image objects and their mutual relations. Therefore, to partition images into sets of useful image objects is a fundamental procedure for successful image analysis or automatic image interpretation. In this sense, image segmentation is critical for subsequent image analysis and understanding. Image segmentation may be defined as the process of subdividing an image into meaningful non-overlapping regions [1]. Image segmentation can be viewed as a clustering problem, which aims to partition the image into clusters such that the pixels within a cluster are as homogenous as possible whereas the clusters among each other are as heterogeneous as possible with respect to a similarity measure. Clustering algorithms can be divided into four main classes: partitioning methods, hierarchical methods, density-based clustering and grid-based clustering. An extensive survey of clustering techniques is described in [2].

VHR Remote sensing image clustering-based segmentation is a complex task as images are noisy and of large size. It is difficult for traditional methods to deal with these images. This type of data has posed a formidable task for finding global optima in most of traditional clustering techniques. This motivates exploring the use of computational intelligence techniques. For many years now, several papers have highlighted the efficiency of approaches inspired from nature [3]. A variety of algorithms inspired from the biological examples by swarming, flocking and herding phenomena. These techniques incorporate swarming behaviours observed in flocks of birds, schools of fish, or swarms of bees, and even human social behaviour.

Swarm Intelligence (SI) is actually a complex multi-agents system, consisting of numerous simple individuals (e.g., ants, birds, etc.), which exhibit their swarm intelligence through cooperation and competition among the individuals. Although there is typically no centralized control dictating the behaviour of the individuals, the accumulation of local interactions in time often gives rise to a global pattern, SI mainly involves two approaches, i.e., Particle Swarm Optimization (PSO) and ant colony optimization (ACO). SI has currently succeeded in solving problems such as traveling salesman problems, data clustering, combination optimization, network routing, rule induction, and pattern recognition [4], However, using SI in remote sensing clustering is a fairly new research area and needs much more work. In [4] PSO was used as a clustering algorithm. The results show that despite k-means is known to be efficient at clustering large data sets, as its computational complexity only grows linearly with the number of data points [5], k-means may converge to solutions that are not optimal, hence PSO outperformed it as well as fuzzy *c*-means and other state-of-the-art clustering algorithms.

In the literature most of the conventional and SI clustering methods are tested on simple scenes, such as low scale grayscale images of rural or suburban sites, where objects of the sensed scenes are quite visible with less shade or occlusion artefacts than in inner cities. Encouraged by the success of PSO in such scenes, in this paper we aim at using the PSO potential in solving complex optimization problems, to handle dense VHR remote sensing images of dense urban areas, the result shows that

PSO algorithm is capable of segmenting such complex images and it outperforms k-means algorithm.

2 Particle Swarm Optimization

Particle Swarm Optimization (PSO) is a population-based evolutionary computation technique developed by [3]. PSO simulates the social behaviour of animals, i.e. birds in a flock or fish in a school. Members of such societies share common goals (e.g., finding food) that are realized by exploring its environment while interacting among them. The popularity of PSO is partially due to the simplicity of the algorithm, but mainly to its effectiveness for producing good results at a very low computational cost [6]. In PSO, each solution can be considered an individual particle in a given D-dimensional search space, which has its own position (x_{id}) and velocity (v_{id}). During movement, each particle adjusts its position by changing its velocity based on its own experience (memory) p_{id}, as well as the experience of its neighbouring particles, until an optimum position is reached by itself and its neighbour. All of the particles have fitness values based on the calculation of a fitness function. Particles are updated by following two parameters called p_{best} and g_{best} at each iteration. Each particle is associated with the best solution (fitness) the particle has achieved so far in the search space. This fitness value is stored, and represents the position called p_{best}. The value g_{best} is a global optimum value for the entire population. The two basic equations which govern the working of PSO are that of velocity vector and position vector given by:

$$v_{id}(t+1) = w\, v_{id}(t) + c_1 r_1(t)(p_{id}(t) - x_{id}(t)) + c_2 r_2(t)(p_{id}(t) - x_{id}(t)) \quad (1)$$

$$x_{id}(t+1) = x_{id}(t) + v_{id}(t+1) \quad (2)$$

The first part of Eq (1) represents the inertia of the previous velocity, the second part is the cognition part and it tells us about the personal experience of the particle, the third part represents the cooperation among particles and is therefore named as the social component. Acceleration constants c_1, c_2 and inertia weight w are the predefined by the user and r_1, r_2 are uniformly generated random numbers in range of [0, 1].

2.1 PSO Clustering

A particle represents a K-cluster centroids. That is, each particle x_i is constructed as $x_i = (m_{i,1},\ldots m_{i,j},\ldots m_{i,d})$ where $m_{i,j}$ refers to the j-th cluster centroid vector of ith particle. Therefore, a swarm represents a number of candidate data clusterings. The quality of each particle is measured using an objective function [7]. There is a matrix representing the assignment of patterns to the cluster of particle i. Each element $z_{i,k,p}$ indicates if pattern z_p belongs to cluster c_k of particle i. The fitness function has an objective to simultaneously minimize the intra-distance between pixels and their cluster means and to maximize the inter-distance between any pair of clusters. The algorithm is composed of the following steps:

1. Initialize each particle to contain N_c randomly selected cluster centroids.
2. For $t = 1$ to t_{max} do

i. For each particle i do
ii. For each data vector z_p
   a. calculate the Euclidean distance $d(z_p, m_{ij})$ to all cluster centroids C_{ij}
   b. assign z_p to cluster C_{ij} such that distance
      $d(z_p, m_{ij}) = \min_{\forall c=1,\ldots N_c} \{ d(z_p, m_{ic}) \}$
   c. calculate the fitness function [7].
iii. Update the global best and local best positions
iv. Update the cluster centroids.

where t_{max} is the maximum number of iterations. The population-based search of the PSO algorithm reduces the effect that initial conditions have, as opposed to the K-means algorithm; the search starts from multiple positions in parallel. However, the K-means algorithm tends to converge faster (after less function evaluations) than the PSO, but usually with a less accurate clustering [4].

2.2 Clustering Validation Measures

These measures are usually used to evaluate to quantitatively evaluate the result of a clustering algorithm[8]. In the following, we briefly explain some quality measures of clustering techniques[7]:

- *Compactness*: samples in one cluster should be similar to each other and different from samples in other clusters. An example of this would be the within-cluster distance and can be calculated by:

$$F_c(m_1, \ldots ..m_k) = \frac{1}{K} \sum_{k=1}^{K} \frac{1}{n_k} \sum_{j=1}^{n_k} d(m_k, y_{jk}) \tag{3}$$

where $d(\cdot)$ is the distance between cluster center, m_k, and y_{jk} which is sample j of cluster k. The objective is to minimize this measurement as possible.

- *Separation*: clusters should be well-separated from each other. It's also known as between-clusters distance. An example of this criterion is the Euclidean distance between clusters centroids. The objective is to maximize the separation between different clusters as possible. Separation is calculated using the following equation:

$$F_s(m_1, \ldots ..m_k) = \frac{1}{K(K-1)} \sum_{j=1}^{K} \sum_{k=j+1}^{K} d(m_j, m_k) \tag{4}$$

- *Combined measure:* This measure is a linear combination of the compactness and separation measures [7]. Having the within-cluster and between-cluster distances defined, we can now construct the combined measure

$$F_{combined} = \omega_1 F_c - \omega_2 F_S \tag{5}$$

where ω_1 and ω_2 are weighting parameters such that $\omega_1 + \omega_2 = 1$.

- *Turi's validity index:* Turi, [9], proposed an index incorporating a multiplier function (to penalize the selection of a small number of clusters) to the ratio between intra-cluster and inter-cluster, the measure is defined as;

$$F_{Turi}(m_1,....,m_K) = (c \times N(2,1)+1) \times \frac{\text{int}\,ra}{\text{int}\,er} \tag{6}$$

where c is a user specified parameter and N(2,1) is a Gaussian distribution with mean 2 and standard deviation of 1. The *intra* term is the average of all the distances between each data point and its cluster centroid and it's used to measure the compactness of clusters as given is Eq. 3 while the inter term is the minimum distance between the cluster centers, this term used to measure the separation of the clusters and is given by:

$$\text{int}\,er = \min\{\|m_k - m_l\|\} \forall k = 1,..K-1, l = k+1,....K \tag{7}$$

The goal is to minimize the Turi's validity index as possible.

3 Experimental Results

The goal here is to compare the performance of the PSO clustering methods and k-means in segmenting VHR remote sensing imagery, and to investigate the PSO ability to segment land-use classes in dense urban areas.

3.1 Dataset

The study area is the city of Kitchener-Waterloo (K-W), Canada. The data was provided by the University Map Library at the University of Waterloo [10] as ortho-rectified aerial images taken in April 2006 at 12 cm spatial resolution by a digital color airborne camera with 8 bit radiometric resolution. We cropped a set of forty test images of size 666*372 from the original image. The cropped test images were chosen for high density urban parts which are highly corrupted by noise. Samples of the test image are shown in Figure. 1.

Fig. 1. Samples of test images that are corrupted by noise

It is difficult to specify any desired number of clusters in the segmentations of remote sensing images, because the ground truth is always not available for the scenes covered by those images. The major objects of interests in urban planning are roads, buildings and green area such as parks. Test images were manually segmented into four land use types (roads, buildings, green area and other). Other represents pixilation which is either difficult to interpret or does not correspond to the objects of interests like building entrance with a very small parking area alongside the road, swimming pools and other small objects in the image. A sample test image is shown in Fig. 2 with its ground truth.

3.2 Experiment Setup

In the applications of VHR remote sensing images in urban studies, road extraction is the most basic and important task. To extract it, the number of clusters is defined empirically to be four clusters. It was chosen by minimize the error between the clustered image and the ground truth images. Although four to seven spectral clusters work well for most of the test images, four clusters have been selected as it gives the best average accuracy for the entire set of the test images.

For each clustering method there are some free parameters need to be tuned in order to assess the best average performance provided by each one of them over the whole set of the test images. K-means doesn't require any parameter tuning. In the PSO clustering algorithm, we carried out different trials with different values for the number of particle, the value was set to 60 particles for all images which is higher then the recommended number, 20 to 30 particles, giving in [11] as the image data is large data set. Increasing the particle number in the algorithm can increase the chance for finding the optimal solution however the algorithm require more time to converge, the inertia weight w is initially set as 1.2 and is reduced by 1% at each generation to ensure good convergence . The acceleration coefficient constants c_1 and c_2 are set as 1.49. These values are chosen based on the results shown in [11].

Throughout this paper, we use f-measure as a quality assessment measure for the mapping between classes (ground truth) and clusters returned using four cluster validity indices. We compare the average performance of PSO clustering and K-means methods. We look at the average over 40 multiple runs for each method and consider the standard deviation. The average execution time of the 40 runs are also compared.

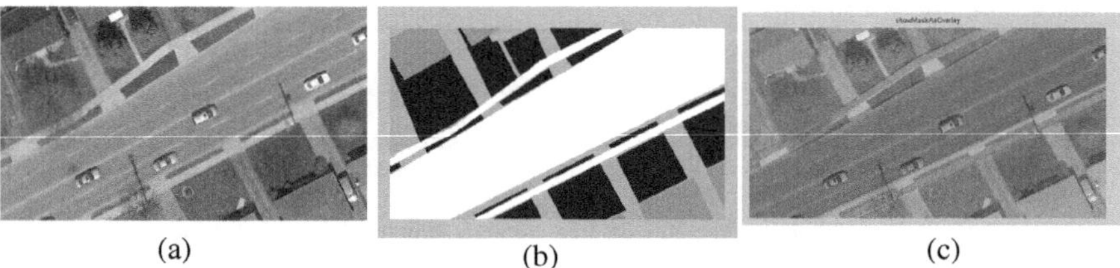

(a) (b) (c)

Fig. 2. Sample of the images and its ground truth: (a) original image, (b) ground truth and (c) ground truth overlaying the original image (right)

3.3 Results

We compare the average performance of PSO and K-means clustering methods focusing on dense urban areas in VHR aerial images. The original image pixel's RGB values are used as spectral feature. The average is taken over 40 for each method. The standard deviation of the error is around 0.1 and 0.2 for all methods. Table 1 shows the results over the 40 test set images using an Intel core 2 Duo T5550 @ 1.83 GHz Processors with 2 MB cache and 3 GB RAM. The table shows the clustering accuracy and the different error measures mentioned in sec 2.2. The result shows the potential of the PSO clustering of aerial images starting from the three RGB bands only. In the experiment we could achieve an average rate of 83% of extracting road areas even in the noisy images of the residential areas.

Table 1. Comparison of clustering accuracies and errors for k-means, PSO clustering methods using different clustering objective functions for road extraction from the aerial images test set

Roads	Clustering Accuracy ↑	*Compactness* ↓	*Separation* ↑	*Combination* ↓	*Turi's index* ↓
PSO-Separation	0.837± 0.081	63.628± 6.703	104.029± 11.487	37.655± 4.422	1.169± 0.203
PSO-Compactness	0.811± 0.078	12.232± 2.456	48.9729± 8.7828	3.166± 0.459	3.050± 1.019
PSO-Combined	0.853± 0.076	21.413± 3.897	87.576± 5.869	3.003± 1.116	2.7883± 0.769
PSO- Trui	0.826± 0.073	32.419± 3.098	55.836± 5.848	8.919± 1.234	1.830± 0.202
K-means	0.754 ± 0.074	6.0837± 0.1474	39.525± 1.022	-0.756± 0.716	0.611± 0.231

4 Conclusion

In this research, which has been motivated by the superiority of PSO over the traditional clustering algorithms in segmenting remote sensing imagery of rural and suburban areas, we tackled the use of PSO in segmenting more complex scenes as VHR remote sensing data in urban areas. The results show that we can extract geographic objects such as roads with 83% accuracy using primitive features as the RGB intensity values of the image pixels.

The next step in this research is to investigate the effect of adding texture and shape descriptors to differentiate between objects with similar spectral signatures such as roads and parking lots.

References

[1] Pal, N.R., Pal, S.K.: A review on image segmentation techniques. Pattern Recognition 9, 1277–1294 (1993)

[2] Jain, A., Duin, R., Mao, J.: Statistical pattern recognition: A review. IEEE Trans. on Pattern Analysis and Machine Intelligence 22, 4–37 (2000)

[3] Kennedy, J., Eberhart, R.C.: Particle swarm optimization. In: Proceedings of IEEE International Conference on Neural Networks (ICNN 1995), vol. 4, pp. 1942–1948. IEEE Service Center, Perth (1995)

[4] Omran, M., Engelbrecht, A.P., Salman, A.: Particle swarm optimization method for image clustering. International Journal of Pattern Recognition and Artificial Intelligence 19(3), 297–321 (2005)
[5] Kotsiantis, S., Pintelas, P.: Recent advances in clustering: a brief survey. WSEAS Transactions on Information Science and Applications 1, 73–81 (2004)
[6] Reyes, M., Coello, C.: Multi-objective particle swarm optimizers: A survey of the state-of-the-art. International Journal of Computational Intelligence Research 3(2), 287–308 (2006)
[7] Ahmadi, A.: Multiple Cooperative Swarms for Data Clustering, PhD Thesis, University of Waterloo (2008)
[8] Halkidi, M., Batistakis, Y., Vazirgiannis, M.: On Clustering Validation Techniques. Intelligent Information Systems Journal 17(2-3), 107–145 (2001)
[9] Ray, S., Turi, R.H.: Determination of Number of Clusters in K-Means Clustering and Application in Colour Image Segmentation. In: Proceedings of the 4th International Conference on Advances in Pattern Recognition and Digital Techniques (ICAPRDT 1999), Calcutta, India, pp. 137–143 (1999)
[10] Tri-Cities and Surrounding Communities Orthomosaics 2006 (computer file). Waterloo, Ontario: The Regional Municipality of Waterloo (2006)
[11] Shi, Y.H., Eberhart, R.C.: Parameter Selection in Particle Swarm Optimization. In: Porto, V.W., Waagen, D. (eds.) EP 1998. LNCS, vol. 1447, pp. 591–600. Springer, Heidelberg (1998)

Image Segmentation under Occlusion Using Selective Shape Priors

Huang Fuzhen and Yang Xuhong

Department of Information & Control Engineering, Shanghai University of Electric Power
Shanghai, 200090, China
{huangfzh,yangxuhong}@shiep.edu.cn

Abstract. In this paper a new method using selective shape priors in a level set framework for image segmentation under occlusion is presented. To solve occluded boundaries, prior knowledge of shape of objects is introduced using the Nitzberg-Mumford-Shiota variational formulation within the segmentation energy. The novelty of our model is that the use of shape prior knowledge is automatically restricted only to occluded parts of the object boundaries. Experiments on synthetic and real image segmentation show the efficiency of our method.

Keywords: Image segmentation, level set, shape priors.

1 Introduction

Image segmentation is a fundamental topic in image processing. Its aim is to partition an image into several parts in each of which the intensity is homogeneous. However, it is often considered a difficult problem due to noise which results in spurious edges and boundary gaps, and occlusions which leads to an overlap of object boundaries. From a variational analysis point of view, level set is considered as a main approach to perform the segmentation [1]. However, classical level set techniques are intensity-based models. They will fail to segment meaningful objects when they are occluded by others or some parts of them are in very low contrast or missing. In fact these situations always happen in practical applications. This hints that shape priors should be incorporated into the segmentation.

There are many work on shape prior segmentation in the literature. Almost all these work are linear combinations of two terms of which one about some specific segmentation functional and the other about shape difference. For example, Leventon et. al. [2] presented a model which incorporates statistical based shape information into Caselles' geometric active contour model. Chen et. al. [3] defined an energy functional depending on the gradient and the average shape of the target object. Cremers et. al. [4] modified the Mumford-Shah's functional by incorporating statistical shape knowledge.

In this paper, we propose an alternative approach consisting in defining a selective shape prior using a variational framework by Nitzberg, Mumford and Shiota (NMS) in [5,6]. The novelty of our model is that the use of shape prior knowledge is

A. Campilho and M. Kamel (Eds.): ICIAR 2010, Part I, LNCS 6111, pp. 89–95, 2010.

automatically restricted only to occluded parts of the object boundaries. That is, the algorithm selectively activates the shape term within the energy functional only for occluded regions. Thus, the evolution of the segmenting level set function for the unoccluded regions is solely driven by image intensity, even though the governing energy functional also includes the shape term. This selective use of local prior shape avoids enforcing shape constraints on regions where the object boundary is clearly defined by image intensity. Finally, this model is used for segmentation on both synthetic and real images.

The remainder of this paper is organized as follows. Section 2 briefly introduces level set method and the NMS model. Our new model is presented in Section 3. Section 4 shows some experimental results and discussions, followed by concluding remarks in Section 5.

2 Preliminaries

2.1 Level Set Theory

Level set [1] is a useful mathematical formulation for implementing efficiently curve propagation. Its central idea is to follow the evolution of a function ϕ whose zero level set always corresponds to the position of a propagating curve. The motion for this evolving function ϕ is determined from a partial differential equation in one higher dimension. The fundamental level set scheme is given by [1]:

$$\frac{\partial \phi}{\partial t} = F\left|\nabla \phi\right| , \tag{1}$$

where ϕ is a surface whose zero level set represents the propagating curve Γ, i.e.:

$$\Gamma(t) = \{\bar{x} \mid \phi(\bar{x}, t) = 0\} , \tag{2}$$

and $\left|\nabla \phi\right|$ denotes the gradient norm of ϕ, F is the speed function that controls the motion of the propagating curve. In general F consists of two terms: an image-based term and a curvature-based term.

Level set representations have many advantages [1]. Firstly, the level set function ϕ always remains a function as long as F is smooth. So topological changes of the propagating curve Γ can be handled naturally. Secondly, the unique, entropy-satisfying weak solution of equation (4) can be obtained relying on viscosity solutions of the associated partial differential equation. Thirdly, the finite difference computational schemes by exploiting numerical solutions of hyperbolic conservation laws are stable and accurate. And finally, intrinsic geometric properties of the propagating curve (e.g., normal, curvature, etc.) can be estimated directly from the level set function and the method can be very easily extended to deal with problems in higher dimensions.

One powerful level set model proposed by Chan and Vese [8] is to minimize the following energy functional

$$
\begin{aligned}
E_{CV}(c_1, c_2, \phi) &= \mu \int_\Omega \delta(\phi)\left|\nabla\phi\right| \mathrm{d}x\mathrm{d}y + v\int_\Omega H(\phi)\mathrm{d}x\mathrm{d}y \\
&+\lambda_1 \int_\Omega \left|I(x,y)-c_1\right|^2 H(\phi)\mathrm{d}x\mathrm{d}y + \lambda_2 \int_\Omega \left|I(x,y)-c_2\right|^2 \left(1-H(\phi)\right)\mathrm{d}x\mathrm{d}y
\end{aligned}, \tag{3}
$$

where $I(x,y)$ is the image intensity, $\mu \ge 0$ □ $v \ge 0$ □ $\lambda_1, \lambda_2 > 0$ are fixed parameters. The Heaviside function H and the one-dimensional Dirac measure δ are defined respectively by

$$
H(z) = \begin{cases} 1, & if\ z \ge 0 \\ 0, & if\ z < 0 \end{cases}, \quad \delta(z) = \frac{d}{dz}H(z). \tag{4}
$$

The parameters c_1, c_2 in (3) are computed as:

$$
c_1 = \frac{\int_\Omega I(x,y)H(\phi)\mathrm{d}x\mathrm{d}y}{\int_\Omega H(\phi)\mathrm{d}x\mathrm{d}y}, \quad c_2 = \frac{\int_\Omega I(x,y)\left(1-H(\phi)\right)\mathrm{d}x\mathrm{d}y}{\int_\Omega \left(1-H(\phi)\right)\mathrm{d}x\mathrm{d}y}. \tag{5}
$$

And its final level set formation is:

$$
\frac{\partial\phi}{\partial t} = \delta(\phi)\left[\mu\kappa - v - \lambda_1\left(I(x,y)-c_1\right)^2 + \lambda_2\left(I(x,y)-c_2\right)^2\right], \tag{6}
$$

where $\kappa = \mathrm{div}\left(\nabla\phi/\left|\nabla\phi\right|\right)$ is the curvature.

This model can detect objects with smooth boundaries or discontinuous boundaries. Moreover, even if the initial image is very noisy, the boundaries can be very well detected and preserved. However, there is still no way to characterize the global shape of an object. Additional information about the object is needed to help the segmentation process, especially those with complex backgrounds.

2.2 Nitzberg-Mumford-Shiota Formulation

In [6,7] Nitzberg, Mumford and Shiota (NMS) proposed a variational framework for the segmentation with depth problem and numerical techniques for minimizing the NMS functional which have been presented in [7,9]. Zhu et al. utilize level set method to minimize the NMS functional in [10]. We briefly review the related NMS formulation for segmentation with depth.

Briefly, segmentation with depth is to determine the ordering of objects in space as well as their boundaries in a given 2D gray scale image I. The ordering of objects refers to the position that one object is farther or nearer an observer than the others. Due to the permission of occlusions between objects, farther objects always consist of two parts: visible and invisible parts. The visible parts are determined by the gray intensity distribution, while the invisible parts should be reconstructed by following some principles. In NMS model, they reconstruct invisible parts of regions by using curvature information along the boundaries. Suppose the 2D image I is composed of n objects $\{O_i\}_{i=1}^n$. And $R_1,\ldots,R_n$ are the regions occupied by the objects inside the image. An occlusion relation '>' is defined on object indices given by $i > j$ when O_i is in front of O_j (from the viewer's perspective). Suppose the objects $O_1, \ldots, O_n$ are listed in order of increasing distance to the observer, so that O_1, O_n are the nearest and farthest objects respectively. Then the visible part A_i of O_i is given by

$$A_1 = R_1,\ A_i = R_i - \bigcup_{j>i} A_j,\ \text{for}\quad i = 2,\dots,n\,, \tag{7}$$

and the background is denoted by $A_{n+1} = \Omega - \bigcup_{j>n+1} A_j$. Then the NMS functional is [6]

$$E = \sum_{i=1}^{n} \{\alpha \int_{\partial R_i \cap \Omega} ds + \beta \int_{\partial R_i \cap \Omega} \phi(\kappa) ds\} + \sum_{i=1}^{n+1} \int_{A_i} (I - c_i)^2 dx\,, \tag{8}$$

where α, β are two nonnegative parameters, and the unknown c_i denote the approximate gray scale intensities of the corresponding objects. Here, the second term is the prior knowledge that the NMS model incorporates to solve occlusions. The function ϕ determines how the curvature information will be incorporated in the functional. The authors in [9] choose ϕ as follows:

$$\phi(x) = \begin{cases} x^2 & |x| \le 1 \\ |x| & |x| > 1 \end{cases}. \tag{9}$$

The level set formulation of NMS functional is [9]

$$\begin{aligned} E = &\sum_{i=1}^{n} \int_{\Omega} (\alpha + \beta\phi(\kappa_i)) |\nabla \psi_i| \delta(\psi_i) dx + \sum_{i=1}^{n} \int_{\Omega} (I - c_i)^2 H(\psi_i) \prod_{j=1}^{i-1} (1 - H(\psi_i)) dx \\ &+ \int_{\Omega} (I - c_{n+1})^2 \prod_{j=1}^{n} (1 - H(\psi_i)) dx \end{aligned}. \tag{10}$$

The unknown intensities c_i are computed as follows:

$$c_i = \frac{\int_\Omega I H(\psi_i) \prod_{j=1}^{i-1} (1 - H(\psi_i) \mathrm{d}x\mathrm{d}y}{\int_\Omega H(\psi_i) \prod_{j=1}^{i-1} (1 - H(\psi_i)) \mathrm{d}x\mathrm{d}y},\quad c_{n+1} = \frac{\int_\Omega I \prod_{j=1}^{n} (1 - H(\psi_j) \mathrm{d}x\mathrm{d}y}{\int_\Omega \prod_{j=1}^{n} (1 - H(\psi_j) \mathrm{d}x\mathrm{d}y}. \tag{11}$$

3 Our New Model Using Selective Shape Priors

In many works on shape prior segmentation in the literature, the prior shape information can be introduced into the level set functional either to the evolution equation [2,4] or directly as a shape energy to the functional [3]. However, just adding a shape term as in these methods means that the shape term might influence boundary shapes even in unoccluded regions, where the boundary is unambiguously defined by image intensity. Hence, we introduce our shape term in a selective manner. That is, the shape term is allowed to take effect only for occluded boundaries. Our model is derived from [11].

For each intersection region of p objects, p>1, define $P = \{\cap_{s=1}^{p} A_s\} - \{\cap_{s=p+1}^{N} A_s\}$ with mean intensity μ_P, then the following shape term,

$$\sum_{s=1}^{p} \int_{P_s} (\mu_P - c_s)^2 \phi(\kappa_s) dx, \tag{12}$$

use shape prior only to boundaries of P that belong to occluded objects, where $P = \{\cap_{t=1, t\neq s}^{p} A_t\} - \{\cap_{s=p+1}^{N} A_s\}$. Firstly, the terms $\phi(\kappa_s)$, that constrain the shape of A_s, are weighted by $(\mu_P - c_s)^2$, which is larger for occluded objects, and is minimal for the object that is in front. Secondly, the shape term in (12) is defined only on P_s, the region that occludes the A_s-boundary of P.

Now for N = 2, for the intersection $P = A_1 \cap A_2$ with mean intensity μ, the local shape term defined in (12) in a level set formulation is

$$\int_{\Omega} \{H(\psi_2)(\mu - c_1)^2 \phi(\kappa_1) + H(\psi_1)(\mu - c_2)^2 \phi(\kappa_2)\} dx. \tag{13}$$

Thus the energy with a local shape term is:

$$\begin{aligned} E = {} & \alpha \int_{\Omega} (|\nabla \psi_1| \delta(\psi_1) + |\nabla \psi_1| \delta(\psi_1)) dx + \beta \int_{\Omega} (\phi(\kappa_1) + \phi(\kappa_2)) dx \\ & + \int_{\Omega} ((I - c_1)^2 H(\psi_1) + (I - c_2)^2 H(\psi_2)) dx + \int_{\Omega} (I - c_3)^2 (1 - H(\psi_1))(1 - H(\psi_2)) dx. \\ & + \lambda \int_{\Omega} \{H(\psi_2)(\mu - c_1)^2 \phi(\kappa_1) + H(\psi_1)(\mu - c_2)^2 \phi(\kappa_2)\} dx \end{aligned} \tag{14}$$

Here, the last term is the shape term used to globally influence the shape of the segmented objects to avoid local minima. β and λ balance the shape terms with λ □ β.

To minimize (14), a finite difference scheme is used to solve the resulting Euler Lagrange equations as in [11].

4 Experimental Results

Experiments with both synthetic image and real image are utilized to demonstrate the performance of our method.

We first evaluate our algorithm on a synthetic image: a triangle (close to an equilateral triangle) occluded by other objects (Fig.1 (a)). The white curve in Fig.1 (a) is the initial curve. The shape prior is an equilateral triangle (Fig. 1 (b)). The final result is shown in Fig.1 (c), where the white curve is the segmentation curve. This example shows that our model can combine the prior shape information to segment an object with a similar shape even though the object is occluded by others or some of its parts are missing.

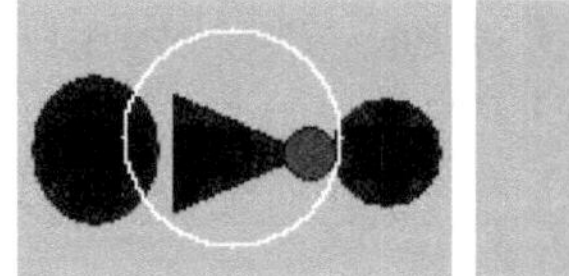

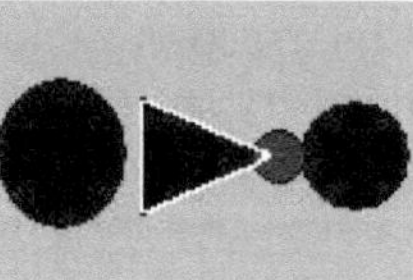

Fig. 1. Segmentation on synthetic image: (a) Original image with initial contour; (b) Prior shape; (c) final result

Our next experiment is carried out on a real image: a hand occluded by other objects (Fig.2 (a)) with cluttered background. The prior shape is a similar hand (Fig.2 (b)). The final results are shown in Fig.2 (c). The example listed here validate that our model is also capable of dealing with real images, i.e., it can segment an object which is similar in shape to the prior one from an image even though the object is occluded by others or has some missing parts.

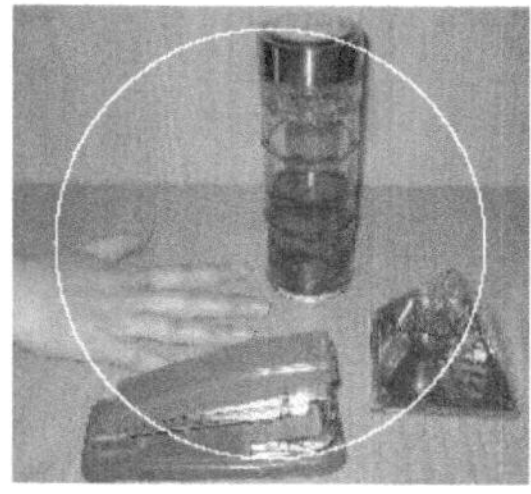

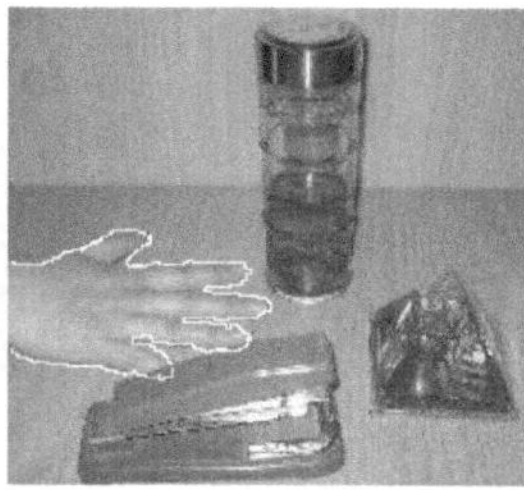

Fig. 2. Segmentation on real image: (a) Original image with initial contour; (b) Prior shape; (c) Final result

5 Conclusions

In this paper, a new model using selective shape priors in a level set framework for image segmentation under occlusion is proposed. Experiments on synthetic and real images show that this model is able to converge to a reference shape, providing robustness to partial occlusion in image segmentation applications. Our future work will be conducted to multiple object segmentation problems.

Acknowledgments. This work is supported by Research Plan of Shanghai Municipal Education Commission (07ZZ147) and Leading Academic Discipline Project of Shanghai Municipal Education Commission (J51301).

References

1. Sethian, J.A.: Level Set methods and Fast Marching Methods: Evolving interfaces in computational geometry, fluid mechanics, computer vision, and materials science. Cambridge University Press, Cambridge (1999)

2. Leventon, M.L., Grimson, W.E.L., Faugeras, O.: Statistical shape influence in geometric active contours. In: Proceedings of the IEEE Computer Society Conference on Computer Vision and Pattern Recognition, pp. 316–323 (2000)
3. Chen, Y., Tagare, H., et al.: Using prior shapes in geometric active contours in a variational framework. International Journal of Computer Vision 50(3), 315–328 (2002)
4. Cremers, D.: Dynamical statistical shape priors for level set-based tracking. IEEE Transactions on Pattern Analysis and Machine Intelligence 28(8), 1262–1273 (2006)
5. Cremers, D., Rousson, M., Deriche, R.: A review of statistical approaches to level set segmentation: integrating color, texture, motion and shape. International Journal of Computer Vision 72(2), 195–215 (2007)
6. Nitzberg, M., Mumford, D.: The 2.1-d sketch. In: Proceedings of the IEEE Conference on Computer Vision (1990)
7. Nitzberg, M., Mumford, D., Shiota, T.: Filtering, segmentation and depth. LNCS. Springer, Heidelberg (1993)
8. Chan, T.F., Vese, L.A.: Active contours without edges. IEEE Transactions onImage Processing 10(2), 266–277 (2001)
9. Esedoglu, S., March, R.: Segmentation with depth but without detecting junctions. Journal of Mathematical Imaging and Vision 18(1), 7–15 (2003)
10. Chan, T., Zhu, W., Esedoglu, S.: Segmentation with depth: A level set approach. UCLA CAM report. 04-49 (2004)
11. Thiruvenkadam, S.R., Chan, T., Hong, R.-W.: Segmentation under occlusions using selective shape prior. UCLA CAM report. 07-30 (2007)

Fusion of Edge Information in Markov Random Fields Region Growing Image Segmentation

Amer Dawoud and Anton Netchaev

School of Computing, University of Southern Mississippi
118 College Drive Box#5106, Hattiesburg MS 39406 USA
{amer.dawoud,anton.netchaev}@usm.edu

Abstract. This paper proposes an algorithm that fuses edge information into Markov Random Fields (MRF) region growing based image segmentation. The idea is to segment the image in a way that takes edge information into consideration. This is achieved by modifying the energy function minimization process so that it would penalize merging regions that have real edges in the boundary between them. Experimental results confirming the hypothesis that the addition of edge information increases the precision of the segmentation by ensuring the conservation of the objects contours during the region growing.

1 Introduction

Image segmentation is a process that decomposes an image into disjoint regions and is a fundamental step for many image-processing tasks such as image understanding and content based image retrieval. In general, image segmentation aims at producing regions that are homogeneous with respect to the extracted features, such as gray level or texture, and have significant different feature values across region boundaries [1]. Many algorithms were proposed over the years. Feature-based approaches such as thresholding [2] and clustering [3] would usually produce noisy results. Many other methods utilize spatial context information explicitly or implicitly. Edge-based segmentation [4], [5], [6] are efficient in describing local behavior but inefficient in producing global meaningful results. Region splitting and merging algorithms [7], [8] have problem with merging and stopping criteria that would usually cause the result to be either over-segmented or under-segmented. Model-based approaches, such as curve evolution [9], [10], [11] and random fields [12] [13] [14] [15], have established mathematical foundation but they require accurate model and the optimization process to be able to converge to a reasonable solution, which are difficult to achieve.

In this paper, image segmentation algorithm is presented, which is based on the work of Yu and Clausi [1]. Our algorithm can be characterized by two aspects: 1) it uses region growing technique in search for optimal solution. 2) it fuses edge information in the energy minimization process in a way that penalizes merging region with real edges in the boundary between them. Next section describes our algorithm in detail, and section 3 shows experimental results demonstrating that our algorithm increases the precision of the segmentation by ensuring the conservation of the objects contours during the region growing.

A. Campilho and M. Kamel (Eds.): ICIAR 2010, Part I, LNCS 6111, pp. 96–104, 2010.

2 Problem Statement and Related Work

This work is an extension to Yu and Clausi's work [1], so it is necessary to distinguish our contribution by briefly introducing their work first and then clearly describing our contribution, which is the fusion of edge information.

2.1 MRF-Based Formulation of Image Segmentation

Let S denote the discrete rectangular lattice on which images are defined. Suppose there are n different classes in the image to be segmented. $\mathbf{X} = \{X_s \mid s \in S\}$ is a set of discrete valued random variables constituting a random field on S, with each variable X_s taking a value in $\{1,\ldots,n\}$ representing the class to which the site s belongs. $\mathbf{Y} = \{Y_s \mid s \in S\}$ is another random field somehow related to $\mathbf{X}$ and the observed image is a realization from $\mathbf{Y}$. Let $\mathrm{x} = \{x_s \mid s \in S\}$ and $\mathrm{y} = \{y_s \mid s \in S\}$ denote the realizations of $\mathbf{X}$ and $\mathbf{Y}$, respectively. The image segmentation is an inverse process that attempts to estimate the best x given the observed image y. With the obtained class labels x, S is segmented to n classes, $\Omega_1, \ldots, \Omega_n$ such that

$$\begin{aligned} &a)\ \Omega_i = \{s|X_s = i,\ s \in S\}, \\ &\text{b) } \textstyle\bigcup_{i=1}^{n} \Omega_i = S, \\ &\text{c) } \forall\ i \neq j\colon \Omega_i \cap \Omega_j = \emptyset. \end{aligned} \tag{1}$$

The image segmentation task can be formulated as a maximum a posterior (MAP) problem for which maximizing the posterior $P(\mathrm{x}|\mathrm{y})$ gives a solution. By the Bayes' rule, this is equivalent to maximizing $p(\mathrm{y}|\mathrm{x})\ P(\mathrm{x})$. Two models are used for analytically representing $p(\mathrm{y}|\mathrm{x})$ (the feature model) and $P(\mathrm{x})$ (the spatial context model). With both the feature model and the spatial context model defined, the MAP formulation of the segmentation task is transformed into minimizing energy

$$E = E_s + E_f \tag{2}$$

where

$$E_s = \sum_{<a,b>\in R} \begin{cases} \beta & if\ x_a \neq x_b \\ 0 & otherwise, \end{cases} \tag{3}$$

where a and b are neighboring sites forming a pair-site clique and β is a positive number. Such a model makes the prior $P(\mathrm{x})$ large if the local neighborhood is dominated by one single class and small otherwise and, hence, is effective in suppressing noisy configurations of class labels. R is the set of all cliques on the entire lattice S.

$$E_f = \textstyle\sum_{s\in S} \frac{1}{2}\ln\left(2\pi\sigma_{x_s}^2\right) + \frac{(y_s - \mu_{x_s})^2}{2\sigma_{x_s}^2}, \tag{4}$$

where μ_i and σ_i^2 are the mean and variance of the pixel values in class i. So the image segmentation problem is formulated as follow as

$$\arg \min_{x_s, s \in S} \{ \sum_{s \in S} \left\{ \frac{1}{2} \ln(2\pi\sigma_{x_s}^2) + \frac{(y_s - \mu_{x_s})^2}{2\sigma_{x_s}^2} \right\} + \beta \sum_{<a,b> \in R} \{1 - \delta(x_a, x_b)\}, \quad (5)$$

where $\delta(.)$ is the Kronecker delta function.

2.2 Extending to Region-Based Segmentation

Finding a solution for (5) represents a combinatorial optimization problem, which could be mathematically intractable. Many combinatorial optimization techniques have been proposed, including iterated conditional mode (ICM) [16], simulated annealing (SA) [17], mean field theory [18], genetic algorithm [19], belief propagation [20], and graph theoretic techniques [21]. To simplify the complexity of the problem, the MRF can be defined on irregular graphs rather than the regular image lattice. This allows the image segmentation problem formulated by (5) to be based on a set of interconnected groups of pixels, with the MRF spatial context model based on a region adjacency graph (RAG) [22]. Here, the labeling is not on single pixels but on regions, where the regions are commonly obtained by a deliberate over-segmentation. Each node in the RAG represents a region and a link between the nodes represents the existence of a common boundary between the regions.

Defined on the RAG, the MRF models the behaviors of the regions in a similar way as for pixels. Let R_i denote node i in the graph and let x_i denote the label for all sites $s \in R_i$. The feature model energy for R_i can be defined as

$$E_f = \sum_{s \in R_i} \left\{ \frac{1}{2} \ln(2\pi\sigma_{x_i}^2) + \frac{(y_s - \mu_{x_i})^2}{2\sigma_{x_i}^2} \right\}, \quad (6)$$

and the MRF pair site clique energy for two neighboring nodes R_i. and R_j is

$$E_s = \sum_{\substack{<s,t> \in C \\ s \in R_i, t \in R_j}} \begin{cases} \beta & if\ x_s \neq x_t \\ 0 & otherwise. \end{cases} \quad (7)$$

Summation of the above energies over the entire RAG gives exactly (5). A combinatorial optimization technique is then applied to RAG nodes instead of pixels. Such a region-based segmentation method is advantageous in computation speed as the number of RAG nodes is usually significantly less than the number of pixels.

3 Proposed Algorithm

3.1 Combing Edge Lines with Watershed Lines

Figure 1 shows the flowchart of the proposed algorithm. The initial step is over-segmentation of input image using watershed [23], which is a well-establish method that is based on the image topology. The magnitude of the gradient is interpreted as elevation information. With successive flooding, watersheds with adjacent catchment basins are constructed. This operation results in image being over-segmented and regions separated by watershed lines as shown in Figure 2 (A).

Next, Canny edge detector [24] is used to detect edges lines. The affect of adjusting the parameter settings of Canny detector, which control how sensitive the detector is, will be discussed in next section. The watershed lines are combined with edge lines, which will divide some of the watershed segmentation regions even into smaller ones. This is demonstrated in Figure 2(B), where the number of regions increased from 4 to 8. Figure 1 shows a real image in A, Detected edges using Canny in B, Watershed over-segmentation in C with number of regions is 642, and the segmentation after combining edge lines with watershed lines where the number of regions is 681.

The purpose of this step is to make sure that the edge lines are coinciding with boundary of the regions. As we will see in the next steps of the algorithm, this is necessary to integrate the edge information in the MRF clique energy.

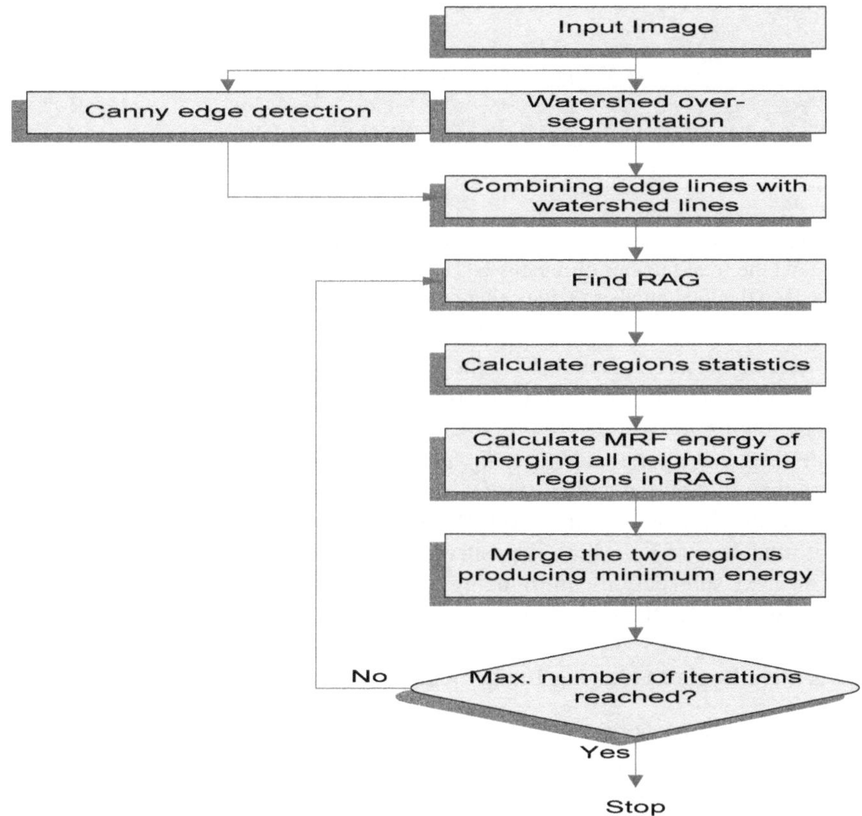

Fig. 1. Algorithm's Flowchart

3.2 Fusion of Edge Information in MRF Energy Minimization

$\mathbf{D} = \{D_s \mid s \in S\}$ is a set of edge random field on S, with each variable D_s taking a value of {0,1} representing a non-edge and an edge site in s, respectively. Let $\mathrm{d} = \{d_s \mid s \in S\}$ denote the realization of **D**.

The MRF pair site clique energy E_s for two neighboring nodes R_i. and R_j is modified from (7) to

$$E_s = \sum_{\substack{<s,t>\in C \\ s\in R_i, t\in R_j}} \begin{cases} \beta_1 & if\ x_s \neq x_t\ and\ d_s = d_t \\ \beta_2 & if\ x_s \neq x_t\ and\ d_s \neq d_t \\ 0 & otherwise. \end{cases} \quad (8)$$

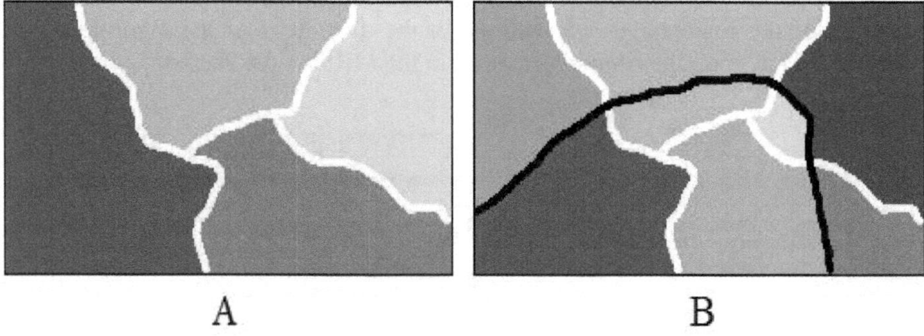

Fig. 2. A) Watershed over-segmentation: the white lines are the watershed lines separating regions, and the total number of regions is 4. B) Combining edge line: black line is a detected edge line, and the total number regions after combining the edge line is 8.

The difference between (7) and (8) is the introduction of β_2, since β_1 is similar to β in (7). Eq. (8) means that for a clique of two neighboring sites to contribute to E_s then they must belong to two different classes. And if one and only one of the sites is an edge site then that clique adds β_2 to E_s, and if both sites are either edge sites or non-edge site then that clique adds β_1 to E_s. The effect of introducing β_2, which should be negative, is that it should reduce the MRF pair site clique energy instead of increasing it as with the case of β_1, which is positive. With regard to region growing (to be discussed in next sub-section), introducing β_2 will promote keeping regions with edge lines passing through the boundaries between them separated.

3.3 Iterative Region Growing Image Segmentation

The algorithm starts merging image regions with the aim of minimizing the overall energy, which is the summation of feature model energy E_f (6) and MRF pair site clique energyE_s (8). This is achieved, as shown in flowchart of Fig. 1, by finding the RAG, and calculating the statistics of all regions μ_{x_i} and $\sigma^2_{x_i}$. Then the differences in energy associated with merging each two neighboring regions, which are represented as an edge in the RAG, are calculated.

Suppose two classes, Ω_i and Ω_j, are being investigated. Let $\Omega_k = \Omega_i \cup \Omega_j$ denote the class obtained by merging the two classes. The energy difference

$$\delta E_{ij} = \sum_{s\in R_k}\left\{\frac{1}{2}\ln\left(2\pi\sigma_{x_k}^2\right) + \frac{\left(y_s-\mu_{x_k}\right)^2}{2\sigma_{x_k}^2}\right\} - \sum_{s\in R_i}\left\{\frac{1}{2}\ln\left(2\pi\sigma_{x_i}^2\right) + \frac{\left(y_s-\mu_{x_i}\right)^2}{2\sigma_{x_i}^2}\right\} - \sum_{s\in R_j}\left\{\frac{1}{2}\ln\left(2\pi\sigma_{x_j}^2\right) + \frac{\left(y_s-\mu_{x_j}\right)^2}{2\sigma_{x_j}^2}\right\} - \sum_{\substack{<s,t>\in C\\ s\in R_i, t\in R_j}}\begin{cases}\beta_1 & if\ x_s \neq x_t\ and\ d_s = d_t\\ \beta_2 & if\ x_s \neq x_t\ and\ d_s \neq d_t\\ 0 & otherwise.\end{cases} \quad (9)$$

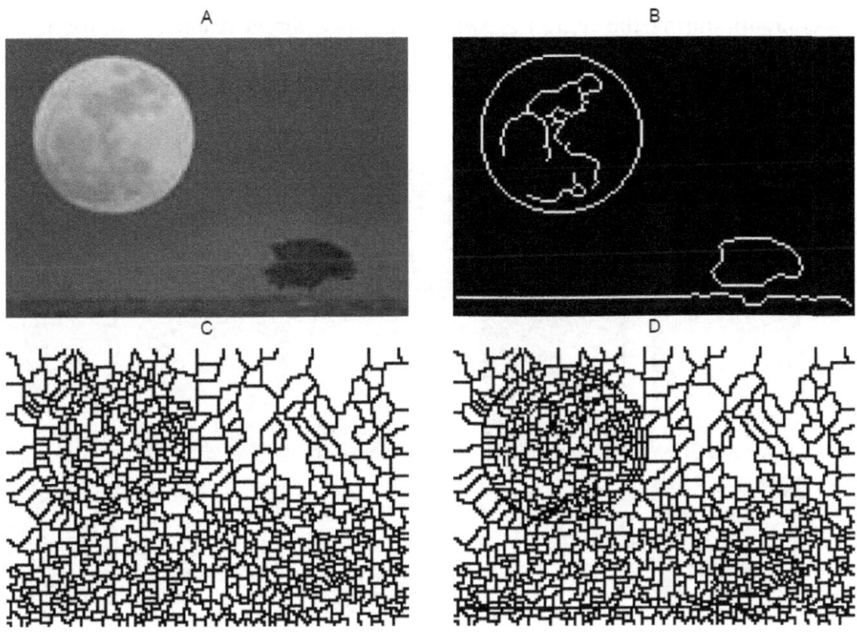

Fig. 3. A) Original image. B) Detected edges using Canny. C) Watershed over-segmentation (number of regions is 642). D) Segmentation after combining Canny edge lines with watershed lines (number of regions is 681).

The two regions producing the lowest δE_{ij}in the RAG are merged. This process is repeated iteratively and regions are merged sequentially, till number of desired classes is reached. In the next iteration, there is no need to do the whole calculations from scratch. The only calculations needed to update RAG and δE_{ij} will be related to new region formed by the merging in pervious iteration, which makes the program run faster.

4 Results

Figure 4 shows experimental results of the image in Figure 3 comparing our algorithm, which fuses the edge information, with Yu and Clausi's algorithm [1]. The segmentation results at various iterations or number of regions (N = 50, 10, 5, 4) are captured. First three columns shows the segmentation results using our proposed algorithm with different β_1 and β_2 settings. The first row of images shows the edge image detected by Canny method at different parameters (sensitivity) settings. Last two columns show the segmentation results without the fusion of edge information (Yu and Cluasi algorithm [1]) with different β settings.

Our algorithm performed better in terms of segmenting the image along the edge lines, which confirms confirming the hypothesis that the addition of edge information increases the precision of the segmentation by ensuring the conservation of the objects contours during the region growing. This is a direct result for introducing that neqative β_2 in the MRF energy function, which makes it difficult to megre regions separated by edge line. Therefore, the settings that controls the sensetivity of the edge

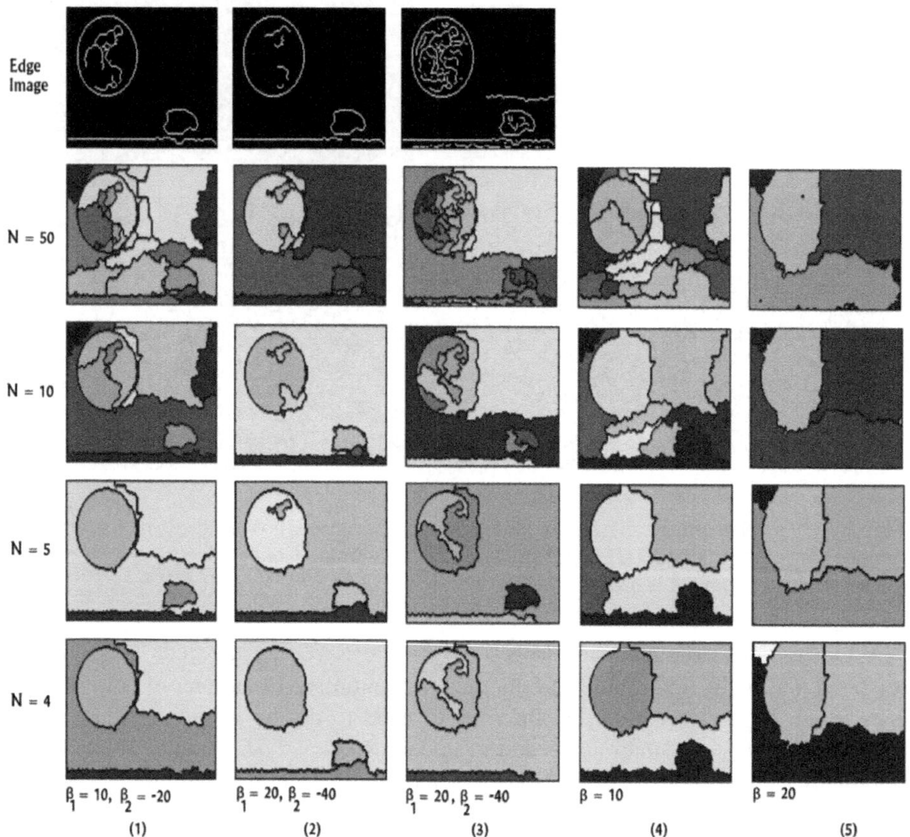

Fig. 4. Segmentation Results for the image shown in Figure 3 at various N (number of regions). First three columns shows the segmentation results using our proposed algorithm with different β_1 and β_2 settings. Last two columns show the segmentation results without the fusion of edge information (algorithm Yu and Cluasi [1]) with different β settings.

detector are also important in achieving a precise segmentation that preserves the object contours.

5 Conclusions

We proposed an algorithm that fuses edge information into Markov Random Fields (MRF) region growing based image segmentation. The idea is to segment the image in a way that takes edge information into consideration. We achieved this by modifying the energy function minimization process so that it would penalize merging regions that have real edge in the boundary between them. Experimental results confirming the hypothesis that the addition of edge information increases the precision of the segmentation by ensuring the conservation of the objects contours during the region growing.

References

[1] Yu, Q., Clausi, D.A.: IRGS: Image Segmentation Using Edge Penalties and Region Growing. Transactions on Pattern Analysis and Machine Intelligence 30, 2126–2139 (2008)

[2] Otsu, N.: A Threshold Selection Method from Gray-Level Histograms. IEEE Trans. Systems, Man, and Cybernetics 9, 62–66 (1979)

[3] Jain, A.K., Dubes, R.C.: Algorithms for Clustering Data. Prentice Hall, Englewood Cliffs (1988)

[4] Ma, W.Y., Manjunath, B.S.: Edgeflow: A Technique for Boundary Detection and Image Segmentation. IEEE Trans. Image Processing 9(8) (2000)

[5] Nguyen, H.T., Worring, M., van den Boomgaard, R.: Watersnakes: Energy-Driven Watershed Segmentation. IEEE Trans. Pattern Analysis and Machine Intelligence 25(3) (March 2003)

[6] Vincent, L., Soille, P.: Watershed in Digital Spaces: An Efficient Algorithm Based on Immersion Simulations. IEEE Trans. Pattern Analysis and Machine Intelligence 13(6), 583–598 (1991)

[7] Adams, R., Bischof, L.: Seeded Region Growing. IEEE Trans. Pattern Analysis and Machine Intelligence 16(6) (June 1994)

[8] Haris, K., Efstratiadis, S.N., Maglaveras, N., Katsaggelos, A.K.: Hybrid Image Segmentation Using Watersheds and Fast Region Merging. IEEE Trans. Image Processing 7(12), 1684–1699 (1998)

[9] Cootes, T.F., Taylor, C.J., Cooper, D.H., Graham, J.: Active Shape Models—Their Training and Application. Computer Vision and Image Understanding 61(1), 38–59 (1995)

[10] Kass, M., Witkin, A., Terzopoulos, D.: Snakes: Active Contour Models. Int'l J. Computer Vision 1(4), 321–331 (1987)

[11] Yezzi, J.A., Tsai, A., Willsky, A.: A Fully Global Approach to Image Segmentation via Coupled Curve Evolution Equations. J. Visual Comm. and Image Representation 13, 195–216 (2002)

[12] Derin, H., Elliott, H.: Modeling and Segmentation of Noisy and Textured Images Using Gibbs Random Fields. IEEE Trans. Pattern Analysis and Machine Intelligence 9(1), 39–55 (1987)

[13] Geman, S., Geman, D.: Stochastic Relaxation, Gibbs Distributions, and the Bayesian Restoration of Images. IEEE Trans. Pattern Analysis and Machine Intelligence 6(6), 721–741 (1984)

[14] Kumar, S., Hebert, M.: Discriminative Random Fields. Int'l J. Computer Vision 68(2), 179–202 (2006)

[15] Won, C.S., Derin, H.: Unsupervised Segmentation of Noisy and Textured Images Using Markov Random Fields. CVGIP: Graphical Models and Image Processing 54(4), 308–328 (1992)

[16] Besag, J.: On the Statistical Analysis of Dirty Pictures. J. Royal Statistical Soc. B 48(3), 259–302 (1986)

[17] Geman, S., Geman, D.: Stochastic Relaxation, Gibbs Distributions, and the Bayesian Restoration of Images. IEEE Trans. Pattern Analysis and Machine Intelligence 6(6), 721–741 (1984)

[18] Andrey, P., Tarroux, P.: Unsupervised Segmentation of Markov Random Field Modeled Textured Images Using Selectionist Relaxation. IEEE Trans. Pattern Analysis and Machine Intelligence 20(3), 252–262 (1998)

[19] Zhang, J.: The Mean Field Theory in EM Procedures for Markov Random Fields. IEEE Trans. Signal Processing 40(10), 2570–2583 (1992)

[20] Cheng, L., Caelli, T.: Unsupervised Image Segmentation: A Bayesian Approach. In: Proc. 16th Int'l Conf. Vision Interface (June 2003)

[21] Boykov, Y., Veksler, O., Zabih, R.: Fast Approximate Energy Minimization via Graph Cuts. IEEE Trans. Pattern Analysis and Machine Intelligence 23(11), 1222–1239 (2001)

[22] Li, S.Z.: Markov Random Field Modeling in Image Analysis. Springer, Heidelberg (2001)

[23] Meyer, F.: Topographic distance and watershed lines. Signal Processing 38, 113–125 (1994)

[24] Canny, F.J.: A computational approach to edge detection. IEEE Trans. Pattern Anal. Machine Intell. 8(6), 679–698 (1986)

Image Segmentation for Robots: Fast Self-adapting Gaussian Mixture Model

Nicola Greggio[1,2], Alexandre Bernardino[2], and José Santos-Victor[2]

[1] ARTS Lab - Scuola Superiore S.Anna, Polo S.Anna Valdera
Viale R. Piaggio, 34 - 56025 Pontedera, Italy
[2] Instituto de Sistemas e Robótica, Instituto Superior Técnico - 1049-001 Lisboa, Portugal
ngreggio@isr.ist.utl.pt

Abstract. Image segmentation is a critical low-level visual routine for robot perception. However, most image segmentation approaches are still too slow to allow real-time robot operation. In this paper we explore a new method for image segmentation based on the expectation maximization algorithm applied to Gaussian Mixtures. Our approach is fully automatic in the choice of the number of mixture components, the initialization parameters and the stopping criterion. The rationale is to start with a single Gaussian in the mixture, covering the whole data set, and split it incrementally during expectation maximization steps until a good data likelihood is reached. Singe the method starts with a single Gaussian, it is more computationally efficient that others, especially in the initial steps. We show the effectiveness of the method in a series of simulated experiments both with synthetic and real images, including experiments with the iCub humanoid robot.

Keywords: image processing, unsupervised learning, self-adapting gaussians mixture, expectation maximization, machine learning, clustering.

1 Introduction

Nowadays, computer vision and image processing are involved in many practical applications. The constant progress in hardware technologies leads to new computing capabilities, and therefore to the possibilities of exploiting new techniques, for instance considered to time consuming only a few years ago. Image segmentation is a key low level perceptual capability in many robotics related application, as a support function for the detection and representation of objects and regions with similar photometric properties. Several applications in humanoid robots [1], rescue robots [2], or soccer robots [3] rely on some sort on image segmentation [4]. Additionally, many other fields of image analysis depend on the performance and limitations of existing image segmentation algorithms: video surveillance, medical imaging and database retrieval are some examples [5], [6].

Two main principal approaches for image segmentation are adopted: Supervised and unsupervised. The latter one is the one of most practical interest. It may be defined as the task of segmenting an image in different regions based on some similarity criterion among each region's pixels. Particularly interesting is the Expectation Maximization algorithm applied to gaussians mixtures which allows to model complex probability distribution functions. Fitting a mixture model to the distribution of the data is

A. Campilho and M. Kamel (Eds.): ICIAR 2010, Part I, LNCS 6111, pp. 105–116, 2010.

equivalent, in some applications, to the identification of the clusters with the mixture components [7].

Expectation-Maximization (EM) algorithm is the standard approach for learning the parameters of the mixture model [8]. It is demonstrated that it always converges to a local optimum. However, it also presents some drawbacks. For instance, EM requires an *a-priori* selection of model order, namely, the number of components (M) to be incorporated into the model, and its results depend on initialization. The more gaussians there are within the mixture, the higher will be the total log-likelihood, and more precise the estimation. Unfortunately, increasing the number of gaussians will lead to overfitting the data and it increases the computational burden. Therefore, finding the best compromise between precision, generalization and speed is a must. A common approach to address this compromise is trying different configurations before determining the optimal solution, e.g. by applying the algorithm for a different number of components, and selecting the best model according to appropriate criteria.

1.1 Related Work

Different approaches can be used to select the best number of components. These can be divided into two main classes: *off-line* and *on-line* techniques.

The first ones evaluate the best model by executing independent runs of the EM algorithm for many different initializations, and evaluating each estimate with criteria that penalize complex models (e.g. the Akaike Information Criterion (AIC) [9] and the Rissanen Minimum Description Length (MDL) [10]). These, in order to be effective, have to be evaluated for every possible number of models under comparison. Therefore, it is clear that, for having a sufficiently exhaustive search the complexity goes with the number of tested models, and the model parameters.

The second ones start with a fixed set of models and sequentially adjust their configuration (including the number of components) based on different evaluation criteria. Pernkopf and Bouchaffra proposed a Genetic-Based EM Algorithm capable of learning gaussians mixture models [11]. They first selected the number of components by means of the minimum description length (MDL) criterion. A combination of genetic algorithms with the EM has been explored.

An example are the greedy algorithms. Applied to the EM algorithm, they usually start with a single component (therefore side-stepping the EM initialization problem), and then increase their number during the computation. The first formulation was originally proposed in 2000, by Li and Barron [12]. Subsequently, in 2002 Vlassis and Likas introduced a greedy algorithm for learning Gaussian mixtures [13]. Nevertheless, the total complexity for the global search of the element to be splitted $O(n^2)$. Subsequently, Verbeek et al. developed a greedy method to learn the gaussians mixture model configuration [14]. However, the big issue in these kind of algorithm is the insertion selection criterion: Deciding when inserting a new component and how can determine the success or failure of the subsequent computation.

Ueda *et Al.* proposed a split-and-merge EM algorithm to alleviate the problem of local convergence of the EM method [15]. Subsequently, Zhang *et Al.* introduced another split-and-merge technique [16]. Merge and split criterion is efficient in reducing number of model hypothesis, and it is often more efficient than exhaustive, random or genetic algorithm approaches. Particularly interesting is the method proposed by Figueiredo and Jain, which uses only merge operations, therefore starting with a high number of mixture parameters, merging them step by step until convergence [17], making then no use of splitting operations. This method can be applied to any parametric mixture where the EM algorithm can be used. However, the higher the number of mixture components is, the more expensive the computation will be. Therefore, since the idea of Figueiredo and Jain starts with a very high number of mixture components, greatly slowing the computation from the first steps.

1.2 Our Contribution

In this paper, we propose an algorithm that automatically learns the number of components as well as the parameters of the mixture model. The particularly of our model is that we approach the problem contrariwise than Figueiredo and Jain did, i.e. by starting from only one mixture component instead of several ones and progressively adapting the mixture by adding new components when necessary. Therefore, in order to accomplish this we needed to define a precise split and stopping criteria. The first is essential to be sure to introduce a new component (and therefore new computational burden) only when strictly necessary, while the second one is fundamental to stop the computation when a good compromise has been obtained (otherwise the algorithm will continue to add components indefinitely, until the maximum possible likelihood is obtained). Our formulation guarantees the following advantages. First, it is a deterministic algorithm; we avoid the different possibilities in the components initializations that greatly affect the standard EM algorithm, or any EM technique that starts with more than one component, by using a unique initialization independently from the input data. Therefore, by applying the same algorithm to the same input data we will get always the same results, Second, it is a low computationally expensive technique - in fact, new components will be added only when strictly necessary.

1.3 Outline

The paper is organized as follows. In sec. 2 we introduce the proposed algorithm. Specifically, we describe the insertion of a new gaussians in sec. 2.4, the initializations in sec. 2.2, the decision thresholds update rules in sec. 2.5, and the stopping criterion 2.6. Furthermore, in sec. 3 we describe our experimental set-up for testing the validity of our new technique and the results. Finally, in sec. 4 we conclude and propose directions for future work.

2 FASTGMM: FAST Self-adapting Gaussian Mixture Model

We distinguish two main important features for our algorithm: The splitting criterion and the stopping criterion. The key issue of our algorithm is looking whether one or more gaussians are not increasing their own likelihood during optimization. In other words, if they are not participating in the optimization process, they will be split into two new gaussians. We will introduce a new concept related to the state of a gaussians component:

- Its age, that measures how long the component's own likelihood does not increase significantly (see sec. 2.1);

Then, the split process is controlled by the following adaptive decision thresholds:

- One adaptive threshold L_{TH} for determining a significant increase in likelihood (see sec. 2.5);
- One adaptive threshold Age_{TH} for triggering the merge or split process based on the component's own age (see sec. 2.5);
- One adaptive threshold S_{TH} for deciding to split a gaussians based on its area (see sec. 2.4).

It is worth noticing that even though we consider three thresholds to tune, all of them are adaptive, and only require a coarse initialization.

These parameters will be fully detailed within the next sections.

2.1 FASTGMM Formulation

Our algorithm's formulation can be summarized within three steps:

- Initializing the parameters;
- Adding a gaussians;
- Updating decision thresholds.

Each mixture component i is represented as follows:

$$\bar{\vartheta}_i = \varrho(w_i, \bar{\mu}_i, \Sigma_i, \xi_i, \Lambda_{last(i)}, \Lambda_{curr(i)}, a_i) \tag{1}$$

where w_i is the *a-priori* probabilities of the class, $\bar{\mu}_i$ is its mean, Σ_i is its covariance matrix, ξ_i its *area*, $\Lambda_{last(i)}$ and $\Lambda_{curr(i)}$ are its last and its current log-likelihood value, and a_i its *age*. Here, we define two new elements, the area (namely, the covariance matrix determinant) and the age of the gaussians, which will be described later.

During each iteration, the algorithm keeps memory of the previous likelihood. Once the re-estimation of the vector parameter $\bar{\vartheta}$ has been computed in the EM step, our algorithm evaluates the current likelihood of each single gaussians as:

$$\Lambda_{curr(i)}(\vartheta) = \sum_{j=1}^{k} log(w_i \cdot p_i(\bar{x}^j)) \tag{2}$$

If a_i overcomes the age threshold Age_{TH} (i.e. the gaussians i does not increase its own likelihood for a predetermined number of times significally - over L_{TH}), the algorithm decides whether to split this gaussians or merging it with existing ones depedending on whether their own single area overcome S_{TH}.

Then, after a certain number of iterations the algorithm will stop - see sec. 2.6. The whole algorithm pseudocode is shown in Fig. 2.1.

Algorithm 2.1. Pseudocode

```
- Parameter initialization;
while (stopping criterion is not met) do
    $\Lambda_{curr(i)}$, evaluation, for $i = 0, 1, \ldots, c$;
    $L(\bar{\vartheta})$ evaluation;
    Re-estimate priors $w_i$, for $i = 0, 1, \ldots, c$;
    Recompute center $\bar{\mu}_i^{(n+1)}$ and covariances $\Sigma_i^{(n+1)}$, for $i = 0, 1, \ldots, c$;
    - Evaluation whether changing the gaussians distribution structure -
    for ($i = 0$ to $c$) do
        if ($a_i > Age_{TH}$) then
            if (($\Lambda_{curr(i)} - \Lambda_{last(i)}$) < $L_{TH}$) then
                $a_i += 1$;
                - General condition for changing satisfied; checking those for each gaussians -
                if ($\Sigma_i > S_{TH}$) then
                    if ($c < maxNumgaussians$) then
                        split gaussians → split ;
                        $c += 1$;
                        reset $S_{TH} \leftarrow \frac{S_{M-INIT}}{ng}$;
                        reset $L_{TH} \leftarrow L_{INIT}$;
                        reset $a_A, a_B \leftarrow 0$, with A, B being the new two gaussians;
                        return
                    end if
                end if
                $S_{TH} = S_{TH} \cdot (1 + \alpha \cdot \xi)$;
            end if
        end if
    end for
end while
```

2.2 Parameters Initialization

At the beginning, S_{TH} will be automatically initialized to the Area of the covariance of all the data set - i.e. the determinant of the covariance matrix relative to the whole data set. The other decision thresholds will be initialized as follows:

$$\begin{aligned} L_{INIT} &= k_{LTH} \\ Age_{INIT} &= k_{ATH} \end{aligned} \tag{3}$$

with k_{LTH} and k_{ATH} (namely, the minimum amount of likelihood difference between two iterations and the number of iterations required for taking into account the lack of a likelihood consistent variation) relatively low (i.e. both in the order of 10, or 20). Of course, higher values for k_{LTH} and smaller for k_{ATH} give rise to a faster adaptation, however adding instabilities.

2.3 Gaussians Components Initialization

The algorithm starts with just only one gaussians. Its mean will be the whole data mean, while its covariance matrix will be those of the whole data set. Of course, one may desire to start with more than one gaussians in case that *a-priori* the gaussians components of the data set are more than one, for sake of convergence speed. In that case means and covariances will be as follows.

2.4 Splitting a Gaussian

If the covariance matrix determinant of the examined gaussians at each stage overcomes the maximum area threshold S_{TH}, then another gaussians is added to the mixture.

More precisely, the original gaussians with parameters $\bar{\vartheta}_{old}$ will be split within other two ones. The new means, A and B, will be:

$$\begin{aligned} \bar{\mu}_A &= \bar{\mu}_{old} + \frac{1}{2}(\Sigma_{i=j})^{1/2} \\ \bar{\mu}_B &= \bar{\mu}_{old} - \frac{1}{2}(\Sigma_{i=j})^{1/2} \quad i,j = \{1,2,\ldots,d\} \end{aligned} \tag{4}$$

where d is the input dimension.

The covariance matrixes will be updated as:

$$\Sigma_{A(i,j)} = \Sigma_{B(i,j)} = \begin{cases} \frac{1}{2}\Sigma_{old(i,j)}, & \text{if } i = j; \\ 0, & \text{othrewise.} \end{cases} \tag{5}$$

The *a-priori* probabilities will be

$$w_A = \frac{1}{2} w_{old} \qquad w_B = \frac{1}{2} w_{old} \tag{6}$$

The decision thresholds will be updated as follows:

$$S_{TH} = \frac{S_{M-INIT}}{ng} \qquad L_{TH} = L_{INIT} \tag{7}$$

where ng_{old} and ng are the previous and the current number of mixture components, respectively. Finally, their ages, a_A and a_B, will be reset to zero.

2.5 Updating Decision Thresholds

The thresholds L_{TH}, and S_{TH} vary at each step with the following rules:

$$\begin{aligned} L_{TH} &= L_{TH} - \frac{\lambda}{ng} \cdot L_{TH} = L_{TH} \cdot (1 - \frac{\lambda}{ng}) \\ S_{TH} &= S_{TH} - \frac{\alpha_{Max}}{ng} \cdot S_{TH} = S_{TH} \cdot (1 - \frac{\alpha_{Max}}{ng}) \end{aligned} \tag{8}$$

with ng is the number of current gaussians, λ, and α_{Max} Using high values for λ, and α_{Max} results in high convergence speed. However, with faster convergence comes significant instability around the optimal desidered point. Following this rules L_{TH} will decrease step by step, approaching the current value of the global log-likelihood increment. This is the same for S_{TH}, which will become closer to the maximum *area* of the gaussians, allowing splitting. This will allow the system to avoid some local optima, by varying its configuration if a stationary situation occurs.

Finally, every time a gaussians is added these thresholds will be reset to their initial value.

2.6 Stopping Criterion

Analyzing the behavior of the whole mixture log-likelihood emerges a common trend: It always acts like a first order system. In fact, it produces a curve with a high derivate at beginning that decreases going on with the number of iterations, reaching the log-likelihood maximum value asymptotically. We know from the theory that the rate at which the response approaches the final value is determined by the time constant. When $t = \tau$ (in our case $i = \tau$), L has reached 63.2% of its final value. When $t = 5\tau$, L has reached 99.3% of its final value. Again, we know from the theory that the time constant τ is the angular coefficient of the output curve at the time $t = 0$.

We know from the EM theory that at each iteration it has to grow, or at least remaining the same. However, spikes during the splitting operations that make the log-likelihood decreasing abruptly are present. Moreover, in order to avoid local optima-like situations, we average the log-likelihood increments by sampling it with a fixed sampling rate (e.g. $T_s = 25$ iterations).

For each $i = n \cdot T_s$, with n an integer number, we store the current log-likelihood within an array. The first time the log-likelihood increment between two consecutive sampled value increases less than 0.7% we store the relative number of iterations $i_{first} = n_{stop}T_s$. Then, we stop after the log-likelihood does not increase over 0.7% for a number of times equal to n_{stop}.

2.7 Computational Complexity Evaluation

Within this section we will use the following convention: ng is the number of the mixture gaussians components, k is the number of input vectors, d is the number of input dimension, and it is the number of iterations.

The computational burden of the EM algorithm is, referring to the pseudocode in tab. 2.1 as follows:

- the original EM algorithm (steps 3 to 6) take $O(k \cdot d \cdot ng)$ for 3 and 6, while step 4 and step 5 take $O(1)$ and $O(k \cdot ng)$;
- our algorithm takes $O(ng)$ for evaluating all the gaussians (step 8 to 26);
- our split (step 15) operation requires $O(d)$.
- the others take $O(1)$.

Therefore, the original EM algorithm takes $O(k \cdot d \cdot ng)$, while our algorithm adds $O(d \cdot ng)$ on the whole, giving rise to $O(k \cdot d \cdot ng) + O(d \cdot ng) = O(k \cdot d \cdot ng + d \cdot ng) = (ng \cdot d \cdot (k+1))$. Considering that usually $d << k$ and $ng << k$ this does not add a considerable burden, while giving an important improvement to the original computation in terms of self-adapting to the data input configuration at best.

3 Experimental Validation

3.1 Experimental Set-Up

To compare our algorithm other EM-based methods we choose three techniques, BIC, AIC, and MDL, as the most common used selection criteria. In order to reduce the artifact of the initialization on the standard EM algorithm, we adopted a standard approach:

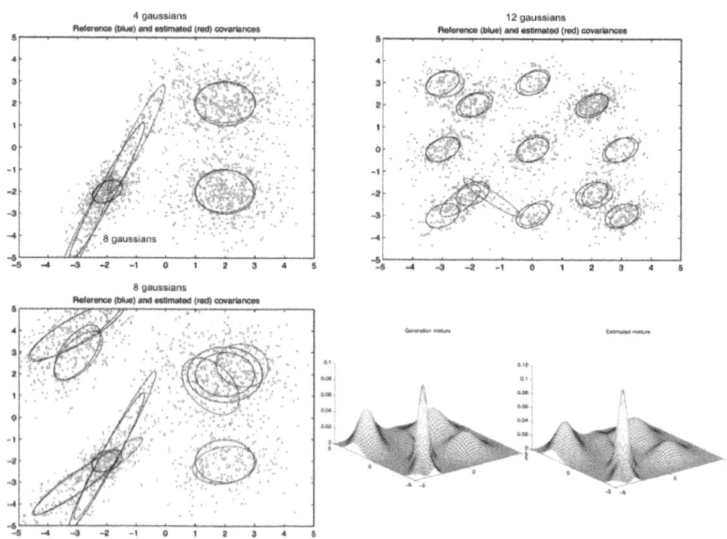

Fig. 1. The 2D representation of the final gaussians mixture generated by our algorithm vs. the real one and the relative log-likelihood outputs as function of the iterations number, for different input mixtures of data (4, 8, 12 gaussians components). Moreover the 8-gaussians case comparison between the generated and computed mixtures is shown on the bottom right.

We selected 10 different initial random conditions, keeping those giving the highest likelihood. The stopping criteria we adopted for the EM computation is the most common used, i.e. it requires that the log-likelihood increment goes below a threshold ϵ. We used $\epsilon = 10 \cdot e^{-5}$. We evaluated our technique's performances by applying it both to synthetic data (artificially generated with a known mixture) and with different kind of pictures, i.e. some well known pictures and some real images (taken by a webcam or by our robotic platform iCub's cameras).

Mixture precision estimation. It is possible to see that FASTGMM usually achieves a higher final log-likelihood than the other techniques, although running more iterations. This suggests a better approximation of the data mixture. However, a higher log-likelihood does not strictly imply that the extracted mixture covers the data better than another one. This is because it is based on the probability of each component, which may be more or less exact, being not deterministic. Nevertheless, it is not a good index on the probability that such mixture would be better.

A deterministic approach is to adopt a unique distance measure between the generation mixture and the evaluated one. In [18] Jensen *et Al.* exposed three different strategies for computing such distance: The Kullback-Leibler, the Earh Mover, and the Normalized L2 distance. The first one is not symmetric, even though a symmetrized version is usually adopted in music retrival. However, this measure can be evaluated in a close form only with mono-dimensional gaussians. The second one also suffers analog problems of the latter. The third choice, finally is symmetric, obeys to the triangle inequality and it is easy to compute, with a comparable precision with the other two. We then used the last one. Its expression states [19]:

$$
\begin{aligned}
z_c N_x(\bar{\mu}_c, \bar{\Sigma}_c) &= N_x(\bar{\mu}_a, \bar{\Sigma}_a) \cdot N_x(\bar{\mu}_b, \bar{\Sigma}_b) \\
&\quad where \\
\bar{\Sigma}_c =&(\bar{\Sigma}_a^{-1} + \bar{\Sigma}_b^{-1})^{-1} \;\; and \;\; \bar{\mu}_c = \bar{\Sigma}_c(\bar{\Sigma}_a^{-1}\bar{\mu}_a + \bar{\Sigma}_b^{-1}\bar{\mu}_b) \\
z_c =&|2\pi\bar{\Sigma}_a\bar{\Sigma}_b\bar{\Sigma}_c^{-1}|^{\frac{1}{2}} \cdot \exp\left\{-\frac{1}{2}(\bar{\mu}_a - \bar{\mu}_b)^T\bar{\Sigma}_a^{-1}\bar{\Sigma}_c\bar{\Sigma}_b^{-1}(\bar{\mu}_a - \bar{\mu}_b)\right\} \\
=&|2\pi(\bar{\Sigma}_a + \bar{\Sigma}_b)|^{\frac{1}{2}} \cdot \exp\left\{-\frac{1}{2}(\bar{\mu}_a - \bar{\mu}_b)^T(\bar{\Sigma}_a + \bar{\Sigma}_b)^{-1}(\bar{\mu}_a - \bar{\mu}_b)\right\}
\end{aligned}
\tag{9}
$$

Therefore, we evaluated the Normalized L2 distance as a measure of synthetic data estimation precision, and we reported our result in tab. 3.3.

3.2 Synthetic Data

In order to evaluate the performance of our algorithm, we tested it by classifying different input data randomly generated by a known gaussians mixture, and subsequently saved to a file. We choose to show the results for 2-dimensional input because they are easier to show than multidimensional ones (for instance, a 2-dimensional gaussians is represented in 2D as an ellipse).

The output of the two algorithms is shown in Fig. 1. Each distribution has a total of 2000 points, but disposed differently. The first one has been generated by a 4 gaussian mixture, the second one by a 8 gaussian mixture and the third one by a 12 gaussian mixture. The generation mixture (blue) and the evaluated one (red) are represented in each subfigure. Finally, the 3D histogram representation of the 8-components generated gaussians mixture data and the estimated one. Due to space limitations, we choose to show only the one that gave rise to the worst log-likelihood estimation plot, i.e. the one with 8 components.

We can see that our algorithm is capable to learn the input data mixture starting from only one component with a good accuracy.

Table 1. Experimental results on synthetic data

Actual number of Gaussian components	Algorithm	Detected number of Gaussian components	Total number of iterations	Final log-likelihood	Normalized L2 distance
4	AIC	4	91	-7403.656573	6.595441
	BIC	4	91	-7405.021887	6.382962
	MDL	6	98	-7460.206259	13.715347
	FASTGMM	4	268	-7405.078438	0.075190
8	AIC	9	120	-8400.626025	34.796101
	BIC	7	91	-8428.323612	18.092732
	MDL	8	111	-8554.125701	22.052649
	FASTGMM	8	650	-8446.063794	6.184175
12	AIC	14	103	-7475.658908	45.687874
	BIC	12	124	-7547.612061	2.811907
	MDL	13	161	-7613.774605	21.293496
	FASTGMM	12	393	-7511.032752	2.658803

3.3 Colored Real Images

Learning the right number of color components (i.e. mixture components) within a colored image is a difficult task. This is because an general image contains several of the three fundamental color combinations. Therefore, it is clear that the number of mixture components needed to represent the image at best rapidly rises up excessively, becoming too high.

The color image segmentation results are shown in Fig. 2. The set of images is divided into two groups: Some general images, on the left (from (1) to (5)), and some images taken by the iCub's cameras, on the right (from (6) to (11)). For each group we show the original images, those obtained with the standard EM algorithm initialized with the BIC/AIC/MDL criteria, and those obtained with our algorithm on the left, in the middle, and on the right, respectively.

Here we will find some differences in the number of mixture components detected by our algorithm and those detected by the BIC/AIC/MDL techniques. Our approach tends to use more components than BIC/AIC/MDL do. This is more evident on the real images (which of course contain more color variations than the artificial ones). In table

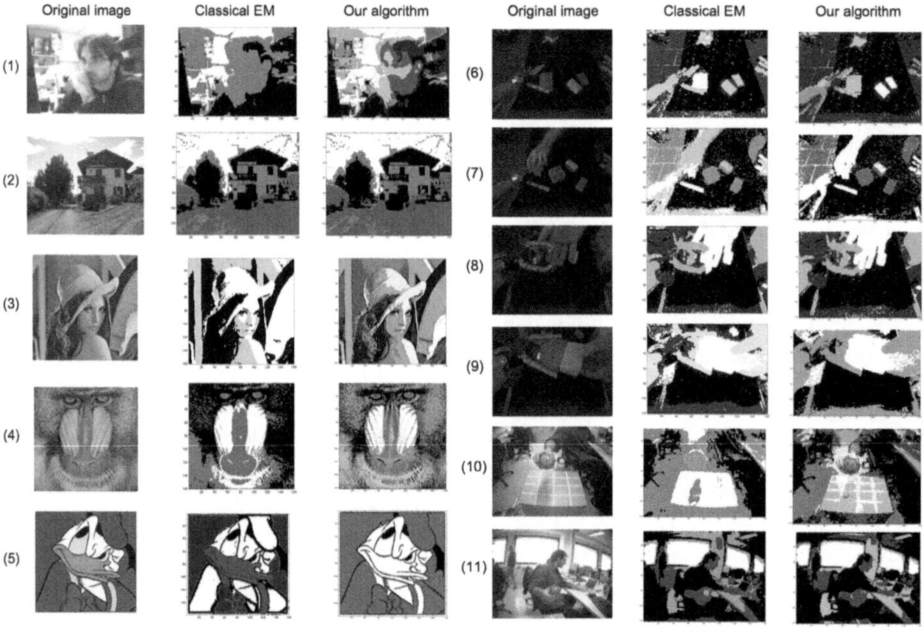

Fig. 2. Color image segmentation results. We divide these images into two groups: Some general images, on the left (from (1) to (5)), and some images taken by the iCub's cameras, on the right (from (6) to (11)). For each group we show the original images, those obtained with the standard EM algorithm initialized with the BIC/AIC/MDL criteria, and those obtained with our algorithm on the left, in the middle, and on the right, respectively.

Table 2. Experimental results on real images

Image (Fig. 2)	Algorithm	Detected number of mixture components	Total number of iterations	Image (Fig. 2)	Algorithm	Detected number of mixture components	Total number of iterations
(1)	AIC	8	85	(6)	AIC	8	234
	BIC	7	91		BIC	7	180
	MDL	7	106		MDL	8	213
	FASTGMM	10	400		FASTGMM	8	350
(2)	AIC	4	120	(7)	AIC	5	92
	BIC	4	91		BIC	4	90
	MDL	4	134		MDL	4	104
	FASTGMM	4	175		FASTGMM	3	150
(3)	AIC	20	213	(8)	AIC	4	92
	BIC	18	192		BIC	4	90
	MDL	18	221		MDL	4	94
	FASTGMM	22	475		FASTGMM	4	175
(4)	AIC	18	145	(9)	AIC	4	78
	BIC	17	126		BIC	3	94
	MDL	16	153		MDL	3	97
	FASTGMM	20	325		FASTGMM	3	150
(5)	AIC	4	86	(10)	AIC	16	121
	BIC	4	93		BIC	16	112
	MDL	4	91		MDL	15	146
	FASTGMM	4	175		FASTGMM	18	300
(11)	AIC	7	131				
	BIC	7	124				
	MDL	6	156				
	FASTGMM	8	350				

3.3 the results of our algorithm and the BIC/AIC/MDL criteria applied to the selected images are shown.

4 Conclusion and Future Work

In this paper we proposed a unsupervised algorithm that learns a finite mixture model from multivariate data on-line. We approached the problem starting from a single mixture component and sequentially *growing* both increases the number of components and adapting their means and covariances. Therefore, its initialization is unique, and it is not affected by different possible starting points like the original EM formulation. Moreover, by starting with a single component the computational burden is low at the beginning, increasing only whether more components are required. We also defined a precise stopping criteria, otherwise the algorithm continues to split indefinitely. Finally, we presented the effectivity of our technique in a series of simulated experiments with synthetic data, artificial, and real images.

- Future work: At the moment we tested our algorithm with synthetic data and static images. As future work, we will improve our algorithm by implementing also a merge technique. So far, it will be possible to remove unused components, too. Our final aim is to apply it to moving objects, online adapting the mixture description with varying input.

Acknowledgements

This work was supported by the European Commission, Project IST-004370 RobotCub and FP7-231640 Handle, and by the Portuguese Government - Fundação para a Ciência e Tecnologia (ISR/IST pluriannual funding) through the PIDDAC program funds and through project BIO-LOOK, PTDC / EEA-ACR / 71032 / 2006.

References

1. Montesano, L., Lopes, M., Bernardino, A., Santos-Victor, J.: Earning object affordances: From sensory motor maps to imitation. IEEE Trans. on Robotics 24(1) (2008)
2. Carpin, S., Lewis, M., Wang, J., Balakirsky, S., Scrapper, C.: Bridging the gap between simulation and reality in urban search and rescue. In: Lakemeyer, G., Sklar, E., Sorrenti, D.G., Takahashi, T. (eds.) RoboCup 2006: Robot Soccer World Cup X. LNCS (LNAI), vol. 4434, pp. 1–12. Springer, Heidelberg (2007)
3. Greggio, N., Silvestri, G., Menegatti, E., Pagello, E.: Simulation of small humanoid robots for soccer domain. Journal of The Franklin Institute - Engineering and Applied Mathematics 346(5), 500–519 (2009)
4. Vincze, M.: Robust tracking of ellipses at frame rate. Pattern Recognition 34, 487–498 (2001)
5. Dobbe, J.G.G., Streekstra, G.J., Hardeman, M.R., Ince, C., Grimbergen, C.A.: Measurement of the distribution of red blood cell deformability using an automated rheoscope. Cytometry (Clinical Cytometry) 50, 313–325 (2002)
6. Shim, H., Kwon, D., Yun, I., Lee, S.: Robust segmentation of cerebral arterial segments by a sequential monte carlo method: Particle filtering. Computer Methods and Programs in Biomedicine 84(2-3), 135–145 (2006)
7. McLachlan, G., Peel, D.: Finite mixture models. John Wiley and Sons, Chichester (2000)
8. Dempster, A., Laird, N., Rubin, D.: Maximum likelihood estimation from incomplete data via the em algorithm. J. Royal Statistic Soc. 30(B), 1–38 (1977)
9. Sakimoto, Y., Iahiguro, M., Kitagawa, G.: Akaike information criterion statistics. KTK Scientific Publisher, Tokio (1986)
10. Rissanen, J.: Stochastic complexity in statistical jnquiry. Wold Scientific Publishing Co., USA (1989)
11. Pernkopf, F., Bouchaffra, D.: Genetic-based em algorithm for learning gaussian mixture models. IEEE Trans. Patt. Anal. Mach. Intell. 27(8), 1344–1348 (2005)
12. Li, J., Barron, A.: Mixture density estimation, vol. 11. NIPS, MIT Press (2000)
13. Vlassis, N., Likas, A.: A greedy em algorithm for gaussian mixture learning. Neural Processing Letters 15, 77–87 (2002)
14. Verbeek, J., Vlassis, N., Krose, B.: Efficient greedy learning of gaussian mixture models. Neural Computation 15(2), 469–485 (2003)
15. Ueda, N., Nakano, R., Ghahramani, Y., Hiton, G.: Smem algorithm for mixture models. Neural Comput. 12(10), 2109–2128 (2000)
16. Zhang, Z., Chen, C., Sun, J., Chan, K.: Em algorithms for gaussian mixtures with split-and-merge operation. Pattern Recognition 36, 1973–1983 (2003)
17. Figueiredo, A., Jain, A.: Unsupervised learning of finite mixture models. IEEE Trans. Patt. Anal. Mach. Intell. 24(3) (2002)
18. Jensen, J.H., Ellis, D., Christensen, M.G., Jensen, S.H.: Evaluation distance measures between gaussian mixture models of mfccs. In: Proc. Int. Conf. on Music Info. Retrieval ISMIR 2007 Vienna, Austria, October 2007, pp. 107–108 (2007)
19. Ahrendt, P.: The multivariate gaussian probability distribution. Tech. Rep. (2005), `http://www2.imm.dtu.dk/pubdb/p.php?3312`

Adaptive Regularization Parameter for Graph Cut Segmentation

Sema Candemir and Yusuf Sinan Akgül

Gebze Institute of Technology, Computer Engineering Department,
Computer Vision Lab., Kocaeli, Turkey
scandemir@bilmuh.gyte.edu.tr; akgul@bilmuh.gyte.edu.tr
http://vision.gyte.edu.tr/

Abstract. Graph cut minimization formulates the segmentation problem as the liner combination of data and smoothness terms. The smoothness term is included in the energy formulation through a regularization parameter. We propose that the trade-off between the data and the smoothness terms should not be balanced by the same regularization parameter for the whole image. In order to validate the proposed idea, we build a system which adaptively changes the effect of the regularization parameter for the graph cut segmentation. The method calculates the probability of being part of the boundary for each pixel using the Canny edge detector at different hysteresis threshold levels. Then, it adjusts the regularization parameter of the pixel depending on the probability value. The experiments showed that adjusting the effect of the regularization parameter on different image regions produces better segmentation results than using a single best regularization parameter.

Keywords: Regularization Parameter, Graph Cut, Image Segmentation.

1 Introduction

The first step of many computer vision systems is the object-background segmentation. For the performance of the advanced steps of the system, the foreground segmentation should be accurate. There are different segmentation techniques in the literature based on clustering [1], curve fitting [2], and energy minimization [3]. The graph cut approach [4,5] is one of the energy based algorithms which solves the object-background segmentation relatively successfully. The algorithm first builds a graph $G = (V, E)$. V consists of set of nodes that correspond to the image pixels. Two additional nodes are also added to V that represent the object and the background terminals. E are the edges that connect the nodes with nonnegative costs. The optimal segmentation is determined by finding the minimum cost cut on the graph through minimizing the graph energy functional. The standard graph energy functional is formulated as,

$$E(f) = \sum_{i \in V} E_d(f_i, d_i) + \lambda \sum_{i,j \in N} E_s(f_i, f_j), \tag{1}$$

A. Campilho and M. Kamel (Eds.): ICIAR 2010, Part I, LNCS 6111, pp. 117–126, 2010.

where V is the nodes on the graph, f_i is the segmentation label and d_i is the a priori data of pixel i, and N represents the neighborhood pixels j of pixel i. The first term in the energy functional is called the data term E_d, which confines the segmentation labels to be close to the observed image. The second term is used for the smoothness which confines the neighboring nodes to have similar segmentation labels.

The data term of the energy formulation is inadequate to obtain a successful segmentation because of the ill-posed nature [6,7] of the segmentation problem. Therefore, energy minimization approaches constrain the solution space by adding a smoothness term. The smoothness term is included in the energy formulation through a regularization parameter λ which determines the degree of the smoothness of the solution. Choosing a suitable λ is important to obtain a meaningful solution. If λ is small, the segmentation will be noisy; on the other hand, if λ is large, the segmentation will not fit the observed data. Figure 1 illustrates the trade-off between the data and smoothness terms on a graph cut minimization. The segmentation with a small regularization parameter (Fig 1.b) produces noisy solutions (grassy regions). If we increase the regularization parameter in order to obtain a noiseless segmentation, this time we lose the details such as the legs and the ears of the horses (Fig 1.c). The better segmentation is obtained with the most suitable regularization parameter (Fig 1.d). However, it still has problems on some parts of the foreground. Note that the ear and the tail regions of the horses are over-segmented (red marked regions); the legs of the horses are under-segmented (blue marked regions).

This paper introduces a new idea that, the trade-off between the data and the smoothness terms should not be balanced by the same regularization parameter for the whole image. For example, the grassy part of the image in Figure 1.a needs higher regularization than the leg part of the horses. However, using the same λ makes the smoothness effect equal on the whole image, and causes over/under-segmented regions even with the most suitable regularization parameter. We propose a method that adaptively changes the regularization parameter of the graph cut minimization depending on the image regions. The method determines the foreground boundary of the image. Then, it adjusts the regularization parameter for each pixel based on the probability of that pixel being part of the boundary.

We introduce the method on interactive graph cut [8,9] which is one of the convenient and widely used graph cut approaches. The user initially marks some pixels as object and some pixels as background to direct the graph cut algorithm. The marked pixels form the object and background intensity distributions. Then, the data and the smoothness energy terms are formulated using these distributions. In order to validate the proposed idea, we implemented a series of experiments on the interactive graph cut minimization by applying our method. The experiments showed that, adjusting the effect of the regularization parameter on different regions of the image produced better segmentation results than using a single best regularization parameter.

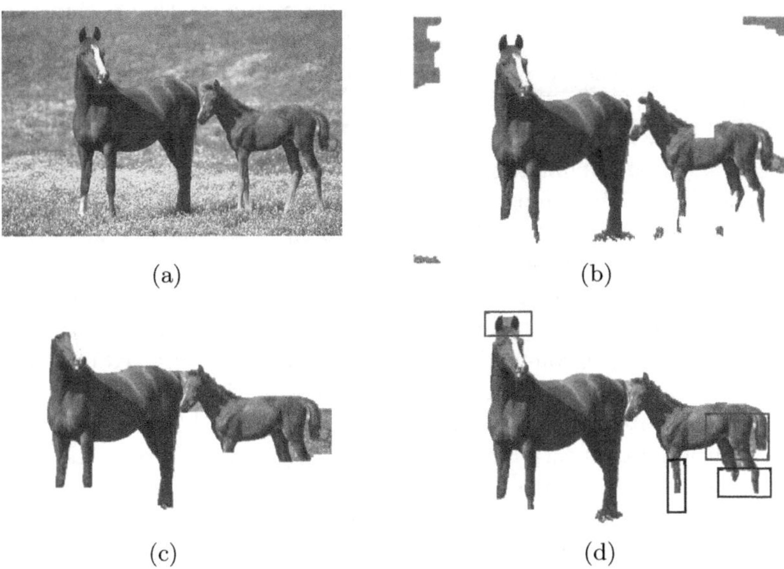

Fig. 1. The illustration of the trade of between the data and the smoothness terms of the graph cut minimization. a) Input image. b) Segmentation is obtained with a small λ, where the smoothness term has small importance in graph energy formulation. c) Segmentation is obtained with a large λ, where the smoothness term has significant importance in the graph energy formulation. d) The best segmentation with the most suitable regularization parameter. Red marked area denote the over segmentations; blue marked area denote the under segmentations.

The rest of the paper is organized as follows: The next section summarizes the related work on the regularization parameter. We introduce the method in Section 3. We show the experimental results in Section 4. Section 5 provides discussions and concluding remarks.

2 Related Work

Obtaining a suitable regularization parameter is as old as the regularization problem [10]. Researchers proposed general techniques such as L-curve method, discrepancy principle, cross-validation principle [11], and U-curve [12]. Besides the general methods, problem-specific techniques are proposed such as for snakes [13] and for image restoration [14]. However, it is recently realized that the suitable regularization parameter depends on statistics of image noise and variation of scene structures [15,16]. Therefore, different image sets need different regularization parameters for an optimal performance. A few papers focused on the optimal choice of regularization parameter from the observed image. Zhang and Seitz [16] proposed a probabilistic mixture model for the λ determination of a stereo pair. Peng and Veksler [17] proposed a regularization parameter selection method for the segmentation problem. Although these approaches estimate suitable parameters, they produce a single regularization parameter for the whole image. Our method on the other hand, claims that the regularization parameter between the energy terms should not be the same for all regions of the image. Different parts of the image needs different regularization effects for a better performance. To the best of our knowledge, our system is the first one to attempt to adaptively regularize the trade-off between the data and the smoothness terms in graph cut segmentation.

As mentioned in the introduction section, using the same λ for the whole image sometimes over-segment the foreground, at other times under-segment such that small foreground details are lost. Vincente and Kolmogorov [18] realized that the graph cut algorithm produces inadequate segmentation for the thin and elongated objects. They proposed a method in which the user marks some additional connectivity priors. Then the algorithm runs the Dijkstra's algorithm for the thin parts of the objects separately. Their work segments the thin parts of the objects successfully. However, it needs additional marking and increases the computational load because of the run-time costs of the Dijkstra's algorithm. Moreover, their approach did not solve the over-segmented part of the segmentation such as the tail part of the horses.

3 Proposed Method

Graph cut approach solves the segmentation problem by minimizing an energy functional which is the linear sum of the problem constraints. The data constraint is inadequate to obtain a unique solution, therefore, the smoothness constraint is included to the energy formulation through a regularization parameter. The general approach is using a single suitable λ for the graph energy formulation, however, we propose that the effect of the regularization should not be the same for the whole image.

We propose a method which arranges the effect of the regularization parameter on different parts of the image. The method first calculates the edge pixels of the image by the Canny edge detector. We run the edge algorithm on the observed image at different hysteresis threshold levels between [0.1-1] (the maximum is 1). Then we calculate the edge probability of each pixel by the linear average of edge maps such that $\overline{I} = \frac{1}{n}\sum_{k=1}^{n} I_k$, where I_k is the binary edge map at hysteresis threshold set k. If pixel i is labeled as an edge pixel for the most of the threshold levels, it has a high probability of being an edge pixel. We illustrated the probability calculation of each pixel in Figure 2. Figure 2.b shows the probabilities of being an edge pixel. In order to decrease the smoothness effect at the near boundary regions, we convolve the probability map with a gaussian kernel (Figure 2.c).

The proposed algorithm modifies the regularization parameter by reducing the smoothness terms of the pixels on the boundary regions. We formulated the proposed idea as

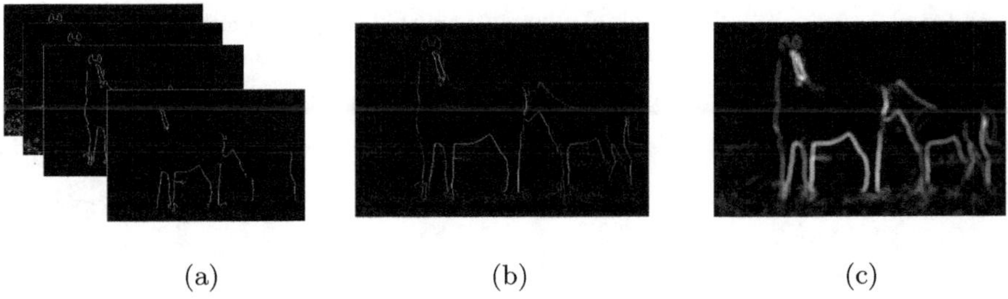

(a) (b) (c)

Fig. 2. a) The edge maps of horse image in fig.1 at different hysteresis threshold levels of Canny edge detector. b) The edge probability map of the horse image is obtained by the linear average of the edge maps at threshold levels between [0.1-1]. c) The probability map is obtained by gaussian smoothing in order to decrease the smoothness effect on the near-boundary regions.

$$E(f) = \sum_{i \in V} E_d(f_i, d_i) + (1 - \overline{I_i})\lambda \sum_{i,j \in N} E_s(f_i, f_j), \tag{2}$$

where $\overline{I_i}$ is the edge probability of pixel i. If the probability of being an edge of the pixel i is high, the regularization parameter will be multiplied with a small value. Therefore, the smoothness effect will decrease for the pixel i. Similarly, if the probability of being an edge is small, we multiplied the smoothness term with a larger value.

4 Experiments

In order to validate the proposed idea, we implemented the interactive graph cut algorithm and segmented the images in Berkeley data set [19]. We first marked some pixels as object and some pixels as background on the image. We constructed the graph structure using the marked pixel histograms. Then we minimized the graph using the energy formulation of Eq. 1 with the regularization parameters between 0 and 99. We calculated the percentage errors of segmentations for each regularization parameter by comparing the obtained labeling with the ground truth segmentation.

We segmented the same images using the proposed approach. We first obtained edge maps of the images using the Canny edge detector at different hysteresis threshold values. Then, we calculated the edge probability of each pixel of the observed image. We minimized the graph structure using the energy formulation of Eq. 2. For a reliable comparison, we used the same graph structure for both approaches. Figure 3 shows the segmentations of both approaches for the best λ values. Note that the tail and the ear regions of the horses are not segmented accurately even using the best λ, which was chosen manually. Some of the leg pixels are also labeled as background (Fig. 3b). The similar corruptions are observed for the other images. The best λ is inadequate to properly segment the region between the wing and the tail of the insect. Some of the

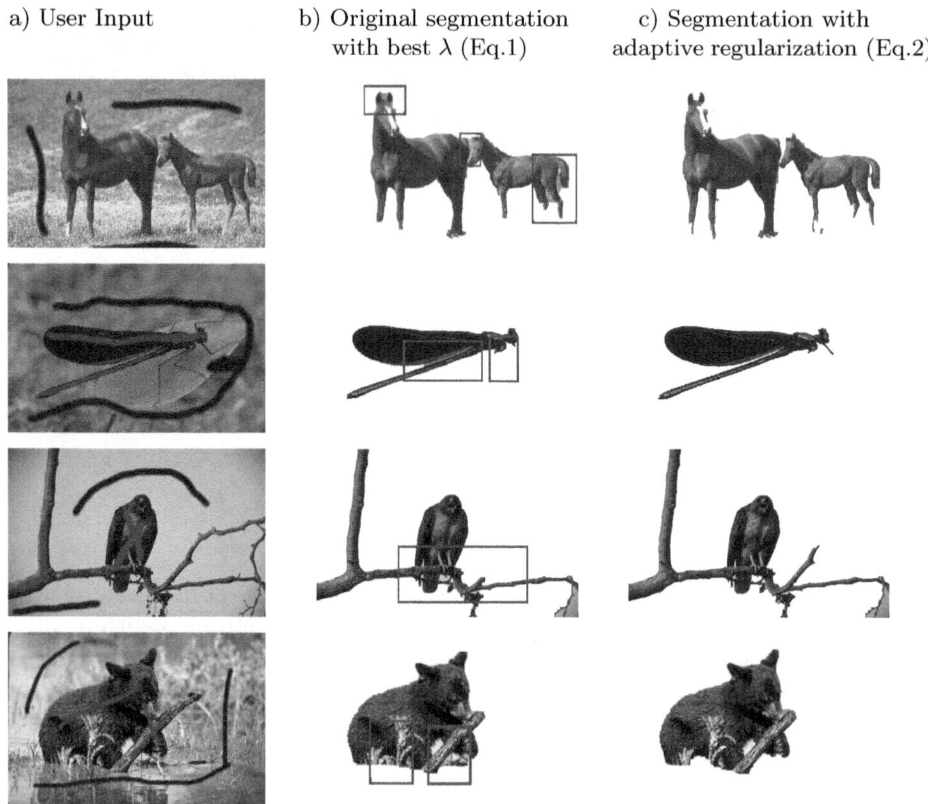

Fig. 3. Comparison of the original graph cut segmentation with the proposed approach. a) User Input. Red marks denote object pixels, blue marks denotes background pixels. b) Original Segmentation with the best λ. Over and under segmented regions are denoted in rectangles. c) Segmentation with adaptive regularization.

background regions, especially at the near boundary parts of the eagle and the bear are segmented as foreground. The proposed approach on the other hand, adaptively adjusts the regularization parameter based on the edge probability. Since the effect of the regularization parameter is decreased on the edge parts of the image, the segmentation solution is not over-smoothed on the thin and elongated parts of the foreground (Fig. 3c).

We illustrate the reversed normalized edge maps in Figure 4 which represent the smoothness proportion of image regions. The dark intensities denote the image parts which should have smaller regularization parameter. The lighter intensities on the other hand should be more regularized in the energy formulation.

We formulated the proposed method based on the boundary knowledge of the observed image. However, the boundary regions cannot be accurately determined. For example, the back parts of the horses in Figure 4.a cannot be determined. In some part of the images, on the other hand, the unnecessary edges

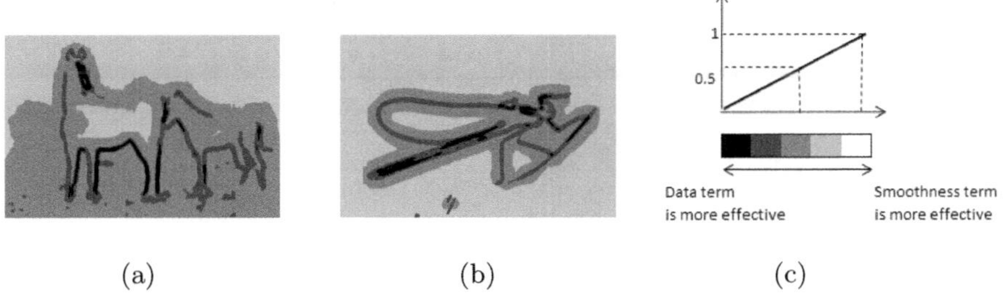

Fig. 4. Illustration of the adaptive regularization values. (a) and (b) are the reversed normalized edge maps of the images in Figure 3. (c) represents the effectiveness of the regularization parameter on the different image regions.

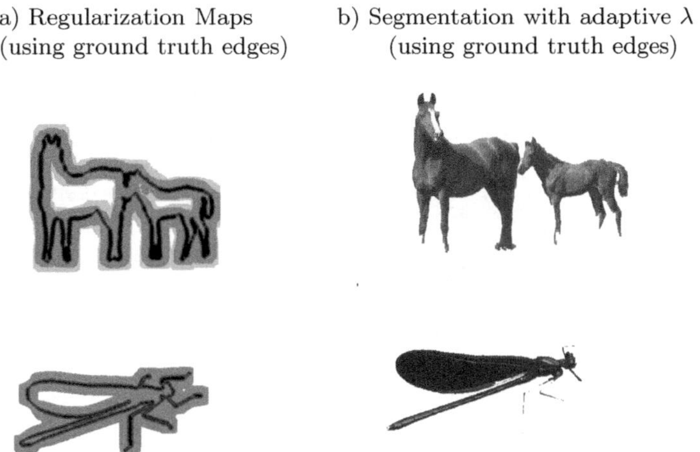

Fig. 5. Segmentation results with the ground truth boundary information

mislead the proposed algorithm (the edge leaf in the insect image Figure 4.b). As a result, the proposed method is influenced by the boundary accuracy. In order to observe the effect of the proposed idea with a better boundary knowledge, we used the ground truth edges of the images. We decrease the effect of the regularization parameter for the boundary and the near-boundary regions of the images manually. Then we used these regularization parameters for the graph cut minimization. The regularization parameters based on the edge maps in Figure 5.a increase the quality of the segmentations (the legs of the horse and the legs of the insect)(Figure 5.b).

We also compared the percentage errors of both approaches using the ground truth segmentation(Figure 6). The red curve denotes the percentage errors of segmentations which are obtained by graph cut minimization in Eq.1. The

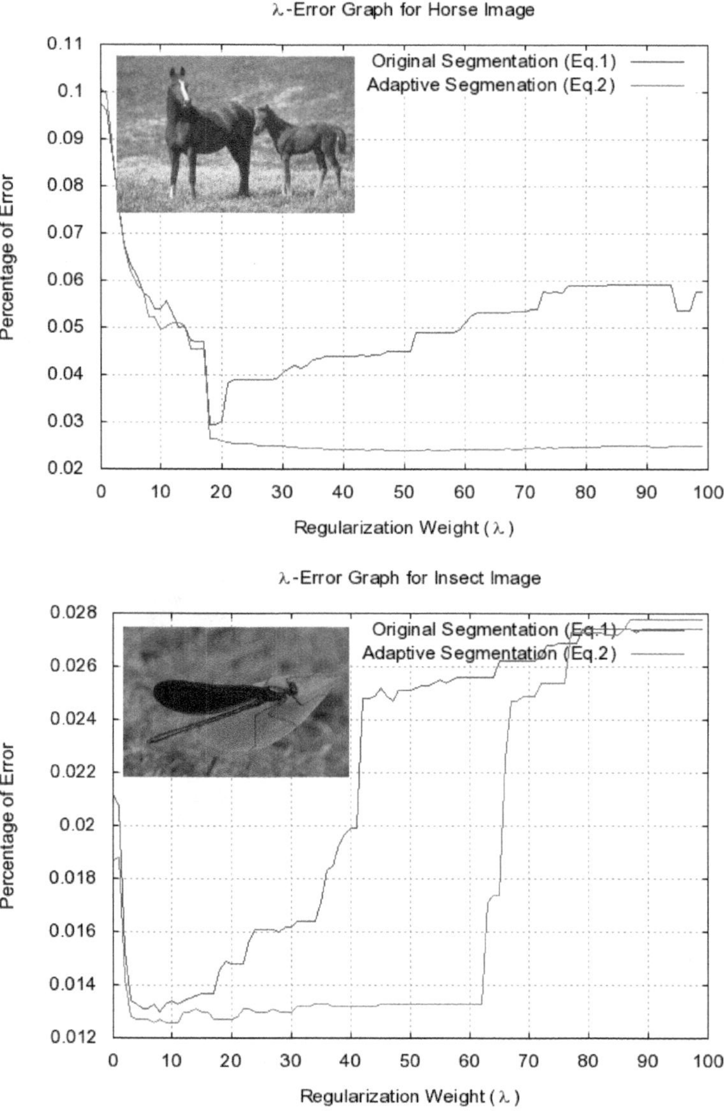

Fig. 6. Comparison of the original graph cut segmentation with the proposed approach for the whole ranges of regularization parameter between 0 and 99

minimum point of this curve is the best λ for the observed image. The green curve on the other hand, denotes the percentage errors of segmentations which are obtained by the proposed approach (Eq.2). Thegraphs show that the proposed method performs better than the original graph cut segmentation for the meaningful ranges of the regularization parameter.

5 Discussion

For the graph cut based segmentation tasks, the trade-off between the data and the smoothness terms should not be balanced by the same λ value for the whole image. The boundary regions should be less regularized than the other regions of the image. We proposed a method which adaptively changes the regularization parameter. The method first determines the boundary regions through the Canny edge detector at different hysteresis threshold levels. The linear average of the edge maps produces the edge probability of each pixel. The proposed method adjusts the effect of the regularization parameter using the probabilities of pixels being part of the boundary. Experimental results showed that the proposed method produces better segmentation results than the original graph cut approach for the best λ.

One of the novelties of this paper is the idea that using adaptive regularization parameters for the different parts of the image improves the segmentation result than using a single regularization parameter. The proposed method is the another novelty of the paper. To the best of our knowledge, our system is the first one to attempt to adaptively regularize the trade-off between the data and the smoothness terms in graph cut segmentation.

The method adjusts the regularization effectiveness depending on the probability of pixels being on the boundary region. If the boundary regions are determined incorrectly, the smoothness effect would not be adjusted properly. In the experiments, we used only the edge knowledge for the boundary probabilities. However, the intensity variations cannot not yield reliable boundary probabilities at all times. Therefore, the proposed method should be improved using the other features of the image for the boundary probability calculation, such as the texture and color features of the image.

References

1. Comaniciu, D., Meer, P.: Mean Shift: A Robust Approach Toward Feature Space Analysis. IEEE Trans. Pattern Anal. Mach. Intell. 24(5), 603–619 (2002)
2. Kass, M., Witkin, A., Terzopoulos, D.: Snakes: Active contour models. International Journal of Computer Vision 1(2), 321–331 (1987)
3. Shi, J., Malik, J.: Normalized Cuts and Image Segmenatation. IEEE Trans. Pattern Anal. Mach. Intell. 22(8), 888–905 (2000)
4. Greig, D.M., Porteous, B.T., Seheult, A.H.: Exact Maximum A Posteriori Estimation for Binary Images. J. Royal Statistical Soc., Series B 51(2), 271–279 (1989)
5. Boykov, Y., Veksler, O., Zabih, R.: Fast approximate energy minimization via graph cuts. IEEE Trans. Pattern Anal. Mach. Intell. 23, 1222–1239 (2001)
6. Hadamard, J.: Sur les problèmes aux dérivées partielles et leur signification physique. Princeton University Bulletin, Princeton (1902)
7. Bertero, M., Poggio, T.A., Torre, V.: Ill-posed problems in early vision. Proceedings of the IEEE 76(8), 869–889 (1988)
8. Boykov, Y., Jolly, M.P.: Interactive graph cuts for optimal boundary and region segmentation of objects in N-D images. In: IEEE Conf. on Computer Vision, vol. 1, pp. 105–112 (2001)

9. Rother, C., Kolmogorov, V., Blake, A.: GrabCut-Interactive foreground extraction using iterated graph cuts. ACM Trans. Graph. 23(3), 309–314 (2004)
10. Tikhonov, A.N., Arsenin, V.A.: Solutions of ill-posed problems. Winston, Washington (1977)
11. Hansen, P.C.: Deconvolution and regularization with toepliz matrices. Numerical Algorithms 29, 323–378 (2002)
12. Krawczyk-Stando, D., Rudnicki, M.: Regularization parameter selection in dicrete ill-posed problems - the use of the U-curve. Int. J. Appl. Math. Comput. Sci. 17(2), 157–164 (2007)
13. Irizarry, R.A.: Choosing smoothness parameters for smoothing splines by minimizing and estimate of risk. Johns Hopkins University, Dept. of Biostatistics Working Papers. Working Paper 30 (2004)
14. Thompson, A.M., Brown, J.C., Kay, J.W., Titterington, D.M.: A study of methods of choosing the smoothing parameter in image restoration by regularization. IEEE Trans. Pattern Anal. Mach. Intell. 13(4), 326–339 (1991)
15. Krajsek, K., Mester, R.: Maximum Likelihood Estimator for Choosing the Regularization Parameters in Global Optical Flow Methods. In: IEEE International Conference on Image Processing, pp. 1081–1084 (2006)
16. Zhang, L., Seitz, S.M.: Estimating optimal parameters for MRF stereo from a single image pair. IEEE Trans. Pattern Anal. Mach. Intell. 29(2), 331–342 (2007)
17. Peng, B., Veksler, O.: Parameter selection for graph cut based image segmentation. In: Proceedings of the British Machine Vision Conference (2008)
18. Vicente, S., Kolmogorov, V., Rother, C.: Graph cut based image segmentation with connectivity priors. In: IEEE Conference on Computer Vision and Pattern Recongnition, pp. 1–8 (2008)
19. `http://www.eecs.berkeley.edu/Research/Projects/CS/vision/`

A New SVM + NDA Model for Improved Classification and Recognition

Naimul Mefraz Khan, Riadh Ksantini,
Imran Shafiq Ahmad, and Boubaker Boufama

School of Computer Science, University of Windsor,
Windsor, ON, Canada N9B 3P4
{khan1126,ksantini,imran,boufama}@uwindsor.ca
http://www.uwindsor.ca

Abstract. Support Vector Machine (SVM) is a powerful classification methodology where the Support Vectors (SVs) fully describe the decision surface by incorporating local information. On the other hand, Nonparametric Discriminant Analysis (NDA) is an improvement over the more general Linear Discriminant Analysis (LDA) where the normality assumption from LDA is relaxed. NDA is also based on detecting the dominant normal directions to the decision surface. This paper introduces a novel SVM + NDA model which combines these two methods. This can be viewed as an extension to the SVM by incorporating some partially global information about the data, especially, discriminatory information in the normal direction to the decision boundary. This can also be considered as an extension to the NDA where the support vectors improve the choice of κ-nearest neighbors ($\kappa - NN$'s) on the decision boundary by incorporating local information. Since our model is an extension to both SVM and NDA, it can deal with heteroscedastic and non-normal data. It also avoids the small sample size problem. Moreover, this model can be reduced to the classical SVM model so that the existing SVM programs can be used for easy implementation. An extensive comparison of the SVM + NDA to the LDA, SVM, NDA and the combined SVM and LDA, performed on artificial and real data sets, has shown the advantages and superiority of our proposed model. In particular, the experiments on face recognition have clearly shown a significant improvement of SVM + NDA over the other methods, especially, SVM and NDA.

Keywords: Linear Discriminant Analysis, Nonparametric Discriminant Analysis, Support Vector Machines, Small Sample Size Problem, Face Recognition.

1 Introduction

In the last decades, a number of powerful linear classifiers have been proposed in the machine-learning community. One of them is the Linear Discriminant Analysis (LDA), whose main goal is to solve the well-known problem of Fisher's

A. Campilho and M. Kamel (Eds.): ICIAR 2010, Part I, LNCS 6111, pp. 127–136, 2010.

linear discriminant criterion. This criterion aims at finding the linear projections such that the classes are well separated, i.e, maximizing the distance between means of classes and minimizing their intraclass variances. The LDA was successfully applied in appearance-based methods for object recognition, such as, face recognition [1].

However, LDA shows poor performance when the underlying assumptions of homoscedasticticity (i.e., data in which classes have equal covariance matrices [8]) and normality (i.e. data with Gaussian distribution [2]) are not satisfied. To relax these assumptions, Fukunaga [2] proposed the Nonparametric DA (NDA) which measures the between-class scatter matrix on a local basis in the neighborhood of the decision boundary. This is done based on the observation that the normal vectors on the decision boundary are the most informative for discrimination [6]. For a data point, these normal vectors are approximated by the $\kappa - NN$'s from the other class in case of a two class classification problem. Therefore, NDA can be considered as a classification method based on the "partially global" characteristics of data which are represented by the $\kappa - NN$'s. However, it is not always an easy task to find a common and appropriate choice of $\kappa - NN$'s on the decision boundary for all class points to obtain the best linear discrimination.

Support Vector Machine (SVM) [10] is another powerful method which emphasizes the idea of maximizing the margin or degree of separation in the training data. SVM tries to find the optimal decision hyperplane using support vectors. The support vectors are the training samples that approximate the optimal decision hyperplane and are the most difficult patterns to classify. In other words, they consist of those data points which are closest to the optimal hyperplane. As SVM deals with a subset of data points (support vectors) which are close to the decision boundary, it can be said that the SVM solution is based on the "local" characteristics of the data. However, SVM does not take into consideration the global or partially global properties of the class distribution on which LDA-based methods (e.g. LDA, NDA) are based.

In this paper, we propose an SVM + NDA classification model which takes into account both the partially global characteristics of data distribution represented by NDA and the local characteristics represented by SVM. Being an extension to both SVM and NDA, this model does not depend on any global distribution pattern of training data. Therefore, it is capable of dealing with heteroscedastic and non-normal data. Moreover, our method combines the discriminatory information represented by the normal vectors to the decision surface and the support vectors which are crucial for accurate classification. Also, our method avoids the small sample size problem, which is a general problem for LDA-based methods (e.g. LDA, NDA) [8]. We have particularly targeted the face recognition problem as an application of interest to our proposed model given that it has become one of the most challenging tasks in the pattern recognition area.

This paper is organized as follows: Section 2 provides the formulation of the classical SVM and NDA method. Section 3 presents the derivation of the novel

SVM + NDA model. We also show that this model is a variation of the classical SVM so that the existing SVM programs can be used. Section 4 provides a comparative evaluation of the SVM + NDA model to LDA, SVM [10], NDA [2] and combined SVM and LDA [11], carried out on a collection of benchmark synthetic, real and face data sets. Finally, Section 5 presents the conclusion.

2 The SVM and NDA Methods

Let $\mathcal{X}_1 = \{x_i\}_{i=1}^{N_1}$ and $\mathcal{X}_2 = \{x_i\}_{i=N_1+1}^{N_1+N_2}$ be the two different classes constituting an input space of $N = N_1 + N_2$ samples or vectors in $\mathbb{R}^M$ and the tags associated with these vectors be represented by $\mathcal{T} = \{t_i\}_{i=1}^{N}$, where $t_i \in \{+1, -1\}$ $\forall i = 1, 2, ..., N$. The goal is to construct an optimal linear separating hyperplane from the training data as represented by the following function:

$$y(x; w) = w^T x + w_0, \tag{1}$$

where x is an input vector and $w = \{w_i\}_{i=1}^{N}$ and w_0 represent the unknown weights to compute.

2.1 Formulation of the NDA

Similar to LDA, NDA tries to find the most discriminative linear projections of the class distributions which can be achieved by maximizing the Rayleigh coefficient (the ratio of the between-class scatter matrix against the within-class scatter matrix) with respect to the weights [8]. Finding the most discriminative projectional direction w^* can be described by the following optimization problem:

$$w^* = \arg \max_{w} \frac{w^T S_{b\kappa} w}{w^T S_w w}, \tag{2}$$

where the within-class scatter matrix is defined as $S_w = \frac{1}{N_1+N_2}(N_1 S_1 + N_2 S_2)$, such that S_1 and S_2 are the covariance matrices for the two classes. $S_{b\kappa}$ is the non-parametric between-class scatter matrix [2] which relaxes the normality assumption of classical LDA. It is constructed on a partially global basis. This formulation is based on the observation that the normal vectors on the decision boundary preserve the classification structure. These normal vectors are approximated by the directions of the lines that connect points between the two classes. For every point, the $\kappa - NN$'s from the other class is considered. The non-parametric scatter matrix is defined as:

$$\begin{aligned} S_{b\kappa} = {} & \frac{1}{(N_1 + N_2)} \sum_{i=1}^{N_1} \omega_i (x_i - M_2^{\kappa}(x_i))(x_i - M_2^{\kappa}(x_i))^T \\ & + \frac{1}{(N_1 + N_2)} \sum_{i=N_1+1}^{N_1+N_2} \omega_i (x_i - M_1^{\kappa}(x_i))(x_i - M_1^{\kappa}(x_i))^T, \end{aligned} \tag{3}$$

where $M_j^{\kappa}(x_i)$ is the mean vector of the $\kappa - NN$'s from class $\mathcal{X}_j$ to sample x_i, $j \in \{1, 2\}$. ω_i is the weighting function to deemphasize samples which are far from the classification boundary and is defined as:

$$\omega_i = \frac{min\{d(x_i, xNN_{1i}^{\kappa})^{\gamma}, d(x_i, xNN_{2i}^{\kappa})^{\gamma}\}}{d(x_i, xNN_{1i}^{\kappa})^{\gamma} + d(x_i, xNN_{2i}^{\kappa})^{\gamma}}, \tag{4}$$

where $d(x_i, xNN_{ji}^{\kappa})$ is the Euclidean distance from x_i to its $\kappa - NN$ from the class $\mathcal{X}_j$. The parameter γ controls how rapidly the value of weighting function falls to zero as we move away from the classification boundary.

Problem (2) can be solved by finding the eigenvalues and eigenvectors of $S_w^{-1} S_{b\kappa}$ [2]. w is formed by the eigenvector corresponding to the largest eigenvalue. However, it is not always an easy task to find the optimal choice of $\kappa - NN$'s for NDA which may be suitable for all data points. The local information crucial for accurate classification is not considered here.

2.2 Formulation of the SVM

In SVM, the decision hyperplane is approximated by two parallel hyperplanes which provide the maximum margin between the two classes. The problem of finding these hyperplanes can be expressed as the following problem:

$$\min_{w \neq 0, w_0} \frac{1}{2} \|w^2\|,$$
$$s.t. \quad t_i(w^T x_i + w_0) \geq 1 \quad \forall i = 1, \ldots N. \tag{5}$$

By transforming this convex optimization problem into its dual problem, the solution can be found as $w = \sum_{i=1}^{N} \alpha_i t_i x_i$ where, equation (5) achieves equality for nonzero values of α_i only. The corresponding data samples are called support vectors. Therefore, SVM considers only those data points which are close to the decision boundary. In this sense, SVM is a local method. On the contrary, NDA incorporates the partially global discriminatory information present in the training data.

3 The SVM + NDA Model

SVM + NDA combines the discriminatory information represented by the normal vectors to the decision surface for the NDA and the support vectors for the SVM. The Model is formulated by different optimization problems in case of linearly separable and non-separable data which we will discuss individually.

3.1 SVM + NDA for Linearly Separable Data

In this case, SVM + NDA is defined by the following optimization problem:

$$\min_{w \neq 0, w_0} \frac{1}{2} w^T (\lambda S_w (S_{b\kappa} + \beta I)^{-1} + I) w,$$
$$s.t. \quad t_i(w^T x_i + w_0) \geq 1 \quad \forall i = 1, \ldots N. \tag{6}$$

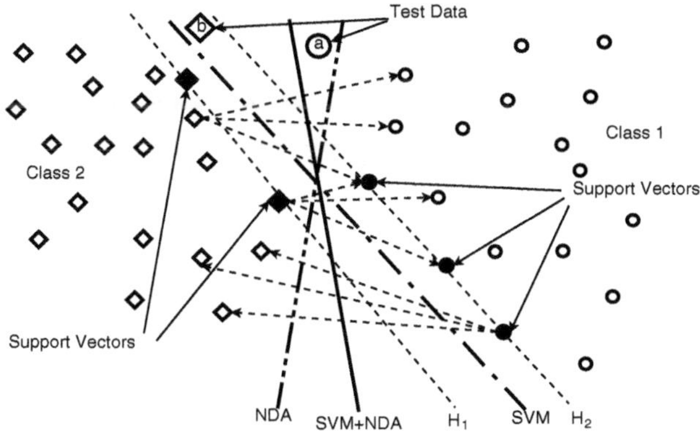

Fig. 1. Geometrical interpretation of SVM + NDA

Here, the term βI represents the regularization matrix to tackle the small sample size problem [8]. β can be any positive scalar value. The idea is to maximize the margin of the separating hyperplane for the two classes and minimize the scatter of data in the normal direction of the hyperplane while incorporating partially global information about the data points simultaneously. Here, the key parameter is λ, which controls the tradeoff. λ can take any value from zero to infinity and is tuned via cross validation. By incorporating the information obtained through NDA in the optimization problem of SVM, we are changing the orientation of the decision hyperplane in such a way that retains the benefit of both SVM and NDA, which may yield better classification accuracy as will be demonstrated in the experimental results section. A rough geometrical interpretation of SVM + NDA compared to the SVM and the NDA in case of a simple two dimensional case can be seen on Figure 1. Here, we see that the hyperplane defined by NDA is obtained by considering $\kappa - NN$'s ($\kappa = 3$) for each data point (for simplicity, only a few of the $\kappa - NN$'s are depicted here). On the other hand, the hyperplane defined by SVM is obtained using two separating hyperplanes H_1 and H_2 which provide the maximum margin. In case of SVM + NDA, we see how the orientation of the decision hyperplane changes in an optimal way by incorporating information from both SVM and NDA. Here, we see that the two (hypothetical) data points for testing a and b are being misclassified by NDA and SVM, respectively, but accurately classified by SVM + NDA.

3.2 Implementation of the SVM + NDA Model

Problem (6) is a convex optimization problem which can be solved by Lagrange undetermined multipliers and using the Karush-Kuhn-Tucker conditions [5]. However, in practice there is an easier way to implement SVM + NDA. According to Lemma 1, our formulation is just a variation of the classical SVM method. Hence, it can be solved using the existing SVM programs, which are widely available [7].

Lemma 1. *The SVM + NDA formulation is equivalent to:*

$$\min_{\hat{w}\neq 0,\hat{w}_0} \frac{1}{2}\|\hat{w}^2\|,$$
$$s.t.\ \ t_i(\hat{w}^T\hat{x}_i + \hat{w}_0) \geq 1 \quad \forall i = 1,\ldots N, \tag{7}$$

where

$$\hat{w} = \Sigma^{1/2}w, \tag{8}$$

$$\hat{x}_i = \Sigma^{-1/2}x_i \quad \forall i = 1,\ldots N \tag{9}$$

and

$$\Sigma = \lambda S_w(S_{b\kappa} + \beta I)^{-1} + I. \tag{10}$$

Proof. Substituting (8-10) into (7) we get (5).

Hence, we can easily use the existing SVM programs for training and testing SVM + NDA. The only pre-calculation we have to do is for $\Sigma^{1/2}$ and $\Sigma^{-1/2}$ which can be achieved by eigenvalue decomposition [4]. Now, the only problem left is to choose a suitable implementation of SVM. We used an SVM implementation for MATLAB by the MathWorks ™[7].

3.3 SVM + NDA for Linearly Non-separable Data

In this case, we use the Soft Margin method described in [10] which leads us to the following optimization problem:

$$\min_{w\neq 0,w_0,C>0} \frac{1}{2}w^T(\lambda S_w(S_{b\kappa} + \beta I)^{-1} + I)w + C\sum_{i=1}^{N}\xi_i,$$
$$s.t.\ \ t_i(w^Tx_i + w_0) \geq 1 - \xi_i,\ \xi_i \geq 0, \quad \forall i = 1,\ldots N. \tag{11}$$

Here, ξ_i denotes the slack variable which measures the degree of misclassification for each data point x_i. C is the regularization parameter, the value of which determines how large or small the penalty factor should be for each misclassified data point. This is again a quadratic optimization problem, and we can reduce it to the classical SVM problem in the same way as described before.

4 Experimental Results

In this section we present a comparison of the SVM + NDA model with the LDA, NDA [2], SVM [10] and SVM + LDA [11] on synthetic, real and face data sets. The value of all the control parameters (λ, κ and regularization parameter C) is tuned properly via cross validation in these experiments.

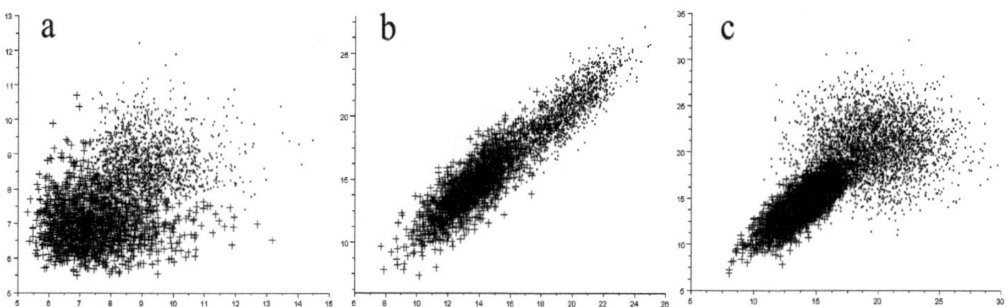

Fig. 2. (a) Gamma distributions (b) Identical Gaussians (c) Different Gaussians

Table 1. Comparison of % classification accuracies on synthetic data sets: best in bold face, second best emphasized

Data Distribution	SVM + NDA	LDA	NDA	SVM	SVM + LDA
Gamma Distributions (a)	**80.4**	77.4	78	79.1	*79.6*
Identical Gaussians (b)	**83.4**	**83.4**	*83.2*	82.4	**83.4**
Different Gaussians (c)	**81.9**	78.9	79.6	*81.2*	*81.2*

4.1 Evaluation and Comparison on Synthetic and Real Data Sets

We have used here synthetic data and a collection of benchmark real data sets to evaluate the SVM + NDA and to compare them to the other four methods in term of classification accuracy. Since SVM + NDA is an extension to the classical SVM and NDA, it makes no assumption on the data. Hence, the synthetic data sets are generated with different distributions (e.g. Gamma distribution and Gaussian distribution) to tease out the advantage of SVM + NDA over methods like LDA or SVM + LDA whose performance depend on the underlying class distributions. The first set of experiments are performed on three synthetic data sets, each of which has two clusters with a total of $N = 3000$ points (1500 points per cluster), generated from Gamma distributions, identical Gaussians and different Gaussians, respectively. The data sets are shown in Figure 2. 10-fold cross validation is used on each data set. Table 1 illustrates the classification accuracies on the synthetic data sets. In the second set of experiments, we have compared the SVM + NDA to the LDA, NDA, SVM and SVM + LDA on eight real data sets obtained from the Benchmark Repository used in [9]: Breast-Cancer, Flare-Solar, German, Heart, Diabetes, Ringnorm, Splice, Thyroid.[1]. 100 partitions into test and training set (about 60%:40%) are generated for each data set(see [9] for details). The results in Table 2 show the average classification accuracy over these 100 runs. From Table 1, we can see that SVM + NDA outperforms the other methods in terms of classification accuracy in all but one case. Even in that case, it is equal to the best method. This strengthens our

[1] The data sets can be obtained from **http://www.first.gmd.de/˜raetsch/**

Table 2. Comparison of % classification accuracies and average computational times (in seconds) on real data sets: best in bold face, second best emphasized

Data set	SVM + NDA	LDA	NDA	SVM	SVM + LDA
Breast-Cancer	**73.8**	71.9	72	*72.9*	*72.9*
Flare-Solar	**66.9**	65.4	65	66.2	*66.5*
German	**76.7**	75.6	74.9	*76.1*	75.9
Heart	**84.9**	83	83.5	*84*	83.6
Diabetes	**76.7**	74.7	74.7	**76.7**	*76.6*
Ringnorm	**75.9**	75	*75.6*	75	74.9
Splice	**84.2**	82.9	83.3	*83.8*	83.2
Thyroid	**90.6**	87.7	*89.4*	88.6	89.2
Avg. time (s)	0.86	**0.09**	*0.15*	0.86	0.86

claim that the information on classification structure can be obtained from two sources, namely the local information represented by the support vectors and the partially global information represented by the normal vectors to the decision boundary. As SVM + NDA combines these two sources, it results in the best classification accuracy. In particular, we see that being an extension to SVM and NDA, our method always yields better results than these two methods. In case of identical Gaussians, we see that LDA and SVM + LDA perform equally to our method. This is expected as LDA is supposed to give the best result possible in case of identical Gaussians. As SVM + NDA is robust and free from underlying assumptions, it can compete with LDA even when the assumptions are satisfied.

From Table 2, it is again obvious that SVM + NDA is superior to all these methods. The inferior performance of NDA compared to LDA in some cases explains the problem of choosing the optimal $\kappa - NN$'s for all data points. SVM + NDA solves this problem by using the support vectors and as a result, outperforms NDA by a considerable margin.

Another interesting observation from these results is the comparison between SVM + NDA and the SVM + LDA method [11]. We can see that the SVM + LDA method comes as the "second best" in most of the cases. Even if SVM + LDA combines the local and global information from SVM and LDA, there is a dependency on LDA here. As a result this method is sensitive to the underlying distributions. But in case of SVM + NDA, the support vectors and the normal vectors to the decision boundary are obtained from SVM and NDA, both of which are free from any underlying assumption. Therefore, as a result, SVM + NDA outperforms SVM + LDA.

In terms of computational complexity, the classical SVM scales with $\mathcal{O}(N^2)$, where N is the number of data points [10]. As for LDA and NDA, the training time scales with $\mathcal{O}(M^3)$ (dominated by the inversion of the within-class scatter matrix [1,2]), where M is the dimension of data points. In SVM + NDA, the inversion of between-class scatter matrix is part of pre-processing and done only once before training. Therefore, the computational complexity of our proposed SVM + NDA method is on par with classical SVM. The last row of table 2 is the training computational times in seconds. As we can see, the required computational time is compatible with the complexity of each method.

4.2 Face Recognition Application

We have also compared the SVM + NDA to the other methods in context of face recognition. This comparison has been carried out on the ORL face database [12] and the Yale face database [3]. The ORL database consists of 400 frontal faces, 10 images each of 40 individuals with variations in pose, illumination, facial expression and facial details. The Yale database consists of 165 images of 15 individuals. There are 11 images per subject, one per different facial expression or configuration. For all these databases, Principal Component Analysis (PCA) is used to project the images onto a reduced subspace. PCA is a standard technique to get rid of redundant information present in patterns. To observe the result of varying PCA dimension on different methods, we have repeated our experiment with projecting all the images onto PCA subspace of $10, 20, 30, \ldots 100$ dimensions each. For each individual (class) of each face database, the number of samples are randomly divided into two equal parts, and both parts are used for training and testing in turns. The classification accuracy is computed as an average of these two runs with one-against-all algorithm.

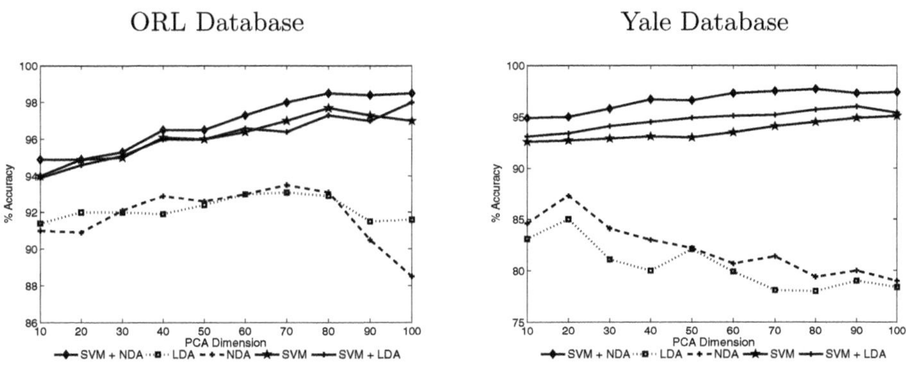

Fig. 3. Classification accuracy vs. PCA dimension for facial databases

From Figure 3, we can see that our proposed method again outperforms all other methods irrespective of the reduced PCA dimension. We also see that LDA and NDA do not perform well in case of high-dimensional data. However, being a variation of classical SVM, SVM + NDA does not suffer from that problem and as we can observe, results in better classification accuracy because of combining all the classification information available through the support vectors and the normal vectors to the decision boundary.

5 Conclusion

We have presented the SVM + NDA classification model which incorporates classification information from two different sources, namely the local information represented by classical SVM using the support vectors and the partially global information represented by NDA using the normal vectors to the decision boundary. SVM + NDA helps improve the choice of $\kappa - NN$'s for NDA using the support vectors. Moreover, since our proposed model is based on SVM and NDA, it is capable of dealing with hetroscedastic and non-normal data. It also resolves the small sample size problem. We have also shown that SVM + NDA is a variation of the classical SVM and can be implemented using the existing SVM programs, which makes it more attractive for real world applications. Our experimental results demonstrate that the SVM + NDA is superior or at least competitive to the LDA, NDA, SVM and the combined SVM and LDA in almost all the cases in terms of classification accuracy. In case of face recognition, the proposed model has provided the best results for the two data sets in case of all PCA dimensions.

In our future work, we intend to investigate extending SVM + NDA to incorporate non-linear decision boundaries. An extension in kernel space of SVM will help to solve the problem of non-linearity and make this method more robust.

References

1. Belhumeur, P.N., Hespanha, J.P., Kriegman, D.J.: Eigenfaces vs. Fisherfaces: recognition using class specific linear projection. IEEE Transactions on Pattern Analysis and Machine Intelligence 19, 711–720 (1997)
2. Fukunaga, K.: Introduction to Statistical Pattern Recognition, 2nd edn. Academic Press, London (2000)
3. Georghiades, A.S.: Yale face database. Center for Computational Vision and Control. Yale University, New Haven (1997)
4. Golub, G.H., Van Loan, C.F.: Matrix Computations, 3rd edn. The Johns Hopkins University Press, Baltimore (1996)
5. Kuhn, H.W., Tucker, A.W.: Nonlinear programming. In: Proceedings of the 2nd Berkeley Symposium, pp. 481–492 (1950)
6. Lee, C., Landgrebe, D.A.: Feature Extraction Based on Decision Boundaries. IEEE Transactions on Pattern Analysis and Machine Intelligence 15(4), 388–400 (1993)
7. MATLAB Bioinformatics Toolbox™. The MathWorks™ (2009)
8. Mika, S., Ratsch, G., Weston, J., Scholkopf, B., Mullers, K.R.: Fisher Discriminant Analysis with Kernels. In: Proceedings of IEEE Neural Networks for Signal Processing Workshop, pp. 41–48 (1999)
9. Ratsch, G., Onoda, T., Muller, K.R.: Soft Margins for Adaboost. Machine Learning 42(3), 287–320 (2000)
10. Vapnik, V.N.: Statistical Learning Theory. John Wiley & Sons, New York (1998)
11. Xiong, T., Cherkassky, V.: A Combined SVM and LDA Approach for Classification. In: Proceedings of the International Joint Conference on Neural Networks, pp. 1455–1459 (2005)
12. Yu, H., Yang, J.: A direct LDA algorithm for high-dimensional data with application to face recognition. Pattern Recognition 34, 2067–2070 (2001)

Incremental Hybrid Approach for Unsupervised Classification: Applications to Visual Landmarks Recognition

Antonio Bandera and Rebeca Marfil

Grupo ISIS, Dpto. Tecnología Electrónica,
Universidad de Málaga, Campus de Teatinos s/n 29071-Málaga, Spain
ajbandera@uma.es, rebeca@uma.es
http://www.grupoisis.uma.es

Abstract. This paper describes a novel approach for incremental subspace learning which combines the best features of the evolving clustering method and the spectral clustering algorithm based on the graph p-Laplacian. The evolving clustering method is employed to classify each input sample into a set of spherically-shaped groups. Then, the spectral clustering algorithm is used to unsupervisedly cluster this reference set, resolving the shape of classes having non-zero covariance. The proposed approach has been applied to the problem of visual landmark recognition, in a mobile robot navigation framework. Experimental results show that the performance of the method is high in terms of error rate.

1 Introduction

Pattern recognition, data mining or time-series prediction are examples of real-world applications where the complete set of training samples could not be provided in advance when building a classifier. Besides, in many of these applications, samples are generally provided little by little and the properties of the data source or the real scenario where they are acquired could be slightly changed as time passes. Therefore, the learning of a system must be also conducted sequentially in an incremental manner. On the other hand, training samples are provided in many cases only when the system does not correctly classify patterns; hence the system is learned incrementally to improve the classification performance. Incremental learning is primarily focused on processing the data in a sequential way so that in the end the classifier is no worse than a hypothetical classifier trained on the batch data [1].

In this paper, we describe an incremental classifier which can perform without an a priori knowledge about the number and shape of the classes which compound the feature space where samples will be represented. This approach combines the evolving clustering method (ECM) [4] and the spectral clustering algorithm based on the graph p-Laplacian [2] as two independent steps. ECM is a distance-based clustering algorithm which includes the input feature vector in the most suitable existing cluster, creating a new one if necessary. As other

A. Campilho and M. Kamel (Eds.): ICIAR 2010, Part I, LNCS 6111, pp. 137–146, 2010.

approaches, it is capable of achieving this goal with no supervision nor previous training. However, contrary to other clustering algorithms, it cannot optimize a classification criterion because it does not memorize information about all samples which were supplied to the classifier. On the other hand, the spectral clustering algorithm is launched when a chunk of training samples has been processed and a new cluster has been created. It mainly avoids to force the shape of the final clusters to a predefined geometric form when there is another shape in the sample density. The proposed classifier works with no previous training process and is specially suitable to be used in applications where computational resources must be minimized.

The remainder of this paper is organized as follows. Section 2 presents the proposed method. Experimental results showing the application of the proposed classifier to visual landmark recognition in a mobile robot framework and revealing the efficacy of the method are described in Section 3. The paper concludes along with discussions and future work in Section 4.

2 Proposed Method

Briefly, the proposed classifier combines two different approaches. Thus, it firstly accomplishes the ECM, a distance-based clustering algorithm. This algorithm can classify the input feature vector into a set of groups or generate a new one if it is necessary. Then, to determine the number and shape of the final clusters, a second step is achieved. This second step of the hybrid approach employs the spectral clustering algorithm based on the graph p-Laplacian. This algorithm merges the data groups generated by the ECM and provides the final groups. Therefore, the shape of these final data groups does not adopt a predefined geometric form. One condition is imposed by this scheme: the ECM must perform an over-classification of the parameter space, i.e. the number of obtained groups at this stage must be greater than the real one. Next subsections briefly deal with the two stages of the proposed method.

2.1 Evolving Clustering Method (ECM)

The ECM is an algorithm for dynamic clustering of an input stream of data [4], where there is no predefined number of clusters. The prototypes are determined such that the maximum distance between an input sample and the closest prototype cannot be larger than a threshold value, T. After create a first group by taking the first input pattern as the prototype $\bar{\mathbf{x}}_1$ and setting to 0 the value of the radius r_1 of this group, the ECM conducts one of the following actions according to its input:

- if the Euclidean distance between the closest prototype to the input sample, $d^{\mathbf{y}\bar{\mathbf{x}}_k}$, is lower than the current radius of this group r_k, the sample $\mathbf{y}$ is included in this group k. Neither a new group is created, nor any existing group is updated;

– if the Euclidean distance between the closest prototype to the input sample is greater than the current radius of this group, then the algorithm will find the group j such as

$$s_j = d^{\mathbf{y}\bar{\mathbf{x}}_j} + r_j = \min\{s_i\}_{i=0}^{n_p}, \tag{1}$$

where n_p is the number of prototypes. If s_j is greater than $2 \cdot T$, then the sample $\mathbf{y}$ does not belong to any existing group. A new group is created and the input sample becomes its prototype $\bar{\mathbf{x}}_{n_p+1} = \mathbf{y}$. The radius r_{n_p+1} is set to 0, and the number of prototypes n_p is increased. Otherwise, the sample $\mathbf{y}$ is included in the group j. This group is now updated by moving its prototype $\bar{\mathbf{x}}_j$ and increasing its radius value r_j. The updated radius, $r_j^{'}$ is set to $s_j/2$, and the prototype $\bar{\mathbf{x}}_j^{'}$ is located on the line that connects the input sample $\mathbf{y}$ and the old prototype, at a distance from $\mathbf{y}$ equal to $r_j^{'}$.

It can be noted that the ECM only needs to store the prototype of every group (*centroid cluster analysis*). Therefore, it is computationally cheap and does not need to store all input samples. On the contrary, all input samples cannot be used to redefine the set of data groups (reference set) when a new sample arrives. Besides, the ECM tends to force observations into spherically-shaped data groups. However, if a low threshold value is employed, then there will be a large number of data groups and the group shapes will be all relatively small, but the resulting data groups will also correctly map the densest part of the feature space. It is interesting to note that the choice of a distance (instead of the classical Euclidean distance) plays an important role allowing to control this biased effect [5].

2.2 Spectral Clustering Based on the Graph p-Laplacian

Given the reference set obtained by the ECM and a similarity function, which should be chosen depending on the domain the data comes on, the data can be transformed into a weighted, undirected graph G. In this graph, vertices are defined by the cluster prototypes provided by the ECM, and the similarity of pairs of prototypes are encoded by positive edge weights. With respect to the edges, there are different strategies to define them, such as the ϵ-neighbourhood graph or the k-nearest neighbour graph [3]. Then, the aim is to divide this graph into subgraphs such that vertices in the same group are similar and vertices in different groups are dissimilar to each other. To achieve that, our approach uses a spectral clustering algorithm. The spectral clustering techniques look for a partition of the graph such that the edges between different groups have a very low weight and the edges within a group have high weight. Besides, the groups should be balanced in the sense that the size of the groups should not differ too much [2].

Although the spectral clustering can divide the graph employing the information provided by the eigenvectors and eigenvalues of its adjacency matrix, it usually works with the Laplacian of the graph adjacency (pairwise similarity) matrix. Let $G = (N, E)$ be an undirected, weighted graph with node set

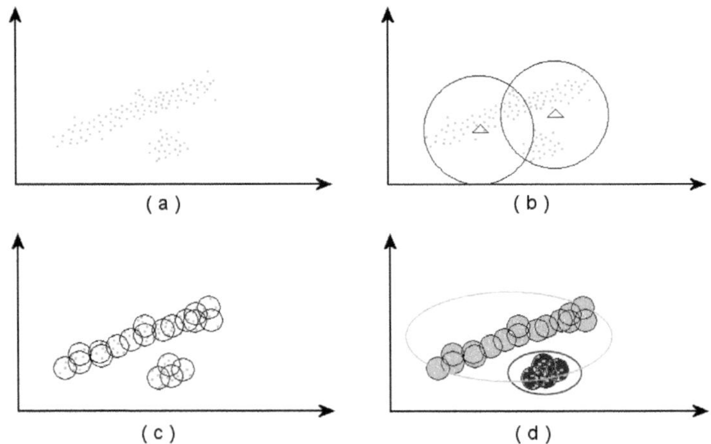

Fig. 1. a) Set of 125 bidimensional samples, b) classification of the set a) using the ECM with a high T value (two data groups), c) classification of the set a) using the ECM with a low T value (20 data groups), and d) classification of the set a) using the proposed hybrid approach

$N = \{n_1, ...n_n\}$ and where each edge between two vertices n_i and n_j has associated a non-negative weight $w_{ij} \geq 0$. The weighted *adjacency matrix* or *similarity matrix* of the graph G is the matrix $W = (w_{ij})_{i,j=1,...n}$. Values w_{ij} equal to zero mean that vertices n_i and n_j are not connected. As the graph is undirected, this matrix is symmetric ($w_{ij} = w_{ji}$). The degree of a vertex $n_i \in N$ is defined as

$$d_i = \sum_{j=1}^{n} w_{ij}. \tag{2}$$

Then, the *degree matrix* D is defined as the diagonal matrix with the degrees $\{d_i\}_{i=1}^{n}$ on the diagonal. Given a subset $A \subset N$, $|A|$ denotes the number of vertices in A and vol(A) is a measure of the size of A defined by the weights of its edges

$$\mathrm{vol}(A) = \sum_{n_i \in A} d_i. \tag{3}$$

A subset $A \subset N$ is connected if any two vertices n_i and n_j in A can be joined by a path $\{n_i...n_m...n_j\}$ such that all intermediate vertices n_m also are in A. The subsets $A_1, ...A_k$ are a *partition* of the graph G if $(A_i \cap A_j)_{i \neq j} = \emptyset$ and $A_1 \cup ... \cup A_k = N$.

The degree and weighted adjacency matrices can be both combined in a matrix which will resume the graph properties. The unnormalized Laplacian matrix is defined as $L = D - W$, and it presents interesting characteristics [6]:

1. it has only real eigenvalues
2. it is positive semidefinite
3. the smallest eigenvalue of L is λ_1=0, with corresponding eigenvector equal to $(1, 1, 1...1)^T$. The multiplicity of this eigenvalue 0 determines the number of connected components of L.

In the literature, there are different variants of graph Laplacians. Thus, in order to normalize the matrix with respect to the number of graph nodes, Chung [8] provides one definition of the normalized Laplacian[1] as

$$L_{norm} = D^{-1/2} L D^{-1/2}. \tag{4}$$

The normalized Laplacian is closely related to the unnormalized one. Thus, if x is an eigenvector of L, then $D^{-1/2}x$ is an eigenvector of L_{norm}. However, the normalization does not imply that the bounds obtained for L_{norm} can be obtained by dividing the eigenvalues of L by the number of graph nodes. Although the two spectra share some global similarities, their properties are different. Chung [8] points out that the eigenvalues of L_{norm} relate well to graph invariants in ways that eigenvalues of L have failed to do. In any case, the number of eigenvalues equal to 0 will be also equal to the number of connected components of the graph.

Then, if $\lambda_1 \leq \lambda_2 \leq ... \leq \lambda_n$ are the eigenvalues of L or L_{norm} in increasing order and repeated according to their multiplicity, then λ_1 will be equal to 0 and λ_2 will be greater than 0 if and only if the graph is connected. The second smallest Laplacian eigenvalue λ_2 of graphs is one of the most important information contained in the spectrum of a graph [6]. It is also known as the Fiedler value [7], and it has been shown that if this value is small, partitioning the graph into two disjoint subsets A_1 and A_2 based on the associated eigenvector will lead to a good *ratio cut* [9]. The ratio cut computes the degree of dissimilarity of a graph partition, and it is defined as

$$\text{ratioCut}(A_1, A_2...A_k) := \sum_{i=1}^{k} \frac{\text{cut}(A_i, \bar{A}_i)}{|A_i|}, \tag{5}$$

where $\text{cut}(A_i, A_j) = \sum_{n \in A_i, m \in A_j} w_{nm}$. Thus, this measure takes into account the total weight of the edges going accross the partitioned parts and the number of vertices of the subsets. The aim is to find balancing partitions, avoiding to cut small subsets of isolated vertices in the graph [7]. In a similar way, the ratio Cheeger cut is defined as

$$\text{CheegerCut}(A, \bar{A}) := \frac{\text{cut}(A, \bar{A})}{\min\{|A|, |\bar{A}|\}}. \tag{6}$$

Both strategies to obtain balanced partitions can be also employed using their corresponding normalized versions, where the term $|A|$ is changed by $\text{vol}(A)$. As

[1] The normalized graph Laplacian is typically defined in two distinct ways: the symmetric matrix ($L_{sym} := D^{-1/2} L D^{-1/2}$) and the random walk ($L_{rw} := D^{-1} L$) [3].

main disadvantage, minimizing these criterion constitutes a NP-hard problem. Spectral clustering provide efficient and robust techniques to approximate the cut of a graph.

In a recent work, Buhler and Hein [2] propose a new generalized version of spectral clustering based on the second eigenvector of the graph p-Laplacian, a nonlinear generalization of the graph Laplacian. The p-Laplacian of a function f on the set of vertices N which can be defined by

$$\sum_{n_j \in N, w_{ij} \geq 0} w_{ij}(f(n_j) - f(n_i))^{[p-1]} \tag{7}$$

where the symbol $a^{[b]}$ denotes a 'power' function that preserves the sign of a ($a^{[b]} = sign(a) \cdot |a|^b$). It can be noted that this p-spectral clustering considers the algorithms based on the previously described Laplacian matrices as special cases for p equal to 2.

Using the Cheeger cut criterion to measure the partition goodness, the normalized generalized spectral clustering algorithm provides a superior performance for $p \to 1$ when compared to standard spectral clustering algorithm [2]. However, there does not exist a multi-partition criterion of the Cheeger cut, which is defined for a bipartition (see Eq. (6)). Hence, the p-spectral clustering algorithm will use a multi-partition version of the ratio cut or the normalized cut [7].

3 Numerical Results and Application to Visual Landmark Recognition

In the mobile robotics community, there exists a recent interest for using object detection and recognition algorithms to provide natural landmarks for the sake of simultaneous robot localization and environment mapping. If a landmark can be defined as a distinct environment feature that the robot can recognize reliably from its sensor observations, then visual landmarks can be associated to 3D surface patches which are significantly different from its surroundings (salient image regions). In our case, we have developed a visual landmark detection approach which has been recently published [10]. This approach looks for image boundaries which delimitate high-contrasted regions of data-dependent shape. In order to describe each landmark, the SIFT descriptor has been extended to characterize the image region defined by an ellipse which have the same first and second moments as the originally shaped region. The orientation of this region is defined by the ellipse's major axis, but it is always assumed that this axis is drawn from top to bottom. That is, we are using a rotation-variant descriptor. Besides, the image region content is blurred and resized to obtain a final descriptor of $(4 \times 4) \times 8 = 128$ components, like the original SIFT descriptor [11].

The experimental evaluation has been conducted on a Pioneer 2AT platform from ActivMedia. The image acquisition system used in the experiments employs a STH-MDCS stereoscopic camera from Videre Design. This is a compact, low-power colour digital stereo head with an IEEE 1394 digital interface. It consists of two 1.3 megapixel, progressive scan CMOS imagers mounted in a rigid body. The camera was mounted at the front and top of the vehicle at a constant orientation, looking forward. Images were restricted to 320 × 240 pixels.

In the trial illustrated in Fig. 2, a set of 528 images was acquired. Although visual landmarks were unsupervisedly detected, they were manually labelled to estimate error rates. In this sense, it must be noted that although the employed approach allows to detect stable landmarks, all of them are not associated to labelled items. Thus, from the total set of 2670 detected landmarks, we only labelled a set of 1237 landmarks. This set of labelled visual landmarks was provided to the incremental classifier in the same order that they were acquired. All training samples were presented only once to learn. Fig. 2 shows the map with some of the detected visual landmarks. A map built previously using a laser scanner and Mapper3 software from ActivMedia Robotics, whose process is based on offline scanmatching techniques applied over the complete set of scans, is shown at Fig. 2a. The trajectory followed by the robot in this trial is also shown in this figure. Fig. 2b illustrates a 3D version of the environment. Landmark locations at this figure are only roughly estimated for illustration purposes.

When a feature vector is provided to the classifier, the ECM tries to include it in a previously defined group, updating the corresponding prototype and cluster radius, but it creates a new one if this process fails. The Euclidean distance has been employed to define the similarity between input feature vectors. As it has been mentioned above, the ECM threshold should be small enough to guarantee over-segmentation. It must be also noted that if this threshold is too small no clusters will be formed and this reduces to spectral clustering on the original data points, a process which will consume large computational resources. Experimental tests have been conducted varying this parameter, and a final value of T equal to 2.0 has been used. When a new cluster is created by the ECM approach, a k-nearest neighbour algorithm was employed to obtain the weighted graph associated to the reference set of prototypes (k=3). The similarity between two vertices is defined by means of the Euclidean distance. Then, the graph is partitioned using the p-spectral clustering. The partitioning is iteratively conducted, and the number of classes is determined by thresholding the last obtained second eigenvalue. When this value is under a given threshold, the partitioning is stopped [3]. Empirical tests have been also used to fix this value to 0.25. With respect to the p value, this test provides a final error rate of 12.0 % when p is equal to 2. This error rate is reduced to 7.8 % when a p value equal to 1.2 is employed (see Fig. 3). It must be noted that if a batch classification is conducted by executing the p-spectral clustering over the whole set of patterns, the final error rate is reduced to 6.7 % (p=1.2).

Finally, once the database of visual landmarks is learned, the recognition system was tested addressing several unsupervised trials. In these tests, a confidence value was employed to determine how confident a recently recognized landmark is that it can match to the closest landmark stored in the database. Thresholding this value, the system is able to classify a landmark as an 'unrecognized' one.

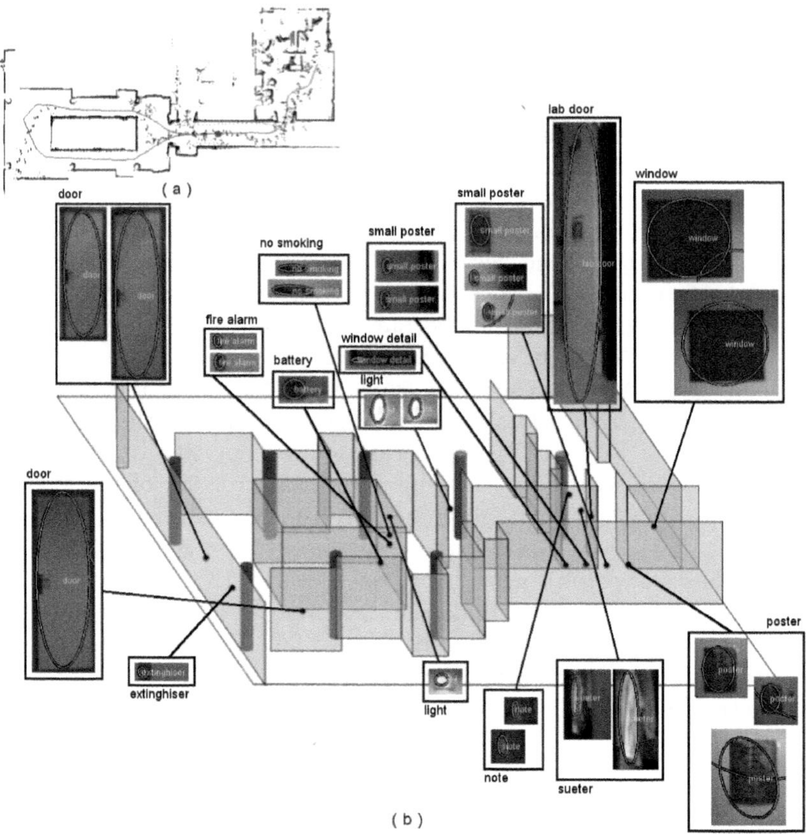

Fig. 2. a) Environment layout; and b) some detected visual landmarks (see text for details)

The confidence value C is defined as

$$C = \min\{C_i\}_{i=0}^{n_p} = \min\{\left(1 - n_p \frac{||\mathbf{y} - \bar{\mathbf{x}}_i||_2}{\sum_{j=0}^{n_p} ||\mathbf{y} - \bar{\mathbf{x}}_j||_2}\right)\}_{i=0}^{n_p}, \tag{8}$$

where C_i displays the similarity of an input sample $\mathbf{y}$ with the i prototype stored in the database, $\bar{\mathbf{x}}_i$. An input sample $\mathbf{y}$ is considered to be classified as 'unrecognized' if the obtained confidence value C is under a fixed threshold, which has been empirically set to 0.2 in our tests. Fig. 4 shows several images acquired in these unsupervised trials. Landmarks and labels, both of them unsupervisedly provided by the system, are overimposed on the images. It can be noted the presence of 'unrecognized' detected landmarks.

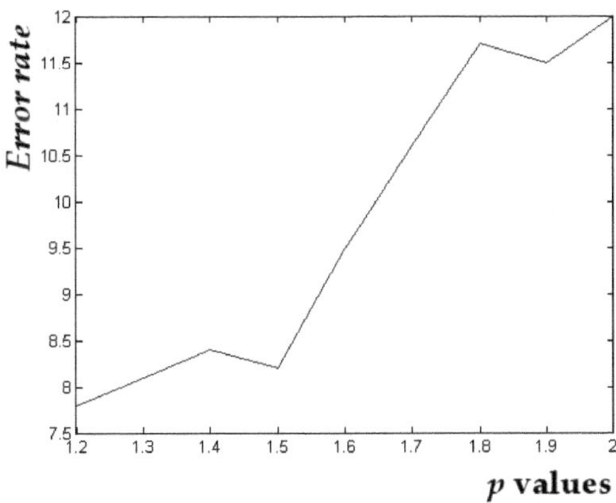

Fig. 3. Results of p-spectral clustering using the normalized ratio cut (patterns were always provided in the same order and all thresholds are fixed except the p value

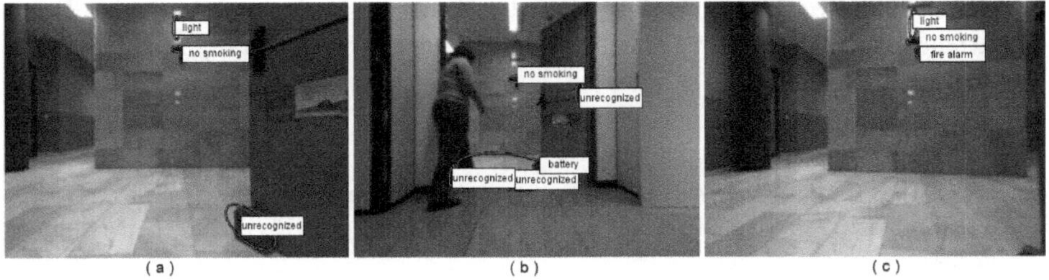

Fig. 4. a-c) Recognized visual landmarks in unsupervisedly performed trials (provided labels are overimposed on the images)

4 Conclusions

This paper has described an incremental classifier which performs two steps. Firstly, it employs a centroid-based, evolving clustering method to reduce the size of the set of stored samples and to obtain a tesselation which maps the densest part of the sample space. The main advantage of this approach is that only cluster prototypes are stored, reducing the storage requirements. This step is able to learn from each pattern without revisiting it. Besides, the time employed to process each pattern do not depend on the number of patterns processed in the past. However, if only the mean and radius are stored, the ECM is very conditioned to look for spherically-shaped clusters. The geometric form of the obtained clusters can be changed if other parameters are employed and updated, but it will be always limited by the chosen shape. The second step of this classifier employs a generalized version of spectral clustering using the graph p-Laplacian [2]. This algorithm divides the reference set by consecutive splitting of clusters, recovering the structure within any elongated cluster.

In a mobile robot navigation framework, this classifier has been tested into a visual landmark recognition system. Experimental results show that the system works correctly although it has no information about the nature of the input patterns. The classifier is specially suitable to deal with applications where a huge number of samples are on-line acquired, and it is necessary to classify them in an unsupervised manner. As main disadvantage, the approach employs a set of thresholds whose values have been empirically set. It is also neccesary to divide the whole reference set into clusters after the ECM creates a new cluster. This issue must be changed and future work will be focused on applying the spectral clustering only to the part of the graph affected by the inclusion of this new vertex.

Acknowledgments

This work has been partially granted by the Spanish Junta de Andalucía under project P07-TIC-03106 and by the Spanish Ministerio de Ciencia y Tecnología (MCYT) and FEDER funds under project no. TIN2008-06196. The code for p-Spectral Clustering has been developed by Thomas Buhler and Matthias Hein, Department of Computer Science, Saarland University, Germany [2].

References

1. Kuncheva, L.: Classifier ensembles for changing environments. In: Roli, F., Kittler, J., Windeatt, T. (eds.) MCS 2004. LNCS, vol. 3077, pp. 1–15. Springer, Heidelberg (2004)
2. Buhler, T., Hein, M.: Spectral Clustering based on the graph p-Laplacian. In: Bottou, L., Littman, M. (eds.) Proc. 26th Int. Conf. on Machine Learning (ICML), pp. 81–88 (2009)
3. von Luxburg, U.: A tutorial on spectral clustering, Tech. Rep. TR-149, Max Planck Inst. Biological Cyb. (2006)
4. Kasabov, N.: Evolving Connectionist Systems: Methods and Applications in Bioinformatics. In: Brain Study and Intelligent Machines. Springer, Berlin (2002)
5. Lomenie, N.: A generic methodology for partitioning unorganised 3D point clounds for robotic vision. In: Proc. 1st Canadian Conf. Computer Robot Vision, pp. 64–71 (2004)
6. Mohar, B.: The laplacian spectrum of graphs. In: Alavi, Y., et al. (eds.) Graph Theory, Combinatorics, and Applications, pp. 871–898 (1991)
7. Shi, J., Malik, J.: Normalized cuts and image segmentation. IEEE Trans. Pattern Analysis Machine Intell. 22, 888–905 (2000)
8. Chung, F.: Spectral graph theory. Am. Math. Soc. (1997)
9. Hagen, L., Kahng, A.: New spectral methods for ratio cut partitioning and clustering. IEEE Trans. Computer-Aided Design 11(9), 1074–1085 (1992)
10. Vázquez-Martín, R., Marfil, R., Núñez, P., Bandera, A., Sandoval, F.: A novel approach for salient image regions detection and description. Pattern Recognition Letters (2009)
11. Lowe, D.: Object recognition from local scale invariant features. In: Proc. 7th Int. Conf. on Computer Vision, pp. 1150–1157 (1999)

Nonlinear Scale Space Theory in Texture Classification Using Multiple Classifier Systems

Mehrdad J. Gangeh[1], Amir H. Shabani[2], and Mohamed S. Kamel[1]

[1] Department of Electrical and Computer Engineering, University of Waterloo, 200 University Avenue West, Waterloo, Ontario, Canada N2L 3G1
{mgangeh,mkamel}@pami.uwaterloo.ca
[2] Department of System Design Engineering, University of Waterloo, 200 University Avenue West, Waterloo, Ontario, Canada N2L 3G1
hshabani@engmail.uwaterloo.ca

Abstract. Textures have an intrinsic multiresolution property due to their varying texel size. This suggests using multiresolution techniques in texture analysis. Recently linear scale space techniques along with multiple classifier systems have been proposed as an effective approach in texture classification especially at small sample sizes. However, linear scale space blurs and dislocates conceptually meaningful structures irrespective of the type of structures exist. To address these problems, we utilize nonlinear scale space by which important geometrical structures are preserved throughout the scale space construction. This adds to the discrimination power of the classification system at higher scales. We evaluate the effectiveness of this approach for texture classification in Brodatz dataset using multiple classifier systems and learning curves. Compared with the linear scale space, we obtain higher accuracy in texture classification utilizing the nonlinear scale space.

Keywords: Multiscale, Nonlinear scale space, texture, multiple classifier systems.

1 Introduction

Texture provides important information in various fields of image analysis and computer vision. It has been used in many different problems including texture classification, texture segmentation, texture synthesis, material recognition, 3D shape reconstruction, color-texture analysis, appearance modeling, , and indexing [1-4].

As texture is a complicated phenomenon, there is no definition that is agreed upon by the researchers in the field [2, 3]. This is one of the reasons that there are various texture descriptors in the literature, each of which tries to model one or several properties of texture depending on the application in hand.

However, most textures show multiresolution property. In the recent years, multiresolution techniques become prevalent in texture analysis due to this intrinsic multiscale nature of textures. Some of the most well-known multiresolution techniques on texture analysis in the literature are: multiresolution histograms [4] including locally

A. Campilho and M. Kamel (Eds.): ICIAR 2010, Part I, LNCS 6111, pp. 147–156, 2010.

orderless images [5], multiresolution local binary patterns [6], multiresolution Markov random fields [7], wavelets [8], Gabor filters [8, 9], multiresolution fractal feature vectors [10], texton based approaches especially those based on MR8 (maximum response 8) filter banks [11], and techniques based on scale space theory [5, 12].

Despite the success of multiresolution techniques in texture analysis, these techniques suffer from high dimensional feature space. This is due to the concatenation of the feature subsets obtained from different scales to be submitted to a classifier. This high dimensional feature space causes that the classifier suffers from the 'curse of dimensionality' [13], i.e., many data samples are required to train the classifier with a reasonable performance. This drawback is not usually revealed in the literature as the results are reported for sufficiently large training set size.

Recently, an alternative approach based on multiple classifier systems (MCS) is proposed that avoids this problem by submitting each feature subset (obtained at a resolution) to a classifier, which is called a base classifiers (BC). Hence, instead of fusion of feature subsets, the decisions made by these BCs are combined. The improvement in the results is especially significant at small sample sizes, which is shown by using learning curves [14].

Linear scale space is used in [14] as multiresolution technique. However, linear scale space suffers from two main restrictions: first, it blurs all the structures in the image without considering their geometrical meaning. This may destroy meaningful structures especially at higher scales. Second, it dislocates the structures in the image, which is due to homogeneous diffusion of the image at all directions irrespective of the structures exist. The first issue is more important in texture classification as we would like to use the information at higher scales to improve the performance of the classification system [14]; vanishing the structures at higher scales may limit this goal.

We propose using nonlinear scale space here to preserve the structures at higher scales and show that this improves the performance of the classification system in comparison to linear scale space, especially at small sample sizes.

2 Scale Space in Texture Classification

In this section, the theoretical background needed for this research is explained. Specifically linear versus nonlinear scale space theory, feature extraction, and multiple classifier systems in the context of multiresolution texture classification are discussed.

2.1 Nonlinear versus Linear Scale Space

A linear scale-space representation of an image can be derived from diffusion equation as given in (1) with constant diffusivity g and (time-like) scale variable s

$$\frac{\partial I}{\partial s} = div\ (g.\nabla I). \tag{1}$$

Using convolution integral, this diffusion equation corresponds to the Gaussian smoothing of the original image I_0 with varying standard deviations. The variance of the Gaussian kernel is, therefore, proportional to the scale parameter ($\sigma^2 = 2s$). In linear diffusion equation, the intensity of each pixel is evolved by the divergence of the radial spatial concentration gradients (∇I) of the surrounding pixels.

Any multiscale signal processing approach that uses linear (Gaussian) scale space filtering suffers from two drawbacks. First, Gaussian smoothing is an isotropic diffusion filtering in which two (or more) regions of different structures might merge as the scale increases. In texture recognition, this side effect may result in blurring of conceptually meaningful structures such as parallel stripes shown in Fig. 1. Consequently, the extracted features at higher scales are less informative and reliable. Second, due to the dislocation of important structures such as edges, any feature extractor has dislocation problem.

To avoid undesirable blurring and dislocation of important structures (e.g., edges) in linear scale space filtering, it is proposed in [15] to control the diffusivity by incorporating the evolving image as a feedback in the smoothing process as follows:

$$\frac{\partial I}{\partial s} = div(g|\nabla I|.\nabla I) \ . \tag{2}$$

In other words, image gradient is used as a measure of edge map. Consequently, an edge-stopping function, like what is given in (3), controls the diffusivity at each direction in this anisotropic filtering scheme

$$g(|\nabla I|) = e^{-(|\nabla I|/k)^2} \ . \tag{3}$$

Stopping the diffusion at the direction of gradients higher than a threshold (k) prevents the sharing of intensity between two (or more) different regions in the image and, hence, avoids their fusion. In this way, as the scale increases, the homogeneous regions smooth more while different regions are still separated.

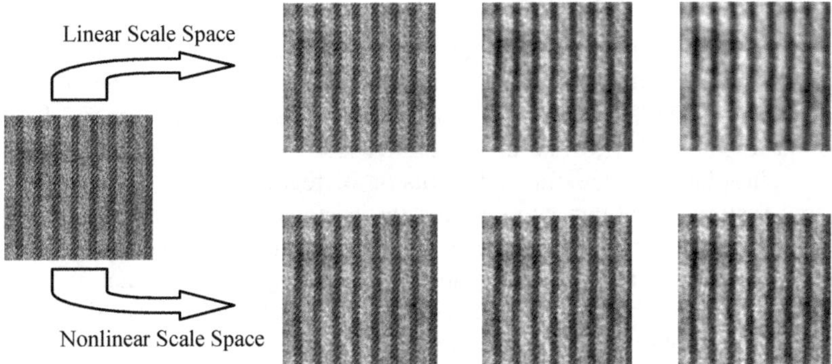

Fig. 1. Linear (*top row*) versus nonlinear (*bottom row*) scale space on texture D11 of Brodatz album (*left texture*). Note that as the scale increases (from left to right), linear scale space fails to preserve small slant patterns while nonlinear scale space can successfully do it.

2.2 Multiscale Feature Spaces

Multiscale feature subsets are obtained by computing the N-jet of derivatives up to the second order at various scales on patches. This means that in (2), I is a patch whose derivative is computed at scale s. Hence, features are computed at each scale and derivative to generate n-dimensional vectors $\mathbf{x}^{(i)} = [x_1, \dots, x_n] \in \mathbb{R}^n, i = 1, \dots, ns \times nd$, where ns and nd are the number of scales and derivatives respectively. This generates $m = ns \times nd$ feature subsets at various scales/derivatives.

These feature subsets can be composed into a single feature space $\mathbf{x} = [\mathbf{x}^{(1)}, \dots, \mathbf{x}^{(m)}]^T$, which is called *distinct pattern representation* (DPR) [16].

2.3 Feature Extraction

As the pixels in each patch are used as the features, the dimensionality of the feature subsets (n) depends on the patch size. As discussed in [14], it is beneficial to the performance of the classification system to increase the patch size at higher scales. The main reason is that at higher scales the coarser structures are emphasized and hence, they should be looked at through larger windows. This increases the dimensionality of feature subsets at higher scales. There are various feature reduction techniques in the literature among which principal component analysis (PCA) is adopted in this research. It is shown in [17] that PCA can have an adaptive feature extraction effect on multiscale texture classification. That is, at higher scales where larger patches are used and, thus, the dimensionality of feature subsets is higher, PCA reduces the dimensionality more than lower scales. The reason is that due to fewer details at higher scales, fewer components are needed to preserve certain fraction of variance of the original space.

By applying PCA to original DPR, a new DPR $\mathbf{y} = [\mathbf{y}^{(1)}, \dots, \mathbf{y}^{(m)}]^T$ is obtained in an uncorrelated space.

2.4 Multiple Classifier Systems

After computation of the DPR in reduced feature space, i.e., $\mathbf{y}$, there are two main approaches for submitting the DPR to the classification system. The common technique (see, for example, [5, 8]) is to fuse the feature subsets and submit the resulting feature space $\boldsymbol{y} \in \mathbb{R}^{k \times m}$ to one classifier[1] $D: \mathbb{R}^{k \times m} \to \Omega$, where $\Omega = \{\omega_1, \dots, \omega_c\}$ is the set of class labels for textures. The fusion of feature subsets generates a high dimensional feature space that can degrade the performance of the classifier D due to the 'curse of dimensionality' [13]. This problem is usually solved in the literature by severe dimensionality reduction of feature subsets, e.g., by computation of moments of histogram [5] or estimation of energy at the output of filter banks [8].

An alternative solution is to submit the DPR to an ensemble of classifiers [14]:

---

[1] For simplicity of the notations, here we assume that each feature subset in reduced space has a dimensionality of k and that there are m feature subsets. However, as mentioned in subsection 2.4, due to adaptive feature reduction effect of PCA, the dimensionality of feature subsets are not necessarily the same.

$$\Gamma = \{D_1, \ldots, D_m\}, \qquad \Gamma: \mathbb{R}^{k\times m} \to \Omega^m \tag{4}$$

where, $D_i: \mathbb{R}^k \to \Omega, i = 1, \ldots, m,$ is the base classifier (BC) trained on each feature subset $\mathbf{y}^{(i)} \epsilon \mathbb{R}^k, i = 1, \ldots, m$. The decisions made by these BCs are subsequently fused to yield a single decision on the class of the pattern submitted for classification. Hence, the problem of finding a classifier $D: \mathbb{R}^n \to \Omega$ is converted into finding an aggregation function $\mathcal{F}$ for combining the outputs of the BCs such that $\mathcal{F}: \Omega^m \to \Omega$.

The outputs of the BCs make a decision matrix, which is also called *decision profile* (DP) as given in (5)

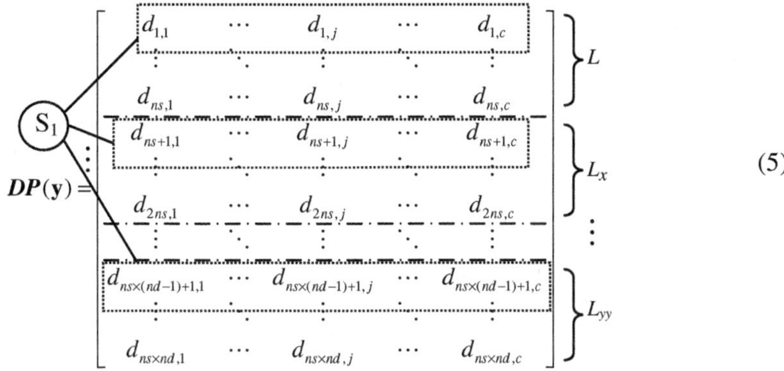

(5)

In (5), each row is the output of one BC and the DP is divided to some submatrices to represent the different derivatives at multiple scales. Here, we assume that the outputs of the BCs are continuous values. That is, each base classifier D_i in the ensemble generates a c-dimensional vector $[d_{i,1}, \ldots, d_{i,c}] \epsilon [0, 1]^c$.

The outputs of the BCs can be combined in one stage. However, in multiscale analysis, it makes sense to group the different derivatives of the same scale in a first stage (as shown in (5) for scale S_1) and then different scales in a second stage to see the effect of each scale on the overall performance of the system. The structure of the proposed two-level classification system is shown in Fig. 2 and can be formulated as:

$$\mu_j(\mathbf{y}) = \mathcal{F}[\mathcal{G}(\mathbf{d}_1^j), \mathcal{G}(\mathbf{d}_2^j), \ldots, \mathcal{G}(\mathbf{d}_{ns}^j)], \tag{6}$$

where $\mathcal{G}: [0,1]^{nd} \to \mathbb{R}$ is the first aggregation function and the vector $\mathbf{d}_i^j$ is defined as:

$$\mathbf{d}_i^j = (d_{i,j}, d_{i+ns,j}, \ldots, d_{i+ns\times(nd-1),j}), \qquad i = 1, \ldots, ns\ . \tag{7}$$

3 Experimental Setup

To evaluate the performance of the system on the classification of small texture patches, we perform some experiments on supervised classification of several test images from Brodatz album. These test images are shown in Fig. 3. All textures used are homogeneous and have a size of 640 × 640 with 8 bit/pixel intensity resolution.

Data Preparation and Preprocessing. There is only one texture image per class in Brodatz album. Hence, to guarantee disjoint training and test sets, each image is divided into two halves. The upper and lower halves are used for the extraction of training and test patches, respectively. The patches of 32 × 32 pixels are then systematically taken from top left to bottom right with some overlap to extract 1769 patches in total from each half. In order to make sure that the classification is based on the texture type not the variations in average intensity or contrast, the patches should be indiscriminable to their mean and variance. To this end, DC cancellation and variance normalization are performed on each patch.

Computation of Multiscale Patches. The N-jet of derivatives up to the second order is used for the computation of multiscale patches. For nonlinear scale space, we set experimentally the edge threshold $k = 10$ and select three scales evenly distributed in 250 iterations of nonlinear diffusion equation (2). This iteration is performed with scale difference $ds = 0.25$ and central finite difference operation. Similar to what is reported in [14], I (zeroth order derivative), I_x, I_y, I_{xx}, I_{xy}, and I_{yy} are computed at multiple scales for each patch.

Construction of Training and Test Sets. As described in subsection 2.4 and also shown in [14], increasing the patch size at higher scales is beneficial to the performance of the classification system. Hence after preprocessing and computing the multiscale patches, the patches of 18 × 18, 24 × 24, and 30 × 30 are taken from the central part of multiscale patches at scales S_1, S_2, and S_3 respectively.

Feature Extraction. Principal component analysis (PCA) is adopted as the feature extraction technique. PCA is computed over all classes in each scale/derivative separately and 95% of original variance is retained in uncorrelated space. As discussed in 2.3, fewer components are needed to retain this percentage of variance at higher scales due to fewer details available at these scales.

Multiple Classifier System. A two-stage parallel combined classifier with the structure shown in Fig. 2 is used in the experiments. Quadratic discriminant classifier (QDC) with regularization at scale S_1 performed the best among the base classifiers (BCs) tested and hence adopted in our experiments. Regularization at scale S_1 is required because the dimensionality of feature subsets at this scale (even after using PCA) is still high and this degrades the performance of the BC at small sample sizes. The mean combiner is used at both stages as it consistently shows good performance comparing to other type of combiners over different sample sizes.

Evaluation. One of the main shortcomings of the papers in multiresolution texture classification is reporting the performance of the system at only a single (usually large) sample size. This keeps the performance of the algorithm in small training set sizes unrevealed. To overcome this problem, we use the learning curves to show the classification error of patches at various sample sizes from 10 to 1500. The experiments are repeated 5 times on different randomized patches in training and test sets and averaged results are reported. The test set size is fixed at 900.

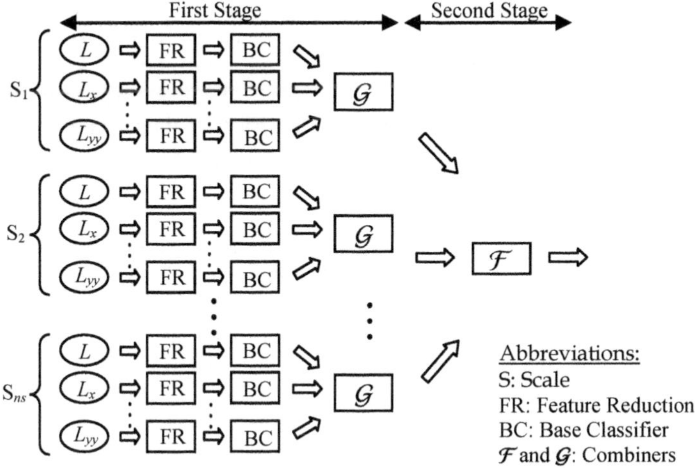

Fig. 2. The structure of proposed two-stage multiple classifier system. In first stage, different derivatives at the same scale are combined. In second stage, different scales are combined.

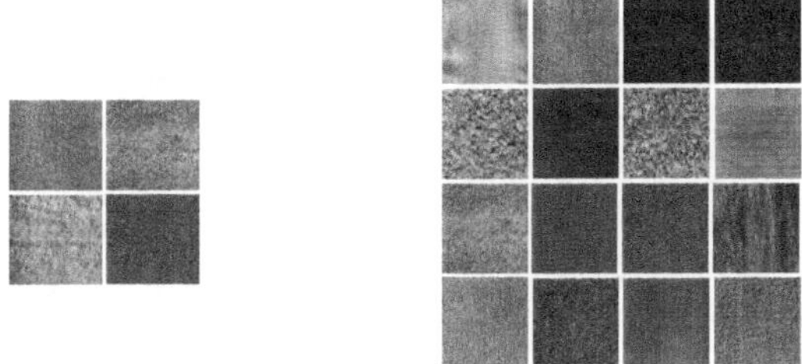

Fig. 3. 4-class (D4, D9, D19, and D57) and 16-class (D3, D4, D5, D6, D9, D21, D24, D29, D32, D33, D54, D55, D57, D68, D77, and D84) problems of Brodatz used in the experiments

4 Results

In this section, we present the results of texture classification using nonlinear scale space and two-stage multiple classifier systems. The performance is compared with linear scale space to show how using nonlinear scale space helps to improve the results, especially at smaller sample sizes.

The results for 4-class and 16-class problems using nonlinear and linear scale space are shown in left and right graphs of Fig. 4, respectively. The top graphs are for 4-class and bottom graphs are for 16-class problems of Brodatz dataset. In each graph, the thick curve displays the overall performance of the classification system, i.e., the output of the second stage in the proposed structure shown in Fig. 2. Thinner curves are the intermediate results, i.e., the outputs of the first stage of classification in Fig. 2, which are the results of combining different derivatives at the same scale.

Comparing the overall performance of left (nonlinear SS) and right (linear SS) graphs in Fig. 4 clearly shows the advantage of using nonlinear over linear scale space. The improvement is especially important at small sample sizes and could be due to this phenomenon that nonlinear SS preserves the structures at higher scale and this adds to the discriminative power of base classifiers at these scales. As can be seen from the graphs in Fig. 4, the performance of combined S_1 and combined S_2 are improved in nonlinear SS comparing to linear one. The overall performance is subsequently improved.

To verify the superiority of combined classifiers over combined feature space (CFS), which is the common technique in the literature, we here compare these two techniques on 4-class problem of Brodatz. Fig. 5 displays these results using CFS. Here, the feature subsets from different scales and derivatives are concatenated and after feature reduction, the combined feature space is submitted to a single QDC with the same regularization parameters as the BCs in Fig. 2. As can be seen from comparing Fig. 5 and top right graph of Fig. 4, high dimensional feature space of CFS, causes that the single classifier suffers from the 'curse of dimensionality' and its performance is significantly degraded especially at smaller sample sizes. If there are many data samples for training, we expect that CFS performs asymptotically as good as MCS. However, in this example this will happen for more than 1500 data samples which is the maximum training set size used in our experiments.

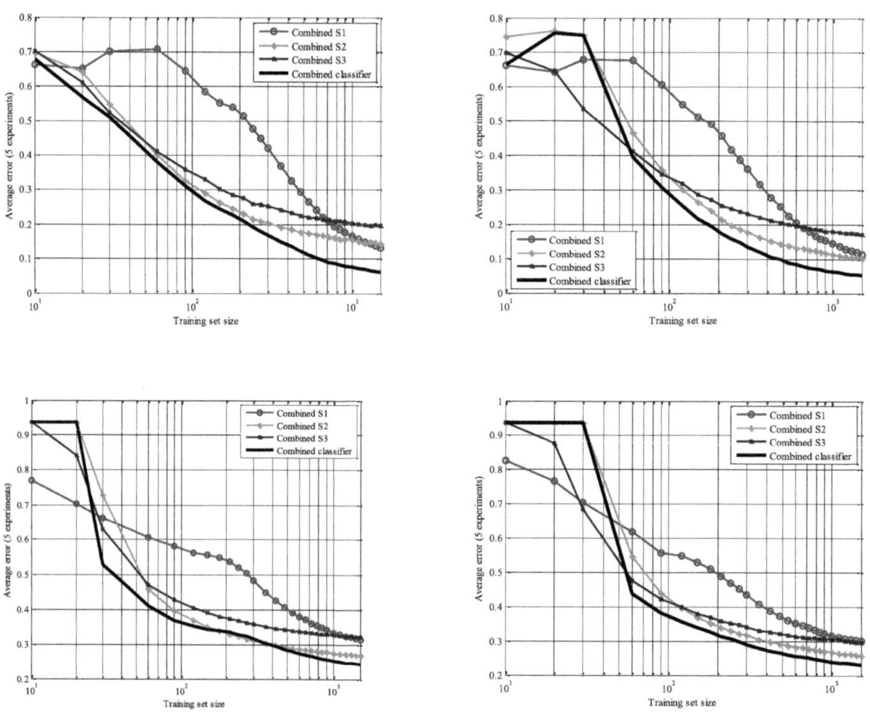

Fig. 4. Learning curves for the classification of 4-class (*top row*) and 16-class (*bottom row*) problems of Brodatz using nonlinear (*left*) and linear (right) scale space texture classification system with the structure proposed in Fig. 2

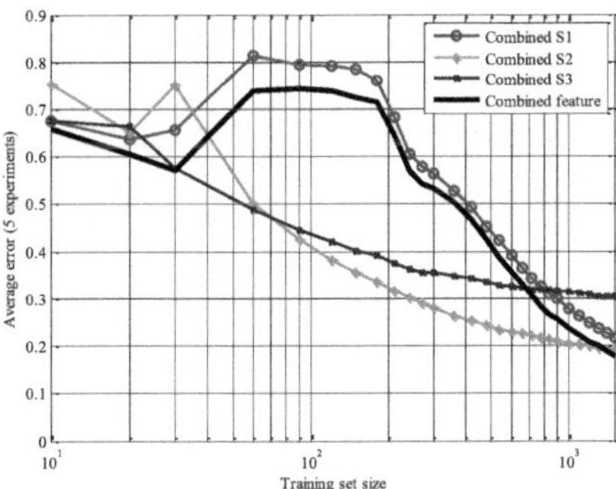

Fig. 5. Learning curves for the classification of 4-class problem of Brodatz using combined feature space (CFS) technique. These curves should be compared with the ones on top right graph of Fig. 4 which are the results for the same problem using multiple classifier systems.

5 Discussion and Conclusion

In this paper nonlinear scale space along with multiple classifier systems are proposed for texture classification. This is to address the problem of linear scale space in blurring and dislocating the important texture structures. Consequently, we obtained improvement in classification of Brodatz texture dataset.

It is shown using learning curves and multiple classifier systems that nonlinear scale space can improve the performance of texture classification system especially at small sample sizes. This is due to more discriminative power available at higher scales of nonlinear scale space comparing to linear one. The improvement of performance at small training set size is important in applications where data acquisition is cumbersome or costly and the number of data samples for training the texture classification system is limited. This is, for example, the case in medical applications such as the diagnosis of lung diseases in high resolution CT [12] or liver diseases in B-scan images of ultrasound [10].

It is also shown that multiple classifier systems improve the performance of texture classification system based on multiresolution techniques comparing to combined feature space.

Acknowledgments. The first author is funded by the Natural Sciences and Engineering Research Council (NSERC) of Canada under Canada Graduate Scholarship (CGS D3-378361-2009). The first author would also like to thank Markus van Almsick from Eindhoven University of Technology, the Netherlands for useful discussions on nonlinear scale space.

References

1. Petrou, M., Sevilla, P.G.: Image Processing Dealing with Texture. John Wiley and Sons, West Sussex (2006)
2. Ahonen, T., Pietikainen, M.: Image Description Using Joint Distribution of Filter Bank Responses. Pattern Recognition Letters 30, 368–376 (2009)

3. Ahonen, T., Pietikainen, M.: Handbook of Texture Analysis. In: Mirmehdi, M., Xie, X., Suri, J. (eds.) Imperial Collage Press, London (2008)
4. Hadjidemetriou, E., Grossberg, M.D., Nayar, S.K.: Multiresolution Histograms and Their Use for Recognition. IEEE Trans. on PAMI 26, 831–847 (2004)
5. van Ginneken, B., ter Haar Romeny, B.M.: Multi-scale Texture Classification from Generalized Locally Orderless Images. Pattern Recognition 36, 899–911 (2003)
6. Ojala, T., Pietikainen, M., Maenpaa, T.: Multiresolution Gray-Scale and Rotation Invariant Texture Classification with Local Binary Patterns. IEEE Trans. on PAMI 24(7), 971–987 (2002)
7. Zhu, S.C., Wu, Y., Mumford, D.: Filters, Random Fields and Maximum Entropy (FRAME): Towards a Unified Theory for Texture Modeling. International Journal of Computer Vision 27(2), 107–126 (1998)
8. Randen, T., Husoy, J.H.: Filtering for Texture Classification: A Comparative Study. IEEE Trans. on PAMI 21, 291–310 (1999)
9. Jain, A.K., Farrokhnia, F.: Unsupervised Texture Segmentation Using Gabor Filters. Pattern Recognition 24, 1167–1186 (1991)
10. Lee, W.L., Chen, Y.C., Hsieh, K.S.: Unsupervised Segmentation of Ultrasonic Liver Images by Multiresolution Fractal Feature Vector. Information Sciences 175(3), 177–199 (2005)
11. Varma, M., Zisserman, A.: A Statistical Approach to Texture Classification from Single Images. International Journal of Computer Vision: Special Issue on Texture Analysis and Synthesis 62, 61–81 (2005)
12. Sluimer, I.C., Prokop, M., Hartmann, I., van Ginneken, B.: Automated Classification of Hyperlucency, Fibrosis, Ground Glass, Solid, and Focal Lesions in High-Resolution CT of the Lung. Medical Physics 33(7), 2610–2620 (2006)
13. Jain, A.J., Duin, R.P.W., Mao, J.: Statistical Pattern Recognition: A Review. IEEE Trans. on PAMI 22(1), 4–37 (2000)
14. Gangeh, M.J., ter Haar Romeny, B.M., Eswaran, C.: Scale Space Texture Classification Using Combined Classifiers. In: Ersbøll, B.K., Pedersen, K.S. (eds.) SCIA 2007. LNCS, vol. 4522, pp. 324–333. Springer, Heidelberg (2007)
15. Perona, P., Malik, J.: Scale space and edge detection using anisotropic diffusion. IEEE Trans. on PAMI 12(7), 629–639 (1990)
16. Kittler, J., Hatef, M., Duin, R.P.W., Matas, J.: On Combining Classifiers. IEEE Trans. Pattern Analysis and Machine Intelligence 20(3), 226–239 (1998)
17. Gangeh, M.J., ter Haar Romeny, B.M., Eswaran, C.: The Effect of Sub-sampling in Scale Space Texture Classification Using Combined Classifiers. In: Int'l Conf. on Intelligent and Advanced Systems (ICIAS), Kuala Lumpur, Malaysia, pp. 25–28 (2007)

The Proof of Completeness of the Graph Method for Generation of Affine Moment Invariants*

Tomáš Suk

Institute of Information Theory and Automation,
Academy of Sciences of the Czech Republic,
Pod vodárenskou věží 4, 182 08 Praha 8, Czech Republic,
suk@utia.cas.cz
http://zoi.utia.cas.cz/suk

Abstract. Features for recognition of affinely distorted objects are of great demand. The affine moment invariants can be generated by a few methods, namely the graph method, the tensor method and the direct solution of the Cayley-Aronhold differential equation. The proof of their equivalence is complicated; it can be derived from the Gurevich's proof for affine tensor invariants. The theme of this paper is this derivation.

1 Introduction

Recognition of objects on images is an important part of many image processing applications. The images are often geometrically distorted and derivation of features invariant to such a distortion is of great demand. The invariants with respect to affine transform are often used, mostly as approximation of a projective distortion. The affine invariants can be computed from various measurements of the image, e.g. as point invariants, differential invariants, Fourier descriptors, etc. The invariants computed from moments play important role among them.

The affine moment invariants can be derived by a few ways. Recently, approximately from beginning of 90's, they are generated automatically, by a computer. There are two groups of methods for this generation, the graph method [1] and computationally equivalent tensor method [2] on one hand and direct solution of the Cayley–Aronhold differential equation [3] on the other hand. The graph method is easier to implement, but with worse computing complexity, while the direct solution of the equation is faster, but less numerically stable.

It is natural to ask the question whether or not all the invariants generated by the graph method are equivalent to that found by means of the Cayley-Aronhold equation, and vice versa. We have used invariants from both methods in pattern recognition for many years and have not found any inequivalence between them, so, the positive answer is likely, but precise proof is difficult. Both the book [2] with the tensor method and the book [4] with a survey of both methods reference only to the Gurevich's proof from [5] (Russian edition [6]) that can be used for derivation of our proof. This derivation is theme of this paper.

* This work has been supported by the grants No. 102/08/1593 and No. 102/08/0470 of the Czech Science Foundation.

A. Campilho and M. Kamel (Eds.): ICIAR 2010, Part I, LNCS 6111, pp. 157–166, 2010.

2 Basic Terms

Affine transformation can be expressed as

$$\begin{aligned}\hat{x} &= q_1^1 x + q_2^1 y + q_3^1 \\ \hat{y} &= q_1^2 x + q_2^2 y + q_3^2,\end{aligned} \tag{1}$$

its Jacobian is $J = q_1^1 q_2^2 - q_2^1 q_1^2$. The geometric moment of the order $p+q$ of an image $f(x,y)$ is defined

$$m_{pq} = \int_{-\infty}^{\infty}\int_{-\infty}^{\infty} x^p y^q f(x,y)\,\mathrm{d}x\,\mathrm{d}y. \tag{2}$$

2.1 Graph Method

The affine moment invariant can be computed as

$$I(f) = \int_{-\infty}^{\infty}\cdots\int_{-\infty}^{\infty} \prod_{k,j=1}^{r} C_{kj}^{n_{kj}} \cdot \prod_{i=1}^{r} f(x_i,y_i)\mathrm{d}x_i\mathrm{d}y_i, \tag{3}$$

where $C_{kj} = x_k y_j - x_j y_k$ is the oriented double area of the triangle, whose vertices are (x_k, y_k), (x_j, y_j), and $(0,0)$ and n_{kj} are some non-negative integers. The number $w = \sum\limits_{k,j} n_{kj}$ is called the weight of the invariant and r is called the degree of the invariant. The maximum order s of moments of which the invariant is composed is called the order of the invariant. Another important characteristic of the invariant is its structure, it is defined by an integer vector $\mathbf{s} = (k_2, k_3, \ldots, k_s)$, where k_j is the total number of moments of the jth order contained in each term of the invariant.

After an affine transform (we consider no translation) it holds $\hat{C}_{kj} = J \cdot C_{kj}$, which means that C_{kj} is a relative affine invariant. The functional (3) can be normalized to translation and scaling to be invariant to the general affine transform. Each such an invariant can be represented by a connected graph, where each point (x_k, y_k) corresponds to one node and each cross-product C_{kj} corresponds to one edge of the graph. If $n_{kj} \geq 1$, the respective term $C_{kj}^{n_{kj}}$ corresponds to n_{kj} edges connecting kth and jth nodes. The problem of derivation of the invariants up to the given weight w is equivalent to generating all connected graphs with at least two nodes and at most w edges.

2.2 Tensor Method

The generation of the affine moment invariants can be expressed in terms of tensors. The moments themselves do not behave under affine transform like tensors, but we can define a moment tensor [7]

$$M^{i_1 i_2 \cdots i_r} = \int_{-\infty}^{\infty}\int_{-\infty}^{\infty} x^{i_1} x^{i_2} \cdots x^{i_r} f(x^1, x^2)\mathrm{d}x^1\mathrm{d}x^2, \tag{4}$$

where $x^1 = x$ and $x^2 = y$. If p indices equal 1 and q indices equal 2, then $M^{i_1 i_2 \cdots i_r} = m_{pq}$. The behavior of the moment tensor under an affine transform

$$\hat{M}^{i_1 i_2 \cdots i_r} = |J|^{-1} q^{i_1}_{\alpha_1} q^{i_2}_{\alpha_2} \cdots q^{i_r}_{\alpha_r} M^{\alpha_1 \alpha_2 \cdots \alpha_r}, \qquad i_1, i_2, \ldots, i_r, \alpha_1, \alpha_2, \ \ldots, \ \alpha_r = 1, 2; \tag{5}$$

i.e. the moment tensor is a relative contravariant tensor with the weight $g = -1$ (in tensor calculus, the affine transform is understood inversely, p^i_α are the coefficients of the direct transform, q^i_α are that of the inverse transform, $J = p^1_1 p^2_2 - p^1_2 p^2_1$ is its Jacobian).

The covariant unit polyvector $\epsilon_{i_1 i_2 \cdots i_n}$ is a skew-symmetric tensor over all indices and $\epsilon_{12 \cdots n} = 1$. The term *skew-symmetric* means that the tensor component changes its sign and preserves its absolute value when interchanging two indices. In two dimensions, it means that $\epsilon_{12} = 1$, $\epsilon_{21} = -1$, $\epsilon_{11} = 0$ and $\epsilon_{22} = 0$. The contravariant unit polyvector (in two dimensions $\epsilon^{i_1 i_2}$) has similar properties except that it is multiplied as contravariant tensor, e.g.

$$\epsilon_{i_1 i_2} \epsilon^{i_1 i_2} = 2. \tag{6}$$

If we multiply the proper number of moment tensors and unit polyvectors so that the number of upper indices at the moment tensors equals the number of lower indices at polyvectors, we obtain a real-valued relative affine invariant, e.g.

$$\begin{aligned} &M^{ij} M^{klm} M^{nop} \epsilon_{ik} \epsilon_{jn} \epsilon_{lo} \epsilon_{mp} = \\ &= 2(m_{20}(m_{21} m_{03} - m_{12}^2) - m_{11}(m_{30} m_{03} - m_{21} m_{12}) + m_{02}(m_{30} m_{12} - m_{21}^2)). \end{aligned} \tag{7}$$

This method is analogous to the graph method. Each moment tensor corresponds to a node of the graph and each unit polyvector corresponds to an edge. The indices indicate, which edge connects which nodes. The graph corresponding to the invariant (7) is on Fig. 1.

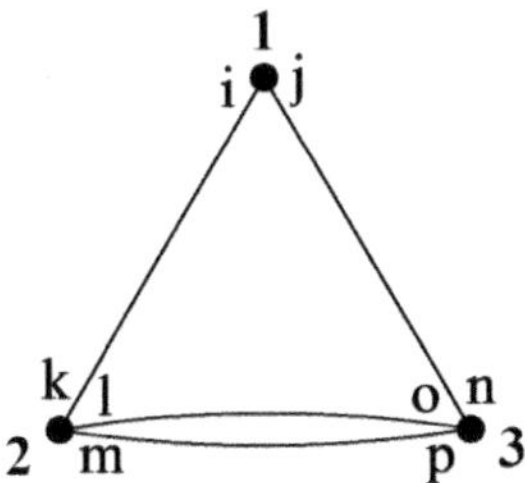

Fig. 1. The graph corresponding to the invariant from (7) and (8)

This invariant can be generated by the graph method as

$$\int_{-\infty}^{\infty} \cdots \int_{-\infty}^{\infty} C_{12} C_{13} C_{23}^2 f(x_1, y_1) f(x_2, y_2) f(x_3, y_3) \mathrm{d}x_1 \mathrm{d}y_1 \mathrm{d}x_2 \mathrm{d}y_2 \mathrm{d}x_3 \mathrm{d}y_3. \tag{8}$$

2.3 Solution of the Cayley-Aronhold Equation

The affine transformation (1) can be decomposed into horizontal and vertical translation, scaling, stretching, horizontal and vertical skewing and possible mirror reflection. Each of these transformations imposes one constraint on the invariants.

The invariance to translation and scaling is provided by the same way in all the methods. We use central moments for translation invariance

$$\mu_{pq} = \int_{-\infty}^{\infty}\int_{-\infty}^{\infty} (x - x_c)^p (y - y_c)^q f(x,y)\, \mathrm{d}x\, \mathrm{d}y, \tag{9}$$

where $x_c = m_{10}/m_{00}$, $y_c = m_{01}/m_{00}$ are the coordinates of the object centroid. The normalization to scaling can be assured by using scale-normalized moments

$$\nu_{pq} = \frac{\mu_{pq}}{\mu_{00}^{((p+q)/2+1)}}. \tag{10}$$

The invariants are supposed to have a form of a linear combination of moment products

$$I = \sum_{j=1}^{n_t} \kappa_j \prod_{k=1}^{r} \nu_{p_{jk},q_{jk}}, \tag{11}$$

where n_t is the number of terms of the invariant. The invariance to stretching can be achieved by using only that products of moments, where the sum of first indices (labeled as p) equals the sum of second indices (labeled as q)

$$\forall j = 1, \ldots, n_t : \sum_{k=1}^{r} p_{jk} = \sum_{k=1}^{r} q_{jk} = w. \tag{12}$$

From the horizontal skew constraint we can derive the Cayley–Aronhold differential equation

$$\sum_p \sum_q p\nu_{p-1,q+1} \frac{\partial I}{\partial \nu_{pq}} = 0. \tag{13}$$

The Cayley–Aronhold differential equation leads to a solution of a system of linear equations for the unknown coefficients κ_j of the invariant.

3 Relation between the Invariants from Both Methods

We have generated two feature sets, one from the graph method, the other as a solution of the Cayley–Aronhold differential equation. The features are homogeneous polynomials of the moments of the same degree, so, the features from one set are linear combinations of the features from the other set and there is the question: are both sets equivalent? If we have some weight limit, then we have finite number of invariants from the graph method, while the number of invariants from the solution of the equations is infinite (each linear combination of the basic solution is also a solution).

All affine moment invariants must satisfy the Cayley–Aronhold differential equation, i.e. if the features generated by the graph method are really affine invariants (and they are), they can be obtained as a solution of the equation. The inverse statement is not so clear, we will formulate it as theorem:

Theorem 1. *All affine moment invariants in the polynomial form (11) can be expressed as linear combinations of some invariants generated by the graph method*

$$I^{(e)} = \sum_{P=1}^{n} c_P I_P^{(g)}. \tag{14}$$

Here $I^{(e)}$ is a general affine moment invariant, e.g. generated as some solution of the Cayley-Aronhold differential equation, and $I_P^{(g)}$, $P = 1, \ldots n$ is a set of invariants generated by the graph method with the same structure as $I^{(e)}$.

Proof. Without loss of generality, we will work with the moments without normalization to translation and scaling, i.e. the geometric moments, because the main question are coefficients of the invariants, not this normalization. The equation (14) can be understood as a system of liner equations for unknown c_P's and we exert to prove that this system has always a solution.

Decomposition. The invariant $I^{(e)}$ can be decomposed into a part of moments B and a part of coefficients K

$$I^{(e)} = K_{i_1 i_2 \cdots i_{2w}} B^{i_1 i_2 \cdots i_{2w}}, \tag{15}$$

where w is the weight of the invariant. The part B can be expressed as a product of moment tensors

$$B^{i_1 i_2 \cdots i_{2w}} = M^{i_1 i_2 \cdots i_{d_1}} M^{i_{d_1+1} i_{d_1+2} \cdots i_{d_1+d_2}} \ldots M^{i_{2w-d_r+1} i_{2w-d_r+2} \cdots i_{2w}}, \tag{16}$$

where r is the degree of the invariant. If $I^{(e)}$ has a structure $(k_2, k_3, \ldots, k_s)$, then k_2 numbers from d_1, d_2, ... , d_r equals 2, k_3 of them equals 3 up to k_s equals s.

The product of moment tensors in (16) contains all possible products of moments with the given structure, so the decomposition (15) is always possible. If some product of moments occurs several times (e.g. m times) in B, then the corresponding components of K must be multiplied by $1/m$. The invariants $I_P^{(g)}$ for each P can be decomposed to the part of coefficients and the part of moments by the same way, the part B is the same for all $I_P^{(g)}$ and $I^{(e)}$, while the part of coefficients of $I_P^{(g)}$ can be expressed as a product of unit polyvectors

$$I_P^{(g)} = \epsilon_{\{i_1 i_2} \epsilon_{i_3 i_4} \cdots \epsilon_{i_{2w-1} i_{2w}\}_P} B^{i_1 i_2 \cdots i_{2w}}, \tag{17}$$

where $\{i_1 i_2 \cdots i_{2w}\}_P$ means P-th permutation of the indices $i_1, i_2, \ldots, i_{2w}$.

Comparison of the decompositions. Equation (14) can then be rewritten as

$$K_{i_1 i_2 \cdots i_{2w}} B^{i_1 i_2 \cdots i_{2w}} = \sum_{P=1}^{(2w)!} c_P \epsilon_{\{i_1 i_2} \epsilon_{i_3 i_4} \cdots \epsilon_{i_{2w-1} i_{2w}\}_P} B^{i_1 i_2 \cdots i_{2w}}. \tag{18}$$

If the moments do not identically equal zero (they identically equal zero for zero image only), then the part B can be omitted

$$K_{i_1 i_2 \cdots i_{2w}} = \sum_{P=1}^{(2w)!} c_P \epsilon_{\{i_1 i_2} \epsilon_{i_3 i_4} \cdots \epsilon_{i_{2w-1} i_{2w}\}_P}. \tag{19}$$

The single equation (18) with variable moments is splitting into system of 2^{2w} linear equations for $(2w)!$ unknown c_P's with constant coefficients. The summation over all permutations of unit polyvector indices is not anything else than summation over all graphs generating invariants with the given structure.

Solvability. Now, we multiply K by the corresponding number of contravariant unit polyvectors. Then we obtain from (19)

$$K_{i_1 i_2 \cdots i_{2w}} \epsilon^{x_1 x_2} \epsilon^{x_3 x_4} \cdots \epsilon^{x_{2w-1} x_{2w}} = \sum_{P=1}^{(2w)!} c_P^* \delta^{x_1}_{\{i_1} \delta^{x_2}_{i_2} \cdots \delta^{x_{2w}}_{i_{2w}\}_P}, \tag{20}$$

where $c_P^* = 2^w c_P$ and $\delta^{i_1}_{i_2}$ is Kronecker delta, $\delta^{i_1}_{i_2} = 1$ if $i_1 = i_2$ and $\delta^{i_1}_{i_2} = 0$ if $i_1 \neq i_2$. The system of equations (20) has 2^{4w} equations for $(2w)!$ unknowns, but many of the equations are linearly dependent, the rank of the system was not increased. Denote it $((2w)! - t)$, where t is some positive integer. Now take the system of equations

$$\sum_{P=1}^{(2w)!} \delta^{x_1}_{\{i_1} \delta^{x_2}_{i_2} \cdots \delta^{x_{2w}}_{i_{2w}\}_P} \lambda_P = 0 \tag{21}$$

with unknowns $\lambda_1, \lambda_2, \ldots, \lambda_{(2w)!}$. The matrices of the systems (20) and (21) are the same, therefore the rank of (21) is also $((2w)! - t)$. That is why the system (21) has t linearly independent solutions

$$\lambda_P = \lambda_P^\sigma, \qquad \sigma = 1, 2, \ldots, t. \tag{22}$$

Now, we can add to the system (21) t equations

$$\sum_{P=1}^{(2w)!} \lambda_P^\sigma \lambda_P = 0 \tag{23}$$

and obtain a system of $2^{4w} + t$ equations. Let the new connected system of equations (21) and (23) has some solution $\lambda_P = \lambda_P^0$, $P = 1, 2, \ldots, (2w)!$. This solution satisfies all the equations (21), therefore it must be a linear combination of the solutions λ_P^σ

$$\lambda_P^0 = \sum_{\sigma=1}^{t} \alpha_\sigma \lambda_P^\sigma, \qquad P = 1, 2, \ldots, (2w)!. \tag{24}$$

The equations (23) must be satisfied for every λ_P, therefore they are satisfied for their arbitrary linear combinations and also for

$$\sum_{\sigma=1}^{t} \alpha_\sigma \sum_{P=1}^{(2w)!} \lambda_P^\sigma \lambda_P = 0. \tag{25}$$

It can be rewritten by (24) in the form

$$\sum_{P=1}^{(2w)!} \lambda_P^0 \lambda_P = 0. \tag{26}$$

It must be satisfied for every λ_P thus also for λ_P^0

$$\sum_{P=1}^{(2w)!} (\lambda_P^0)^2 = 0, \tag{27}$$

i.e. $\lambda_1^0 = \lambda_2^0 = \cdots = \lambda_{(2w)!}^0 = 0$. It means the only solution of the connected system of equations (21) and (23) is zero and therefore its rank is $(2w)!$. A relation of the form

$$\sum_{P=1}^{(2w)!} \lambda_P^\sigma \delta_{\{i_1}^{x_1} \delta_{i_2}^{x_2} \cdots \delta_{i_{2w}\}_P}^{x_{2w}} = 0 \tag{28}$$

corresponds to each of the solutions (22). Let $p_{x_1 x_2 \cdots x_{2w}}$ be an arbitrary tensor of covariance $2w$. Since

$$\delta_{i_1}^{x_1} \delta_{i_2}^{x_2} \cdots \delta_{i_{2w}}^{x_{2w}} p_{x_1 x_2 \cdots x_{2w}} = p_{i_1 i_2 \cdots i_{2w}}, \tag{29}$$

then we obtain from (28)

$$\sum_{P=1}^{(2w)!} \lambda_P^\sigma p_{\{i_1 i_2 \cdots i_{2w}\}_P} = 0. \tag{30}$$

The components of the tensor $p_{x_1 x_2 \cdots x_{2w}}$ can be selected quite arbitrarily and in spite of it each component on the left-hand side of (30) equals zero. From it

$$\sum_{P=1}^{(2w)!} \lambda_P^\sigma c_P^* \delta_{\{i_1}^{x_1} \delta_{i_2}^{x_2} \cdots \delta_{i_{2w}\}_P}^{x_{2w}} = 0, \qquad \sigma = 1, 2, \ldots, t. \tag{31}$$

The equality (31) gives t independent linear relations between the unknown coefficients c_P^*. If we add (31) to the equations (20), we obtain a system (A) of $2^{4w} + t$ equations in the coefficients c_P^*. The matrix of this system coincides with the matrix of the connected system (21) and (23). Consequently, the rank of the system (A) is $(2w)!$ and one may select from it $(2w)!$ equaitons in such a way that the determinant formed by their system (B) is non-zero; the system (B) involves all the t equations (31) and $((2w)! - t)$ equations of the system (20) obtained from certain definite values of the indices $x_1, x_2, \ldots, x_{2w}$. Solving the system (B) we express the left-hand side of (20) in the form of linear combinations of the right-hand sides of the system (B), i.e. again in the form of the right-hand sides of (20). It means (20) has always a solution. □

Notes: The solution of (19) is not unique, since we can add to the right-hand side of (20) any linear combination of the left-hand sides of (28). We supposed two-dimensional space here, but the proof can be generalized for arbitrary number of dimensions.

4 Example

For illustration, we created the simplest affine moment invariants up to the third order by the graph method

$$I_1 = (\mu_{20}\mu_{02} - \mu_{11}^2)/\mu_{00}^4$$

with weight $w = 2$ and structure $\mathbf{s} = (2)$,

$$I_2 = (-\mu_{30}^2\mu_{03}^2 + 6\mu_{30}\mu_{21}\mu_{12}\mu_{03} - 4\mu_{30}\mu_{12}^3 - 4\mu_{21}^3\mu_{03} + 3\mu_{21}^2\mu_{12}^2)/\mu_{00}^{10}$$

with weight $w = 6$ and structure $\mathbf{s} = (0, 4)$,

$$I_3 = (\mu_{20}\mu_{21}\mu_{03} - \mu_{20}\mu_{12}^2 - \mu_{11}\mu_{30}\mu_{03} + \mu_{11}\mu_{21}\mu_{12} + \mu_{02}\mu_{30}\mu_{12} - \mu_{02}\mu_{21}^2)/\mu_{00}^7$$

with weight $w = 4$ and structure $\mathbf{s} = (1, 2)$,

$$\begin{aligned} I_4 = (&-\mu_{20}^3\mu_{03}^2 + 6\mu_{20}^2\mu_{11}\mu_{12}\mu_{03} - 3\mu_{20}^2\mu_{02}\mu_{12}^2 - 6\mu_{20}\mu_{11}^2\mu_{21}\mu_{03} - 6\mu_{20}\mu_{11}^2\mu_{12}^2 \\ &+12\mu_{20}\mu_{11}\mu_{02}\mu_{21}\mu_{12} - 3\mu_{20}\mu_{02}^2\mu_{21}^2 + 2\mu_{11}^3\mu_{30}\mu_{03} + 6\mu_{11}^3\mu_{21}\mu_{12} \\ &-6\mu_{11}^2\mu_{02}\mu_{30}\mu_{12} - 6\mu_{11}^2\mu_{02}\mu_{21}^2 + 6\mu_{11}\mu_{02}^2\mu_{30}\mu_{21} - \mu_{02}^3\mu_{30}^2)/\mu_{00}^{11} \end{aligned}$$

with weight $w = 6$ and structure $\mathbf{s} = (3, 2)$,

$$\begin{aligned} I_5 = (&\mu_{20}^3\mu_{30}\mu_{03}^3 - 3\mu_{20}^3\mu_{21}\mu_{12}\mu_{03}^2 + 2\mu_{20}^3\mu_{12}^3\mu_{03} - 6\mu_{20}^2\mu_{11}\mu_{30}\mu_{12}\mu_{03}^2 \\ &+6\mu_{20}^2\mu_{11}\mu_{21}^2\mu_{03}^2 + 6\mu_{20}^2\mu_{11}\mu_{21}\mu_{12}^2\mu_{03} - 6\mu_{20}^2\mu_{11}\mu_{12}^4 + 3\mu_{20}^2\mu_{02}\mu_{30}\mu_{12}^2\mu_{03} \\ &-6\mu_{20}^2\mu_{02}\mu_{21}^2\mu_{12}\mu_{03} + 3\mu_{20}^2\mu_{02}\mu_{21}\mu_{12}^3 + 12\mu_{20}\mu_{11}^2\mu_{30}\mu_{12}^2\mu_{03} \\ &-24\mu_{20}\mu_{11}^2\mu_{21}^2\mu_{12}\mu_{03} + 12\mu_{20}\mu_{11}^2\mu_{21}\mu_{12}^3 - 12\mu_{20}\mu_{11}\mu_{02}\mu_{30}\mu_{12}^3 \\ &+12\mu_{20}\mu_{11}\mu_{02}\mu_{21}^3\mu_{03} - 3\mu_{20}\mu_{02}^2\mu_{30}\mu_{21}^2\mu_{03} + 6\mu_{20}\mu_{02}^2\mu_{30}\mu_{21}\mu_{12}^2 \\ &-3\mu_{20}\mu_{02}^2\mu_{21}^3\mu_{12} - 8\mu_{11}^3\mu_{30}\mu_{12}^3 + 8\mu_{11}^3\mu_{21}^3\mu_{03} - 12\mu_{11}^2\mu_{02}\mu_{30}\mu_{21}^2\mu_{03} \\ &+24\mu_{11}^2\mu_{02}\mu_{30}\mu_{21}\mu_{12}^2 - 12\mu_{11}^2\mu_{02}\mu_{21}^3\mu_{12} + 6\mu_{11}\mu_{02}^2\mu_{30}^2\mu_{21}\mu_{03} \\ &-6\mu_{11}\mu_{02}^2\mu_{30}^2\mu_{12}^2 - 6\mu_{11}\mu_{02}^2\mu_{30}\mu_{21}^2\mu_{12} + 6\mu_{11}\mu_{02}^2\mu_{21}^4 - \mu_{02}^3\mu_{30}^3\mu_{03} \\ &+3\mu_{02}^3\mu_{30}^2\mu_{21}\mu_{12} - 2\mu_{02}^3\mu_{30}\mu_{21}^3)/\mu_{00}^{16} \end{aligned}$$

with weight $w = 9$ and structure $\mathbf{s} = (3, 4)$, The corresponding graphs are on Fig. 2. All other invariants were eliminated as linearly dependent. The invariant I_5 has dependent absolute value

$$|I_5| = \sqrt{-4I_1^3I_2^2 + 12I_1^2I_2I_3^2 - 12I_1I_3^4 - I_2I_4^2 + 4I_3^3I_4},$$

therefore it was omitted from the set. Its sign can be used for recognition of an object from its mirror reflection.

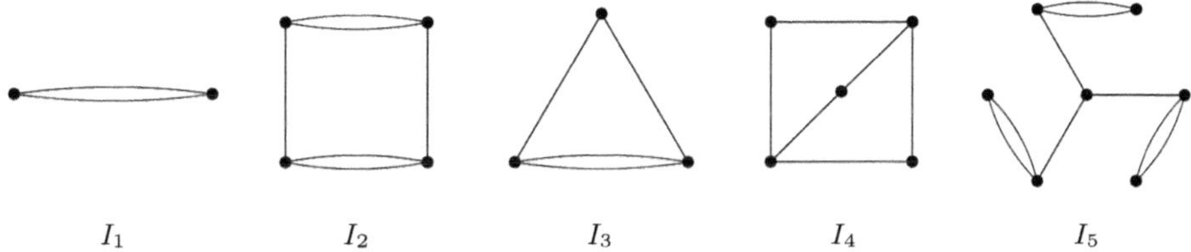

Fig. 2. The generating graphs of the invariants I_1, I_2, I_3, I_4 and I_5

The same invariants were also generated as a solution of the Cayley-Aronhold differential equation except $I_4' = -(I_4 + 6I_1I_3)$ instead of I_4. So, we have two sets of features for recognition of affinely distorted objects, $\{I_1, I_2, I_3, I_4\}$ and $\{I_1, I_2, I_3, I_4'\}$, there is one-to-one mapping between them and features from one set are linear combinations of that from the other set. The presented proof means the same situation is in all higher orders.

5 Numerical Experiment

The goal of this experiment is to show the behavior of the affine moment invariants. We have photographed a series of cards used in a game called mastercards

Fig. 3. The mastercards: Girl, Old scratch, Tyre-ride, Room-bell, Fireplace, Winter cottage, Spring cottage, Summer cottage, Bell and Star

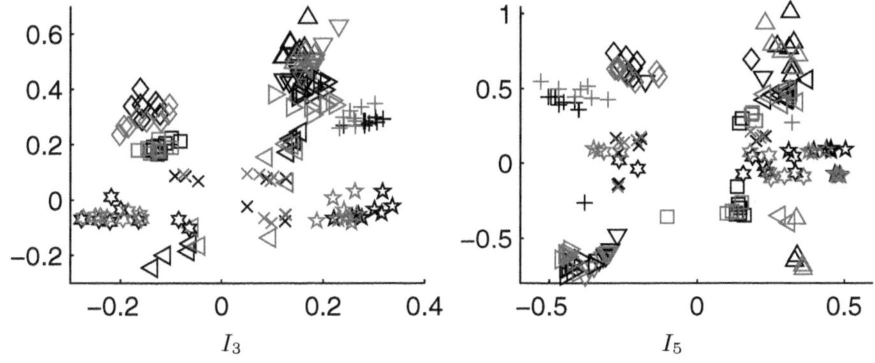

Fig. 4. The feature space of I_3 and I_5 of red (horizontal axis) and blue (vertical axis) channels. Legend: ▽ Girl, △ Old scratch, □ Tyre-ride, ◇ Room-bell, ▷ Fireplace, × Winter cottage, ◁ Spring cottage, + Summer cottage, ✩ Bell and ✡ Star. A card from each pair is expressed by the black symbol while the other card is expressed by the magenta (gray) symbol.

(also pexeso), where the objective is to find the same pairs of cards turned face-down. Cards from each of the ten pairs are shown on Fig. 3.

First, we used the feature set $\{I_1, I_2, I_3, I_4\}$ for each color channel separately, i.e. 12 features. The result was 3 errors from 140 cases, i.e. error rate 2.1%. Then we added I_5 to the feature set and the result was 17 errors, i.e. the error rate worsened to 12.1%. It is an illustration, how a dependent invariant can worsen the recognition. The feature space of the independent invariant I_3 is on Fig. 4a, that of the dependent invariant I_5 is on Fig. 4b.

6 Conclusion

The proved theorem means the features from all the methods mentioned in the paper are equivalent. The suitable method for the generation of the affine moment invariants can be chosen freely, only on the base of the computational aspects as complexity of programming, computing complexity, memory demands and numerical precision.

References

1. Suk, T., Flusser, J.: Graph method for generating affine moment invariants. In: Proceedings of the 17th International Conference on Pattern Recognition ICPR 2004, pp. 192–195. IEEE Computer Society, Los Alamitos (2004)
2. Reiss, T.H.: Recognizing Planar Objects Using Invariant Image Features. LNCS, vol. 676. Springer, Berlin (1993)
3. Suk, T., Flusser, J.: Affine moment invariants generated by automated solution of the equations. In: Proceedings of the 19th International Conference on Pattern Recognition ICPR 2008. IEEE Computer Society, Los Alamitos (2008)
4. Flusser, J., Suk, T., Zitová, B.: Moments and Moment Invariants in Pattern Recognition. Wiley, Chichester (2009)
5. Gurevich, G.B.: Foundations of the Theory of Algebraic Invariants, Nordhoff, Groningen, The Netherlands (1964)
6. Gurevich, G.B.: Osnovy teorii algebraicheskikh invariantov. OGIZ, Moskva, The Union of Soviet Socialist Republics (1937)
7. Cyganski, D., Orr, J.A.: Applications of tensor theory to object recognition and orientation determination. IEEE Transactions on Pattern Analysis and Machine Intelligence 7(6), 662–673 (1985)

A Novel Human Motion Recognition Method Based on Eigenspace

Abdunnaser Diaf[1], Riadh Ksantini[1], Boubakeur Boufama[1], and Rachid Benlamri[2]

[1] University of Windsor, Windsor, ON N9B 3P4, Canada
{diaf,ksantini,boufama}@uwindsor.ca
[2] Lakehead University, Thunder Bay, ON P7B 5E1, Canada
rbenlamr@lakeheadu.ca

Abstract. This paper proposes a novel and robust appearance-based method for human motion recognition based on the eigenspace technique. This method has three main advantages over the existing appearance-based methods. First, the Linear Discriminant Analysis (LDA) is used for dimensionality reduction and eigenspace generation, while preserving maximum separability between classes. Second, by combining a novel centering technique with an incremental procedure, the motion data becomes more concise, expressive, and less confused. Third, data storage is greatly enhanced by using a directed acyclic graph (DAG) structure based on Euclidean distance between projected data. The method is rigorously trained and tested using KTH dataset which contains a large number of motion videos partitioned into six human motions. The experimental results are very promising yielding an average recognition rate of 94.17%.

1 Introduction

Human motion recognition has gained great attention by researchers in the computer vision area. This attention has been triggered by the interest in many applications, such as, motion recognition in surveillance systems, robotics, wireless interfaces and interactive environments. In the literature, two main types of methods for human motion recognition are available: model-based methods and appearance-based methods. Since model-based methods require much more computation complexity to build models, the appearance-based methods have acquired priority over model-based methods. In the appearance-based methods, the image sequence is considered as the primitive component, and is transformed into a suitable form for the purpose of learning as well as recognition. There are many techniques available for human motion analysis and recognition in the context of the appearance-based methods. For example, hidden markov model, motion energy and history images, dynamic time warping and many others. Comparatively to the techniques mentioned above, the eigenspace technique, used in [1, 2, 3, 4], does not require geometrical calculation or partial segmentation of the models. Therefore, it is much more easily adaptable and computationally less expensive [1]. However, the existing appearance-based methods

A. Campilho and M. Kamel (Eds.): ICIAR 2010, Part I, LNCS 6111, pp. 167–175, 2010.

based on the eigenspace technique have used the Principal Component Analysis (PCA) for dimensionality reduction and eigenspace generation. The PCA may discard dimensions that contain important discriminative information [5]. Moreover, these methods have relied on the standard XOR compression technique for motion discrimination [2, 4]. This technique can lead to a strong overlap between the motions, especially, when the human moves in different directions in the scene. Finally, some of these methods [2] have employed the binning and basic B-trees techniques for motion indexing. The performance of these techniques degrades in high dimensions. To overcome these drawbacks, we propose a novel appearance-based method based on the eigenspace technique. This method has three main advantages over the existing methods. First, we use the Linear Discriminant Analysis (LDA) for dimensionality reduction and eigenspace generation. In fact, unlike the PCA, LDA seeks to find the projection directions along which the classes are best separated [5]. Second, instead of the XOR compression technique, we combine a centering technique with an incremental procedure for motion compression to make data more concise and expressive, and then to avoid motion overlapping. Third, data indexing and retrieving are significantly enhanced using a DAG structure based on Euclidean distance between projected data, instead of the B-trees technique. This paper is organized as follows. Section 2 presents the preprocessing operations. Section 3 details the training phase. Section 4 describes the recognition process. The implementation and experiments are given in Section 5.

2 Preprocessing Operations

Preprocessing operations are crucial for achieving efficient high level processing such as training and recognition tasks. Two essential preprocessing operations are considered here: (1) Background subtraction, where each image frame is subtracted from a background model [6]. The output of this step are binary images with white representing the removed background and black the human-body(see Fig. 1). (2) Centering Human-body Blobs, by performing this operation on each frame, we can narrow down and restrict human motion to a very limited and expressive area. Centering the human body provides many advantages. First, the resulting data will be concise and expressive (see Fig. 2), second, the centering will help to get rid of the static portions and keeps only the parts of the body, where the motion is concentrated. As a consequence, the subsequent steps will be faster and more reliable. The algorithm below describes the centering process:

input <V:background-subtracted binary video>
  for each image $I_i \in V$, *do*
    using a vertical scan line moves from left to right in I_i, *do*
      find $h =$ *max. height of the blob, and* $x = XCoor(h)$;
    using two vertical scan lines move towards x *in* I_i, *do*
      find x_1 *and* x_2 *where blob height* $>= T \times h$;
    $C = x_1 + \frac{x_2 - x_1}{2}$; *// C: blob centre*

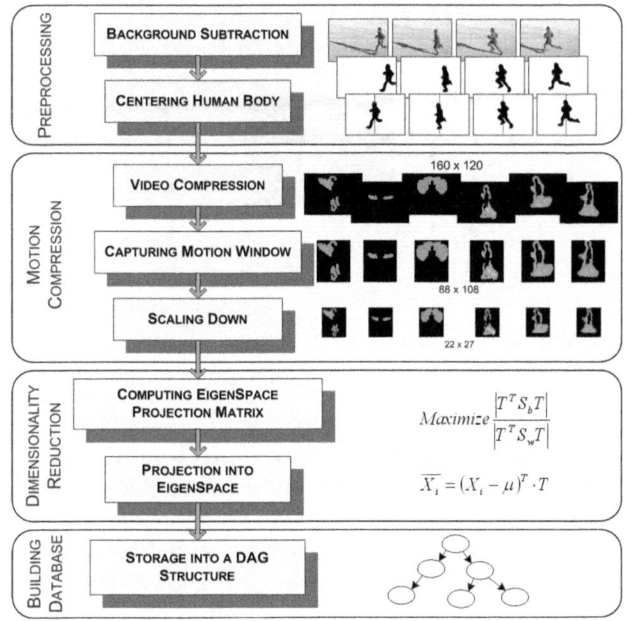

Fig. 1. The training stage

shift the blob horizontally by $[\frac{width(I_i)}{2} - C]$ *pixels;*
next I_i*;*
output <*V:binary video*>

3 Learning Human Motion

Recognition systems cannot recognize anything unless they have learned it. In our proposed method, the learning stage is built up on the data samples produced from the preprocessing stage. Each sample used for learning is a blob-centered background-subtracted binary video. The main goal of this stage is to reduce the dimensionality of the training samples to their minimum possible dimensionality as long as discrimination between classes can be achieved. This stage consists of three main steps, (1)video compression, (2)projection into an eigenspace and, (3)data storage (see Fig. 1).

3.1 Generating Motion Intensity Images (MIIs)

Given the output from our preprocessing stage, centered background-subtracted binary blobs, the goal here is to collapse(compress) these images into a single image, an MII. We have used an incremental procedure for this image compression where, only the differences between consecutive frames are considered. This type of compression provides more details about slow and fast moving portions

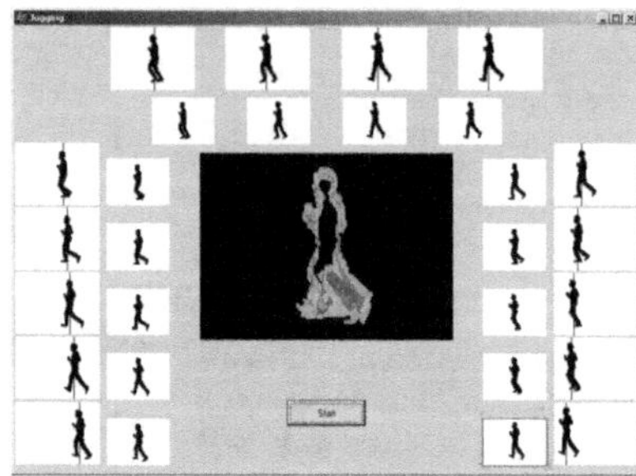

Fig. 2. Generating the MII from a binary video

of the blob. The middle colored image of Fig. 2 shows clearly how an MII of a jogging motion appears. Motion intensity varies from light to heavy and can easily be noticed through colors from green to red respectively. In the same figure, the outer set of binary images represents an input motion video starts from the lower-left frame. The vertical line in each frame indicates the horizontal blob center. The corresponding blob-centered video is shown by the intermediate set of binary images. The following algorithm describes how an MII is created:

input <V :background-subtracted binary video>
  reset C;
  starting with the 2nd *image* $I_2 \in V$, *do*
    for each pixel $P_j \in I_i$, *do*

$$P_j(C) = P_j(C) + \begin{cases} 1, & if\ P_j(I_i) \neq P_j(I_{i-1}) \\ 0, & otherwise \end{cases}$$

  for each pixel $P_j \in C$, $P_j(C) = \frac{P_j(C)}{f} \times 100;$
output <C: motion intensity image>

where f is the number of frames in V and C is the MII.
Standard XOR compression techniques have been used for motion discrimination [2] when the body remains at the same location. However, when the body moves in different directions, an incremental method is more appropriate.

3.2 Capturing and Scale-Normalizing Motion Window

As CPU-time is always a major issue, the size of images are typically reduced to their minimum as long as there is no negative effect on the outcome. The most popular methods for image scaling are Nearest Neighbor, Linear Interpolation and Cubic Interpolation. The latter is often preferred over the first two when quality is important [7]. As it can be seen from Fig 2, the human motion is concentrated at the center of the MII. Therefore, most of the MII area shows no motion information or some scattered noise. In our case, each MII image is a $W \times H$ matrix that can be reduced to a new $w \times h$ MII, where $w < W$ and $h < H$. We have scaled down our MIIs by a factor of 4 using the cubic interpolation method, hence reducing our data size by more than 16 times.

3.3 Dimensionality Reduction and Eigenspace

Significant improvements can be achieved by mapping the data into a lower-dimensionality space in both recognition and computational aspects. Many recognition systems have employed the eigenspace technique for dimensionality reduction. There are two common algorithms used for constructing eigenspaces, PCA and LDA. The existing recognition systems using the eigenspace technique have employed PCA [2] [4]. We have analyzed both PCA and LDA in order to determine the best suitable for human motion recognition. We have found that LDA preserves the most of the class discriminatory information and seeks to find directions along which classes can be best separated [5]. As we have seen, human-blob centering has helped a great deal in obtaining very expressive motion representations. Hence, LDA is the better algorithm for dimensionality reduction in our case. It seeks the best separation between classes by taking into account the scatter within-classes as well as the scatter between-classes. It tries to maximize the ratio of between-classes matrix (S_b) to within-classes matrix (S_w), $S_w^{-1}S_b$ [5]. Eigenspace projection matrix can then be constructed by choosing the eigenvectors with the largest R eigenvalues of $S_w^{-1}S_b$, where R is the desired reduced dimensionality. These eigenvectors provide the directions of the maximum discrimination.

Methodology: Let W be the total number of elements in each MII, C be the number of classes, μ_c be the mean feature vector for class c, N_c be the number of training samples from class c, and N be the total number of training samples from all classes, $N = \sum_{c=1}^{C} N_c$, then:

1. Each MII image $I_n (n = 1, .., N)$, is formed as a vector $x_n = (p_1, .., p_W)$.
2. A super matrix $X_c (c = 1, .., C)$ is constructed for each class from all its training vectors, $X_c = (x_1, ..., x_{N_c})$.
3. A mean vector μ_c is computed for each super matrix, $\mu_c = \frac{1}{N_c}\sum_{n=1}^{N_c} x_n$, and then subtracted from its super matrix, $X_c = X_c - \mu_c$.
4. A super mean matrix is constructed from all mean vectors, $M = (\mu_1, \mu_2, .., \mu_C)$
5. A mean-of-means vector μ is computed out from all mean vectors, $\mu = \frac{1}{C}\sum_{c=1}^{C} \mu_c$, and then subtracted from the super mean matrix, $M = M - \mu$.
6. The scatter within-classes and scatter between-classes matrices, S_w and S_b, are computed using Eq. 1.
7. Finally, the transformation matrix T is constructed by choosing the eigenvectors, $e_r (r = 1, .., R)$, with the largest R eigenvalues of $S_w^{-1}S_b$.

$$S_w = \sum_{c=1}^{C} X_c^T X_c, \qquad S_b = M^T M \tag{1}$$

The dimensionality of each MII can be drastically reduced and projected as a point on the eigenspace G using Eq. 2.

$$g_n = (x_n - \mu)^T \cdot T \tag{2}$$

By labeling and projecting all training data into G, a global eigenspace for all classes is built.

3.4 Storage into a DAG Structure

Before any data storage, we have integered the original indices of data samples in order to lower the computational cost of subsequent arithmetic and logical operations. We have achieved this by shifting the floating point d digits to the right ($d \geq 6$) then, replacing the number with the integer closest to it. A DAG [8] structure G is built based on Euclidean distances between projected data. Within each node $n \in G$, there are three data fields, an index consisting of R sub-indices along each eigenspace axis, a label indicating its class and, a set of pointers pointing to its children. For constructing the database, a bubble scenario has been used where the closest point to the center of all projected training data will be taken as a starting root. Each other data point will be connected to the root directly or indirectly through a closer point. The following algorithm describes this process:

input <*E:eigen-space distribution,* $G = \phi$*:DAG*>
 find the closest point $c \in E$ *to the distribution center;*
 remove c *from* E *and make it the root of* G*;*
 while $E \neq \phi$*, do*
  for each leaf node $n \in G$*, do*
   for each projected point $p \in E$*, do*
    if [1:] p *is closer to* n*'s parent than other leafs' parents*
    and [2:] p *is closer to* n *than* n*'s siblings*
    and [3:] *no intermediate point* $t \in E$ *satisfies 1, 2,*
    and $d(p,t) + d(t,n) > T \times d(p,n)$*: where* $T > 1$ *then*
     remove p *from* E *and make it a child node of* n*;*
output <*G:DAG*>

4 Human Motion Recognition

By storing the projected training data in a DAG structure, the system is learned and ready to be used for motion recognition. This is the most important part from the user's point of view as it is responsible for giving the final decisions. The accuracy level of the decisions plays a central role in evaluating recognition systems. This stage is very similar but much faster than the learning stage. The unknown human motion will simply be recognized as the retrieved nearest neighbor's class.

Given an unknown video as input, the following steps will take place in order: (**i**) Background subtraction; (**ii**) Human-blob centering; (**iii**) Monochrome-video compression generating a primary MII; (**vi**) Capturing and then scaling down the motion window producing a final MII; (**v**) Dimensionality reduction of the MII by forming it into a vector $U = (p_1, p_2, .., p_W)$, and then projecting it into the eigenspace using Eq. 2.

As our DAG is well-structured for finding nearest neighbors, the search will be fast, accurate and simple. The unknown-projected motion, as a point $\overline{u}$, is taken and, its coordinates are integered into u. Finding the nearest neighbor in our DAG structure is done based on finding a node whose Euclidean distance to the unknown motion index is the smallest. The unknown motion is simply recognized as its closest neighbor's class. Once the compound query index is ready, its closest training neighbor can be retrieved from the DAG using the algorithm below:

```
input < G:DAG, u:CompoundIndex>
  NN = root(G);   // Nearest Neighbor
  MinD = distance(u, NN);   // Min. Euclidean distance
  CurrNode = NN;
  while children(CurrNode) ≠ φ do
    find a child node ch with min. d_ch = distance(u, ch);
    if d_ch < MinD then
      NN = ch;   MinD = d_ch;
    endif;
    CurrNode = ch;
  endwhile;
output<NN's ClassLabel>
```

5 Implementation Results

Our system has been implemented in C++ on a 1.6GH PC. KTH dataset from [9] is mainly used for both training and testing of our method.

Table 1. Confusion matrix for testing 150 new videos

InputVideo→ Classification↓	Boxing	Hand clapping	Hand waving	Jogging	Running	Walking
Boxing	100%	0%	0%	0%	0%	0%
HandClapping	0%	100%	0%	0%	0%	0%
HandWaving	0%	0%	100%	0%	0%	0%
Jogging	0%	0%	0%	80%	12.5%	2.5%
Running	0%	0%	0%	15%	87.5%	0%
Walking	0%	0%	0%	5%	0%	97.5%

The obtained results emphasize the effectiveness of the proposed human motion recognition system when compared to methods reported in [2, 9, 4]. In our implementation, each MII is a 160×120 matrix yielding 19,200 elements. An 88×108 window is used to capture the motion within each MII and becomes a new MII of 9,504 elements. Each new MII is scaled down to 22×27, reducing our data size about 32 times, with almost no data loss (see Fig. 1). Data is then projected drastically into an eigenspace of 5 axes using LDA, while preserving maximum class separability. We have used more than 30,000 images representing 625 short motion videos to train our system. Those videos are captured for 25

Table 2. Confusion matrix for testing 775 videos

InputVideo→ Classification↓	Boxing	Hand clapping	Hand waving	Jogging	Running	Walking
Boxing	100%	0%	0%	0%	0%	0%
HandClapping	0%	100%	0%	0%	0%	0%
HandWaving	0%	0%	100%	0%	0%	0%
Jogging	0%	0%	0%	94.28%	3.38%	0.71%
Running	0%	0%	0%	4.29%	96.62%	0%
Walking	0%	0%	0%	1.43%	0%	99.29%

different people having 6 different motions in different environments and clothing. The motions are boxing, hand-clapping, hand-waving, jogging, running and, walking. We have first used 8,000 images representing 150 new motion videos and have obtained the recognition results shown as a confusion matrix in table 1. Table 2 shows the resulting confusion matrix obtained from the performed tests using the whole set of videos in our possession - 775 videos. These two tables show an excellent recognition rate with only a few confusions. As it can be seen, there is a clear separation between each of boxing, hand-clapping, and hand-waving, and all the others. Walking and running has also a clear separation. On the other hand, jogging has some confusion with both running and walking. Although this can be seen as a shortcoming, one has to realize that jogging is a motion that falls between walking and running. It becomes especially confusing when some actors jog too slow/fast as it is here.

6 Conclusions

We have proposed a novel and efficient appearance-based method based on eigenspace for human motion recognition. Thanks to the combination of the centering technique with the incremental procedure, the motion discrimination has significantly improved. Moreover, the eigenspace is generated and the data dimensionality is drastically reduced using the LDA, while preserving maximum class separability. Hence, the recognition process is very fast and efficient, especially, when it is enhanced by the DAG structure. Future work will be dedicated to make the recognition method robust to more complicated motions and to improve the LDA discriminatory power using Kernel space.

References

1. Rahman, M., Ishikawa, S.: Human motion recognition using an eigenspace. Pattern Recognition Letters 26, 687–697 (2005)
2. Eftakhar, S., Tan, J., Kim, H., Ishikawa, S.: An efficient approach to human motion recognition employing large motion-database structure. In: SICE 2008, August 2008, pp. 2239–2243 (2008)

3. Murase, H., Sakai, R.: Moving object recognition in eigenspace representation: gait analysis and lip reading. Pattern Recognition Letters 17(2), 155–162 (1996)
4. Ogata, T., Tan, J., Ishikawa, S.: High-speed human motion recognition based on a motion history image and an eigenspace. IEICE - Transactions on Information and Systems E89-D(1), 281–289 (2006)
5. Yu, H., Yang, J.: A direct lda algorithm for high-dimensional data - with application to face recognition. Pattern Recognition 34(10), 2067–2070 (2001)
6. Mcivor, A.M.: Background subtraction techniques. In: The 15th International Conference on Image and Vision Computing, Auckland, New Zealand, November 2000, pp. 147–153 (2000)
7. Lin, T., Chen, S., Truong, T., Chen, C., Lin, C.: Still image compression using cubic spline interpolation with bit-plane compensation, September 2007, vol. 6696. SPIE, San Jose (2007)
8. Bondy, J., Murty, U.: Graph Theory (Graduate Texts in Mathematics), 1st edn., August 2008. Springer, Heidelberg (2008)
9. Schuldt, C., Laptev, I., Caputo, B.: Recognizing human actions: A local svm approach. In: Proceedings of the 17th International Conference on Pattern Recognition, August 2004, vol. 3, pp. 32–36 (2004)

Human Body Pose Estimation from Still Images and Video Frames

Amar A. El-Sallam and Ajmal S. Mian

School of Computer Science and Software Engineering
University of Western Australia, 35 Stirling Highway Crawley, WA 6009
{elsallam,ajmal}@csse.uwa.edu.au

Abstract. This paper presents a marker-less approach for human body pose estimation. It employs skeletons extracted from 2D binary silhouettes of videos and uses a classification method to partition the resultant skeletons into five regions namely, the spine and four limbs. The classification method also identifies the neck, the head and the shoulders. Using the center of mass principles, a model is fitted to the body parts. The spine is modeled with a 2nd order curve while each limb is modeled by two intersected lines. Finally, the model parameters represented by a reference point and two angles belonging to the lines are estimated and the pose is reconstructed. The proposed approach can estimate body poses from single images as well as multiple frames and is considerably robust to occlusions. Unlike existing methods, our approach is computationally efficient and can track human motion while correcting for pose errors using multiple frames. The proposed approach was tested on real videos from MuHAVi and MAS databases and gave promising results. [1]

1 Introduction

Recently, there has been intensive interest in the estimation of human body pose from images and video. Applications of human pose estimation are many including visual surveillance, image retrieval, human computer interaction, sign language, and animation. Pose estimation is the process of estimating human articulated poses with the aim to identify the posture of the human body. This gives information about the action being performed by a person. Pose estimation can be done using a single image, a stereo image pair, or a video sequence. Human pose estimation is very challenging because of the large number of possible articulated deformations and different body scales. Moreover, factors like loose clothing, self occlusion can significantly deteriorate the data[1,2,3]. Several algorithms have been proposed in the literature for pose estimation. These approaches can be broadly divided into two main categories namely analytic algorithms and learning-based algorithms [4,5,6]. Analytic algorithms are used when the geometry of the body is known or can be determined i.e. the projected image of the body is assumed to be a well-known function of a person's pose. In

[1] This research is sponsored by ARC Discovery Grant DP0881813.

A. Campilho and M. Kamel (Eds.): ICIAR 2010, Part I, LNCS 6111, pp. 176–188, 2010.

this case, if a set of control points of the body such as corners, edges or other feature points are identified, it is possible to determine the 3D body pose coordinates from their corresponding 2D image ones in an analytical way. Learning based algorithms use machine learning techniques to map 2D image features to the respective 3D pose representation. The mapping is learned offline from a sufficiently large training data of images of the human body, with different and known poses. During online phase, the learnt model is used to estimate the 3D body pose from 2D image features.

In existing literatures, the most effective methods for estimating/tracking a person's pose are based on markers. Some of these methods use sensing devices e.g. electro-mechanical/magnetic (data gloves), or light emitting diodes attached to the body [8,9,10]. However, these methods are very expensive, can hinder the real body posture, require complex setup and calibration in order to provide precise pose information and are not user friendly. Moreover, they cannot be used for surveillance applications. In [11] a single hypothesis approach for pose tracking is proposed using a Kalman filter, but it invariably fails because human body exhibits multiple motions for different limbs. This requires the simultaneous tracking of multiple body parts. To achieve this, particle filters have been employed which maintain multiple hypotheses [12], however a large number of particles are required in order to estimate the full range of poses. Furthermore, both of the previous methods have to be initialized by external means at the beginning and every time the tracking is lost e.g. due to occlusions. In [3], a 3D skeleton-based body pose recovery method using multiple views is developed. However, the 3D skeleton obtained from the 3D hull was very noisy and as a result a computationally expensive probabilistic approach was employed, even though it was not efficient in minimizing the skeleton noise. In [13], a learning-based method is conducted for 3D human body pose recovery using single images and monocular image sequences. The method however requires complex computations and a large library of labeled poses. In [16], a multicamera silhouette-based power spectral method is proposed, the method however uses training data and space-time volume of poses and sensitive to the frame rate.

In this paper, we propose a marker-less approach that uses skeletons of silhouettes extracted from images or video sequences. Our method does not require labeled training data and is computationally efficient. It can track human motion and correct for errors using pose estimates from previous frames guided by a regression model. The proposed approach was tested on real data comprising still images and video sequences from the MuHAVi and MAS databases and achieved promising results. A demo of the results is provided as supplementary material with the paper.

2 Data Model

The model used in this work consists of 15 body points connected with a skeleton as shown in Fig 1. These points correspond to the head, neck, top-left arm (TL), top-right arm (TR), bottom-left leg (BL) and bottom-right leg (BR).

We assume a prior segmentation of the image and work with 2D silhouettes extracted from an image or a video sequence [1,13,3]. The data used in our experiments is provided by [14,15] which include Multicamera Human Action Video Data (MuHAVi) and Manually Annotated Silhouette Data (MAS) (with prior background subtraction). Our aim is to estimate the human pose by fitting the skeleton model to the silhouettes Fig. 1. The proposed approach should be able to identify different body poses within given error bounds and should be able to deal with still images, video frames and be robust to self occlusions.

3 Human Body Pose Identification

The proposed approach proceeds as follows: (i) an initial estimate of skeleton is first obtained from the 2D binary silhouettes, (ii) skeleton parts/data that describe the body spine, arms and legs are identified and separated into different sub-skeletons (partitions), (iii) each sub-skeleton is then modeled by lines or curves (depending on the natural shape of the human body parts), and (iv) finally, using these lines/curves, the associated body points depicted in the model in Fig. 1 are identified and the pose is reconstructed. Details about the procedure are explained in the following sections.

3.1 Skeletonization Process

Unlike existing methods [11]-[13], where the raw 2D image or the edge maps are used for human pose classification, we only use body skeleton which makes our method computationally very efficient and able to classify human poses more effectively. The proposed skeletonization algorithm is summarized in Table 1.

Table 1. Proposed Skeletonization Algorithm

Step 1: read the silhouette image/frame, define I_o and choose a binary image threshold $0.5 \leq \tau \leq 0.9$
Step 2: crop out the silhouette I from I_o and let $I_n = I$
Step 3: find the blurred silhouette $I_b = I * *G$, where G is a Gaussian point spread function (PSF) of size $\ell \times \ell$ and variance σ^2
Step 4: find the borders/edges of I_b, $I_d = \sqrt{(I_b * *H)^2 + (I_b * *H^T)^2}$ where H^T is the transposed of the horizontal sobel PSF H
Step 5: find the the binary borders $I_{BW}(x,y) = \begin{cases} 1 & I_d(x,y) \geq \tau \\ 0 & otherwise \end{cases}$, then determine $I_n = I_n - I_{BW}$
Step 6: find the shrinked silhouette $I(x,y) = \begin{cases} 1 & I_n(x,y) \geq \tau \\ 0 & otherwise \end{cases}$
Step 7: remove from the shrinked silhouette any pixel of value equal to '1' that is surrounded by N_z zero pixels and do the same of a zero pixel surrounded by N_z '1' pixels
Step 8: repeat steps 3 to 6 until the error between the current and previous shrinked silhouettes reaches ϵ (a small value).

To demonstrate the skeletonisation algorithm see the examples shown in Fig. 3 for $\tau = 0.85$, $\ell = 9$, $\sigma^2 = 3$ and $N_z = 5$ and $\epsilon = 0.01$ which can be adjusted based on the captured image/frame size.

3.2 Classifying the Estimated Skeletonized Image

To correctly classify the skeletonized data/image obtained using the previous section, we employ combinations of the following methods: (i) center of mass fundamentals, (ii) curve fitting, and (iii) pixel histograms. To initiate the classification process, first we divide the skeletonized frame into four different partitions, (P_{11}) for right arm, (P_{12}) for left arm, (P_{21}) for right leg, and (P_{22}) for left leg. These partitions can be done equally or differently using mathematical approaches that use body weight or shape. In the curve fitting-based method, each partition can be fitted to an nth order polynomial with a typical value of $n = 2$ or 3. Because each arm or leg can be approximated by only two lines, then an accurate approach is to find the two best fit lines for the considered n^{th} order polynomial. In the center of mass based method (CM), each partition's curve is divided into N_p different segments, the CM of each segment is then obtained, and finally the two best fit lines for the N_p segments are identified. Center of Mass of a system of N particles each of which has a mass m_i located a distance $r_i = \sqrt{x_i^2 + y_i^2}$ from an origin is defined by,

$$CM = \frac{\sum_{i=1}^{N} r_i m_i}{\sum_{i=1}^{N} m_i} \tag{1}$$

where x_i and y_i, $i = 1, 2, \ldots, N$ define the cartesian locations of $m_i = 1$.

3.3 Estimated Skeleton Partitions

Since we are dealing with binary images, then the value of the mass m_i will be either 1 or 0. In this case CM takes the form,

$$CM = \frac{\sum_{i=1}^{N} r_i}{N} \tag{2}$$

To separate each body part correctly, reference points are required. In recent pose classification techniques such as the one in [13], points belonging to all of the body edges were used as references, thus the computational complexity was high. In this work however we use few reference points to classify different human body poses. The first is the body curve of equilibrium which defines a curve that passes through the (x, y) coordinates around which the body masses $(m_i = 1, i = 1, 2, \ldots, N)$ are in equilibrium. The equilibrium curve, as an example, is shown in "cyan" in Fig. 3. Mathematically, a coordinate $y = y_o$ at $x = x_o$ in this curve can be determined by,

$$y_o = mean\{\#(I_o(x_o, y) = 1)\}\forall y \tag{3}$$

where $\#()$ define the number of times the statement inside the parenthesis is true. The second important reference is the center of mass for the entire body define $CM_o = \{CM_{ox}, CM_{oy}\}$ (see Fig. 1) which will be defined using the formulae,

$$CM_{ox} = \frac{\sum_{i=1}^{N} x_i}{N}, \quad CM_{oy} = \frac{\sum_{i=1}^{N} y_i}{N} \tag{4}$$

CM_o, indicated by red stars and circles in Fig. 3, together with the curve of equilibrium are used as pilots to divide the skeleton into five main partitions.

3.4 Identification of Spine

The spine of the body is the most important reference for pose estimation. In this work, the spine is identified using the above determined curve of equilibrium along the location of CM_o. This is done by first moving CM_o (red star) to the nearest skeleton coordinates (red circle) Fig. 3(b)(c). Then all points belonging to the curve of equilibrium are shifted towards the closest skeleton points. Since the body weight/width is larger than that of any nearby arm, this will guarantee that the curve of equilibrium will converge to the skeleton part representing the body and not the arms. Moreover, the neck and the intersection point between the two legs shown in green in Figs. 1, 3(b)(c) are identified by shifting a window of proper dimensions along the new line of equilibrium (green). At each shift, all pixels inside the circle are summed and recorded. The two shifts around CM_o that have the maximum number of pixels, (i) define the neck location and (ii) the legs intersection location (see Fig. 2). This can also be interpreted as finding the distribution of pixels (histogram) along the shifted curve of equilibrium and the two locations that have the highest histograms around CM_o are the required ones. In cases where any of the two locations is close to CM_o, the body weight represented by the sum of pixels can be used. To correct for the location. For example, see the false neck location in Fig. 2, to correct this location, the body weight starting from this false location (above CM_o in the direction of the equilibrium curve) is calculated and compared with the total body weight. Based on human body standards [17] and using several conducted experiments, the weight of the head plus the neck contributes a certain portion of the body weight. Based on that, the weight associated with this false location is compared with these standards and the false location is disregarded, i.e. a new search for the correct neck which excludes this false location is considered. We use the formula,

$$x_n = 0.9(1 - W_n)/W_t \ \ x_f \tag{5}$$

to provide the final correct x coordinate for the false x_f neck coordinate. The y_n correct coordinate is the corresponding y for the coordinate x_n in the shifted curve of equilibrium, where W_n is the body weight associated with the identified neck, and W_t is the total body weight. Once the two locations, define $p_n = \{x_n, y_n\}$ and similarly p_b (see Fig. 2) are accurately identified, the spine location is found by fitting all points belong to the shifted curve of equilibrium and between the two identified locations to a 2nd order curve, $c_o(x) = a_2x^2 + a_1x + a_o$ as shown in Fig. 3(d). The values of a_2, a_1, a_o are calculated using least squares estimation techniques.

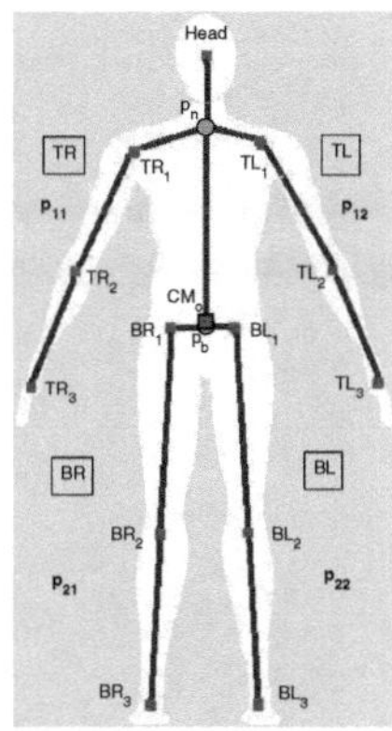

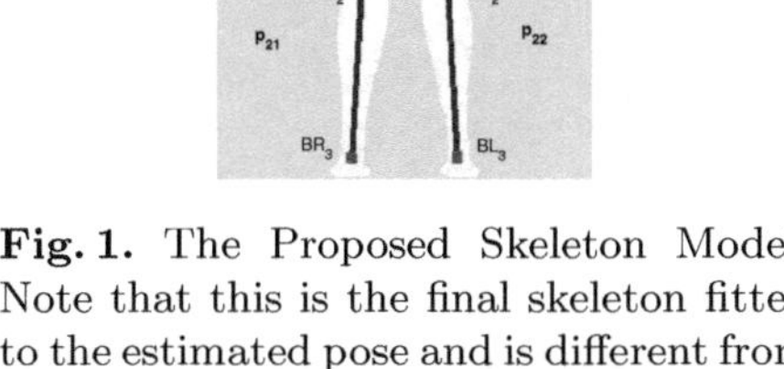

Fig. 1. The Proposed Skeleton Model. Note that this is the final skeleton fitted to the estimated pose and is different from the outcome of the initial skeletonization.

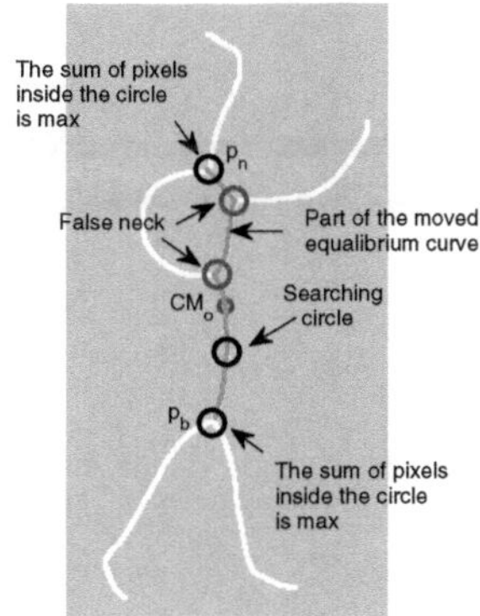

Fig. 2. Estimation of the correct neck position and the intersection point of the legs using a histogram-based search approach.

3.5 Identification of the Arm Portions

To identify the arm poses, i.e. points $\{TR_1, TR_2, TR_3\}$ and $\{TL_1, TL_2, TL_3\}$, we first use the 2nd order curve $c_o(x)$ identified above for the spine as a reference. We then generate two different 2nd order curves $c_r(x)$ (right) $c_l(x)$ (left) along the spine curve (Fig. 3(d)). Again, using human body standards [17] together with several experiments, the two curves are located at a distance equal to $1/4W$ at the location p_b and $2/4W$ at the neck location p_n, where W is the body width at the intermediate location CM_o, p_b, i.e.,

$$\begin{aligned} c_r(x) &= c_o(x) - \frac{1/4W}{(x_{p_b} - x_{p_n})}(x - x_{p_n}) - 1/4\ W \\ c_l(x) &= c_o(x) + \frac{1/4W}{(x_{p_b} - x_{p_n})}(x - x_{p_n}) + 1/4\ W \end{aligned} \tag{6}$$

The skeleton data between the two curves are set to zero, see Fig. 3(e). Finally, the skeleton data behind the curve $c_r(x)$ is used to represent the partition p_{11} (TR), and the skeleton data in front of the curve $c_l(x)$ is used to represent the partition p_{12} (TL), the lower boundaries for the two partition is set to $p_b - W/2$ as shown in Fig. 3(f).

3.6 Identification of the Leg Portions

The leg partitions p_{21} and p_{22} are identified according to the following steps: (i) CM_o and a portion of the original curve of equilibrium (in cyan) which starts from the location p_b down to the last skeleton point are used for referencing, (ii) the location CM_o is considered as a common body point for the two legs, (iii) the skeleton data between CM_o and p_b are set to zero as shown in Fig. 3(e) and, (iv) finally the right and left skeleton data for p_{21} and p_{22} are identified as shown in Fig. 3(f).

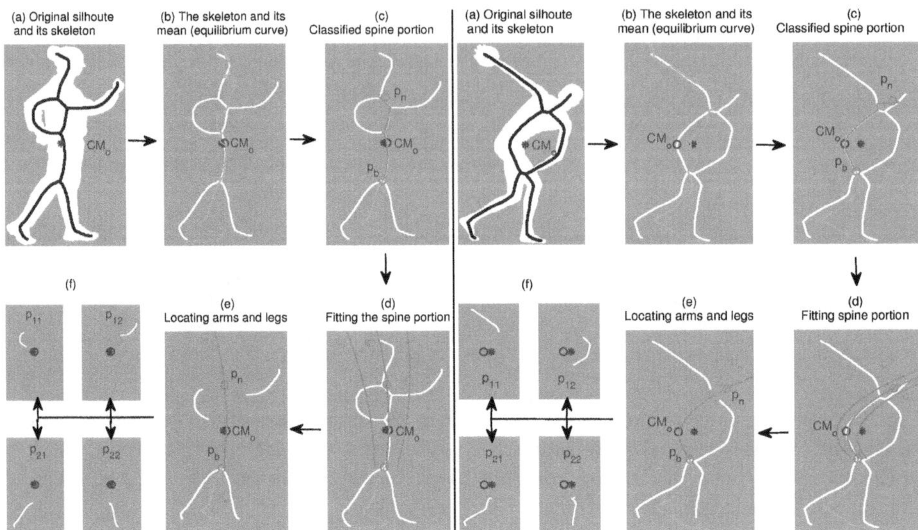

Fig. 3. Two examples explaining the sequence of steps in the approach: (a) Skeletonizing the silhouettes (b) Estimating CM and equilibrium curve (c) Detecting the spine (d) Fitting a 2nd order curve to the spine (e) Locating and (f) separating the limbs

3.7 Challenging Poses

In real situations, human body can take infinite number of poses. In many of them, the previously described technique can be easily used to isolate each arm and leg portion. However, since we are dealing with 2D silhouettes, other pose cases for example when arm(s) or leg(s) is/are occluded or merged within the body image, will be difficult to handle. Many existing approaches fail to deal with such situations. In this paper, we attempt to deal with such type of difficult poses as follows:

- when an arm (or both) are hidden, we use portion of the body curvature (edges) to represent the unknown pose of the hidden arm (e.g. Fig 4). This portion will then be shifted to the mid-distance between it and p_n. In cases of video sequences, we will also consider further correction to this arm pose case by considering information from the previous arm poses.
- when the two legs are close (fully overlapped, or occlude each other), their skeleton in this case will be a single line. In this situation the two legs will be assigned the same line as shown in Fig 4. In case of on-line pose tracking in videos, further correction will also be considered using information from previous leg poses. Information from future poses can also be incorporated for off-line cases or during online pose estimation but with a time lag.
- distances between the point p_n, and the closest skeleton points of each separated arm (see Fig. 4) are employed to correct for shoulder locations. This can happen when a small part of an arm(s) is available. If this distance contributes more than 50% of the arm length, then the person is mostly moving

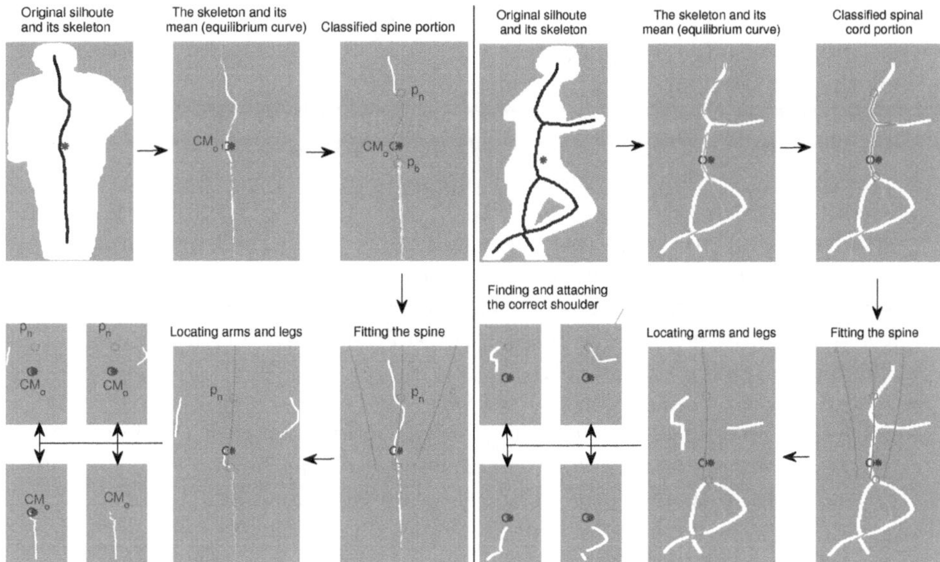

Fig. 4. Two examples of special body poses showing occlusions and ambiguity in the locations of the limbs. The body curvature is used to correct for the ambiguities to a certain extent.

right or left and the shoulder location associated with arm is the point p_n, see Fig. 4 in the examples.

3.8 Estimation of the Overall Body Pose

Using the above procedure, we are now able to isolate the body arms and legs and identify the spine. Our target then is to fit each one of these parts to a model that can correctly (fully) estimate its pose. To achieve this, a 2nd order curve is fitted to the spine. The curve starts at CM_o and ends at the neck point p_n. It can be clearly seen that the spine is a portion of $c_o(x)$. The head location define $c_o(p_h)$ is found by extending $c_o(x)$ further than the neck point p_n using a length related to human body standards for average neck length [17].

Since the arms and the legs can only be formed into two intersected lines, we consider a two-lines (a single reference and two angles) model to represent their poses. To clarify our two-lines model, let us consider the identified right arm of the lady silhouette in Fig. 3. The arm partition is separated and detailed in Fig. 5. The proposed model can be described as follows:

- use Eqn. 6 to find the center of mass for the arm, define CM_r
- use CM_r as a reference to divide the skeletonized arm into N_p sub-partitions of equal weight (Fig. 5). We use an odd number for $N_p \geq 5$ to enforce a common intersecting point between the arm portions.
- find the center of mass for each sub-partition, $CM_1, CM_1, \ldots, CM_{N_p}$

- fit the first CM_{n_p} points, $n_p = 1, \ldots, (N_p + 1)/2$ to a line and determine its angle w.r.t the horizontal θ_{11}^1
- fit the last overlapped CM_{n_p} points, $n_p = (N_p + 1)/2, \ldots, N_p$ to another a line and determine its angle w.r.t the horizontal θ_{11}^2. For accurate results we use a common point $n_p = (N_p + 1)/2$ for both lines.
- find the intersection between the two lines
- record the outputs $\{TR_1, TR_2, TR_3, \theta_{11}^1, \theta_{11}^2\}$ (see Fig. 5). It can be clearly seen that $\{TR_1, \theta_{11}^1, \theta_{11}^2\}$ and basic knowledge about the length of a human arm are enough to fully describe each arm's pose
- follow the same procedure above for the skeletonized leg case, but consider the total body center of mass CM_o as one of the CM_{n_p}. The feet are excluded when determining the leg CM_{n_p} points.
- record the same results for the two legs
- reflect the overall identified results in the model depicted in Fig. 6 where L_a, L_b, L_c, L_d are settings for the arms and legs lengths
- compare the estimated pose in Fig. 6 with the original body pose

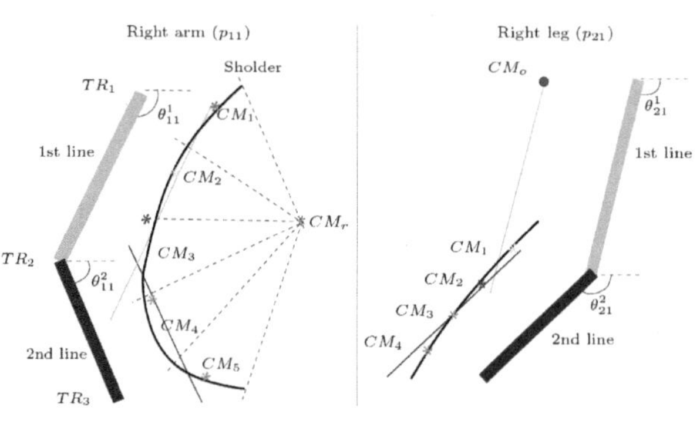

Fig. 5. Identifying the shoulder and fitting the obtained Center of Mass points to the two-lines model then estimating the two angles required for the pose reconstruction process

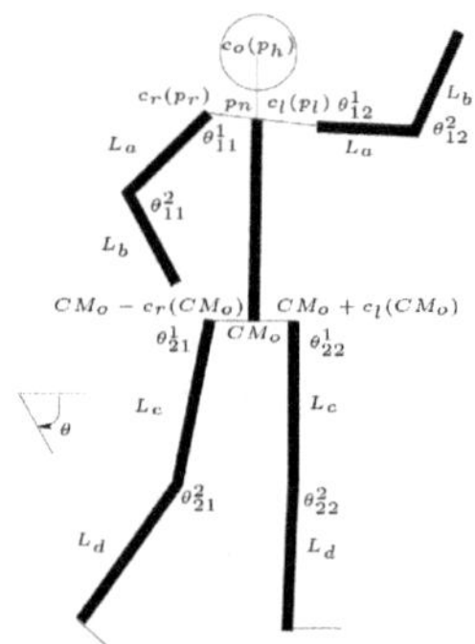

Fig. 6. The final reconstructed pose after reflecting all identified and estimated body pose parameters

4 Experimental Results and Discussion

Experiment 1: In this experiment, we apply the proposed approach to estimate 8 real human body poses using their stand alone silhouette images. The skeleton is obtained using $\ell = 9$, $\sigma^2 = 3$ and $N_z = 5$ and $\epsilon = 0.01$ (approximately 20 iterations). The number of segments used to represent each arm and leg is $Np = 15$, then the two-line model is applied and the required body points and angles are determined then reflected in Fig. 6. As shown in Fig. 7, 8, our method was successful in estimating body poses where self-occlusion is absent. For silhouette images with partial occlusions, the proposed method was able

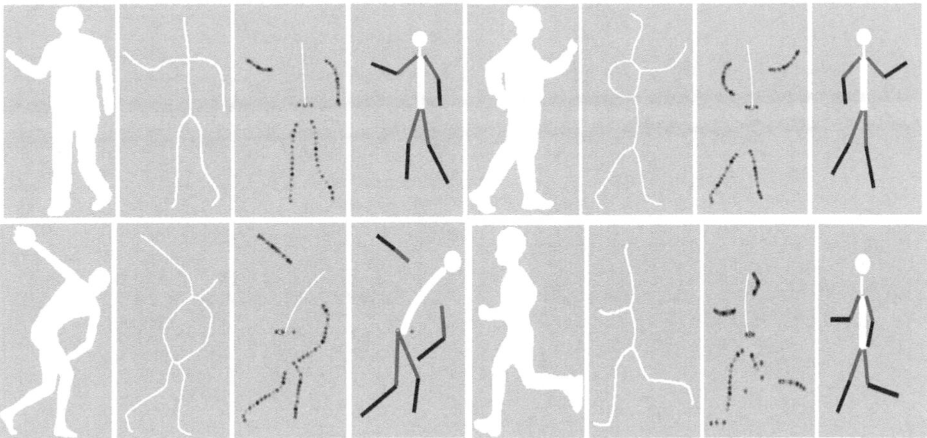

Fig. 7. Four examples showing the reconstruction of some human body poses where ambiguities are minimal to average. In these cases the reconstructed poses can be seen very close to the original ones.

to recover the pose partially, but for cases with excessive self occlusions, the method attempted to estimate the best body pose match using the body edges. Some mis-match pose cases between the actual pose and the estimated pose also present, however they are not very significant. Note that pose estimation using single still images in the presence of occlusions is one of the biggest challenges in computer vision. However, our approach achieved promising results in this scenario with minimal computational complexity compared to [13,16]. A Matlab implementation of the approach takes approximately 3 seconds on a 1.73Ghz machine with 2.5G to reconstruct a pose from single image.

Experiment 2: In this experiment, we apply the proposed approach to estimate body poses from real video representing a person walking and turning changing the direction of movement. The video has 466 frames of silhouette images [14]. Skeletons are obtained as before but with $Np = 5$ segments. Unlike the single image case, a correction is applied to the estimated pose to eliminate mis-identified poses. This is done by smoothing the change in the estimated pose between successive frames. More precisely, if the difference in the estimated angles is greater than a certain threshold then the current pose angle is determine through an auto regressive (AR) process of order 3. The estimated and the corrected angles for this video sequence is depicted in Fig. 9 [2]. For video frames that have no or insignificant occlusion, the maximum difference (error) in degrees between a real pose and an estimated pose is found to be less than 8° degrees. In this experiment, the total number of fully mis-classified poses was about 10/466. This number is represented by the spikes in Fig. 9, however, when the person

[2] The estimated and corrected poses for the video are provided as supplementary material with the paper.

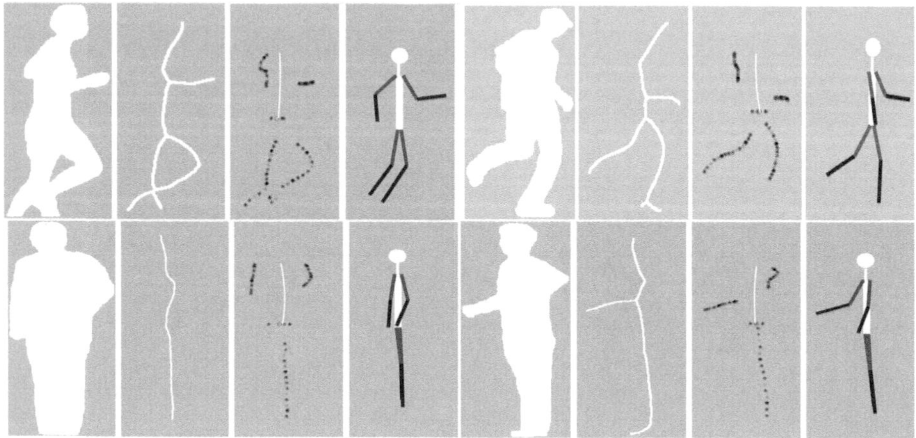

Fig. 8. Four examples showing the reconstruction of some human body poses where ambiguities are average to sever. In these cases some of the reconstructed poses are close to the original ones, while others with sever ambiguities, the approach attempted to identify the best fit pose configuration to them.

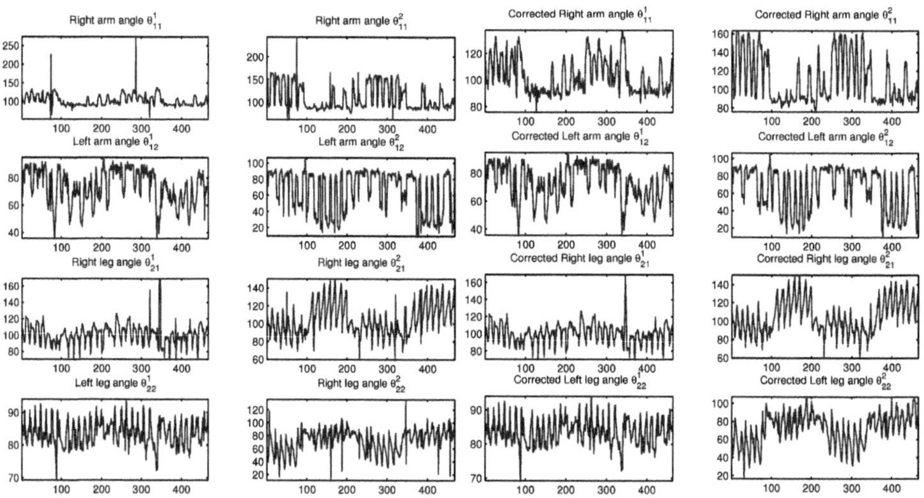

Fig. 9. The estimated limb angles w.r.t the frame index of a real video representing a person walking and turning changing the direction of movement. Correction for for errors using pose estimates from previous frames guided by a regression model can be seen effective.

is turning or his/her hands are merged within the body, there are partial pose mismatches between the original pose the estimated one.

5 Conclusion

We presented a marker-less approach for human body pose estimation from still images and from video sequences. The proposed approach is computationally very efficient. Unlike existing approaches, where the raw image or silhouettes are used to recognize the pose, our approach only uses a limited number of points obtained from an automatically estimated body skeleton. The skeletons are obtained using a fast filtering algorithm applied to 2D binary silhouettes of images or video frames. The proposed approach can deal with some of the challenging scenarios involved in single image based pose estimation and can correct for pose errors in video sequences using temporal information. The proposed approach gave promising results when tested with real data comprising still images and video sequences from the MuHAVi and MAS databases.

References

1. Huang, S.S., Fu, L.C., Hsiao, P.Y.: Silhouette-based human pose estimation using reversible jump Markov chain Monte Carlo. IEEE Electronics letters 42(10), 575–577 (2006)
2. Sigal, L.: Measure locally, reason globally: Occlusion sensitive articulated pose estimation. In: IEEE Conference Computer Vision and Pattern Recognition, pp. 2041–2048 (2006)
3. Menier, C., Boyer, E., Raffin, B.: 3D Skeleton-Based Body Pose Recovery. In: 3rd International Symposiom on 3D Data Processing, Visulization and Transmission, pp. 389–396 (2006)
4. Ioffe, S., Forsyth, D.A.: Probabilistic methods for finding people. International J. Computer Vision 43(1), 45–68 (2001)
5. Shapiro, L.G., Stockman, G.C.: Computer Vision. Prentice-Hall, Englewood Cliffs (2001)
6. Felzenszwalb, P.F., Huttenlocher, D.P.: Distance transforms of sampled functions. Technical Report TR2004-1963, Cornell Computing and Information Science (2004)
7. Lan, X., Huttenlocher, D.P.: Beyond trees: Common factor models for 2d human pose recovery. In: IEEE International Conference on Computer Vision, pp. 470–477 (2005)
8. Erol, A., Bebis, G., Nicolescu, M., Boyle, R.D., Twombly, X.: Vision-based hand pose estimation: A review. Computer Vision and Image Understanding 108(1-2), 52–73 (2007)
9. Sturman, D.J., Zeltzer, D.: A survey of glove-based input. IEEE Computer Graphics and Applications 14(1), 30–39 (1994)
10. Foxlin, E.: Motion tracking requirements and technologies. In: Stanney, K.M. (ed.) Handbook of Virtual Environments: Design, Implementation, and Applications, pp. 163–210. Lawrence Erlbaum Associates, Hillsdale (2002)

11. Rohr, K.: Towards model-based recognition of human movements in image sequences. Graphical Models and Image Processing 25(1), 49–115 (1994)
12. Deutscher, J., Blake, A., Reid, I.: Articulated body motion capture by annealed particle filtering. In: IEEE Conference on Computer Vision and Pattern Recognition, pp. 126–133 (2000)
13. Agarwal, A., Triggs, B.: Recovering 3D Human Pose from Monocular Images. IEEE Transactions on Pattern Analysis and Machine Intelligence 28(1), 44–58 (2006)
14. Ragheb, H., Velastin, S., Remagnino, P.: Multicamera Human Action Video Data (MuHAVi) and Manually Annotated Silhoutte Data (MAS), http://dipersec.king.ac.uk/MuHAVi-MAS
15. Ragheb, H., Velastin, S., Remagnino, P.: ViHASi: virtual human action silhouette data for the performance evaluation of silhouette-based action recognition method. In: The 1st ACM Workshop on Vision Networks for Behaviour Analysis, pp. 77–84 (2008)
16. Ragheb, H., Velastin, S., Remagnino, P., Ellis, T.: Human Action Recognition Using Robust Power Spectrum Feature. In: IEEE International Conference on Image Processing, pp. 753–756 (2008)
17. Osterkamp, L.K.: Current Perspective on Assesment of Human Body Proportions of Relevance to Amputees. Journal of the American Dietetic Association 95(2), 753–756 (1995)

3D Human Action Recognition Using Model Segmentation

Sang Min Yoon[1] and Arjan Kuijper[2]

[1] GRIS, TU-Darmstadt
[2] Fraunhofer IGD
Fraunhoferstrasse 5, Darmstadt, Germany

Abstract. This paper addresses a learning-based human action recognition system from multiple images based on integrating features of segmented 3D human body parts such as face, torso, and limbs. The innovation of our proposed 3D human action recognition system consists of three parts: (1) 3D reconstruction of the target object by tracking the position of a target object in a scene to voxelize the accurate 3D human model, (2) Human body model segmentation into several human body parts using ellipsoidal models in the space of second-order three dimensional diffusion tensor fields, and (3) Classification and recognition of human actions from features of the segmented human model using Multiple-Kernel based Support Vector Machine. Experimental results on a set of test volume data show that our proposed method is very efficient to visualize and recognize the human action using few parameters which are independent to partial occlusion, dimension, and viewpoint.

1 Introduction

Human action recognition systems [1-3], which are defined to understand basic human actions from images, have a long history in computer vision. They give rise to many applications, such as automated surveillance systems, smart home applications, video indexing and browsing, and human-computer interaction. Human action recognition from a single 2D image is heavily studied by numerous researchers, but still a challenging issue due to the partial occlusion, clutter, dependence of viewpoint, and pose ambiguity in the image. In multiple camera environment, the numbers of observables which can be used for robust 3D human action recognition are extended and its recognition ratio is more reliable than single camera based methods and independent of viewpoint. However, these systems are much more complex and difficult in a high-dimensional and multi-modal space.

Even though there are some approaches for human action recognition from multiple images or 3D model recognition systems, human action recognition system from 3D reconstruction of multiple images combined with a segmentation technique is new to our knowledge. We propose to solve the problems of 3D human action recognition by focusing on adequate feature extraction and separation of the human body model into several human body parts from 3D

A. Campilho and M. Kamel (Eds.): ICIAR 2010, Part I, LNCS 6111, pp. 189–199, 2010.

reconstructed models, and measuring the similarity using the features which are extracted from three-dimensional second order diffusion tensor fields.

The contribution of our proposed 3D human action recognition system can be summarized as:

1. A photo-realistic 3D reconstruction methodology by tracking the center of gravity of the target object (section 3.1).
2. The ellipsoidal representation of 3D reconstructed human models whose scale and rotation are determined by normalized eigen-features (section 3.2), followed by a 3D model segmentation into several parts which have similar tensorial characteristics used as features to recognize human actions (section 3.2).
3. Human action recognition using Multiple Kernel based Support Vector Machine (section 3.3).

2 Related Work

2.1 3D Model Reconstruction

The topic of 3D scene reconstruction with multiple images has been investigated and produced numerous results in the area of computer vision. The 3D reconstruction research started early on from a stereo vision based reconstruction technique [4]. Kang et al. [5] developed a method of multi-view reconstruction from images to overcome the large occlusions. Hence, they are usually not suitable for a full 3D scene reconstruction. Image based visual hull reconstruction (IBVH) [6] is a real-time 3D scene reconstruction technique from multiple view images. Seitz et al. [7] presented voxel coloring method to reconstruct the concave objects which cannot be solved by IBVH.

2.2 3D Segmentation

Most 3D model segmentation techniques are based on polygon meshes which are flexible enough to approximate an arbitrary shape. Chen et al. [8] surveyed and implemented the 3D segmentation of 3D model segmentation methodologies such as K-means, graph cuts, hierarchical clustering, primitive fitting, random walks, core extraction, tubular multi-scale analysis, critical point analysis, spectral clustering, and so on. The number of segments and segment area are different according to the characteristic of features and clustering methods.

2.3 3D Human Action Recognition

2D/3D human action recognition systems can be largely separated into four categories. First, structural methods [9] use parameterized models describing geometric configurations and relative motions of parts in the motion patterns. This method provided explicit locations of parts which led to advantages for application of HCI and motion animation, but this approach requires a large number of free parameters that have to be estimated. Second, appearance-based methods using template features [10] need a lower degree of freedom than those of structural approach, but they rely on either spatial alignment, or spatial-temporal registration of image sequences prior to reconstruction. The statistical approach [11] was proposed to overcome the difficulty of finding corresponding features between models and structures in test images of structural and appearance based methods. Lastly, the event-based motion interpretation method [12] are popularly used for human action recognition.

3 Our Approach

Most of the above studies are based on computing local space-time gradients or other intensity based features and might thus be unreliable in the cases of low quality video, motion discontinuities and motion aliasing. To overcome these problems, we will explain our new approach for photo-realistic 3D reconstruction from multiple images, 3D model segmentation, and multiple kernel based human action understanding.

3.1 Photo-Realistic 3D Reconstruction Based on 3D Boundary Tracking

We propose a photo-realistic 3D reconstruction methodology from multiple images which have camera calibration data and appearance model of target object.

From numerous previous photo-realistic 3D reconstruction techniques from multiple images, IBVH [6] and voxel coloring [7] promised an efficient 3D reconstruction technique in real-time. Nevertheless, the IBVH algorithm is very dependent on the number of images used, on the position of each viewpoint considered, on the camera's calibration quality, and on the complexity of object's shape. Voxel coloring takes much time to voxelize the whole environment.

Our proposed 3D reconstruction methodology continuously tracks the 3D boundary of the target object and then reconstructs the radiance or color at the surface points by projecting every voxel which is within tracked 3D boundary. Using this method, we do not need to voxelize the hole 3D scene and check the color consistency for the meaningless voxels.

We consider a scene observed by n calibrated static cameras and we focus on the state of one voxel at position V chosen among the positions of the 3D lattice used to discretize the scene. Here we model how knowledge about the occupancy state of voxel V influences image formation, assuming a static appearance model which is extracted from kernel density estimation based background subtraction [13]. As the target object moves in the 3D scene, we track the 3D position of the target object and extract the candidate 3D region in a 3D scene. Figure 1 shows the concept of our approach. We continuously track the center of gravity $g_1, g_2, ..., g_n$ of the appearance model in each image and calculate the G points in the 3D scene which is obtained by intersection of n 3D lays. We extract the 3D lattice by combining the silhouette images of the target object to be reconstructed with camera calibration information to set the visual rays in 3D space

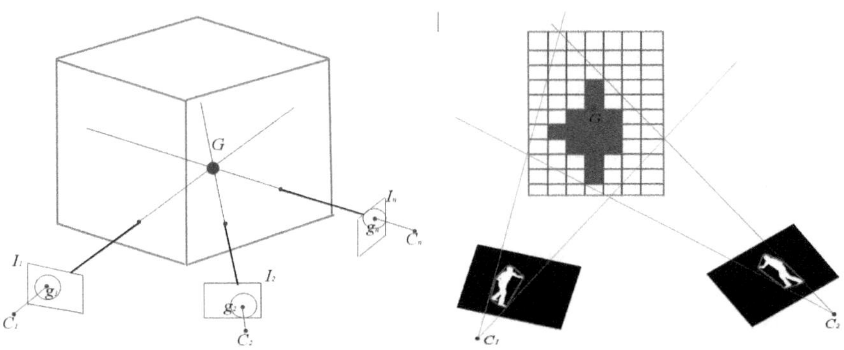

Fig. 1. 3D lattice configuration by tracking the 3D boundary of a target object and voxel carving using color consistency check within 3D lattice. The 3D lattice is determined by intersection of convex cones. The voxels which are painted with "green" are survived voxels by using the color consistency check.

for all silhouette points. They define a generalized cone in which the same object is presented. The 3D lattice in a whole scene is determined by its intersection of these cones. Within a 3D lattice, we used the photo-consistency measure to determine if a certain voxel V belongs to the object being reconstructed or not. Figure 2 shows the 3D reconstructed object from different viewpoints using our method. Figure 2(a) shows the input images in various viewpoint, Figure 2(b) is the 3D lattice which is built by the intersection of convex cones and its carved 3D volume data in a 3D lattice, and Figure 2(c) visualizes 3D model which is reconstructed by using our approach.

(a)Input images from multiple viewpoints (b)3D lattice and 3D reconstructed object within the 3D lattice (c)3D reconstructed human body model

Fig. 2. Multiple images and its reconstruction procedure

3.2 3D Human Model Segmentation in Diffusion Tensor Fields

Among variety of 3D surface representation methodologies, the ellipsoidal representation of 3D deformable human models is very efficient and effective to visualize and recognize its actions using few parameters. It allows us to automatically segment the decomposed human model into several body parts by measuring the similarity between neighbor voxels for the component-based 3D human action recognition.

In the following, we describe the two stages in this process: the computation of the ellipsoidal representation of voxels of the 3D human model and the segmentation procedure.

Ellipsoidal model based 3D human model representation. The tensorial maps, defined as topological representations of 3D symmetric, second-order tensor fields, contain and provide more information than other spaces to measure the similarity between neighbor regions. The generalized symmetric, second-order three dimensional diffusion tensor fields are defined as follows:

$$T = \begin{pmatrix} T_{xx} & T_{xy} & T_{xz} \\ T_{yx} & T_{yy} & T_{yz} \\ T_{zx} & T_{zy} & T_{zz} \end{pmatrix}, \tag{1}$$

where $T_{xy} = T_{yx}, T_{xz} = T_{zx}, T_{yz} = T_{zy}$ because the tensor is a symmetric positive definite matrix. This matrix can be reduced to its principal axes by solving the characteristic equation:

$$(T - \lambda \cdot \mathbf{I})\mathbf{e} = \mathbf{0}, \tag{2}$$

where $\mathbf{I}$ is the identity matrix, λ are the eigenvalues of the tensor and e are the eigenvectors. In each pixel, the tensor can be represented by an ellipsoid, where the main axis lengths are proportional to the eigenvalues $\lambda(\lambda_1 > \lambda_2 > \lambda_3 > 0)$.

Evaluating tensor ellipsoidal geometry and their properties are faciliated with an intuitive domain that spans all possible tensor shapes. Such a domain is afforded by the geometric anisotorpy metrics of Kindlmann [14]. Given the non-negative tensor eigenvalues $\lambda_1, \lambda_2, \lambda_3$, the metrics quantify the certainty with which a tensor may be said to have a given shape like:

$$c_l = \frac{\lambda_1 - \lambda_2}{\lambda_1 + \lambda_2 + \lambda_3},\ c_p = \frac{2(\lambda_2 - \lambda_3)}{\lambda_1 + \lambda_2 + \lambda_3},\ c_s = \frac{3\lambda_3}{\lambda_1 + \lambda_2 + \lambda_3}.$$

The three metrics add up to unity, and define a barycentric parametrization of a triangular domain, with the extremes of linear, planar, and spherical shapes at the three corners. Figure 3 represents a 3D ellipse whose main hemiaxis length is proportional to the square root of eigenvalues λ and the direction correspond to the respective eigenvector e of each tensor. Each voxel within 3D reconstructed human body model is replaced by its corresponding ellipsoidal model which is shown in Figure 3.

3D model segmentation. The degree of anisotropy using the ellipsoidal decomposed 3D model can be quantified in a single number called a diffusion anisotropy index and it is represented as *fractional anisotropy* (FA). The FA representation method of the tensorial elements geometrically characterizes the shape of 3D ellipsoid of each voxel and is defined as follows.

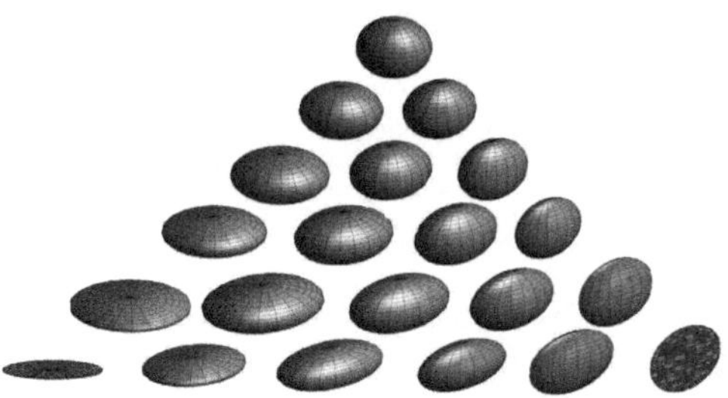

Fig. 3. Ellipsoidal representation of each voxel according to its eigenvalues and eigenvectors

$$FA = \sqrt{\frac{3[(\lambda_1 - \lambda_{avg})^2 + (\lambda_2 - \lambda_{avg})^2 + (\lambda_3 - \lambda_{avg})^2]}{2 \times (\lambda_1^2 + \lambda_2^2 + \lambda_3^2)}} \quad (3)$$

where λ_{avg} is the average of λ_1, λ_2, and λ_3. The FA is used as feature to measure the similarity of neighbor voxels to segment the 3D model.
The 3D model segmentation procedure is as follows:

STEP0 : Initially, the numbers of subregions of the human body model is equal to the number of voxels of the 3D human model. Calcuate $FA_{ij} = \sqrt{(FV_i - FV_j) \times (FV_i - FV_j)}$, where FV_i and FV_i are FA of voxel i and j, respectively.

STEP1 : Progressively merge the neighbor voxels if FA_{ij} is less than threshold and recalculate the average FA of the merged subregions, $FA_{sub} = \frac{1}{n}\sum_{k=1}^{n} FA_k$, where the merged sugregion which have n voxels. The voxels i and j are splitted if FA_{ij} exceeds the threshold.

STEP2 : Repeat **STEP1** until there are no subregions whose the FA_{sub} is less than threshold.

Figure 4 shows our segmented human actions from unlabeled volume data to segmented subregions through the characteristic of ellipse. Figure 4(a) shows the 3D reconstructed human action from multiple images using our proposed method, Figure 4(b) shows the ellipsoidal decomposition of the 3D model, and Figure 4(c) visualizes the 3D segmentation by measuring the similarity between neighbor voxels.

3.3 3D Human Action Classification Using Multiple Kerrnel Based Support Vector Machine

In this section, we describe how Support Vector Machine (SVM) is used for efficient classification of highly variant human motions. SVM is primarily a classier

(a) 3D reconstructed human body model of boxing, jogging, and walking (b) Superquadric representation

(c) 3D segmentation of human body

Fig. 4. 3D model segmentation from volume data of boxing, jogging, and walking

method that performs classification tasks by constructing hyperplanes in a multidimensional space that separates cases of different labels. SVM supports classification tasks and handles multiple continuous and categorical variables.

The performance of different classifiers applied in object detection and recognition systems have been evaluated and compared in the area of pattern recognition and data learning. Bazzani concluded that the Support Vector Machine (SVM) performs better than the Multi-Layer Perception (MLP) for a small number of training data [15]. Papadopoulos [16] has also shown that SVM achieves a higher accuracy rate than Neural Network. Having evaluated the SVM, Kernel Fisher Discriminant (KFD), Relevance Vector Machine (RVM), Feedfoward Neural Network (FNN), and committee machines.

In $\{x_i, y_i\}_{i=1}^{l}, x \in R^m$ where l the number of training features, each x is then mapped to a $\Phi(x)$ and y_i is separated into human actions like boxing, jogging, and walking. The non-linear SVM maps the training samples from the input space into a higher-dimensional feature space via a mapping function Φ and construct a hyperplane defined as $w\Phi(x) + b = 0$ to separate examples from the classes. $\{x_i, y_i\}_{i=1}^{l}$ in the kernel-induced feature space is related to the kernel function K which intuitively computes the similarity between examples in SVM. The standard SVM [17] tries to find a hyperline that has large margin and small training error.

Instead of having a single kernel (SK) K, suppose that we have a set of M base kernels $K_1, K_2, ..., K_M$ with corresponding kernel-induced feature maps $\Phi_1, ..., \Phi_M$. The MK-SVM [18] is extended from the SK-SVM as follows:

$$min_{w,b,\xi}\frac{1}{2}(\Sigma_{k=1}^{M}||w_k||)^2 + C\sum_{i=1}^{l}\xi_i \quad (4)$$

s.t. $y_i(\Sigma_{k=1}^{M} w_k^T \Phi_k(x_i) + b) \succeq 1 - \xi_i, \xi_i \succeq 0, i = 1, 2, , ..., l$, where $w = \{w_1, w_2, ..., w_M\}$ which is the weight for component Φ_k, and ξ is the non-negative slack variables. The regularization parameter C determines the trade-off between the maximization margin $\frac{1}{||w||^2}$ and the minimum experience risk.

The eigen-features x, given by FA_{sub}, which are extracted from segmented human body parts from last section, are used for the learning based human action recognition system. The FA represents the characteristic of the segmented subregion of human body and it is used for classification of human action recognition instance.

4 Experiments

We setup our proposed methodology with a Pentium 4 and a CUDA which is a technology for GPU computing from NVIDIA Geforce 8200. Our experiments are separated in three categories: (1) 3D reconstruction, (2) 3D segmentation in the space of diffusion tensor fields, and (3) action recognition results from MK-SVM technique. Table 1 shows the average running time of our procedure. In our approach the 3D segmentation clearly consumes most time.

Table 1. Average running times for 3D action recognition from multiple images using 128x128x128 dimensional human body model

Category	time(ms)
3D Reconstruction	153
3D Segmentation	1286
3D Action classification	28

4.1 3D Reconstruction

First, we tested our algorithm on the HumanEva dataset[1]. The images came from the HumanEva database which contains 7 calibrated video sequences We first reconstructed the target object from multiple images by the tracking based 3D reconstruction procedure using the provided camera calibration data and the statistics of the background modeling with voxel size 128x128x128 and 64x64x64 to compare the recognition ratio by changing its dimension. Figure 5 visualizes the difference between our proposed tracking based 3D reconstruction method and the original voxel coloring method. The proposed 3D reconstruction method provides more detail than the original voxel coloring method.

[1] http://vision.cs.brown.edu/humaneva/

(a) Our proposed 3D reconstruction method (b) Original voxel carving method

Fig. 5. Comparison of 3D reconstruction between our proposed and original voxel coloring method

4.2 3D Model Segmentation and Action Recognition

Next, we conducted our proposed MK-SVM based 3D human action recognition and compared it with K-Nearest Neighbor(KNN) classification, and a single kernel based SVM(SK-SVM). The HumanEva dataset provides various human motions for four different people. We reconstructed the 3D human model of boxing, jogging and walking actions and trained the tensorial features. Table 2 is the human action recognition matrix of boxing, walking, and jogging actions by changing the dimension of 3D human body model from 64x64x64 to 128x128x128. The acceptance ratio of each action in dimension of 128x128x128 is better than in dimension of 64x64x64, especially for jogging.

Table 2. 3D human action recognition ratio using HumanEav Dataset of 128x128x128 and 64x64x64 dimension

Human action	Training	Testing	Boxing	Walking	Jogging	Boxing	Walking	Jogging
Boxing	600	1200	94.2	1.9	3.9	92.7	2.2	5.1
Walking	600	1200	6.7	84.7	8.6	5.8	83.8	10.4
Jogging	600	1200	7.5	10.8	81.7	5.1	13.5	81.4

(a) 128x128x128 dimension (b) 64x64x64 dimension

Table 3. Comparison of human actio recognition system using KNN and single-kernel SVMs to compare with our proposed MK-SVM based human action recogntion system

Human action	Traning	Testing	Boxing	Walking	Jogging	Boxing	Walking	Jogging
Boxing	600	1200	86.2	6.8	8.4	87.4	4.4	7.2
Walking	600	1200	9.3	75.8	12.9	11.6	74.8	13.6
Jogging	600	1200	7.8	12.5	79.7	4.8	14.3	80.

(a) KNN classification method. (b) SK-SVM classification method

Analyzing the errors of not correctly recognized actions we found that, for example, the action of "boxing" goes to "walking" category because the main human upper body parts arms and torso are too close to each other and could not be segmented as separate parts.

We also tested the 3D motion action recognition system in an experimental environment using four IEEE 1394 cameras using our proposed method. We

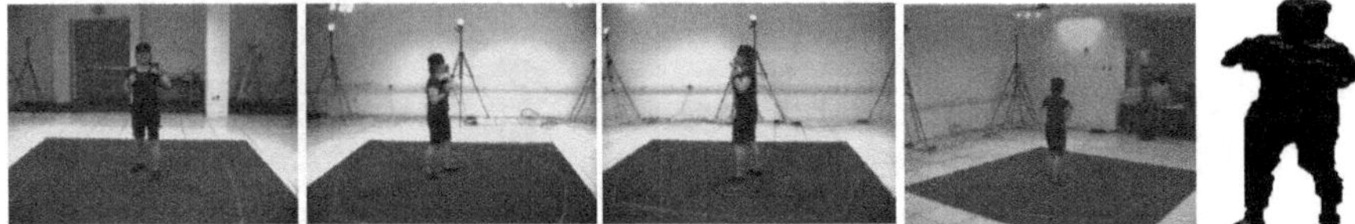

Fig. 6. Example of a wrong classification of our proposed human action recognition system

Table 4. 3D human action recognition matrix in our experimental environment

Human action	testing	Boxing	Walking	Jogging
Boxing	500	83.2	5.4	10.4
Walking	500	13.8	72.7	16.1
Jogging	500	9.5	16.3	74.2

tested 500 3D human models (500 images from 4 cameras which have 640x480 resolution) in a large environment (4m x 15m). The human action recognition ratio of walking and jogging is lower than the acceptance ratio of boxing, because their motions are is more similar than that of boxing.

Unfortunately, there is no public dataset for 3D human model based action recognition system, so we could not directly compare our approach and others. The state of the art in human action recognition from 2D images [19] is based on the Weizman and KTH human action dataset. It is not able to correctly recognize the actions jogging, walking, and running satisfactorily, while our proposed approach using the 3D reconstructed model has a balanced human action recognition ratio and overcome the dependency of viewpoints.

5 Discussion

In this paper, we have presented a novel 3D human action recognition technique whose properties come from segmented human body parts' eigen-features. Our system extends existing 3D reconstruction methods in the view of photo-realistic 3D reconstruction, and eigen-feature based 3D model recognition technique whose properties are come from diffusion tensor fields. Our system is also very efficient and robust in partial occlusion and clutter of 3D model as shown in Section 4.

Future work will focus on efforts to reduce processing time. The approach will also be extended to a 3D model detection and retrieval system.

References

1. Weinland, D., Ronfard, R., Boyer, E.: Free Viewpoint Action Recognition using Motion History Volumes. In: Computer Vision and Image Understanding (2006)
2. Blank, M., Gorelick, L., Shechtman, E., Irani, M., Basri, R.: Actions as Space-Time Shapes. In: Proceeding of ICCV (2005)

3. Schuldt, C., Laptev, I., Caputo, B.: Recognizing Human Actions: A Local SVM Approach. In: Proceeding of CVPR (2004)
4. Marr, D.C., Poggio, T.: A computational theory of human stereo vision. Proceeding of the Royal Society of London B204, 301–328 (1979)
5. Kang, S.B., Szeliski, R., Chai, J.: Handling occlusions in dense multi-view stereo. In: Proceeding of Computer Vision and Pattern Recognitio, pp. 103–110 (2001)
6. Matusik, W., Buehler, C., Raskar, R., Gortler, S.J., McMillan, L.: Image based visual hulls. In: SIGGRAPH 2000 Proceedings, pp. 369–374 (2000)
7. Seitz, S.M., Dyer, C.R.: Photorealistic scene reconstruction by voxel coloring. In: Proceeding of Computer Vision and Pattern Recognition, pp. 1067–1073 (1997)
8. Chen, X., Golovinskiy, A., Funkhouser, T.: A Benchmark for 3D Mesh Segmentation. ACM Transactions on Graphics 28(3) (2009)
9. Sminchisescu, C., Triggs, B.: Kinematic jump processes for monocular 3D human tracking. In: Proceeding Computer Vision and Pattern Recognition, pp. 69–76 (2003)
10. Ekin, A., Tekalp, A.M., Mehrotra, R.: Automatic soccer video analysis and summarization. IEEE Transaction on Image Processing 12(7), 796–807 (2003)
11. Doretto, G., Cremers, D., Favaro, P., Soatto, S.: Dynamic texture segmentation. In: Proceeding of Ninth International Conference on Computer Vision, pp. 1236–1242 (2003)
12. Rao, C., Yilmaz, A., Shah, M.: View-invariant representation and recognition of actions. International Journal of Computer Vision 50(2), 203–226 (2002)
13. Han, B., Comaniciu, D., Davis, L.S.: Sequential kernel density approximation through mode propagation: applications to background modeling. In: Proceeding of ACCV (2004)
14. Kindlmann, G.: Superquadric Tensor Glyphs. In: Joint EUROGRAPHICS - IEEE TCVG Symposium on Visualization (2004)
15. Bazzani, A.: An SVM classifier to separate false signals from microcalcifications in digital mammograms. Phys. Med. Biol. 46, 1651–1663 (2001)
16. Papadopoulos, A., Fotiadis, D.I., Likas, A.: Characterization of clustered microcalcifications in digitized mammograms using neural networks and support vector machines. Artificial Intelligence in Medicine 34(2), 141–150 (2004)
17. Scholkopf, B., Smola, A.: Learning with Kernels: Support Vector Machines, Regularization, Optimization, and Beyond (Adaptive Computation and Machine Learning). MIT Press, Cambridge (2002)
18. Hu, M., Chen, Y., Kwok, J.T.-Y.: Building Sparse Multiple-Kernel SVM Classifiers. IEEE Transactions on Neural Networks (2009)
19. Wang, Y., Mori, G.: Max-Margin Hidden Condtional Random Fields for Human Action Recognition. In: Proceeding of CVPR (2009)

Image-Based Grasping Point Detection Using Boosted Histograms of Oriented Gradients*

Leonidas Lefakis[1,3], Horst Wildenauer[2], Manuel Pascual Garcia-Tubio[3], and Lech Szumilas[4]

[1] IDIAP Research Center, Martigny, Switzerland
[2] Institute for Computer Aided Automation, Vienna University of Technology
[3] Automation and Control Institute, Vienna University of Technology
[4] Research Industrial Institute for Automation and Measurement, Warzawa

Abstract. In this paper, we describe the components of a novel algorithm for the detection of grasping points from monocular images of previously unseen objects. A basic building block of our approach is the use of a newly devised descriptor, capable of representing grasping point shape and appearance by the use of histograms of oriented gradients in a semi-local manner. Combined with boosting our method learns discriminative grasp point models for new objects from a set of annotated real-world images. The method has been extensively evaluated on challenging images of real scenes, exhibiting largely varying characteristics concerning illumination conditions, scene complexity, and viewpoint. Our experiments show that the method, despite these variations, works in a stable manner and that its performance compares favorably to the state-of-the-art.

1 Introduction

Building affordable and scalable platforms, capable of interacting with real environments represents a tempting goal for robotics research. In this context, solutions based solely on visual sensory input are moving more into the center of interest. On one side there are economical considerations to reduce prices by avoiding expensive sensors. On the other hand, vision input already contains rich information to harvest for the task of reasoning about an observed scene, and ultimately manipulating its content by grasping and moving objects.

In our work, we focus on mining monocular vision input to detect potential points for grasping of previously unseen objects. Recently, Saxena et al. [16,17] presented a promising approach capable of grasping previously unseen objects purely based on vision. Their local, texture and colour based, grasp point representation is learned from artificially created images of object examples [15]. Grasp points are separately searched for in two input images provided by a stereo system and only image locations with high confidence of being a grasp

* This work was partly supported by the European Union project GRASP (FP7-215821).

A. Campilho and M. Kamel (Eds.): ICIAR 2010, Part I, LNCS 6111, pp. 200–209, 2010.

point are triangulated to infer the 3D-position were the object can be grasped. Thus the need for reconstructing the object's 3D shape is avoided.

Bohg and Kragic [5] find grasping points by describing the global object shape using shape context [3]. As shape context is known to perform poorly in cluttered scenes [19], the work assumes high quality figure ground segmentation. In practice this is achieved with help of an active stereo system and approximate knowledge of object placement. In [4] their approach is directly compared to Saxena's work drawing upon the latter's training database - reporting significantly improved performance. However, this database of artificially created images presents objects on homogeneous background, thereby greatly simplifying the task of figure-ground segmentation. Furthermore, performance was quantified utilizing metrics devised for binary classification - an assessment which, as we will detail later in the paper, is not suitable for the detection task at hand.

Our approach is motivated by the existence of similar semi-local parts in objects that themselves have rather dissimilar shapes. A typical example is the presence of handles in a large variety of objects ranging from scissors to jugs. In that sense, our method is similar in spirit to the one proposed in [16]. However, by encoding more semi-local information around grasp points, we arrive at grasp point representations which, as we will show experimentally, are able to ignore image clutter to a larger extent. Here, our contribution is twofold: a) We devised a novel image descriptor based on radially configured histograms of oriented gradients, facilitating efficient grasp point detection in real scenes. The descriptor is simple to implement, and can be easily extended to include more visual cues such as color or texture. b) In contrast to preceding work utilizing artificially created data, we demonstrate that discriminative grasp point representations can be learned from images of real scenes. In experiments on a challenging data set, we show that our method is able to significantly outperform the state-of-the-art.

The remainder of the paper is organized as follows: A sketch of the entire method is given in Sec. 2. Sec. 3 describes the image-based representation of grasp points followed by an outline of how these representations are learned in Sec. 4. The process of detecting grasp points in a novel image is detailed in Sec. 5. We present our evaluation in Sec. 6 and conclude the paper with Sec. 7.

2 Method Overview

The proposed method is composed of two main steps: 1) Grasp point representations are learned from annotated images in a discriminative fashion by means of boosting. For this, our descriptor based on radially configured histograms of oriented gradients is employed. 2) In the detection phase, the learned models are densely scanned over a range of scales of an input image. Then, the mean-shift algorithm is employed to detect the modes of the resulting scale-space response maps, yielding both the scale and position of potential grasp points. A typical result obtained with our mehod is depicted in Fig. 1.

Fig. 1. Detected grasp points (blue circles) and detector responses (right image). Note the zoom view of the bottle neck in the lower right of the left image.

Fig. 2. Illustration of the descriptor. Probes (red circles) are radially arranged around the center (red dot). Each probe pools the gradient strength separately from underlying orientation channels and stacks them in a histogram. To avoid clutter, only $K = 3$ concentric probe rings on $O = 4$ orientation channels are shown.

3 Grasp Point Model / Descriptor

Our representation of grasp points is an adoption of Carmichael and Hebert's [6] shape descriptor using a circular arrangement of edge probes. Each of these probes captures the density of the underlying edge image by weighted integration within a gaussian-shaped receptive field. Borrowing the idea from [20], we extend the descriptor to operate on oriented gradient responses instead of edges. Having an input image $\mathbf{I}$, we compute a number C of *blurred orientation channels* $\mathbf{G}_o^{\sigma_p} = G_{\sigma_p} * \mathbf{G}_o$, $o = 1 \ldots C$, one for each discrete orientation. Here, $\mathbf{G}_o = \frac{\partial \mathbf{I}}{\partial o}$ is the image gradient with the derivative taken w.r.t. orientation o. G_{σ_p} denotes a Gaussian kernel with standard deviation σ_p and $*$ stands for convolution. Using the blurred orientation channels, probe values at image location (x,y) for orientation o can now be efficiently obtained by simply accessing $\mathbf{G}_o^{\sigma_p}(x, y)$ which equals the pooled oriented gradient density at that position. By stacking all channel-values for one probe location into a vector, a C-dimensional histogram $\mathbf{p}$ of oriented gradients is obtained.

Surrounding a probe at the query position, additional probes are located on K concentric circles with radii $r_k = k\sigma_p$, $k = 1..K$. Each circle is populated with an increasing number of $6k$ evenly spaced probes resulting in a total number of $N_p = 3K(K + 1) + 1$ probe positions, see Fig. 2.

From this, our descriptor is constructed by stacking all probe histograms into one vector and normalizing it to unit length to achieve a certain degree of illumination invariance. Overall, the representation has three parameters: The size of a probe σ_p, the number of circles K, and the number of orientation channels C. Its total dimensionality is given by $C(3K^2 + 3K + 1)$.

4 Learning

We employ boosting to learn a discriminative visual representation of grasp points from annotated training examples. In the context of object detection and image classification, boosted classifiers have been widely adopted and have been empirically shown to achieve excellent performance [14,18]. Here, we utilize the GentleBoost algorithm to build a so-called strong classifier by iteratively combining the outputs of weak learners. In our case, the weak learners are defined as regression stumps [11] built from individual probe-based gradient histograms. Following the idea of Laptev [12], at each boosting round weighted Linear Discriminant Analysis (wLDA) is conducted on the vectors formed by the bins of the orientation histograms for each probe position in the descriptor. The histogram-vectors are then projected onto the normal $\mathbf{w}$ of the discriminant and regression stumps are fitted to the resulting scalars.

After M rounds of boosting, the final classifier has the form of:

$$H = \sum_{m=1}^{M} a_m(\mathbf{w}_m^T \mathbf{p} > th_m) + b_m, \tag{1}$$

where a_m, b_m, th_m are the parameters of the best weak classifier, and $\mathbf{w}_m$ is returned by wLDA - all at round m. $\mathbf{p}$ is the C dimensional feature vector described in Sec. 3.

4.1 Training Procedure

In order to train the boosted grasp point detector, positive and negative examples of grasp points are extracted from training set annotations. Positive examples are obtained by scaling the annotated grasp regions in each image to the canonical scale and computing the descriptor at the central point of the grasping region.

Given the high dimensionality of the data and the relatively low number of positive examples compared to the enormous space of negative examples to sample from, we further augment the set of positive examples by employing jittering. Small amounts of noise are added by randomly re-scaling and rotating the image, and translating the grasp point position in small ranges [12].

To obtain the negative examples we use two different methods: a) For each training image we extract descriptors from positions chosen randomly from points not on the grasping region. b) For positions close to the grasping region but which do not constitute grasp points, the classifier is often not able to construct adequate discriminative models based on the randomly chosen negative examples, and thus returns false detections during testing. To counter this, we shape the classifier response by providing additionally negative examples positioned near the grasping region [18]. In particular, we use positions located on circles centered at the grasp points, with a radius 1.5 times of that of the grasping region.

5 Detection

Next, we describe how grasp points are found by a sliding window approach, often utilized in image-based object detection frameworks of which the famous Viola-Jones detector[21] is probably the most prominent example. Specifically, to find grasp points of different sizes, we scan images in a range of predefined scales $\{s^k\}, k = 1 \ldots K$. For an image at scale s^k, we proceed as follows:

1. Gradients are computed and their energies are distributed over C different channel images according to their orientation. Between adjacent orientation bins, i.e. at the same image position, linear interpolation is used to arrive at smooth estimates for the channels. The resulting maps are then smoothed by a Gaussian kernel to obtain blurred channel images $\mathbf{G}_o^{\sigma_p}$, see Sec. 3.
2. At each image position (x, y), the boosted classifier is evaluted on the descriptor values extracted by accessing the blurred orientation maps at the radially configured probe positions centered on it.

Repeating the above for each scale, we obtain the strong classifiers confidence $H(x, y, s^k)$ which we convert to the posterior probabilities of a grasp point presence at image position (x, y) and scale s^k using the logistic transform proposed in [10]:

$$P(grasp\_point_{(x,y,s^k)}) = \frac{1}{1 + e^{-H(x,y,s^k)}} \tag{2}$$

For a confidence map computed in such way, we refer the reader to Fig. 1. To find the set of grasp point detections, mean-shift mode estimation is adopted as described by Shotton et al. [18]. Mean-shift efficiently locates the local maxima of the underlying probability distribution, and delineates the associated basin of attraction thereby effectively supressing nearby weaker maxima. Location and scale of a detected grasp point are given by the respective mode, while the confidence in the detection is obtained from the probability density estimate at the mode's location.

Fig. 3. Example images from the dataset consisting of mugs, bottles, and Martini glasses

Fig. 4. Illustration of grasp point and object annotation. Bounding boxes (red) and grasp points (black-white dots).

6 Evaluation

6.1 Dataset

To emulate a challenging testing scenario, we compiled a dataset containing images of 3 object categories taken in realistic settings. The collection consists of 630 images, of which 210 were images of mugs, 210 of bottles, and 210 of Martini glasses. 30 of the mug images and 30 bottle images were taken from the database of Ferrari et al. [9], the remainder was found by a Google image search. The images exhibit viewpoint changes, considerable background clutter and in many cases more than one object instance and class are present. The number of annotated object instances totaled 720. Examples are depicted in Fig. 3.

Grasp points are represented by circular regions giving position and approximate scale of the relevant structure. Two grasp points were selected for each mug - one at the top of the handle and one in the middle. Martini glass grasp points are located at the upmost part of the shaft, bottles were annotated by the top of the neck. Overall, 956 grasp points have been annotated. In addition,

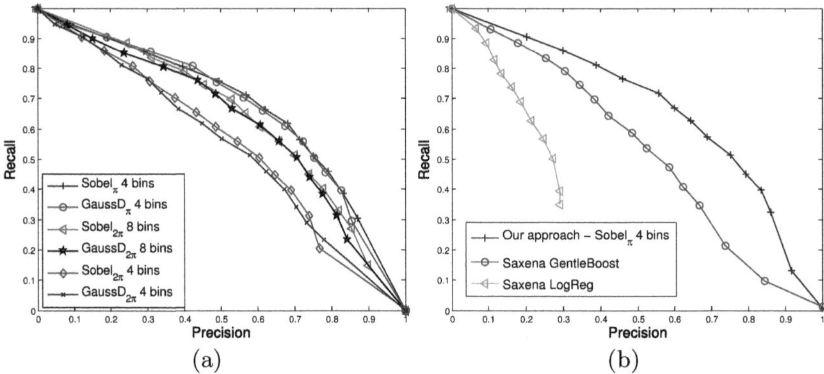

Fig. 5. Evaluation of grasp point detection performance. (a) Precision-recall curves for different variants of our approach. (b) Our algorithm (blue crosses) versus Saxena's [16] method (red circles and green triangles).

each object instance is provided with a bounding box and a label of the object category, designating the class of the associated grasp points. Note that the geometric information provided by the bounding box is not used in the current approach. In our future work however, we plan to integrate this as means of delineating the class specific image context around the grasping point. Fig. 4 shows examples of annotated object instances and grasp points.

6.2 Procedure

The dataset is split into two equally sized sets for training and testing. From the training set, grasp point models are learned using the position and size given by the annotation. During training, images are rescaled such that each grasp point attains a canonical radius of 7 pixels before extracting the descriptor.

Test images were not rescaled and grasp points exhibit a scale range of roughly 3× from smallest to largest. The detection procedure returns the positions, sizes, and confidences of grasp point presence at the respective locations. Given a minimum confidence threshold, resulting detections are regarded as correct if the circular region of the inferred grasp point r_{inf} agrees sufficiently with the ground truth grasping point r_{gt}. This is checked using the symmetric overlap criterion $\frac{Area(r_{gt} \cap r_{inf})}{Area(r_{gt} \cup r_{inf})}) > 0.25$ similar to [1].

In contrast to [4], we compare detection performance by means of precision-recall (PR) curves [8,13] rather than receiver-operating-characteristics (ROC) which have been designed for binary classification tasks. The fundamental problem is that the number of negatives used in ROC's false positive rate is not clearly defined for the detection task we are facing. See [2] for a more thorough discussion on this matter.

Table 1. PR-AUC values for our approach using different orientation quantisation and gradient operators. The subscripts π and 2π denote the polarity-ignoring and non-ignoring filter versions respectively.

	4 bins	8 bins
Derivatives	PR-AUC	PR-AUC
$GaussD_{\pi}$	0.6627	0.6164
$GaussD_{2\pi}$	0.5251	0.6112
$Sobel_{\pi}$	**0.6656**	0.6325
$Sobel_{2\pi}$	0.5441	0.6290

6.3 Results

In order to study the influence of histogram granularity and the particular choice of gradient computation, we compared Gaussian derivatives and the Sobel operator in two variants: Orientation estimation in the full 4-quadrant range, and ignoring the gradient direction by mapping its orientation in the range from 0 to π, i.e. bright to dark image transitions have the same orientation as dark to bright. Additionally, orientations were quantisized into $C = 4$ and $C = 8$ bin histograms (channel images). During all tests reported here, the remaining descriptor parameters (see Sec.3) were set to $\sigma_p = 5$ and $K = 5$, determined by cross-validation over the training set.

The results of these experiments are depicted in Fig. 5 (a), the corresponding area-under-curve values (PR-AUC) [18] are listed in Tab. 1. Note that we omitted plots of $Sobel_{\pi}$ and $GaussD_{\pi}$ for 8 bins to reduce clutter. One can see that the Sobel filter consistantly outperforms Gaussian derivatives and that ignoring gradient polarity has the edge over its counterpart. This is in accordance with [7]. Overall, the best PR-AUC of 0.6656 was obtained by the polarity-ignoring Sobel operator ($Sobel_{\pi}$) using orientation quantization into 4 channels. Fig. 6 shows some example detections taken from the test set.

In addition, we compared our method with the approach suggested in [16]. There, a descriptor based on Laws masks was used to encode texture over multiple scales. Since experiments revealed a poor performance (PR-AUC of 0.3460) of the proposed logistic regression algorithm, to have a fairer comparison we also present the improved results (PR-AUC of 0.5249) obtained using our GentleBoost-based learning framework. As can be seen from the precision-recall curves depicted in Fig. 5 (b), the proposed semi-local detector achieves significantly higher performance.

Finally, we tested our algorithm on images showing novel object classes, not contained in the training set, with semi-local structures similar to those learned during training. Thus the handles on the jar, though belonging to a quite dissimilar object than the mugs, were detected as they resemble the mug handles. The same effect can be seen in the case of scissors. Furthermore as can be observed in case of the flowers, the detector is able to detect similarities which are perhaps not immediately apparent - the similarity of a flower stem to a martini glass

Fig. 6. Detection examples showing successful detections (red) and false positives (blue)

Fig. 7. Grasp point detections for object classes not contained in the training set showing meaningful detections (red)

shaft. These examples illustrate that the descriptor is capable of capturing the shape similarity of image structures leading to meaningful detections of grasping regions.

7 Conclusions

We presented a method for detecting grasp points in monocular images of newly seen objects, based on learning grasp points from images containing three object classes and four distinct grasping types. Extensive tests have shown that our approach based on boosted histograms outperforms the state-of-the-art. We were able to demonstrate that the approach is capable of capturing grasping relevant information, achieving promising results on familiarly shaped objects from classes not contained in the training set.

Current work focuses on incorporating more monocular image cues into our descriptor in order to examine their influence on the detection rate. A more careful investigation of the blurring scale and extensions to automatically determine the best size of the the semi-local descriptor representation is also of interest.

References

1. http://www.pascal-network.org/challenges/VOC
2. Agarwal, S., Awan, A., Roth, D.: Learning to detect objects in images via a sparse, part-based representation. IEEE Transactions on Pattern Analysis and Machine Intelligence 26(11), 1475–1490 (2004)
3. Belongie, S., Malik, J., Puzicha, J.: Shape context: A new descriptor for shape matching and object recognition. In: NIPS, pp. 831–837 (2000)
4. Bhog, J., Kragic, D.: Learning grasping points with shape context. Robotics and Autonomous Systems (2009) (in Press)
5. Bohg, J., Kragic, D.: Grasping familiar objects using shape context. In: 14th Intl. Conference on Advanced Robotics, Munich, Germany (2009)
6. Carmichael, O., Hebert, M.: Shape-based recognition of wiry objects. IEEE Trans. on Pattern Analysis and Machine Intelligence 26(12) (2004)
7. Dalal, N., Triggs, B.: Histograms of oriented gradients for human detection. In: CVPR, pp. 886–893 (2005)
8. Davis, J., Goadrich, M.: The relationship between precision-recall and roc curves. In: ICML 2006: Proceedings of the 23rd International Conference on Machine Learning, pp. 233–240. ACM, New York (2006)
9. Ferrari, V., Jurie, F., Schmid, C.: From images to shape models for object detection. International Journal of Computer Vision (2009)
10. Friedman, J., Hastie, T., Tibshirani, R.: Additive logistic regression: a statistical view of boosting. Annals of Statistics 28, 2000 (1998)
11. Kevin, A.T., Murphy, K.P., Freeman, W.T.: Sharing features: Efficient boosting procedures for multiclass object detection. In: CVPR, pp. 762–769 (2004)
12. Laptev, I.: Improving object detection with boosted histograms. Image and Vision Computing (27), 535–544 (2009)
13. Murphy, K.P., Torralba, A.B., Eaton, D., Freeman, W.T.: Object detection and localization using local and global features. In: Toward Category-Level Object Recognition, pp. 382–400 (2006)
14. Opelt, A., Pinz, A., Fussenegger, M., Auer, P.: Generic object recognition with boosting. T-PAMI 28(3), 416–431 (2006)
15. Saxena, A.: Stanford synthetic object grasping point data, http://ai.stanford.edu/~asaxena/learninggrasp/data.html
16. Saxena, A., Driemeyer, J., Kearns, J., Ng, A.Y.: Robotic grasping of novel objects. In: Schölkopf, B., Platt, J., Hoffman, T. (eds.) Advances in Neural Information Processing Systems, vol. 19, pp. 1209–1216. MIT Press, Cambridge (2007)
17. Saxena, A., Driemeyer, J., Ng, A.Y.: Robotic grasping of novel objects using vision. The Intl. Journal of Robotics Research 27(2), 157–173 (2008)
18. Shotton, J., Blake, A., Cipolla, R.: Multiscale categorical object recognition using contour fragments. T-PAMI 30(7), 1270–1281 (2008)
19. Thayananthan, A., Stenger, B., Torr, P.H.S., Cipolla, R.: Shape context and chamfer matching in cluttered scenes. In: IEEE Conference on Computer Vision and Pattern Recognition, pp. 127–133 (2003)
20. Tola, E., Lepetit, V., Fua, P.: Daisy: An efficient dense descriptor applied to wide baseline stereo. IEEE T-PAMI 99(1) (2009)
21. Viola, P., Jones, M.: Robust real-time object detection. International Journal of Computer Vision (2001)

Efficient Methods for Point Matching with Known Camera Orientation*

João F.C. Mota and Pedro M.Q. Aguiar

Institute for Systems and Robotics / IST, Lisboa, Portugal
{jmota,aguiar}@isr.ist.utl.pt

Abstract. The vast majority of methods that successfully recover 3D structure from 2D images hinge on a preliminary identification of corresponding feature points. When the images capture close views, *e.g.*, in a video sequence, corresponding points can be found by using local pattern matching methods. However, to better constrain the 3D inference problem, the views must be far apart, leading to challenging point matching problems. In the recent past, researchers have then dealt with the combinatorial explosion that arises when searching among $N!$ possible ways of matching N points. In this paper we overcome this search by making use of prior knowledge that is available in many situations: the orientation of the camera. This knowledge enables us to derive $\mathcal{O}(N^2)$ algorithms to compute point correspondences. We prove that our approach computes the correct solution when dealing with noiseless data and derive an heuristic that results robust to the measurement noise and the uncertainty in prior knowledge. Although we model the camera using orthography, our experiments illustrate that our method is able to deal with violations, including the perspective effects of general real images.

1 Introduction

Methods that infer three-dimensional (3D) information about the world from two-dimensional (2D) projections, available as ordinary images, find applications in several fields, *e.g.*, digital video, virtual reality, and robotics, motivating the attention of the image analysis community. Using single image brightness cues, such as shading and defocus, researchers have proposed methods that work in highly controlled environments, like laboratories, but result sensitive to the noise and are unable to deal with more general scenarios. Consequently, the effort of the past decades was mainly on the exploitation of a much stronger cue: the motion of the brightness pattern between images. In fact, the image projections of objects at different depths move differently, unambiguously capturing the 3D

* J. Mota is also affiliated with the Dep. of Electrical and Computer Engineering, Carnegie Mellon University, Pittsburgh PA, USA. This work was partially supported by Fundação para a Ciência e Tecnologia, under ISR/IST plurianual funding (POSC program, FEDER), grant MODI-PTDC/EEA-ACR/72201/2006, and grant SFRH/BD/33520/2008 (CMU-Portugal program, ICTI).

A. Campilho and M. Kamel (Eds.): ICIAR 2010, Part I, LNCS 6111, pp. 210–219, 2010.

shape of the scene. This lead to the so-called 3D Structure-from-Motion (SfM) methods.

SfM splits the problem into two separate steps: **i)** 2D motion estimation, from the images; **ii)** inference of 3D structure (3D motion of the camera and 3D shape of the scene), from 2D motion. Usually, the 3D shape of the scene is represented in a sparse way, by a set of pointwise features, thus the 2D motion is represented by the corresponding set of trajectories of image point projections. When dealing with video sequences, consecutive images correspond to close views, and those trajectories can be obtained through tracking, *i.e.*, by using *local* motion estimation techniques. However, since very distinct viewpoints are required to better constrain the 3D inference problem, in many situations there is the need to process a single pair of distant views. In this scenario, the 2D motion estimation step **i)**, *i.e.*, the problem of matching pointwise features across views, becomes very hard and, in fact, the bottleneck of SfM (step **ii)** has been extensively studied and efficient methods are available [1]).

Researchers have then addressed the problem of computing point correspondences in a *global* way, by incorporating the knowledge that the feature points belong to a 3D rigid object. However, the space of correspondences to search grows extremely fast: considering N feature points, there exist $N!$ ways to match them. Due to this combinatorial explosion, only sub-optimal methods have been proposed to solve the problem, see, *e.g.*, [2], for an iterative approach that strongly depends on the initialization. Curiously, in the simpler scenario of dealing with noisy observations of *geometrically equal* point clouds, the optimal solution can be efficiently obtained as the solution of a convex problem [3]. The challenge in SfM is that the point clouds from which we must infer the correspondences have *distinct shape* because they are different 2D projections of the (unknown) 3D shape.

In this paper, we overcome the difficulty pointed out in the previous paragraph by using as prior knowledge the orientation of the camera. In fact, in many situations, that knowledge is available from camera calibration or can be computed without using feature points and their correspondences. For example, in scenarios where many edges are aligned with three orthogonal directions, *e.g.*, indoor or outdoor urban scenes, the orientation of the camera can be reliably obtained from the vanishing lines of a single image, see, *e.g.*, [1], or even directly from the statistics of the image intensities [4]. We show how the knowledge of camera orientation simplifies the problem, enabling us to derive an algorithm of complexity $\mathcal{O}(N^2)$. We prove that this algorithm computes the optimal set of correspondences for the orthographic camera projection model in a noiseless scenario and propose a modified version that results robust to uncertain measurements and violations of orthography.

2 Problem Formulation

Consider the scenario of Fig. 1, where two cameras C_1 and C_2 (or, equivalently, the same camera in two different positions) capture two different views of the world. As usual when recovering SfM, we assume that a set of N feature points was extracted from each of the images, and their coordinates in the image plane are represented by

$$I_1 := \begin{bmatrix} x_1^{(1)} & x_2^{(1)} & \cdots & x_N^{(1)} \\ y_1^{(1)} & y_2^{(1)} & \cdots & y_N^{(1)} \end{bmatrix}, \quad I_2 := \begin{bmatrix} x_1^{(2)} & x_2^{(2)} & \cdots & x_N^{(2)} \\ y_1^{(2)} & y_2^{(2)} & \cdots & y_N^{(2)} \end{bmatrix}, \tag{1}$$

where the superscript (i) indexes the points to C_i, for $i = 1, 2$. Each feature point has 3D coordinates (X_n, Y_n, Z_n), with respect to some fixed coordinate frame. Let that frame be attached to C_1 such that: **1)** the axes X and Y are parallel to the axes x and y of the camera frame; **2)** the optical center of the camera C_1 is aligned with the axis Z (see Fig. 1). The major challenge when attempting to recover $\{(X_n, Y_n, Z_n), n = 1, \ldots, N\}$ from I_1 and I_2 is the correspondence problem. In fact, we do not know the pairwise correspondences between the columns of I_1 and I_2 in (1) because there is not a "natural" way to automatically order the feature point projections. Although estimating this ordering leads to a combinatorial problem whose solution, in general, becomes a quagmire for large N, we show in this paper that, when the relative orientation of the cameras is known and the perspective projection is well approximated by the orthographic projection model, an efficient solution can be found.

Consider the orthographic model of a camera [1]: $\mathbf{x} = P\mathbf{X}$, where $\mathbf{X} \in \mathbb{P}^3$ and $\mathbf{x} \in \mathbb{P}^2$ are, respectively, the homogeneous coordinates of the points in space and in the image plane. The matrix $P \in \mathbb{R}^{3\times 4}$ is given by

$$P = \begin{bmatrix} R & t \\ 0_3^T & 1 \end{bmatrix}, \tag{2}$$

where $R \in \mathbb{R}^{2\times 3}$ contains the first two rows of a 3D rotation matrix, $t \in \mathbb{R}^2$ is a translation vector and 0_3 is the zero vector in $\mathbb{R}^3$. With the choice of reference

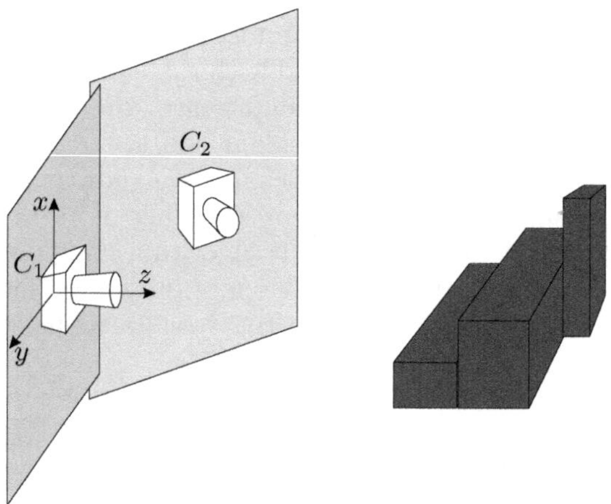

Fig. 1. Our scenario, with a choice for the reference frame

frame of the previous paragraph, it is straightforward to see that camera C_1 captures the first two coordinates of the feature points, *i.e.*, that $(x_n^{(1)}, y_n^{(1)}) = (X_n, Y_n), n = 1, \ldots, N$. Naturally, camera C_2 captures projections that depend on the relative position of the cameras, the 3D coordinates of the points, and their correspondences:

$$\begin{bmatrix} I_2 \\ 1_N^T \end{bmatrix} = \begin{bmatrix} R & t \\ 0_3^T & 1 \end{bmatrix} \begin{bmatrix} X_1 & X_2 & \cdots & X_N \\ Y_1 & Y_2 & \cdots & Y_N \\ Z_1 & Z_2 & \cdots & Z_N \\ 1 & 1 & \cdots & 1 \end{bmatrix} \Pi, \tag{3}$$

where $1_N \in \mathbb{R}^N$ has all its entries equal to 1, and $\Pi \in \mathbb{R}^{N \times N}$ is a permutation matrix, *i.e.*, a matrix with exactly one entry equal to 1 per row and per column and the remaining entries equal to 0 (when we multiply a matrix M by Π, we get a matrix with the same entries of M but with the columns arranged in a possibly different order).

By using (3), we obtain the model relating the projections of the feature points in images I_1 and I_2 with all the unknowns:

$$I_2 = \left[\hat{R} I_1 + \hat{r} Z^T + t 1_N^T \right] \Pi, \tag{4}$$

where $Z = [Z_1, Z_2, \ldots, Z_N]^T$ and R was decomposed as $R = [\hat{R}, \hat{r}]$, with $\hat{R} \in \mathbb{R}^{2\times 2}$ and $\hat{r} \in \mathbb{R}^{2\times 1}$. When the relative orientation of the cameras is known (which, as discussed in the previous section, occurs in several practical situations), *i.e.*, when $\hat{R}$ and $\hat{r}$ are known, the problem becomes to find a permutation matrix Π, a set of 3D point depths $\{Z_1, \ldots, Z_N\}$, and a translation vector t that solve (4). In general, the problem is hard due to the huge cardinality of the set of all $N \times N$ permutation matrices: $N!$.

3 Closed-Form Solution for Translation

The choice of the reference frame in the previous section leaves one degree of freedom: we can place the frame at any point along the axis Z. We now choose this position in such a way that the problem is simplified: let it be such that $\sum_{n=1}^{N} Z_n = 1_N^T Z = 0$, *i.e.*, that the plane XY contains the center of mass of the feature points.

Multiplying both sides of (4) by 1_N and simplifying, we get

$$I_2 1_N = \left[\hat{R} I_1 + \hat{r} Z^T + t 1_N^T \right] 1_N \tag{5}$$

$$= \hat{R} I_1 1_N + N t. \tag{6}$$

where (5) uses the fact that $\Pi\, 1_N = 1_N$ (permutation of a vector with all equal entries) and (6) uses equalities $Z^T 1_N = 0$ (from the choice of reference frame) and $1_N^T 1_N = N$. From (6), we see that the solution for the translation vector t does not depend on the remaining unknowns (Π, Z):

$$t = \frac{1}{N}\left(I_2 - \hat{R}I_1\right) 1_N. \tag{7}$$

By removing the (now known) translation from the problem, *i.e.*, by replacing the solution (7) in (4) (and using $1_N^T \Pi = 1_N^T$), we get

$$I_2 = \left[\hat{R}I_1 + \hat{r}Z^T\right] \Pi + \frac{1}{N}\left(I_2 - \hat{R}I_1\right) 1_N 1_N^T. \tag{8}$$

To simplify notation, we re-define our observations by introducing matrices $\tilde{I}_1$ and $\tilde{I}_2$, both computed from known data:

$$\tilde{I}_2 := I_2 - \frac{1}{N}\left(I_2 - \hat{R}I_1\right) 1_N 1_N^T, \qquad \tilde{I}_1 := \hat{R}I_1. \tag{9}$$

With these definitions, problem (4) is re-written as

$$\tilde{I}_2 = \left[\tilde{I}_1 + \hat{r}Z^T\right] \Pi, \tag{10}$$

where the unknowns are the depths $Z_1, \ldots, Z_N$, in Z, and the correspondences, coded by Π.

4 Optimal Solution for Noiseless Data

We first present an efficient algorithm to compute the solution to our problem when there is no noise, meaning that there exists at least one pair (Z, Π) that solves (10).

Naturally, the solution for the permutation matrix Π is given by the association of each column of $\tilde{I}_1$ with a column of $\tilde{I}_2$, for the correct value of Z. Let column n of $\tilde{I}_1$ (resp. $\tilde{I}_2$) be represented by $[\tilde{X}_n, \tilde{Y}_n]^T$ (resp. $[\tilde{x}_n, \tilde{y}_n]^T$) and consider the error E_{ij} of associating column j of $\tilde{I}_1$ with column i of $\tilde{I}_2$, *i.e.*,

$$E_{ij} = \min_{Z_j} \left[\tilde{x}_i - \tilde{X}_j - \hat{r}_1 Z_j\right]^2 + \left[\tilde{y}_i - \tilde{Y}_j - \hat{r}_2 Z_j\right]^2, \tag{11}$$

where $\hat{r} = [\hat{r}_1, \hat{r}_2]^T$. The minimizer Z_j^* solving (11) is straightforwardly obtained in closed-form:

$$Z_j^* = \frac{\hat{r}_1(\tilde{x}_i - \tilde{X}_j) + \hat{r}_2(\tilde{y}_i - \tilde{Y}_j)}{\|\hat{r}\|^2}. \tag{12}$$

Our algorithm, detailed and analyzed in the sequel, computes for each column i of $\tilde{I}_2$, the column j^* of $\tilde{I}_1$ that minimizes error E_{ij} (11) with respect to j (without noise, for each i there exists at least one j^* such that $E_{ij^*} = 0$). In the algorithm description below, the $N \times N$ permutation matrix Π is simply parameterized by a $N \times 1$ vector perm: the jth column of Π has entry perm_j equal to 1 (and, obviously, the others equal to zero); also, $|S|$ denotes the cardinality of set S and $S_1 \backslash S_2$ the set of elements of S_1 that do not belong to S_2.

Algorithm 1

Inputs *Matrices* $\tilde{I}_1$ *and* $\tilde{I}_2$*, organized into the corresponding sets of columns* $\mathcal{B}_1 = \{[\tilde{X}_1, \tilde{Y}_1]^T, \dots, [\tilde{X}_N, \tilde{Y}_N]^T\}$ *and* $\mathcal{A} = \{[\tilde{x}_1, \tilde{y}_1]^T, \dots, [\tilde{x}_N, \tilde{y}_N]^T\}$*, and vector* $\hat{r}$*.*

Procedure *For* $i = 1, \dots, N$ $(N = |\mathcal{A}|)$

- *For all* $j = 1, \dots, |\mathcal{B}_i|$*, compute* Z_j^* (12) *and* E_{ij} (11)*;*
- $j^* = \arg\min_j E_{ij}$*;*
- $\text{perm}_{j^*} = i$*,* $Z_{j^*} = Z_{j^*}^*$*;*
- $\mathcal{B}_{i+1} = \mathcal{B}_i \backslash [\tilde{X}_{j^*}, \tilde{Y}_{j^*}]^T$*.*

Outputs *Vectors* perm *and* Z*.*

Algorithm 1 consists of N loops where, in each loop, a column of $\tilde{I}_2$ is assigned to a column of $\tilde{I}_1$. Each assignment requires a search over, at most, N possibilities. It is then clear that our algorithm has complexity of $\mathcal{O}(N^2)$, in particular, we obtain the total number of floating point operations (flops) as $7N^2 + 7N - 14$. Before proving optimality of Algorithm 1, we interpret it in a geometric way. Defining each possible "displacement" $\tilde{I}_1 \rightarrow \tilde{I}_2$ as $a_{ij} := [\tilde{x}_i - \tilde{X}_j, \tilde{y}_i - \tilde{Y}_j]^T$, the cost minimized in (11) can be written as $\|a_{ij} - Z_j \hat{r}\|^2$. So, for each column $[\tilde{x}_i, \tilde{y}_i]^T$ of $\tilde{I}_2$, our algorithm searches the column $[\tilde{X}_j, \tilde{Y}_j]^T$ of $\tilde{I}_1$ that minimizes $\|a_{ij} - Z_j \hat{r}\|^2$ for all possible values of Z_j. Since this expression achieves its minimum (zero) when a_{ij} is collinear with $\hat{r}$ (which we synthetically denote by $a_{ij}//\hat{r}$), Algorithm 1 assigns pairs of columns such that their difference is "as parallel as possible" to $\hat{r}$. This collinearity is a re-statement of the fact that epipolar lines are parallel in an orthographic stereo pair [1] (more generally, the trajectories of image projections of a rigid scene can be represented in a rank 1 matrix [5]).

Theorem 1 (Optimality of Algorithm 1). *If there exists at least one pair* (Z, Π)*, such that* (10) *holds, then the outputs of Algorithm 1 determine a pair* $(\bar{Z}, \bar{\Pi})$ *that solves* (10)*.*

Proof. Suppose the pair (Z^*, Π^*) is such that (10) holds. For each $i = 1, \dots, N$, there exists one and only one k such that

$$\Pi_{ki}^* = 1 \tag{13}$$

(because Π^* is a permutation matrix). We now denote by $j^*(i)$ the assignment produced by Algorithm 1, *i.e.*, we make explicit the dependence of j^* on i. Obviously, if $j^*(i) = k$ for all $i = 1, \dots, N$, then the algorithm returned an optimal solution. So, for the remaining of the proof, we assume there is an index i such that $j^*(i) \neq k$. We will see that, even in this case, (10) holds for the solution provided by the algorithm, because $E_{ij^*(i)} = 0$, for all i.

A simple way to complete the proof is using contradiction. Assume i is the smallest index such that $E_{ij^*(i)} > 0$ (obviously $j^*(i) \neq k$). If $E_{ij^*(i)} > 0$, then $[\tilde{X}_k, \tilde{Y}_k]^T \notin \mathcal{B}_i$ (at the ith loop). Thus, there exists an index l $(1 \leq l < i)$ such that $[\tilde{x}_l, \tilde{y}_l]^T // [\tilde{X}_k, \tilde{Y}_k]^T$ (because $E_{lj^*(l)} = 0$ for all $1 \leq l < i$). According to the assignment defined by (13), we have $[\tilde{X}_k, \tilde{Y}_k]^T // [\tilde{x}_i, \tilde{y}_i]^T$, thus $[\tilde{x}_l, \tilde{y}_l]^T // [\tilde{x}_i, \tilde{y}_i]^T$.

Also, since Π^* is a permutation matrix, there exists an index m ($1 \leq m \leq N$), such that $\Pi^*_{ml} = 1$, or, equivalently, such that $[\tilde{X}_m, \tilde{Y}_m]^T // [\tilde{x}_l, \tilde{y}_l]^T$, thus, $[\tilde{X}_m, \tilde{Y}_m]^T // [\tilde{x}_i, \tilde{y}_i]^T$. We now consider two cases: **1)** if $[\tilde{X}_m, \tilde{Y}_m]^T \in \mathcal{B}_i$, there is a contradiction because $E_{im} = 0$; **2)** if $[\tilde{X}_m, \tilde{Y}_m]^T \notin \mathcal{B}_i$, it is straightforward to find a vector $[\tilde{X}_{m'}, \tilde{Y}_{m'}]^T \in \mathcal{B}_i$ such that $[\tilde{X}_{m'}, \tilde{Y}_{m'}]^T // [\tilde{x}_i, \tilde{y}_i]^T$, by performing steps like the ones above, which brings us back to case **1)**.

5 Approximate Solution for Noisy Data

In practice, not only the knowledge of the camera orientation is uncertain but also the feature point projections are noisy. Since Algorithm 1 is based on the collinearity of a vector that depends on the camera orientation ($\hat{r}$) with vectors that depend on the feature point projections ($[\tilde{x}_i - \tilde{X}_j, \tilde{y}_i - \tilde{Y}_j]^T$), its behavior is sensitive to disturbances affecting these vectors. We now propose a modification of this algorithm, which results robust not only to the noise but also to violations of the orthographic projection model.

From model (10) we note that the clouds of points in $\tilde{I}_1$ and $\tilde{I}_2$ differ by $\hat{r}Z^T$. Since $\hat{r}$ contains entries of a rotation matrix, thus with magnitude smaller than 1, in practice, the patterns of points in $\tilde{I}_1$ and $\tilde{I}_2$ will almost coincide when the depth of the scene is not too large (more rigorously, when $\hat{r}Z^T$ is negligible if compared to the minimum distance between points), even if the corresponding points in I_1 and I_2 are very distant (see an insightful example in Fig. 4). This motivated us to use the matching criterion of minimizing the Euclidean distance between points in $\tilde{I}_1$ and $\tilde{I}_2$,

$$E'_{ij} = \left\| \begin{bmatrix} \tilde{x}_i \\ \tilde{y}_i \end{bmatrix} - \begin{bmatrix} \tilde{X}_j \\ \tilde{Y}_j \end{bmatrix} \right\|^2, \tag{14}$$

rather than the less robust collinearity implicit in (11).

Algorithm 2

Inputs *Matrices* $\tilde{I}_1$ *and* $\tilde{I}_2$*, organized into the corresponding sets of columns* $\mathcal{B}_1 = \{[\tilde{X}_1, \tilde{Y}_1]^T, \ldots, [\tilde{X}_N, \tilde{Y}_N]^T\}$ *and* $\mathcal{A} = \{[\tilde{x}_1, \tilde{y}_1]^T, \ldots, [\tilde{x}_N, \tilde{y}_N]^T\}$*, and vector* $\hat{r}$*.*

Procedure *For* $i = 1, \ldots, N$ ($N = |\mathcal{A}|$)
- *For all* $j = 1, \ldots, |\mathcal{B}_i|$*, compute* E'_{ij} (14);
- $j^* = \arg\min_j E'_{ij}$;
- $\text{perm}_{j^*} = i$, $Z_{j^*} = Z^*_{j^*}$ (12);
- $\mathcal{B}_{i+1} = \mathcal{B}_i \backslash [\tilde{X}_{j^*}, \tilde{Y}_{j^*}]^T$.

Outputs *Vectors* perm *and* Z.

Our experiments, some of them singled out in the following section, demonstrate that Algorithm 2 successfully infers correct feature point correspondences when dealing with real images. In spite of correctly determining correspondences, the accuracy of the depth estimates in Z strongly depends on the magnitude of the components of $\hat{r}$. In fact, assuming the correspondences are known, for example, $\Pi = I_{N\times N}$ (for simplicity), model (10) becomes $\tilde{I}_2 - \tilde{I}_1 = \hat{r}Z^T$, making clear that the accuracy in the estimation of Z depends not only on the accuracy of the measurements $(\tilde{I}_1, \tilde{I}_2, \hat{r})$ but also on the magnitude of the components of $\hat{r}$. In particular, we obtain an upper-bound for the depth estimation error as $\rho_Z = \max|\tilde{I}_2 - \tilde{I}_1| / \min|\hat{r}|$. Naturally, when the ratio ρ_Z is large, we can still use our algorithm to estimate the correspondences between the feature points (the bottleneck of the problem), whose accuracy is not affected by ρ_Z, and then use a standard algorithm to recover SfM, eventually using a larger set of images to reduce ambiguity, see, *e.g.*, [1].

6 Experiments

To test the algorithms with ground truth, we synthesized data. In particular, we generated the 3D world as a set of 50 points randomly distributed in $[-200, 200]^3$ and relative orientations between the cameras by specifying random rotation matrices. Then, we synthesized measurements according to the model in expression (3), for random permutation matrices. As expected, according to our theoretical derivation of Section 4, Algorithm 1 always produced the correct result: it successfully recovered the permutation, *i.e.*, the correct correspondences between the points, and their depth. To test robustness to disturbances, we then ran experiments by considering inaccurate knowledge of camera orientation and noisy feature point projections. As anticipated in Section 5, we observed that Algorithm 2 results more robust than Algorithm 1. The plot in Fig. 2 illustrates this point by showing the average number of wrong correspondences as functions of the (white Gaussian) measurement noise standard deviation (st.dv.). Note that, even for noise st.dv. of 5 pixels, Algorithm 2 almost always recovers totally correct correspondences. In what respects to depth estimation accuracy, the magnitudes of the errors were smaller than the magnitudes of the measurement noise.

We tested our algorithms with real images. Two examples are shown in Fig. 3, which contains the two pairs of images with feature points superimposed. Note that, in both examples, corresponding features are far from being close to each other, preventing thus the usage of "local" methods. We used standard calibration techniques to compute camera orientation [6] and then run our algorithms. The plots in Fig. 4 provide insight over our approach: while the feature point projections of corresponding features in I_1 and I_2 are in general far apart, their "versions" in $\tilde{I}_1$ and $\tilde{I}_2$ are close. As a consequence, Algorithm 2 recovered the correct correspondences in both cases. We emphasize that these examples strongly depart from the assumed orthographic projection (see the perspective effects between the pairs of images in Fig. 3), thus, that our approach is able to deal with a wide range of real life scenarios.

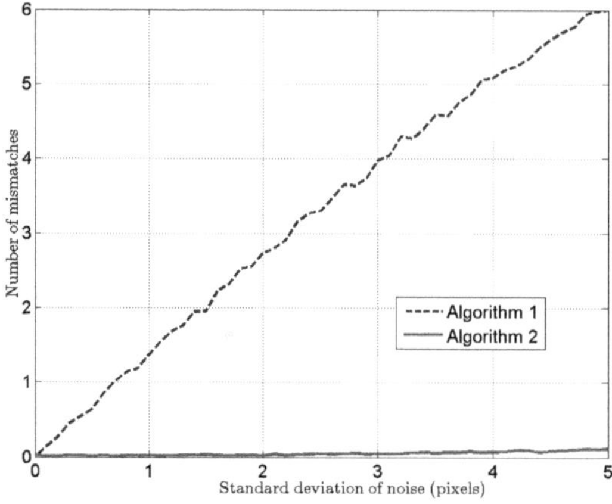

Fig. 2. Number of incorrect correspondences, for a 3D world of 50 points, as functions of the noise power (mean over 1000 runs)

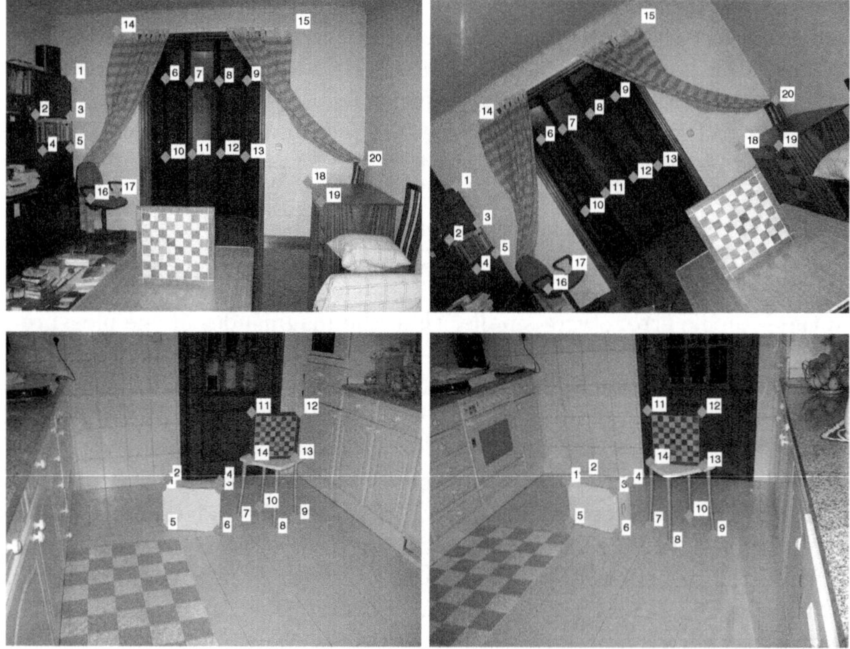

Fig. 3. Two pairs of real images with feature points superimposed

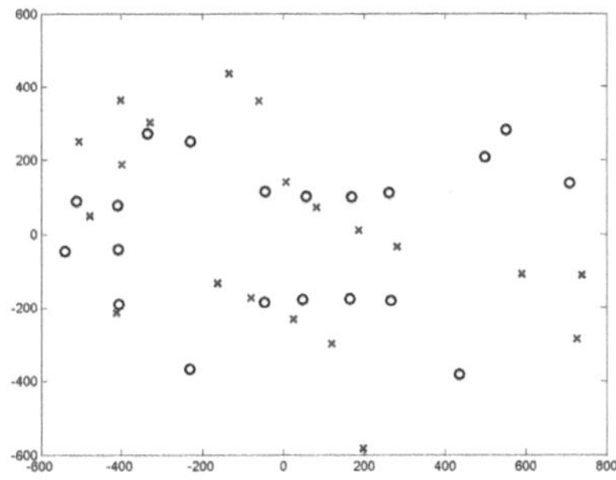
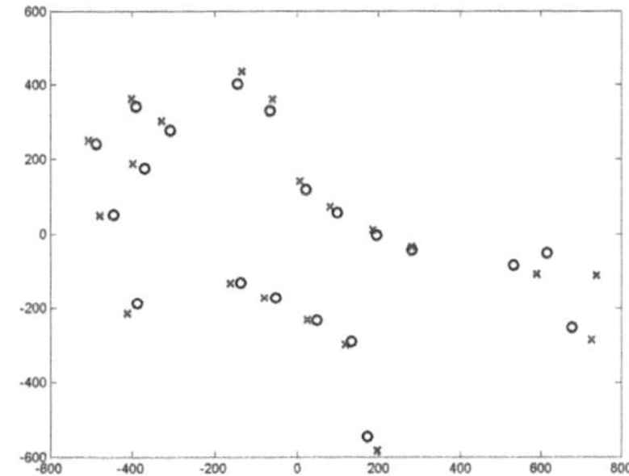

Fig. 4. Left: feature point coordinates in I_1 and I_2, extracted from the pair of images in the top of Fig. 3 (the blue circles are from the left image and the red crosses from the right one). Right: corresponding entries of $\tilde{I}_1$ and $\tilde{I}_2$, computed from known data, see (9).

7 Conclusion

We proposed efficient algorithms for finding simultaneously the correspondences between points in two images and their depth in the 3D world. Our approach is based on the facts that, in many situations, the relative orientation of the cameras is available, or can be easily inferred, and the camera model can be approximated by an orthographic projection. The resulting complexity is $\mathcal{O}(N^2)$, where N is the number of feature points (compare with $N!$, the number of possible correspondences). We prove the optimality of a first algorithm when dealing with noiseless data and develop a modified version that results more robust to uncertainty in the measurements.

References

1. Hartley, R., Zisserman, A.: Multiple View Geometry In Computer Vision. Cambridge University Press, Cambridge (2003)
2. Dellaert, F., Seitz, S., Thorpe, C., Thrun, S.: Structure from motion without correspondence. In: IEEE Conf. on Computer Vision and Pattern Recognition, Hilton Head SC, USA (2000)
3. Shivaswamy, P., Jebara, T.: Permutation invariant SVMs. In: Int. Conf. on Machine Learning, Pittsburgh PA, USA (2006)
4. Martins, A., Aguiar, P., Figueiredo, M.: Orientation in Manhattan. IEEE Trans. on Pattern Analysis and Machine Intelligence 27(5) (2005)
5. Aguiar, P., Moura, J.: Rank 1 weighted factorization for 3D structure recovery: Algorithms and performance analysis. IEEE Trans. on Pattern Analysis and Machine Intelligence 25(9) (2003)
6. Bouguet, J.: Camera calibration toolbox for Matlab (2008), `http://www.vision.caltech.edu/bouguetj/calib_doc`

Real-Time Scale Invariant 3D Range Point Cloud Registration

Anuj Sehgal[1], Daniel Cernea[2], and Milena Makaveeva[3]

[1] Indian Underwater Robotics Society,
E-118 Nar Vihar Part-I, Sector 34,
201301 Noida, India
anuj@iurs.org
[2] University of Kaiserslautern, Department of Computer Science,
67653 Kaiserslautern, Germany
cernea@informatik.uni-kl.de
[3] Jacobs University Bremen, Computer Science, Campus Ring 1,
28759 Bremen, Germany
mmakaveeva@jacobs-university.de

Abstract. Stereo cameras, laser rangers and other time-of-flight ranging devices are utilized with increasing frequency as they can provide information in the 3D plane. The ability to perform real-time registration of the 3D point clouds obtained from these sensors is important in many applications. However, the tasks of locating accurate and dependable correspondences between point clouds and registration can be quite slow. Furthermore, any algorithm must be robust against artifacts in 3D range data as sensor motion, reflection and refraction are commonplace. The SIFT feature detector is a robust algorithm used to locate features, but cannot be extended directly to the 3D range point clouds since it requires dense pixel information, whereas the range voxels are sparsely distributed. This paper proposes an approach which enables SIFT application to locate scale and rotation invariant features in 3D point clouds. The algorithm then utilizes the known point correspondence registration algorithm in order to achieve real-time registration of 3D point clouds.

1 Introduction

Due to the relative inexpensiveness and multiple benefits available from representing the viewed environment in 3D images, sensors and stereo cameras that are able to provide 3D point cloud data are becoming increasingly popular. 3D point cloud data representation is extremely important in various fields such as archaeology, geology, oceanography and lately even in robotics, where 3D point clouds are increasingly utilized for mapping and localization of robots in a 3D environment.

However, in all these application domains it is rare to obtain a single representation of the data [1] and as such multiple frames of point clouds have to be obtained and registered with respect to each other in order to construct a

A. Campilho and M. Kamel (Eds.): ICIAR 2010, Part I, LNCS 6111, pp. 220–229, 2010.

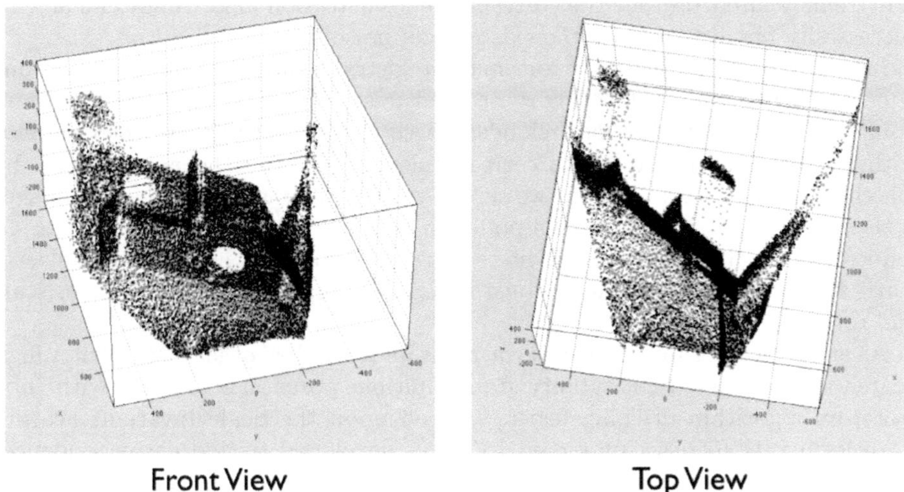

Fig. 1. Point cloud obtained from an IR-Ranger

composite map or scene. This composite data can be then further utilized for localization, analysis or visualization purposes.

Full automation of the registration process of range image 3D point clouds is a topic of active research and most systems still rely upon user input in order to determine the initial transformation. Additionally, the algorithms are highly processor intensive [2] making real-time registration of these point clouds a non-trivial effort. Furthermore, range point clouds provide another challenge as compared to intensity images in the form of noise that may be present within the returned data, causing false artifacts to appear in the point clouds or making the point cloud too sparse, with not enough usable information within it [3]. For example, range image point cloud data obtained from IR rangers can be highly noisy because of spurious readings resulting from ambient light and also as a result of the surface property of the target object; if the object is dark the range data would be erroneous since infra-red light is absorbed by darker colors. Though it is possible that this erroneous data may provide matchable features in some applications, in all our tests there were no matches located within such areas. Furthermore, the 3D shape formation errors induced by this data can lead to the registered point clouds appearing highly deformed.

A rendering of one such frame of a point cloud is provided in Figure 1. The noisy nature of this data is clearly visible in this representation. The point cloud frame consists of a number of boxes stacked upon each other; the boxes have a large circular black section painted in the middle. In the front view, this section appears as a white empty space, but if a top view is obtained it becomes clear that this feature is still present, but displays much further behind the box. This erroneous result is introduced because the black color absorbs the IR ranging beacon and as such the farthest possible distance is assumed for that region. It is extremely important for the correspondence detection algorithm to be able to successfully function despite the existence of noise [4].

In order to enable real-time automatic registration of these point clouds, our approach depends upon locating robust features, invariant to scale, rotation and point of view within the point clouds. These robust features can then be used with a high certainty to locate correspondences between the point clouds, by matching these features within two sets of voxels. Registration of the two images is then carried out using a known point correspondences algorithm. However, to reduce the effect of the possible noise in range data, it is necessary to select a feature descriptor that is very robust and, more importantly, invariant to scale and rotation changes.

As such, to meet our goals of locating a high number of features with a high degree of certainty, repeatability from multiple poses and in data with high noise, an algorithm utilizing features based upon the Scale Invariant Feature Transform (SIFT) descriptor model [5] was developed to find correspondences between the point clouds. The SIFT features are highly robust in that they are orientation invariant and are applicable at multiple scales. The SIFT feature detector algorithm is able to generate a large number of localized features with a relatively low computational cost. The detected SIFT features in the point clouds can be matched with a high degree of certainty and repeatability, from multiple poses and without respect to scale.

The following sections of the paper present information related to the SIFT algorithm and then proceed to describe the approach used to find SIFT features in 3D point clouds. The correspondence detection algorithm and the registration method used are also discussed. Some results, test and performance data obtained using the approach are presented and discussed along with the conclusions that are drawn from the results.

2 The SIFT Feature Detector

Robust detection of features in a scene are necessary in order to find correspondences within a point cloud so as to expedite the registration process, which normally can be computationally expensive. The features provided by the SIFT algorithm are local and invariant to image scale and rotation, thereby making them quite robust. These features are also robust in response to changes in illumination and minor changes in viewpoint while, being highly distinctive [6]. The SIFT algorithm is implemented in four stages that provide a result in form of multiple feature descriptors that are represented as a 128-element vector to achieve scale and rotational invariance.

The first stage of the algorithm is where all possible points of interest, known as key-points, are detected. In order to achieve this, the input data is successively convolved with Gaussian filters at different scales, and then the difference of successive Gaussian-blurred images are taken. The local extremum points that exist within the Difference of Gaussians (DoG), an approximation to the Laplacian, at multiple scales are then accepted as the key points. Once the initial set of candidate key points is obtained from the DoG images, they are analyzed within their own neighborhood and adjacent scales, to determine whether they are a local maxima or minima. Furthermore, the second step discards the key-point coordinates that are located in noisy space. This is achieved by eliminating candidates that lie in a region of low contrast or on the edges.

The third step achieves invariance to rotation by assigning each key-point one or more orientations. To compute the orientation of a point in a scale-invariant manner, the Gaussian-smoothed image corresponding to the scale from which the key-point was originally derived is taken and an orientation and gradient magnitude assigned to it. Magnitude and direction calculations for the key-points are performed for every pixel in the neighborhood and an orientation histogram is generated with 36 bins, each bin covering 10 degrees. Once the histogram is fully populated the orientations with the highest peaks and those that are within 80% of the highest peaks are assigned to the key point.

The final step in the SIFT algorithm actually computes the descriptor vector that can be used to identify and further match each key point. This step is extremely similar to the orientation assignment method. The feature descriptor is computed as a set of orientation histograms on a pixel neighborhood of size 4 times 4. The histograms are relative to the key point orientation and the orientation data is derived from the image that corresponds to the key point's scale. The representations now contain 8 bins, each leading to the derivation of a SIFT feature vector that contains 128 elements. This vector may be used to perform image matching or pattern recognition.

3 The Registration Algorithm

The aim of the work presented is to be able to enable automatic real-time registration of the 3D point clouds. Currently, the most popular method for registration is the Iterative Closest Point (ICP) algorithm or some derivative of the same [1]. The ICP algorithm and most of its derivatives are computationally expensive, giving rise to the necessity of being able to perform registration based upon pre-located correspondences from a fewer set of points, in order to speed up the overall performance of the registration process. However, this approach requires that the pre-computed correspondences between the point clouds be calculated quickly, while also ensuring their accuracy between frames that could have changing rotation, translation and scaling. In order to achieve this goal a three-step algorithm that uses the SIFT feature descriptor to describe key points in point clouds is designed. The three steps of the algorithm (data preprocessing; SIFT descriptor generation and feature matching to locate correspondences; and registration of point clouds) are discussed in the following subsections.

3.1 Data Preprocessing

The SIFT feature detector is designed to function only with 2D datasets and as such, in order to extract SIFT features from the range data in the point cloud,

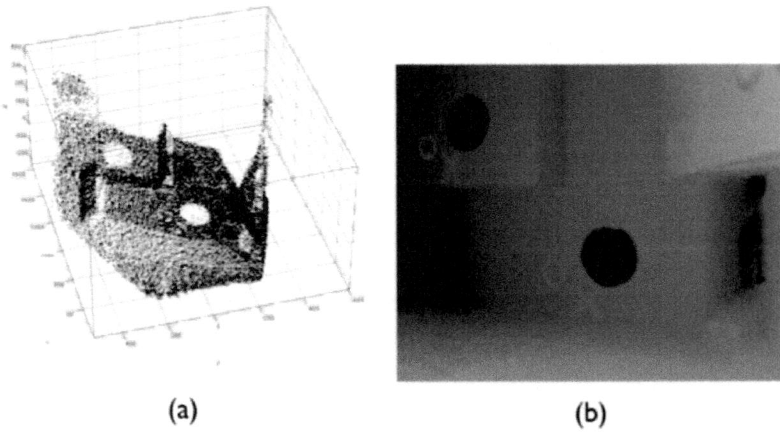

Fig. 2. Square-root scaled image of a point cloud; (a) The point cloud; (b) Square-root scaled range data

this information is square root scaled to fit between 0-255. The Euclidean distance to each individual (x, y, z) coordinates within the point cloud is calculated from the origin $(0, 0, 0)$ and scaled using square root scaling. An image representation of this range data square root scaling can be seen in Figure 2. This data was derived from a Swiss IR Ranger mounted on a mobile robot.

Upon performing the square-root scaling, the data is passed through a PNG converter in order to obtain images to which the SIFT operator is applied. The SIFT feature detector requires continuous points in the neighborhood of a pixel to function. Voxels in a 3D range point-cloud are not densely located, and as such the SIFT detector cannot be extended to 3D range point-clouds directly. This necessiates the square-root scaling step before the SIFT feature detector can be used.

3.2 SIFT Feature Detector and Matching

The SIFT feature detector is built using OpenCV [7] to follow closely the SIFT algorithm from [5,6]. The SIFT algorithm takes as input a PNG image corresponding to the square-root scaled representation of the point cloud and computes the 128-element vectors for every identified feature key point.

Upon obtaining the SIFT feature descriptors from the square root scaled images for the two point clouds to be registered, correspondences between the (x, y, z) coordinates in the point clouds is obtained by searching for matching SIFT descriptors, using the RANSAC algorithm [8]. The RANSAC algorithm selects a set of feature pairs randomly and computes the set of all feature pairs conforming to the implied transformation. A support set is rejected if it results in a size that is below a certain threshold.

Figure 3 shows matches found between the scaled point cloud images. The results in Figure 3 make it appear as though the the number of corresponding

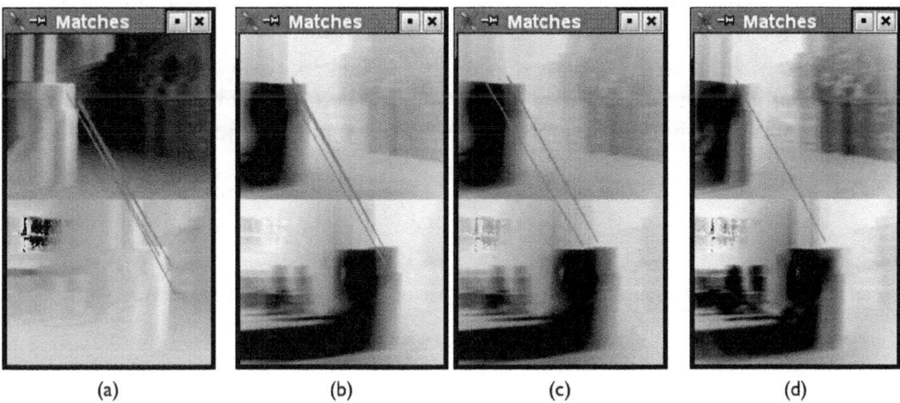

Fig. 3. SIFT descriptor based matches for two frames from a 3D point cloud data set. Each subfigure represents different types of point cloud data from the same frame (a) Range point cloud (b) Intensity point cloud (c) Range & Intensity combined point cloud (using intensity as another dimension in scaling) (d) Another Range and Intensity combined point cloud (obtained by multiplying intensity and distance before scaling).

matches is not very high. However, the data depicted in this figure is from a Swiss IR Ranger mounted on a robot that is moving swiftly, causing it to be blurry. This gives rise to a limited set of features to match, but our approach functions by successfully performing registration between images as long as at least three correspondences are located. Since a relatively high number of correspondences are located even in noisy and blurry data, the SIFT feature detector appears to function robustly on scaled range images as well.

Once the matches are found on the basis of the RANSAC algorithm, correspondences between the 3D point clouds are easily derivable since the corresponding location of each key point in the square-root scaled data is known within the point cloud as well. The set of resultant correspondences can now be further used with the chosen registration algorithm.

3.3 Point Cloud Registration

Registration is necessary in order to be able to compare or integrate the data from different measurements. This step provides the relative rotation, translation and scale of the two 3D point clouds being compared. The popular ICP algorithm is memory and processor intensive, thereby being unsuitable for real-time applications [9]. However, if a known points correspondence algorithm is used, this can considerably speed up the registration performance.

As such, for the purpose of speeding up the registration step and owing to the robustness of the SIFT features, the known points correspondence registration algorithm based on quaternions is utilized in our approach. Every corresponding point in the range point clouds can represent a quaternion with $c = 0$ and x, y

and z coordinates given by the respective coordinates of the point in the 3D point cloud.

By comparing data from two consecutive point clouds, we are able to retrieve the quaternion of the rotation matrix $\breve{u}^*$ from the eigenvector corresponding to the maximum positive eigenvalue of the 4x4 matrix N shown below:

$$N = \sum_{i=1}^{n} \bar{\Gamma}(\breve{r}_{l,i})^T \Gamma(\breve{r}_{r,i}) \tag{1}$$

Further, the rotation matrix can be obtained from the rotation quaternion by,

$$R = \bar{\Gamma}(\breve{u}^*)^T \Gamma(\breve{u}^*) \tag{2}$$

Having obtained the rotation matrix R, the translation quaternion and scale factor are calculated using,

$$r^*_{l/r} \equiv \bar{r}_r + sR_{r/l}(\bar{r}_l) \tag{3}$$

$$s^* = \frac{\sum_{i=1}^{n} (r'_{r,i})^T R_{r/l}(r'_{l,i})}{\sum_{i=1}^{n} \left\| r'_{l,i} \right\|^2} \tag{4}$$

The matrix calculations required for the registration step were performed using the GSL library [10] and several extensions were written in order to calculate the eigenvectors, eigenvalues, vector normalization and etc.

We also further extended our work in order to utilize registration to derive the roll, pitch and yaw between the consecutive frames. This can be extremely useful in robotics since it provides a method to derive odometry by using only range sensor data, rather than depending upon sensors like GPS, which may not function under certain conditions. After obtaining the rotation matrix, calculation of the respective roll, pitch and yaw is a straightforward task of selecting the appropriate row/column pairs from the rotation matrix. Having obtained translation and the yaw, pitch and roll the robot odometry is available. This can be used in localization and mapping tasks commonly performed by robots.

4 Testing and Results

The test dataset used to evaluate the overall algorithm was obtained from a Swiss IR Ranger mounted on a mobile robot. The dataset consists of 290 frames of point clouds. The (x, y, z) location within the point cloud corresponds to the measured distance in millimeters. This data is retrieved frame by frame and supplied to the software running on a Linux platform in order to register the two point clouds. The test system used was a SuSe Linux installation on a platform with 512MB RAM and a 1 GHz AMD Athlon64 CPU.

The robustness of the SIFT features and their ability to have more than a single orientation at a particular point can cause the RANSAC algorithm to

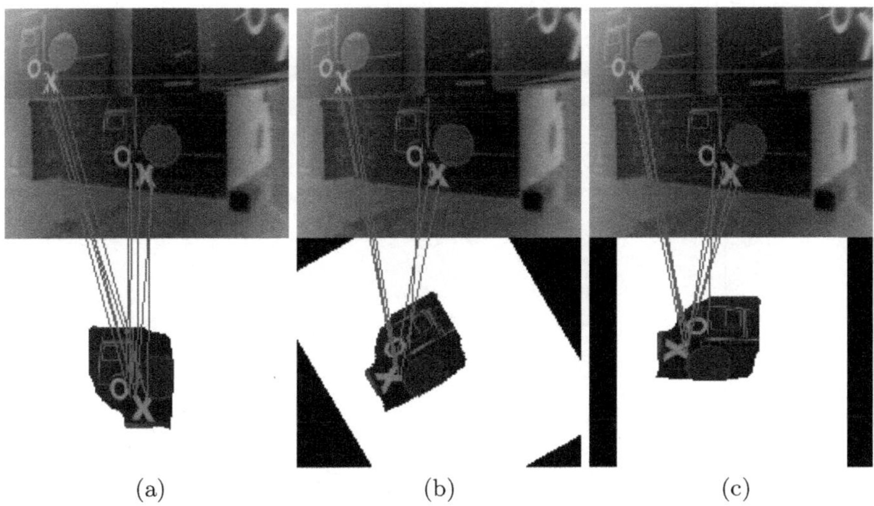

(a) (b) (c)

Fig. 4. Results of matching a template; (a) Template without rotation; (b) 60° rotated template; (c) 90° degree rotated template

successfully find more than one correspondence of a feature in the adjacent frame. This is especially useful in case a particular pattern or object needs to be located within a particular point cloud. This may be achieved by having a reference template point cloud of a particular object and then using our approach to register the target and template point clouds.

Figure 4 shows the results of such a template matching experiment. In this case the template used was a subset of a cardboard box placed in front of the Swiss IR Ranger. The target point cloud was obtained by stacking multiple boxes of this type and then registering the point cloud to the template. The lower half of all the images in Figure 4 show the template and the upper half are the target point cloud representation. The template data was also rotated in order to ensure that the features from the template could still be matched in the target point cloud. As is clear from Figures 4(a), (b) and (c), our approach is able to find multiple correspondences between the template and the target point cloud, irrespective of the rotation of the template data and angular position of the targets within the point clouds.

We also ran an experiment to test the ability of our approach to provide robot odometry using the method described in Section 3.3. In order to do so, each consecutive 3D point cloud was registered with the previous one and a rotation, translation and scaling were derived, which also provided us with the yaw, pitch and roll. The obtained yaw, pitch, roll, and translation values were compared with those provided by the on-board sensors by plotting a route map for the robot as predicted by both data sources and also plotting the yaw, pitch and roll in a similar fashion. Since the robot was moving on a 2D surface, the

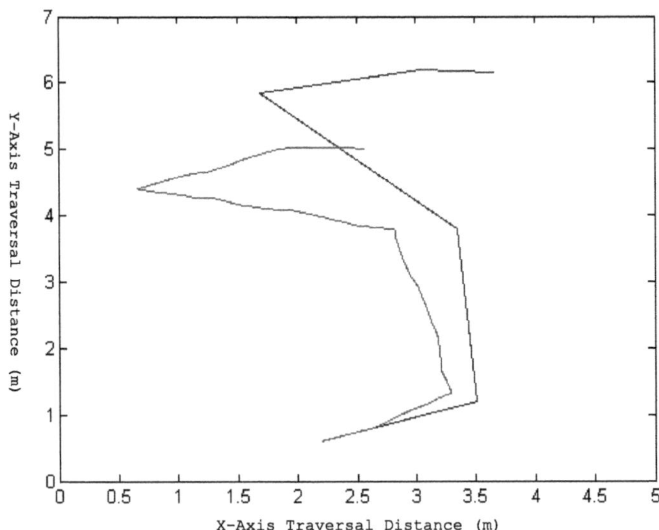

Fig. 5. Motion of the robot as predicted by the odometry derived from 3D point cloud registration vs. robot motion as obtained from navigational sensors (IMU, gyroscope and compass). Red line is from point clouds matching, while blue is the recorded robot odometry.

roll data was always constant. However, the odometry derived from the 3D point clouds closely matched that provided by the other sensors.

Figure 5 shows the path taken by our test robot. While the shape of the positional odometry derived from the 3D point cloud data is similar to the actual path, there is quite a lot of deviation between the two. However, this deviation can be explained by the fact that some of the frames that had multiple correspondences, like those in Figure 4, were not considered in the final result since the multiple locations of the correspondences led to errors. In the figure, the odometry seems more accurate, but it represents the ideal path, which does not consider wheel-spin. The difference between the resulting endpoints can reduced by applying a Kalman filter on the SIFT matching results. As such, these results could further be improved by applying some filtering, however, they clearly demonstrate the effectiveness of using 3D point cloud data for obtaining odometry information as well. The obtained results can at least be used as a basis for understanding robot trajectory.

Lastly, the other important performance criterion is the run-time performance of the algorithm in finding the correspondences between point clouds and then performing the corresponding registration. In our tests, the software was able to achieve a frame rate of 6.36 fps. We are confident that this could further be improved by adding optimizations methods, which were omitted in this version for the sake of simplicity and testing. However, this frame rate is within the range for real-time performance.

5 Conclusion

In our work we porposed a new feature extraction method for 3D data by applying the SIFT feature descriptor to 3D point clouds and scaling the data sets into images, which the SIFT descriptor could work with. The point clouds could then be registered based upon the correspondences of the feature points. This implementation makes it clear that the SIFT descriptor retains its robustness even when utilized with range data since the correspondences obtained appear to be visually correct. Moreover, even in highly noisy IR range data, multiple correspondences are successfully found in case of template matching.

The algorithm performs quite well in locating correspondences between point clouds and registering them with near real time performance. Furthermore, the preliminary experimental results suggest that even odometry data derived from calculating the relative translation, rotation and scaling between successive point clouds is close to being accurate, however, it may need further filtering and an implementation of Kalman filters to have the error component removed.

References

1. Bendels, G., Degener, P., Wahl, R., Koertgen, M., Klein, R.: Image-based registration of 3d-range data using feature surface elements. In: Proceedings of The 5th International Symposium of Virtual Reality, Archaeology and Cultural Heritage, VAST 2004 (2004)
2. Callieri, M., Cignoni, P., Ganovelli, F., Montani, C., Pingi, P., Scopigno, R.: Vclab's tools for 3d range data processing. In: Proceedings of the 1st EURO-GRAPHICS Workshop on Graphics and Cultural Heritage, Brighton, UK (November 2003)
3. Cernea, D.: Graphical methods for online surface fitting on 3d range sensor point clouds. Master's thesis, Jacobs University Bremen, Germany (August 2009)
4. Schall, O., Belyaev, A., Seidel, H.P.: Robust filtering of noisy scattered point data. In: Proceedings of Point-Based Graphics on Eurographics/IEEE VGTC Symposium, June 2005, pp. 71–144 (2005)
5. Lowe, D.: Distinctive image features from scale- invariant keypoints. International Journal of Computer Vision 60(2) (November 2004)
6. Lowe, D.: Object recognition from local scale-invariant features. In: Proceedings of International Conference on Computer Vision, Corfu, Greece (September 1999)
7. Pisarevsky, V., et al.: Opencv, the open computer vision library (2008), `http://mloss.org/software/view/68/`
8. Fischler, M.A., Bolles, R.C.: Random sample consensus: a paradigm for model fitting with applications to image analysis and automated cartography. ACM Commun. 24(6), 381–395 (1981)
9. Zhang, Z.: Iterative point matching for registration of free-form curves and surfaces. Int. J. Comput. Vision 13(2), 119–152 (1994)
10. Gough, B. (ed.): GNU Scientific Library Reference Manual, 2nd edn. Network Theory Ltd. (2003)

On-Board Monocular Vision System Pose Estimation through a Dense Optical Flow*

Naveen Onkarappa and Angel D. Sappa

Computer Vision Center, Edifici O, Campus UAB
08193 Bellaterra, Barcelona, Spain
{naveen,asappa}@cvc.uab.es

Abstract. This paper presents a robust technique for estimating on-board monocular vision system pose. The proposed approach is based on a dense optical flow that is robust against shadows, reflections and illumination changes. A RANSAC based scheme is used to cope with the outliers in the optical flow. The proposed technique is intended to be used in driver assistance systems for applications such as obstacle or pedestrian detection. Experimental results on different scenarios, both from synthetic and real sequences, shows usefulness of the proposed approach.

1 Introduction

During the last decade on-board vision has gained popularity in the automotive applications due to the increase of traffic accidents in modern age. According to the World Health Organization, every year almost 1.2 million people are killed and 50 million are injured in traffic accidents worldwide [1]. A key solution to this is the use of intelligent vision systems that are able to predict dangerous situations and anticipate accidents; these systems are usually referred in the literature as advanced driver assistance systems (ADAS). They help the driver by providing warnings, assisting to take decisions and even taking automatic evasive actions in extreme cases. Some common examples are *lane departure warning*, *pedestrian protection systems* and *adaptive cruise control*.

On-board vision systems can be classified into two different categories: *monocular* or *stereo*. Although each one of them has its own advantages and disadvantages both approaches have a common problem: real-time estimation of on-board vision system pose—position and orientation—, which is a difficult task since: (*a*) the sensor undergoes motion due to the vehicle dynamics, and (*b*) the scene is unknown and continuously changing.

In general, monocular based approaches tackle the camera pose problem by using the prior knowledge of the environment as an extra source of information. For instance, Coulombeau and Laurgeau [2] assume that the road observed on

* This work has been partially supported by the Spanish Government under project TRA2007-62526/AUT; research programme Consolider-Ingenio 2010: MIPRCV (CSD2007-00018); and Catalan Government under project CTP 2008ITT 00001.

A. Campilho and M. Kamel (Eds.): ICIAR 2010, Part I, LNCS 6111, pp. 230–239, 2010.

images has a constant known width; Liang et al. [3] assume that the vehicle is driven along two parallel lane markings, which are projected to the left and to the right of the image; Bertozzi et al. [4] assume that the camera's position and orientation remain constant through the time. Obviously the performance of these methods depends on fulfillment of assumptions, which in general cannot be taken for granted.

On the other hand, stereo based approaches have also used prior knowledge of the scene to simplify the problem and to speed up the whole process by reducing the amount of information to be handled. For instance, [5] proposes to reduce the processing time by computing 3D information only on edge points (e.g., lane markings on the image). Similarly, the edge based v-disparity approach proposed in [6], for an automatic estimation of horizon lines and later used for applications such as obstacle or pedestrian detection (e.g., [7],[8]), only computes 3D information over local maxima of the image gradient. A different stereo vision based approach has been proposed in [9]. It uses dense depth maps and is based on the extraction of a dominant 3D plane that is assumed to be the road plane. Camera's position and orientation are directly computed, referred to that plane. A recent work [10] proposes a novel paradigm that is based on raw stereo images provided by a stereo head. This paradigm includes a stochastic technique to track vehicle pose parameters given stereo pairs arriving in a sequential fashion. In [10], the assumption is that the selected region only contains road points, as well as the road surface is assumed to be a plane.

The current work proposes a novel approach for estimating camera's position and orientation for monocular vision systems, which are finally represented as a single value. It is based on a dense optical flow estimated by means of the TV-L^1 formulation. Previous approaches rely on local formulations: a technique based on optical flow with template matching scheme was used in [11], while a maximum likelihood formulation over small patches was introduced in [12].

The main advantage of the proposed approach with respect to other monocular based approaches is that it does not require feature extraction neither imposes restrictive assumptions. The advantage with respect to the previous optical flow based approaches is that the current one is based on an accurate variational dense optical flow formulation. Finally, since it is based on a monocular vision, a system cheaper than stereo based solutions can be reached.

The remainder of this paper is organized as follows. Section 2 briefly introduces the TV-L^1 formulation used to compute dense optical flow, together with the proposed adaptation to reduce the processing time or to increase the accuracy of the flow estimation. The proposed approach is presented in Section 3. Experimental results on different sequences/scenarios are presented in Section 4. Finally, conclusions are given in Section 5.

2 TV-L^1 Optical Flow

State of the art in optical flow techniques unveil that varational techniques give dense estimation with more accuracy as compared to other approaches. TV-L^1

is a variational optical flow technique proposed in [13] that gives dense flow field. In the current work, an improved version [14] is used, which is briefly presented in this section. As the initial formulation of the variational method proposed by Horn and Schunck [15], the formulation in [14] also involves an optical flow constraint and a regularization term but both of them with L^1 norm. The TV-L^1 optical flow is obtained by minimizing the following energy function:

$$E = \int_{\Omega} \{ \alpha \underbrace{|I_1(\boldsymbol{x} + \boldsymbol{u}(\boldsymbol{x})) - I_0(\boldsymbol{x})|}_{Data\ Term} + \underbrace{|\nabla \boldsymbol{u}|}_{Regularization} \} \, d\boldsymbol{x}, \tag{1}$$

where I_0 and I_1 are two images; $\boldsymbol{x} = (x_1, x_2)$ is the pixel location within a rectangular image domain $\Omega \subseteq \mathbf{R}^2$; and $\boldsymbol{u} = (u_1(\boldsymbol{x}), u_2(\boldsymbol{x}))$ is the two dimensional displacement field. The parameter α weighs between data term and regularization term. The objective is to find the displacement field $\boldsymbol{u}$ that minimizes the energy function in (1). The regularization term $|\nabla \boldsymbol{u}|$ with L^1 norm is called total variation regularization. Replacing these data and regularization terms with L^2 norm lead us to the original Horn and Schunck formulation [15]. Since the terms in (1) are not continuously differentiable, the energy function can be minimized using dual formulation for minimizing total variation as proposed in [16] and adapted to optical flow in [13]. Linearizing I_1 near to $(\boldsymbol{x} + \boldsymbol{u}_0)$, where $\boldsymbol{u}_0$ is a given flow field, the whole data term is denoted as an image residual $\rho(\boldsymbol{u}) = I_1(\boldsymbol{x} + \boldsymbol{u}_0) + \langle \nabla I_1, \boldsymbol{u} - \boldsymbol{u}_0 \rangle - I_0(\boldsymbol{x})$. Then, by introducing an auxiliary variable $\boldsymbol{v}$, the data term and regularization term in (1) can be rewritten as indicated in (2), making easier the minimization process. Without loss of generality, in the two-dimensional case, the resulting energy can be expressed as:

$$E = \int_{\Omega} \left\{ \alpha \, |\rho(\boldsymbol{v})| + \sum_{d=1,2} (1/2\theta)(u_d - v_d)^2 + \sum_{d=1,2} |\nabla u_d| \right\} d\boldsymbol{x}, \tag{2}$$

where θ is a small constant, such that $\boldsymbol{v}$ is a close approximation of $\boldsymbol{u}$; and d indicating the dimension takes value as 1 and 2. This convex energy function is optimized by alternative updating steps 1 and 2 for $\boldsymbol{u}$ and $\boldsymbol{v}$:

Step 1. By keeping $\boldsymbol{u}$ fixed, $\boldsymbol{v}$ is computed as:

$$\min_{\boldsymbol{v}} \{ \alpha \, |\rho(\boldsymbol{v})| + \sum_{d=1,2} (1/2\theta) \, (u_d - v_d)^2 \}, \tag{3}$$

Step 2. Then, by keeping v_d fixed for every d, u_d is computed as:

$$\min_{u_d} \int_{\Omega} \{ 1/2\theta (u_d - v_d)^2 + |\nabla u_d| \} d\boldsymbol{x}. \tag{4}$$

Equation (4) can be solved for each dimension using the dual formulation. The solution is given by:

$$u_d = v_d - \theta \, \mathbf{div} \boldsymbol{p}_d, \tag{5}$$

where the dual variable $\boldsymbol{p} = [p_1, p_2]$ for a dimension d is iteratively defined by

$$\tilde{\boldsymbol{p}}^{n+1} = \boldsymbol{p} + \tau/\theta(\nabla(v_d + \theta \, \mathbf{div} \boldsymbol{p}^n)), \tag{6}$$

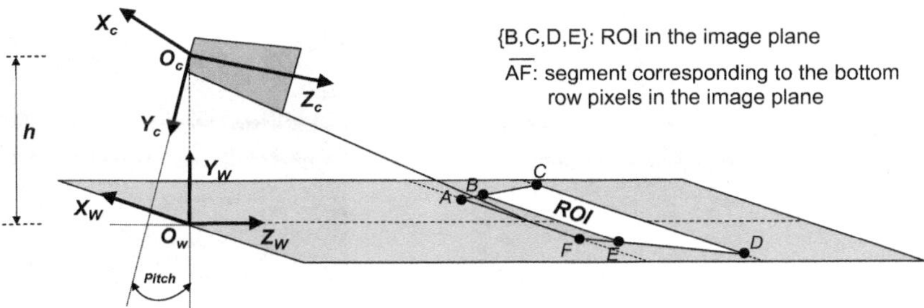

Fig. 1. Camera coordinate system (X_C, Y_C, Z_C) and world coordinate system (X_W, Y_W, Z_W)

$$\boldsymbol{p}^{n+1} = \tilde{\boldsymbol{p}}^{n+1} / \max(1, \, |\tilde{\boldsymbol{p}}^{n+1}|), \tag{7}$$

where $\boldsymbol{p}^0 = \mathbf{0}$ and the time step $\tau \leq 1/4$.

The solution of equation (3) is a simple thresholding step since it does not involve derivative of $\boldsymbol{v}$, and is given by:

$$\boldsymbol{v} = \boldsymbol{u} + \begin{cases} \alpha\theta\nabla I_1 & \text{if} \quad \rho(\boldsymbol{u}) < -\alpha\theta|\nabla I_1|^2 \\ -\alpha\theta\nabla I_1 & \text{if} \quad \rho(\boldsymbol{u}) > \alpha\theta|\nabla I_1|^2 \\ -\rho(\boldsymbol{u})\nabla I_1/|\nabla I_1|^2 & \text{if} \quad |\rho(\boldsymbol{u})| \leq \alpha\theta|\nabla I_1|^2 \end{cases} \tag{8}$$

In this optical flow method, the structure-texture blended image that is robust against sensor noise, illumination changes, reflections and shadows as explained in [14] is used. Additionally, in the current implementation an initialization step is proposed for reducing the CPU time or increasing the accuracy. This step consists in using the optical flow computed between the previous couple of frames as initial values for the current couple instead of initializing by zero.

3 Proposed Approach

Before detailing the approach proposed to estimate the monocular vision system pose, the relationships between the coordinate systems (world and camera) and the camera parameters, assuming a flat road are presented.

3.1 Model Formulation

Camera pose parameters are computed relative to a world coordinate system (X_W, Y_W, Z_W), defined for every frame, in such a way that: the $X_W Z_W$ plane is co-planar with the current road plane. Figure 1 depicts the camera coordinate system (X_C, Y_C, Z_C) referred to the road plane. The origin of the camera coordinate system O_C is contained in the Y_W axis—it implies a $(0, t_y, 0)$ translation of the camera w.r.t. world coordinate system. Hence, since *yaw* angle is not considered in the current work (i.e., it is assumed to be zero), the six camera pose parameters[1] $(t_x, t_y, t_z, yaw, roll, pitch)$ reduce to just three $(0, t_y, 0, 0, roll, pitch)$,

[1] A 3D translation and a 3D rotation that relates O_C with O_W.

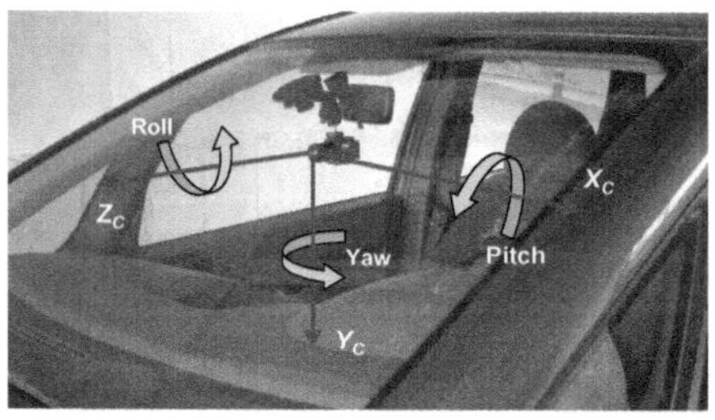

Fig. 2. (*left*) On-board camera with its corresponding coordinate system. (*right*) Horizon line (r_{HL}) estimated by the intersection of projected lane markings.

denoted in the following as (h, Φ, Θ) (i.e., camera height, roll and pitch). Figure 2($left$) shows the onboard camera used for testing the proposed approach.

Among the parameters (h, Φ, Θ), the value of the roll angle (Φ) will be very close to zero in most situations, since when the camera is rigidly mounted on the car, a specific procedure is followed to ensure an angle at rest within a given range, ideally zero, and in regular driving conditions this value scarcely varies (more details can be found in [9]). Finally, the variables (h, Θ) that represents the camera pose parameters are encoded as a single value, which is the *horizon line* position in the image plane (e.g., [17],[18]). The horizon line corresponds to the back-projection of a point, lying over the road at an infinite depth. Assuming the road can be modelled as a plane, let $ax + by + cz + h = 0$ be the road plane equation and h the camera height, see Fig. 1 (since $(h \neq 0)$ the plane equation can be simplified dividing by $(-h)$). Let $P_i(0, y, z)$ be a point lying over the road plane at an infinite depth z from the camera reference frame with $x = 0$; from the plane equation the y_i coordinates of P_i corresponds to $y_i = \frac{1-cz_i}{b}$. The backprojection of y_i into the image plane when $z_i \to \infty$ defines the row coordinate of the horizon line r_{HL} in the image. It results into:

$$r_{HL} = r_0 + f\frac{y_i}{z_i} = r_0 + \frac{f}{z_i b} - f\frac{c}{b} \ , \tag{9}$$

where f denotes the focal length in pixels, r_0 represents the vertical coordinate of the camera principal point, and z_i is the depth value of P_i. Since $(z_i \to \infty)$, the row coordinate of the horizon line in the image is finally computed as $r_{HL} = r_0 - f\frac{c}{b}$. Additionally, when lane markings are present in the scene, the horizon line position in the image plane can be easily obtained by finding the intersection of these two parallel lines, see Fig. 2($right$).

3.2 Horizon Line Estimation

In the current work, a RANSAC based approach is proposed to estimate the horizon line position. It works directly in the image plane by using the optical flow vectors computed between two consecutive frames. The TV-L^1 optical flow [14] with a minor modification as explained in the previous section is used. The flow vectors within a rectangular region centered in the bottom part of the

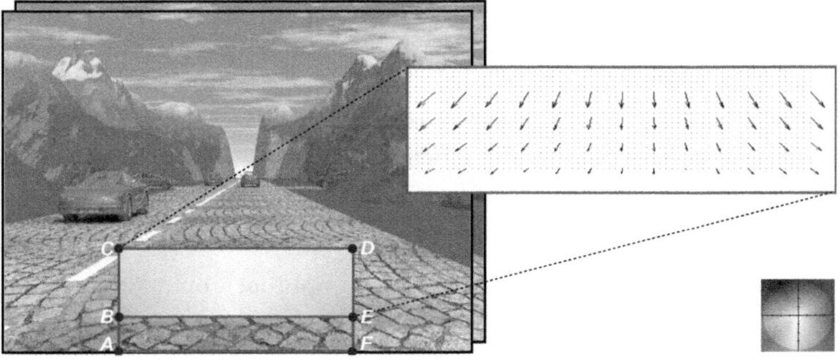

Fig. 3. A couple of consecutive synthetic frames illustrating the rectangular free space {A,C,D,F}, containing the ROI {B,C,D,E} from which computed flow vectors are used for estimating horizon line position. (*top – right*) Enlarged and sub-sampled vector field from the ROI. (*bottom – right*) Color map used for depicting the vector field in the ROI.

image are used instead of considering the flow vectors through the whole image. The specified region is a rough estimation of the minimum free space needed for a vehicle moving at 30km/h to avoid collisions—rectangle defined by the points $\{A, C, D, F\}$, in Fig. 3. Note that, at a higher speed this region should be enlarged. Actually from this rectangular free space only the top part is used (rectangular ROI defined by the points $\{B, C, D, E\}$ in Fig. 3), since the flow vectors at the bottom part (image boundary) may not be as accurate as required. Figure 3 presents a couple of synthetic frames with the optical flow computed over that ROI; an enlarged and sub-sampled illustration of these flow vectors is given in the top-right part.

Let $\boldsymbol{u}$ be the computed flow field corresponding to a given ROI $\{B, C, D, E\}$. This vector field can be used for recovering the camera motion parameters through a closed form formulation (e.g., [11] and [12]). However, since it could be noisy and contains outliers, a robust RANSAC based technique [19] is proposed for computing the horizon line position. It works as follow:

Random sampling: *Repeat the following three steps K times*

1. Draw a couple of vectors, $(\boldsymbol{u}^1, \boldsymbol{u}^2)$ from the given ROI where $\boldsymbol{u}^1 = (u_1^1, u_2^1)$ and $\boldsymbol{u}^2 = (u_1^2, u_2^2)$.
2. Compute the point (S_x, S_y) where these two vectors intersect.
3. Vote into the cell $\boldsymbol{C}_{(i,j)}$, where $i = \lfloor S_y \rfloor$ and $j = \lfloor S_x \rfloor$ and (i, j) lie within the image boundary.

Solution:

1. Choose the cell that has the highest number of votes in the voting matrix $\boldsymbol{C}$. Let $\boldsymbol{C}_{(i,j)}$ be this solution.
2. Set the sought horizon line position r_{HL} as the row i.

Fig. 4. Horizon line computed by the proposed approach on a synthetic sequence

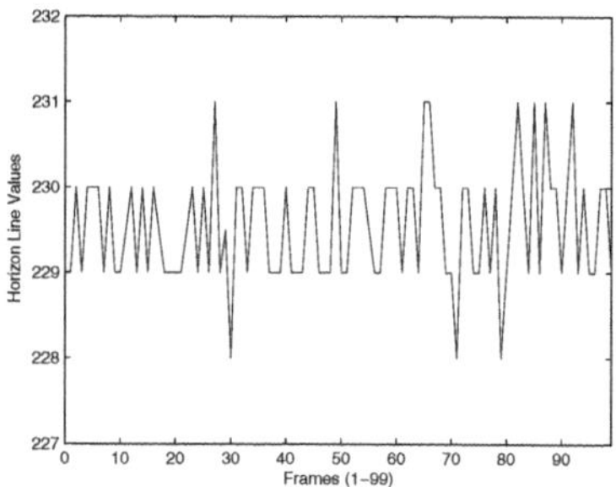

Fig. 5. Plot of variations in horizon line in a sequence of 100 frames

4 Experimental Results

The proposed technique has been tested on several synthetic and real video sequences. Firstly, a synthetic sequence (gray scale sequence-1 in set 2 of *enpeda* [20]) was used for validating the proposed approach. Figure 4 shows some frames with the horizon line computed by the proposed technique. Note, that in this case, since a perfect flat road without any vehicle dynamics, camera pose almost remains constant (horizon line variation through this synthetic sequence is presented in Fig. 5). On the contrary, horizon line undergoes large variations in Fig. 6. This synthetic sequence (gray scale sequence-2 in Set 2 of *enpeda* [20]) contains uphill, downhill and flat road scenarios. Figure 7(*left*) presents the variations of horizon line for the whole sequence. Figure 7(*right*) depicts the pitch angle variation from the ground-truth data. The similarity between these two plots confirms the effectiveness of the presented approach. The sequences in Fig.4, and Fig.6 are of a resolution of 480×640 pixels, and the ROI contains 96×320 pixels placed above 48 pixels from the bottom of the image.

Figure 8 shows a frame from a real sequence (Intern-On-Bike-left sequence in set 1 of *enpeda* sequences [20]) with the horizon line estimated by the proposed approach. The variation of the horizon line over a set of 25 frames of that sequence is presented in Fig. 8(*right*). Additionally, few different real frames, with horizon line estimated by the proposed approach, are shown in Fig.2(*right*) and Fig. 9. Notice that the horizon lines estimated by intersecting the projected lane markings (dotted lines) also coincide with those obtained by the proposed

Fig. 6. Horizon lines computed by the proposed approach on a synthetic video sequence illustrating different situations: uphill, downhill and flat roads

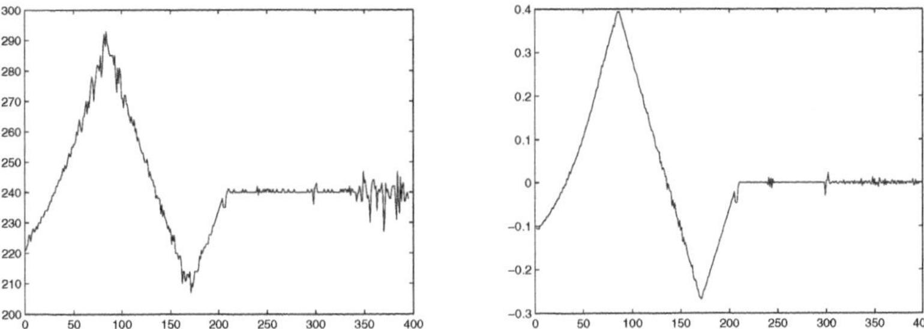

Fig. 7. (*left*) Variations in horizon line position over a sequence of 396 frames. (*right*) Pitch angle variations from the ground-truth.

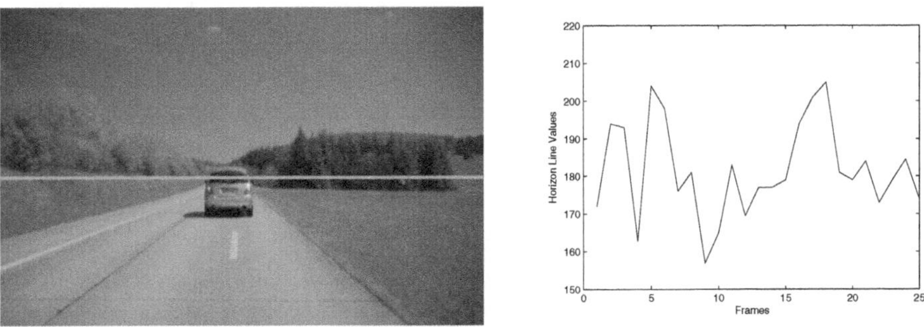

Fig. 8. Horizon line for a real sequence and its variations for 25 frames

approach, in spite of the fact that some frames contain outliers (see lane barriers in the top-left frame in Fig. 9). The video frames in Fig. 9 are captured at a resolution of 480×752 pixels at about 30fps. The value of K is empirically determined and the better value is about half of the total number of flow vectors in the specified ROI. The specified ROI contains 96×376 pixels and is placed above 48 pixels from the bottom of the image.

Fig. 9. Real video frames with the horizon lines estimated by the proposed approach (note that they correspond with the intersections of the projected lane markings)

5 Conclusions

A robust technique for pose estimation of an on-board monocular vision system has been presented. It uses dense flow field from a state of the art variational optical flow technique that is robust against common obstructions in real traffic such as shadows, reflections and illumination changes. The proposed modified initialization step to the optical flow estimation has the advantage to be more accurate or less computation time. The camera pose parameters estimation is modelled as a horizon line estimation problem and has been solved using a RANSAC based approach that is robust against outliers in the flow field. The proposed approach is validated on both synthetic and real sequences. With the advance in the real-time implementation of optical flow algorithms and particularly, for our problem of estimating the flow vectors only in the specified region instead of the whole image, the proposed approach can be implemented on real applications with real-time performance.

References

1. Peden, M., Scurfield, R., Sleet, D., Mohan, D., Hyder, A., Jarawan, E., Mathers, C.: World Report on road traffic injury prevention. World Health Organization, Geneva (2004)
2. Coulombeau, P., Laurgeau, C.: Vehicle yaw, pitch, roll and 3D lane shape recovery by vision. In: Proc. IEEE Intelligent Vehicles Symposium, Versailles, France, pp. 619–625 (2002)

3. Liang, Y., Tyan, H., Liao, H., Chen, S.: Stabilizing image sequences taken by the camcorder mounted on a moving vehicle. In: Proc. IEEE Int. Conf. on Intelligent Transportation Systems, Shangai, China, pp. 90–95 (2003)
4. Bertozzi, M., Broggi, A., Carletti, M., Fascioli, A., Graf, T., Grisleri, P., Meinecke, M.: IR pedestrian detection for advaned driver assistance systems. In: Proc. 25th. Pattern Recognition Symposium, Magdeburg, Germany, pp. 582–590 (2003)
5. Nedevschi, S., Vancea, C., Marita, T., Graf, T.: Online extrinsic parameters calibration for stereovision systems used in far-range detection vehicle applications. IEEE Trans. on Intelligent Transportation Systems 8(4), 651–660 (2007)
6. Labayrade, R., Aubert, D., Tarel, J.: Real time obstacle detection in stereovision on non flat road geometry through 'V-disparity' representation. In: Proc. IEEE Intelligent Vehicles Symposium, Versailles, France, pp. 646–651 (2002)
7. Bertozzi, M., Binelli, E., Broggi, A., Del Rose, M.: Stereo vision-based approaches for pedestrian detection. In: Proc. IEEE Int. Conf. on Computer Vision and Pattern Recognition, San Diego, USA (2005)
8. Labayrade, R., Aubert, D.: A single framework for vehicle roll, pitch, yaw estimation and obstacles detection by stereovision. In: Proc. IEEE Intelligent Vehicles Symposium, Columbus, OH, USA, pp. 31–36 (2003)
9. Sappa, A., Dornaika, F., Ponsa, D., Gerónimo, D., López, A.: An efficient approach to on-board stereo vision system pose estimation. IEEE Trans. on Intelligent Transportation Systems 9(3), 476–490 (2008)
10. Dornaika, F., Sappa, A.: A featureless and stochastic approach to on-board stereo vision system pose. Image and Vision Computing 27(9), 1382–1393 (2009)
11. Suzuki, T., Kanade, T.: Measurement of vehicle motion and orientation using optical flow. In: Proc. IEEE Int. Conf. on Intelligent Transportation Systems, Tokyo, Japan, pp. 25–30 (1999)
12. Stein, G., Mano, O., Shashua, A.: A robust method for computing vehicle egomotion. In: IEEE Intelligent Vehicles Symposium, Dearborn Michigan, USA, pp. 362–368 (2000)
13. Zach, C., Pock, T., Bischof, H.: A duality based approach for realtime TV-L^1 optical flow. In: Proc. 29th Annual Symposium of the German Association for Pattern Recognition, Heidelberg, Germany, pp. 214–223 (2007)
14. Wedel, A., Pock, T., Zach, C., Cremers, D., Bischof, H.: An improved algorithm for TV-L1 optical flow. In: Proc. of the Dagstuhl Motion Workshop, Dagstuhl Castle, Germany, pp. 23–45 (2008)
15. Horn, B.K.P., Schunk, B.G.: Determining optical flow. Artificial Intelligence 17, 185–203 (1981)
16. Chambolle, A.: An algorithm for total variation minimization and applications. J. Math. Imaging Vis. 20(1-2), 89–97 (2004)
17. Zhaoxue, C., Pengfei, S.: Efficient method for camera calibration in traffic scenes. Electronics Letters 40(6), 368–369 (2004)
18. Rasmussen, C.: Grouping dominant orientations for ill-structured road following. In: Proc. IEEE Int. Conf. on Computer Vision and Pattern Recognition, Washington, USA, pp. 470–477 (2004)
19. Fischler, M., Bolles, R.: Random sample consensus: A paradigm for model fitting with applications to image analysis and automated cartography. Graphics and Image Processing 24(6), 381–395 (1981)
20. Vaudrey, T., Rabe, C., Klette, R., Milburn, J.: Differences between stereo and motion behaviour on synthetic and real-world stereo sequences. In: Proc. Image and Vision Computing New Zealand, Christchurch, New Zealand, pp. 1–6 (2008)

II-LK – A Real-Time Implementation for Sparse Optical Flow

Tobias Senst, Volker Eiselein, and Thomas Sikora

Technische Universität Berlin
Communication Systems Group
{senst,eiselein,sikora}@nue.tu-berlin.de
http://www.nue.tu-berlin.de

Abstract. In this paper we present an approach to speed up the computation of sparse optical flow fields by means of integral images and provide implementation details. Proposing a modification of the Lucas-Kanade energy functional allows us to use integral images and thus to speed up the method notably while affecting only slightly the quality of the computed optical flow. The approach is combined with an efficient scanline algorithm to reduce the computation of integral images to those areas where there are features to be tracked. The proposed method can speed up current surveillance algorithms used for scene description and crowd analysis.

Keywords: Lucas-Kanade, optical flow, fast implementation, integral images, optimization, real-time.

1 Introduction

Computation of optical flow is a common topic in the computer vision community whose applications range from motion estimation to point tracking. There are many different approaches to compute optical flow, among them the classical Lucas-Kanade method [12].

Introduced in 1981, it still has many applications ([1]) and is a popular method to compute the movement of sparse feature points from one video frame to the next. This is often used for tracking objects or persons directly as e.g. in [11] or [9]. As another approach based on the idea of individual motion, [5] builds trajectories from sparse feature tracks which are afterwards clustered to obtain the number of persons in a scene. Similarly, yet in a different context, crowds are described in [14] by their pointwise motion which yields information on their activity and on abnormal events. Other applications are 3D pose and camera parameter estimation as e.g. in [10]. The Lucas-Kanade method also inspired other more recent algorithms as e.g. [7], [6] or [3].

Formulating brightness constancy between two points in consecutive images leads to an equation in two unknowns and cannot be solved as such. This is commonly known as the *aperture problem.* As a solution, the Lucas-Kanade method assumes constant flow for a window around the current pixel and solves

A. Campilho and M. Kamel (Eds.): ICIAR 2010, Part I, LNCS 6111, pp. 240–249, 2010.

the corresponding system of equations in an iterative manner. We recapitulate this process shortly in section 2. While computing the solution, it is necessary to repeatedly sum the pixel values around the features which is a considerable computational effort. This complexity can be reduced by using integral images as shown in section 2.2. To take full advantage of these, we propose to modify the Lucas-Kanade method by using another linearization in the image difference term. Thus, we are able to compute all sums of image pixels by means of integral images which accelerates the algorithm remarkably. In section 3 we explain an extension to reduce computational complexity in the case of tracking of sparse features. In section 4 we show evaluations of the proposed method and the possible benefit in terms of speed-up.

2 Modifying the Lucas-Kanade Method by Integral Images (II-LK)

Let $I^{A}(x,y)$ and $I^{B}(x,y)$ be two consecutive grayscale images of a video sequence. I^A is an image at a discrete time t and I^B the consecutive image at time $(t+1)$. To compute the optical flow of an image domain $\Omega \subset \mathbb{R}^2$ Lucas and Kanade [12] formulated the well-known *brightness constancy assumption* $I^A(x,y,t) = I^B(x+u, y+v, t+1)$ in combination to the assumption that all pixels in domain $\Omega = w_x \times w_y$ are subject to a constant movement. Formulating this into an error energy term yields

$$\epsilon = \sum_{\Omega} \left(I^{A}(x,y) - I^{B}(x+u, y+v) \right)^2 \rightarrow min. \tag{1}$$

Minimizing the term is done via a Newton-Raphson iteration scheme which is linearized using a Taylor expansion. Using

$$\begin{aligned} \mathbf{h} = (u,v)^T, \mathbf{G} = \textstyle\sum_{\Omega} \nabla I \cdot \nabla I^T, \\ \mathbf{b}_{k-1} = \textstyle\sum_{\Omega} (I^{A}(x,y) - I^{B}(x+u_{k-1}, y+v_{k-1})) \cdot \nabla I^T \end{aligned} \tag{2}$$

with $\mathbf{h}$ being the motion vector, $\mathbf{G}$ the gradient matrix and $\mathbf{b}$ the mismatch vector, the iteration scheme for the k-th iteration is given by

$$\mathbf{h}_k = \mathbf{h}_{k-1} + \mathbf{G}^{-1} \cdot \mathbf{b}_{k-1}. \tag{3}$$

Now it is possible to apply an iterative scheme to solve the equation and obtain the image displacement vector $\mathbf{h}$ for the window Ω. This implies computing repeatedly the sums of pixel values and gradient values within the domain Ω. Computing motion vectors for a whole image results in a computational effort of $\mathcal{O}(N w_x w_y)$ with N being the number of pixels. As a consequence of overlapping domains there are regions where constant values are computed twice or even more often. In the next section we present a method to reduce this effort.

2.1 Integral Images Reduce Computational Overhead

Integral images or summed area tables are image representations used for fast computation of region sums. Introduced in [8], they are often used to accelerate learning algorithms (e.g. in [13]). The values of an integral image, denoted $\hat{I}$, contain the sum of the original pixel values I from the upper left corner to their position. They can be computed in one pass over the image using the following formal description

$$s(x, y) = s(x-1, y) + I(x, y) \tag{4}$$
$$\hat{I}(x, y) = s(x, y) + \hat{I}(x, y-1) \tag{5}$$

with the boundary condition $s(x, -1) \equiv 0$ and $\hat{I}(-1, y) \equiv 0$. Using integral images, a sum of a rectangular region given by four points A (upper left), B (upper right),C (lower left), D (lower right) can be computed by

$$\sum_{\Omega} I = \hat{I}(A) - \hat{I}(B) - \hat{I}(C) + \hat{I}(D). \tag{6}$$

In the case of not using integral images, the computational complexity for the sum depends on the window size with $\mathcal{O}(w_x \times w_y)$. Using integral images it reduces to $\mathcal{O}(4)$ for the sum and, as shown above, $\mathcal{O}(N)$ (with N being the number of pixels) to create the integral image of the current frame.

Within the iteration scheme (3) we can directly substitute building the gradient matrix $\mathbf{G}$ using the integral images $\hat{I}_{I_u^2}, \hat{I}_{I_u I_v}, \hat{I}_{I_v^2}$. Provided that we use a non-weighted window, the gradient matrix

$$\mathbf{G} = \begin{bmatrix} \sum_{\Omega} I_u \cdot I_u & \sum_{\Omega} I_u \cdot I_v \\ \sum_{\Omega} I_u \cdot I_v & \sum_{\Omega} I_v \cdot I_v \end{bmatrix} = \begin{bmatrix} G_{uu} & G_{uv} \\ G_{uv} & G_{vv} \end{bmatrix} \tag{7}$$

for a window Ω with its vertices $A, B, C, D \in \Omega$ can be built rapidly by

$$\begin{aligned} G_{uu} &= \hat{I}_{I_u^2}(A) - \hat{I}_{I_u^2}(B) - \hat{I}_{I_u^2}(C) + \hat{I}_{I_u^2}(D) \\ G_{vv} &= \hat{I}_{I_v^2}(A) - \hat{I}_{I_v^2}(B) - \hat{I}_{I_v^2}(C) + \hat{I}_{I_v^2}(D) \\ G_{uv} &= \hat{I}_{I_u I_v}(A) - \hat{I}_{I_u I_v}(B) - \hat{I}_{I_u I_v}(C) + \hat{I}_{I_u I_v}(D) \end{aligned} \tag{8}$$

2.2 Modification of the Lucas-Kanade Method

Using (7) we enhance the computation of the gradient matrix by using integral images. However, due to the iterative window shifts the computation of $\mathbf{b}$ in (3) is difficult to compute by integral images. We propose therefore an extension of (3) which uses a second linearization of the image difference term and allows us to use integral images in the whole equation. Approximating the shifted image by a first-order Taylor-expansion results in

$$I^{B}(x + u_{k-1}, y + v_{k-1}) = I^{B}(x, y) + \mathbf{h}_{k-1} \cdot \nabla I + \epsilon. \tag{9}$$

The mismatch vector $\mathbf{b}$ can then be approximated by

$$\tilde{\mathbf{b}}_{k-1} = \sum_{\Omega} \left(I^A(x,y) - I^B(x,y) - \mathbf{h}_{k-1} \cdot \nabla I \right) \cdot \nabla I^T \tag{10}$$

and the final iteration scheme by

$$\mathbf{h}_k \approx \mathbf{h}_{k-1} + \tau \cdot \mathbf{G}^{-1} \cdot \tilde{\mathbf{b}}_{k-1}. \tag{11}$$

Now all sums of pixel values in this formula can be computed by using integral images and benefit from the huge speed-up they provide. Therefore, we will refer to this as the *I*ntegral *I*mage *L*ucas-*K*anade method (II-LK). However, the proposed linearization introduces additional approximation errors. In practice it has been shown that a parameter $\tau \leq 1$ enhances the robustness of the solution by avoiding overshooting. The approximated mismatch vector $\tilde{\mathbf{b}}$ can now be computed with the four integral images

$$\begin{aligned}
B_{Au} &= \hat{I}_{I^A \cdot I_u}(A) - \hat{I}_{I^A \cdot I_u}(B) - \hat{I}_{I^A \cdot I_u}(C) + \hat{I}_{I^A \cdot I_u}(D) \\
B_{Av} &= \hat{I}_{I^A \cdot I_v}(A) - \hat{I}_{I^A \cdot I_v}(B) - \hat{I}_{I^A \cdot I_v}(C) + \hat{I}_{I^A \cdot I_v}(D) \\
B_{Bu} &= \hat{I}_{I^B \cdot I_u}(A) - \hat{I}_{I^B \cdot I_u}(B) - \hat{I}_{I^B \cdot I_u}(C) + \hat{I}_{I^B \cdot I_u}(D) \\
B_{Bv} &= \hat{I}_{I^B \cdot I_v}(A) - \hat{I}_{I^B \cdot I_v}(B) - \hat{I}_{I^B \cdot I_v}(C) + \hat{I}_{I^B \cdot I_v}(D)
\end{aligned} \tag{12}$$

and

$$\begin{aligned}
\tilde{b}_{u_{k-1}} &= B_{Au} - B_{Bu} - u_{k-1} \cdot G_{uu} - v_{k-1} \cdot G_{uv} \\
\tilde{b}_{v_{k-1}} &= B_{Av} - B_{Bv} - u_{k-1} \cdot G_{uv} - v_{k-1} \cdot G_{vv}
\end{aligned} \tag{13}$$

so that the final displacement is given by

$$\begin{aligned}
u_k &= u_{k-1} + \tau \frac{\tilde{b}_{u_{k-1}} \cdot G_{vv} - \tilde{b}_{v_{k-1}} \cdot G_{uv}}{G_{uu} \cdot G_{vv} - G_{uv} \cdot G_{uv}} \\
v_k &= v_{k-1} + \tau \frac{\tilde{b}_{v_{k-1}} \cdot G_{uu} - \tilde{b}_{u_{k-1}} \cdot G_{uv}}{G_{uu} \cdot G_{vv} - G_{uv} \cdot G_{uv}}
\end{aligned}. \tag{14}$$

In section 4, we show that the additional approximation error is acceptable in surveillance applications, where runtime is an important criterion. A larger window size increases the advantage of integral images because in this case the classical approach has a complexity depending on Ω compared to a constant complexity of integral images.

Yet, if one only needs to compute optical flow vectors for some points in a large image, the creation of integral images might be more costly than the benefit that they bring. To cope with this problem, we propose in section 3 a scanline algorithm which permits to identify the regions where an integral image is needed thus reduces the overall computational cost.

3 Tracking Sparse Features

For tracking tasks it is common to select only a certain number of features to track. A classical approach to find easily trackable features is the "Good

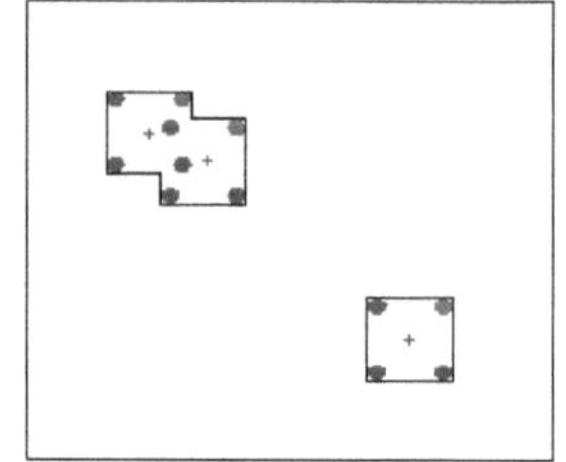

Fig. 1. Left: Example for "Good Features To Track" (Picture from PETS 2006). Right: marker scheme for features: (cross: feature point, blue: 2, green: -2, violet: 1, red: -1) and area for which integral images have to be built.

Features To Track" method [15]. Its main idea is to select good feature points by choosing the highest minimal eigenvalues of the covariance matrix **G** of the points, computed by

$$\lambda_{min} = \frac{G_{uu} + G_{uv}}{2} - \sqrt{\frac{(G_{uu} - G_{vv})^2}{2} - G_{uv}^2}. \qquad (15)$$

As shown in (6) integral images can be easily used to reduce the computational overhead in this method. So the minimal eigenvalues can be computed in one pass over the image. This is an even bigger advantage because the same integral images that are needed for feature selection can afterwards be used to compute the flow vectors for this frame. Therefore it is not necessary to compute these integral images twice.

Given a set $\mathcal{S}$ of feature points to track, an important point is that the computational benefit integral images provide is dependent on the window size and the number of features $|\mathcal{S}|$. For a small $|\mathcal{S}|$, their gain might be less than the additional effort to build them. We therefore propose an algorithm to compute integral images only in image regions covered by feature domains.

For every window Ω, four labels $l \in$ { *2 ("top left"), -2 ("top right"), 1 ("bottom left"), -1 ("bottom right")*} are stored in a mask to represent its vertices. The output of the scanline algorithm is a mask indicating if a pixel lies within a feature window or not. Therefore, a variable n_H is used to store the number of connected windows on the current line. The number of overlapping windows on the current column (see. figure 2) is directly stored in the mask which is parsed from top-left to bottom-right. If a "top-left"-label is found, this label is propagated to the same column in the next line.

For all newly found "top left"- ["top right"-] labels, the propagated label is increased [decreased] by one. Accordingly, for the labels "bottom left" ["bottom right"], the propagated label is decreased [increased]. Finding a "left"-label, n_H is increased whereas for a "right"-label it is decreased. So the pixels for which an integral image pixel has to be created are identified by $n_H > 0$. Formally:

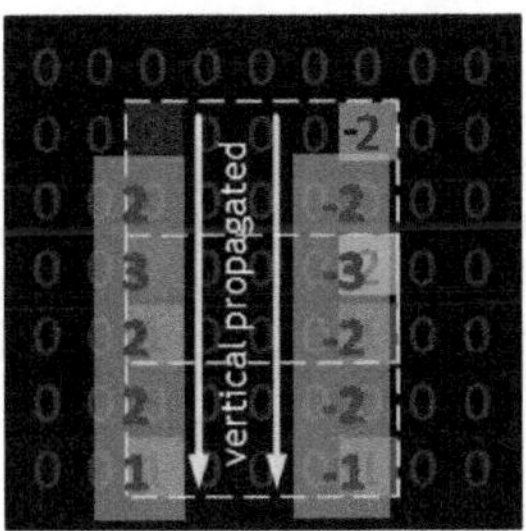

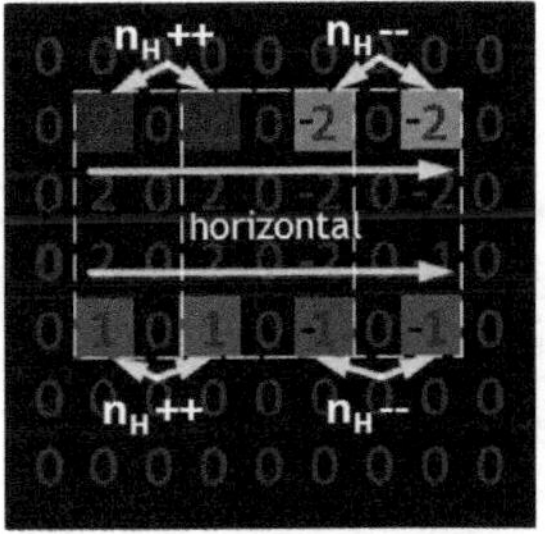

Fig. 2. Left: Vertical propagation scheme for scanline algorithm: propagated values (red) depend on feature window labels (gray). Right: Horizontal parsing scheme for scanline algorithm: n_H (white) is increased or decreased depending on feature window labels (gray).

$$s(x,y) = \begin{cases} s(x-1,y) + I(x,y), & \text{if } n_H > 0 \\ 0, & \text{otherwise} \end{cases} \tag{16}$$

$$\hat{I}(x,y) = \begin{cases} s(x,y) + I(x,y-1), & \text{if } n_H > 0 \\ 0, & \text{otherwise} \end{cases} \tag{17}$$

In this way it is possible to build the integral images only for the image regions which have to be processed.

4 Experimental Results

In this section we evaluate the execution time and accuracy of our algorithm and compare the results to the public implementation of S. Birchfield's KLT-Tracker[1] and to the very fast implementation in Intel's OpenCV library[2] described in [4]. OpenCV is one of the standard vision libraries and known to provide fast implementations for many algorithms. Our implementation is based on the Birchfield algorithm and not optimized for specific CPU architectures as OpenCV. We give the results for these three implementations to demonstrate the performance gain II-LK provides compared to the standard Lukas-Kanade-method and to see this in relation with the highly optimized OpenCV method.

The numerical evaluation was performed for different, benchmark sequences and for sequences from the PETS workshop 2006. No color information was used by either algorithm. All tests were conducted on a PC with an 3.00 Ghz Intel Core2Duo CPU using a C/C++-implementation, one thread and a $\tau = 0.25$. The runtime was measured by the Microsoft©Visual Studio 2008 profiler. To compare the quality of the computed motion vectors, we used ground truth values from [2].

[1] Available at http://ces.clemson.edu/ stb/klt

[2] Available at http://sourceforge.net/projects/opencvlibrary

Fig. 3. (Left) Image detail from the PETS 2006 dataset for comparison of our method with OpenCV. (Middle) Optical Flow result (Birchfield): good overall result. In homogeneous regions, outliers can reach high values and are then filtered out which leads to regions without result vectors. (Right) Optical Flow result (II-LK): overall comparable results. Outliers are not filtered as their size is limited through the introduced linearization.

Figure 3 illustrates a visual difference in the results of II-LK and the standard KLT-Tracker: By using the introduced second linearization in (9), image gradients are not recalculated. This leads to smaller outlier errors which are not filtered out as in the standard KLT algorithm. Visually, the results are not very different than the ones obtained by the standard KLT which shows the applicability of this algorithm for e.g. crowd analysis in surveillance scenarios.

The results obtained for the *PETS 2006* sequence (720×576 pixels) can be found in table 1. Our method outperforms the standard KLT-Tracker implementation and as well OpenCV's highly optimized implementation especially for a large number of feature points. Comparing the number of feature points and the time needed for the computation of their motion vectors shows that the runtime per feature for our method decreases for higher numbers of features, which is due to the fact that in this case windows around feature points tend to overlap. This and the scanline algorithm in our method reduce the corresponding effort of building integral images.

Table 1. Execution time and time per feature for the *PETS 2006* sequence (Values in ms, window size $\Omega = 17 \times 17$) for different numbers of features (N)

Run-time per frame (per feature) in msec

# Features	OpenCV	KLT	II-LK
345280	2249 (0.007)	43470 (0.126)	**713 (0.002)**
86320	850 (0.010)	15726 (0.182)	**292 (0.003)**
21580	210 (0.010)	3977 (0.184)	**97 (0.004)**
9657	95 (0.010)	1813 (0.187)	**63 (0.006)**
5395	54 (0.010)	1044 (0.194)	**50 (0.009)**

Table 2. Execution time for different window sizes (21580 feature points, Values in ms). Compared to the other candidates, the performance of "II-LK" changes only slightly with an increasing window size which is due to the constant time for summing up the windows around features.

Run-time per frame in msec

Ω	OpenCV	KLT	II-LK
25x25	368	7648	**136**
21x21	285	5705	**112**
17x17	210	3977	**97**
13x13	143	2519	**84**
7x7	90	1359	**73**

Table 3. Evaluation results: color code scheme proposed in [2] shows the visual performance of the algorithms. Discarded points are labeled black. EP: relative endpoint errors (window $\Omega = 17 \times 17$), DP: percentage of discarded points ($\overline{0}$ - rounded to zero). Overall, II-LK performs comparable to KLT. As visible in the "Urban3" sequence, it generates less outliers than the KLT algorithm.

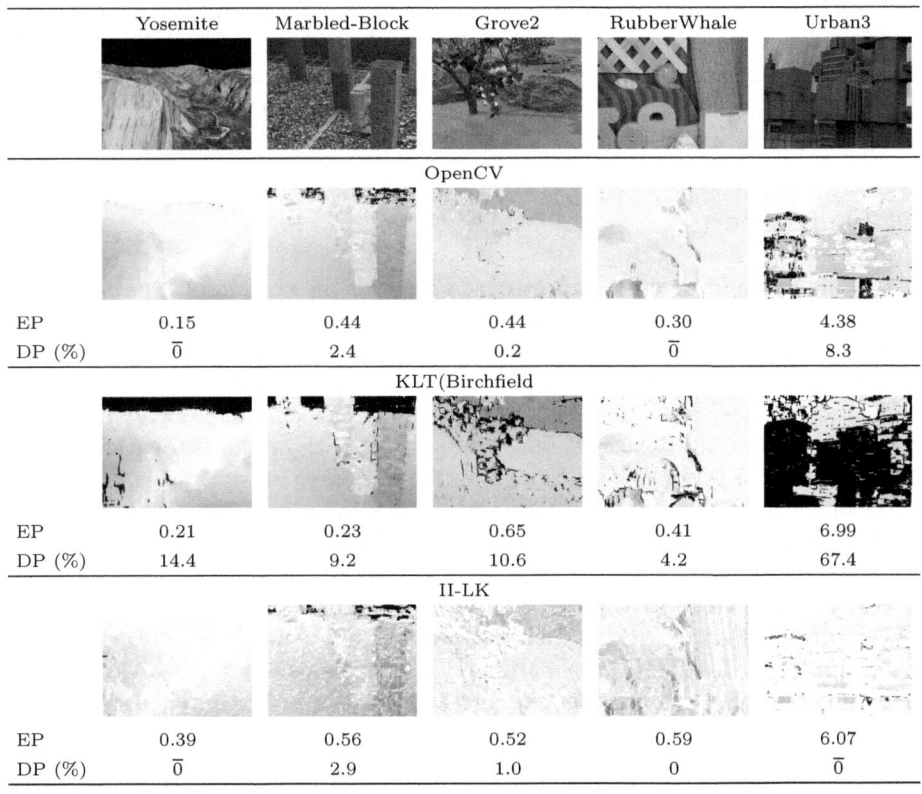

	Yosemite	Marbled-Block	Grove2	RubberWhale	Urban3
OpenCV					
EP	0.15	0.44	0.44	0.30	4.38
DP (%)	$\overline{0}$	2.4	0.2	$\overline{0}$	8.3
KLT(Birchfield					
EP	0.21	0.23	0.65	0.41	6.99
DP (%)	14.4	9.2	10.6	4.2	67.4
II-LK					
EP	0.39	0.56	0.52	0.59	6.07
DP (%)	$\overline{0}$	2.9	1.0	0	$\overline{0}$

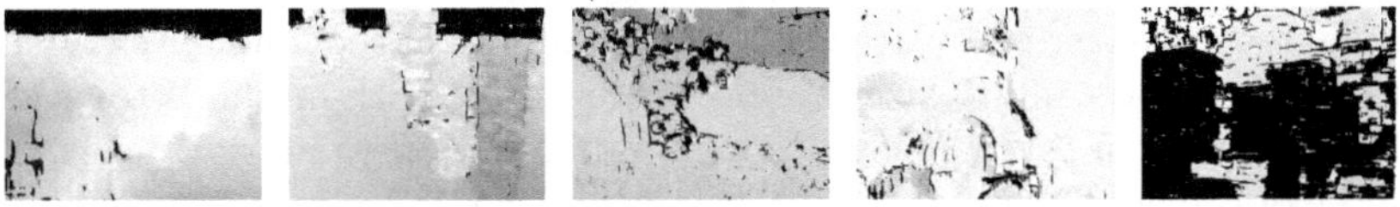

Table 2 shows the runtime that the algorithms needed for different window sizes. The time needed by the KLT-Tracker and OpenCV rises strongly with the window size. Thanks to the constant time integral images need for computing the sums over Ω, our algorithm shows a better performance. Still, the needed time grows slightly because the regions in which integral images are computed grow with the window size.

Table 3 shows the relative endpoint errors of the different methods. Normally the KLT algorithm discards bad matches (outliers) automatically, so for a more accurate comparison, we allow all methods to discard their bad matches. This is done by setting a threshold of 200% of the maximal absolute value of the ground truth data.

Compared to KLT, OpenCV's runtime is superior, but II-LK is still much faster as seen in table 1 and 2. As our algorithm is based on the Birchfield implementation, we focus on a comparison between these two.

The error of e.g. 0.39 pixels in the "Yosemite" sequence compared to Birchfield's 0.21 pixels means that our algorithm expects an average feature 0.18 pixels further away from its actual position than the standard KLT method. This is due to the introduced linearization. On the other hand, II-LK discards less points and can improve the robustness of the results in homogeneous regions. By comparing the results of the "Urban3" sequence, it can be seen, that II-LK still returns many acceptable values e.g. inside the buildings where the KLT tracker produces outliers. Overall, the accuracy loss is acceptable (see also figure 3) for applications which do not focus on subpixel accuracy, as [5] or [14].

5 Conclusion

In this paper we present an approach to speed up the tracking of sparse feature points. By modifying the classical Lucas-Kanade feature tracker by a second linearization, integral images can be used in all parts of the equation. This method is combined with an efficient scanline algorithm to compute integral images only in the regions where they are needed. Depending on the number of features to be tracked and the window size around the features, both methods together allow a huge speed-up of several magnitudes compared to a standard tracker. Compared to the optimized OpenCV code, our method needs also much less time. Applying similar optimizations to our code as they are in OpenCV will further enhance the system's speed.

We propose this algorithm for surveillance applications needing motion vectors and feature trajectories which can be used for crowd description or crowd behaviour analysis. Those applications can be speeded up and can thus track more points which will further enhance their results. With the upcoming of high-definition video content the need for fast optical flow algorithms in the surveillance domain becomes more important and we propose this algorithm as an adequate solution.

For the future, we are working towards an optimization of our method to obtain a superior runtime behaviour and more accurate flow vectors by adapting the minimization scheme.

References

1. Baker, S., Matthews, I.: Lucas-Kanade 20 Years on: A Unifying Framework. International Journal of Computer Vision 56, 221–255 (2004)
2. Baker, S., Scharstein, D., Lewis, J.P., Roth, S., Black, M.J., Szeliski, R.: A Database and Evaluation Methodology for Optical Flow. In: IEEE 11th International Conference on Computer Vision, pp. 1–8 (2007)
3. Birchfield, S., Pundlik, S.J.: Joint tracking of features and edges. In: IEEE Computer Society Conference on Computer Vision and Pattern Recognition, pp. 1–6. IEEE Computer Society, Alaska (2008)
4. Bouguet, J.Y.: Pyramidal Implementation of the Lucas Kanade Feature Tracker - Description of the Algorithm. Technical report, Intel Corporation, Microprocessor Research Labs (1999)
5. Brostow, G.J., Cipolla, R.: Unsupervised Bayesian Detection of Independent Motion in Crowds. In: IEEE Computer Society Conference on Computer Vision and Pattern Recognition, vol. 1, pp. 594–601 (2006)
6. Bruhn, A., Weickert, J., Feddern, C., Kohlberger, T., Schnörr, C.: Variational optical flow computation in real time. IEEE Transactions on Image Processing 14, 608–615 (2005)
7. Bruhn, A., Weickert, J., Schnörr, C.: Combining the Advantages of Local and Global Optic Flow Methods. In: Proceedings of the 24th DAGM Symposium on Pattern Recognition, pp. 454–462. Springer, London (2002)
8. Crow, F.C.: Summed-area tables for texture mapping. In: SIGGRAPH 1984: Proceedings of the 11th annual conference on Computer graphics and interactive techniques, pp. 207–212. ACM, New York (1984)
9. Ercan, A.O., Guibas, L.J.: Object tracking in the presence of occlusions via a camera network. In: IPSN 2007: Proceedings of the 6th International Conference on Information Processing in Sensor Networks, pp. 509–518. ACM Press, New York (2007)
10. Hong, H.S., Chung, M.J.: 3D pose and camera parameter tracking algorithm based on Lucas-Kanade image alignment algorithm. In: International Conference on Control, Automation and Systems, ICCAS, pp. 548–551 (2007)
11. Landgraf, T., Rojas, R.: Tracking honey bee dances from sparse optical flow fields. Technical report (2007)
12. Lucas, B.D., Kanade, T.: An Iterative Image Registration Technique with an Application to Stereo Vision. In: 7th International Joint Conference on Artificial Intelligence, pp. 674–679. William Kaufmann, Vancouver (1981)
13. Paul, V., Michael, J.: Robust Real-time Object Detection. International Journal of Computer Vision 57, 137–154 (2004)
14. Saxena, S., Brémont, F., Thonnat, M., Ma, R.: Crowd Behavior Recogniton for Video Surveillance. In: Blanc-Talon, J., Bourennane, S., Philips, W., Popescu, D., Scheunders, P. (eds.) ACIVS 2008. LNCS, vol. 5259, pp. 970–981. Springer, Heidelberg (2008)
15. Shi, J., Tomasi, C.: Good Features to Track. Technical report, Cornell University (1993)

High Accuracy Optical Flow Method Based on a Theory for Warping: 3D Extension

Weixin Chen and John L. Barron

Dept. of Computer Science
The University of Western Ontario
London, Ontario, Canada, N6A 5B7
barron@csd.uwo.ca

Abstract. This paper describes the implementation and qualitative and quantitative evaluation of a 3D optical flow algorithm, whose derivation is based on the 2D optical flow method published by Brox et al. [ECCV2004]. The optical flow minimizes an energy function built with three assumptions: a brightness constancy assumption, a gradient constancy assumption, and a smoothness assumption. Brox et al. minimize the 2D version of this function using a robust estimator, which make the functional convex and guarantees convergence to a single solution. They propose a numerical solution based on nested fixed points iterations and use this scheme within a coarse-to-fine warping strategy in a 2D hierarchical image pyramid. In our 3D extension, our solution requires the regularization of a 3D function based on 3D extensions of their assumptions in a 3D hierarchical volume pyramid. We solve the corresponding Euler-Lagrange equations iteratively using nested iterations. We present 3D quantitative results on three sets of 3D sinusoidal data (with and without motion discontinuities), where the correct 3D flow is known. We also present a qualitative evaluation on the 3D flow computed using gated MRI beating heart sequence.

Keywords: 2D and 3D optical flow, regularization, brightness constancy constraint, gradient constancy constraint, Horn and Schunck smoothness constraint, hierarchical volume pyramid, 3D reverse warping, quantitative and qualitative error analysis.

1 Introduction

Starting with the pioneering work of Horn and Schunck (1981) [1] and Lucas and Kanade (1981) [2], many methods have been proposed to estimate optical flow. Brox et al. [3] presented a variational model for computing 2D optical flow that integrates three constraints: brightness constancy and global smoothness (both proposed by Horn and Schunck [1]) and gradient constancy. They employed an image warping technique and showed that warping is a good way to handle large displacements in a multi-resolution pyramid. Their published quantitative results are still the best for the Yosemite image sequence. Indeed, they obtain near perfect flow for the clouds in the Yosemite images by using heavy smoothing [4]. These clouds are fractal-based deformable objects and should not lend themselves to an accurate flow recovery. Later, Papenberg et al. [5] extended this method and investigated additional constancy constraints. These only make the flow slightly better so we ignore them in our algorithm.

A. Campilho and M. Kamel (Eds.): ICIAR 2010, Part I, LNCS 6111, pp. 250–262, 2010.

This paper presents the design, implementation and analysis of a 3D extension of the 2-frame 2D Brox et al. algorithm in MatLab to see if we can obtain good 3D flow using 3D volumetric data sequences [6]. We quantitatively evaluate our algorithm on three datasets of synthetic sinusoidal data (with both continuous and discontinuous motions) and we also perform qualitative evaluation on a gated MRI cardiac dataset [7,8,6].

2 3D Optical Flow Algorithm

Brox et al.'s algorithm [3] uses several constraints which we extend to 3D. The **Grayvalue Constancy Constraint** requires the grayvalue distribution about a voxel move with that voxel. That is, $I(x,y,z,t) = I(x+u, y+v, z+w, t+1)$, where I denotes a 4D volumetric dataset and $(u, v, w, 1)$ is the displacement vector between an image (when we say image from now on we mean 3D image or volume) at times t and $t+1$. Performing a 1st order Taylor series expansion of $I(x, y, z, t)$, yields the **optical flow constraint** (sometimes called the **motion constraintt**) equation in 3D, $\nabla I \cdot (u, v, w) + I_t = 0$, where $\nabla I = (I_x, I_y, I_z)$ is the spatial intensity gradient at (x, y, z). The **Gradient Constancy Constraint** is a constraint that ensures I is invariant under small grayvalue changes by requiring $\nabla I(x,y,z,t) = \nabla I(x+u, y+v, z+w, t+1)$. These constraints only consider displacement of a voxel locally without taking into account the interaction between neighboring voxels. Therefore, if the gradient vanishes somewhere or if the flow can only be computed in the direction normal to the gradient (this is the aperture problem), the model presented so far will not work. As a result, we need to include a spatial Horn and Schunck like **Smoothness Constraint** to attenuate these problems and give a globally smooth flow field. This constraint requires $|\nabla u|^2 + \nabla v|^2 + |\nabla w|^2$ be as near 0 as possible. Finally, a hierarchical pyramid of the images is constructed with smoothing and downsampling using a small reduction factor, $\eta \approx 0.95$. We use MatLab's function **imresize** to build the pyramid from 2 original Gaussian blurred images ($\sigma = 1.3$). The use of a pyramid allows faster motions to be computed using a coarse to fine strategy (motions are slower the further up the pyramid you go), where at each pyramid level, the flow between the 2 images is computed and use to reverse warp the 2nd image into the 1st image: if the flow is good, the 1st image and the warped second image will closely agree. Warping, effectively, removes the motion from the 2nd image.

2.1 The Energy Function

The grayvalue and gradient constraints are measured by the energy:

$$E_{Data}(u,v,w) = \int_{\Omega} \Psi \left(|I(x+u, y+v, z+w, t+1) - I(x,y,z,t)|^2 \right.$$
$$\left. + \gamma \, |\nabla I(x+u, y+v, z+w, t+1) - \nabla I(x,y,z,t)|^2 \right) dxdydzdt. \tag{1}$$

Here Ψ does robust estimation by ensuring the function is convex (a single global solution results) and $\Psi(s^2) = \sqrt{s^2 + \epsilon}$. Brox et al. use $\epsilon = 0.001$. A smoothness term that models the assumption of piecewise smoothness is:

$$E_{smooth}(u,v,w) = \int_{\Omega} \Psi \left(|\nabla u|^2 + |\nabla v|^2 + |\nabla w|^2 \right) dxdydzdt. \tag{2}$$

Here $\nabla = (\partial_x, \partial_y, \partial_z)$ is the spatio-temporal gradient. The total energy is the weighted sum between the data term and the smoothness term:

$$E(u, v, w) = E_{Data} + \alpha E_{smooth}, \tag{3}$$

with regularization parameter $\alpha > 0$. Now the computational task is to find the values u, v and w that minimizes this function. Brox et al. derive the Euler-Lagrange equations (which are nonlinear) and then use a nested iteration scheme to minimize these. The velocity at iteration $a+1$, $(u^{a+1}, v^{a+1}, w^{a+1})$, is computed from the velocity at iteration a as (u^a, v^a, w^a) + (du^a, dv^a, dw^a), where (du^a, dv^a, dw^a) is the iteration improvement velocity vector. The idea is to initially set (du^a, dv^a, dw^a) to $(0, 0, 0)$, compute Ψ'_{Data} and Ψ'_{Smooth} using these values at the start of the outer iterations and then to iteratively refine (du^a, dv^a, dw^a) in an inner iteration to best satisfy these Ψ' values. Then these new (du^a, dv^a, dw^a) values are used to update Ψ'_{Data} and Ψ'_{Smooth} and modify (u^a, v^a, w^a) for the next step in the outer iteration, after which (du^a, dv^a, dw^a) are zero to $(0, 0, 0)$ again and then iteratively re-computed in another inner iteration using these new Ψ' values. This outer iteration and all the inner iterations terminate when either a specified maximum number of iterations is reached (typically 100 or 200) or the difference between the computed vector between two adjacent iterations is less than a threshold TOL (we use 10^{-3} in this paper).

Processing starts at the highest level in the pyramid. This computed flow field is then projected down one level in the pyramid (see details in Section 2.3) and used to inverse warp the second image into the first image (see detail in Section 2.4). The idea is that the measured motion between the 1st and 2nd images is "removed". Of course, the flow field used to do the warping is not precisely correct. Flow is now computed between the 1st image and the inverse warped 2nd image and this correction flow field is added to the projected flow field from the higher pyramid level to obtain the new flow field for that level. Projection and warping are continually performed until the final level in the pyramid (the original image) is reached. Note that this processing allows "fast" motion to be handled because at the higher levels in the pyramid the motion is slowed by the image size reduction and the warping calculation prevents the motion from increasing in magnitude as processing descends the pyramid. Full mathematical development, including all the equations and their derivations and pseudo code for the nested iterations algorithm are available in a M.Sc. thesis [6].

2.2 Intensity Differentiation

Brox et al. used 4 point central differences to compute spatial derivatives I_x, I_y and I_z using the kernel $(0.0866, -0.6666, 0.0, 0.6666, -0.0866)$. Second order derivatives are computed using the same kernel applied on the first order derivatives. Temporal derivatives are computed as simple voxel differences. In image processing nomenclature, this is called a 2 point difference and is known to be a poor approximation to a temporal derivative.

2.3 Projecting Velocities between Adjacent Levels

Note that the size (including the width, the height and the depth) of images at different pyramid levels are not the same. Therefore, we use MatLab function `imresize` to use resample the new flow field (with rescaling by multiplication of this new flow field by expansion factor $\frac{1}{\eta}$) to go from a coarser level to the next finer level in the pyramid.

2.4 3D Inverse Warping

Given a flow field for an image $I(t)$, $\boldsymbol{v}(t) = (u(t), v(t), w(t))$ and the 2nd image $I(t+1)$ in the sequence we can compute the 1st image using inverse warping. That is, for floating point image location $(i + u(t), j + v(t), k + w(t))$, one can use tri-linear interpolation via MatLab function `interp3` on the grayvalues at the 8 surrounding neighbourhood integer image locations in the 2nd image to compute the grayvalue at integer location (i, j, k) in the 1st image. The reverse nature of this calculation ensures each pixel in the interpolated 1st image gets a grayvalue (if $(i+u(t), j+v(t), k+w(t))$ is outside the image boundaries the interpolated value is set to 0).

3 Generation of 3D Sinusoid Volume Datasets

The main advantage of sinusoidal sequences is that both the flow fields and the 1st and 2nd order derivatives can be computed precisely, allowing quantitative analysis of the algorithm's performance. We generate three sinusoid volume datasets: **sinL**, **sinR** and **sin**. Each volume contains 31 slices of 256×256 data (unsigned shorts in the range [0-4095], i.e. 12 bits). The **sin** data is a combination of **sinL** and **sinR** which have different velocities in its left part and right part. We generate **sinL** and **sinR** using:

$$sin(\boldsymbol{k}_1 \cdot \boldsymbol{p} + \omega_1 t) + sin(\boldsymbol{k}_2 \cdot \boldsymbol{p} + \omega_2 t) + sin(\boldsymbol{k}_3 \cdot \boldsymbol{p} + \omega_3 t), \tag{4}$$

where $\boldsymbol{k}_i = (k_{ix}, k_{iy}, k_{iz})$ are the spatial frequencies, $\boldsymbol{p} = (x, y, z)$ are 3D spatial coordinates, ω_i is the temporal frequency and t is the temporal coordinate. We use $\boldsymbol{v}_L = (3, 2, 1)$ for the **sinL** sequence and $\boldsymbol{v}_R = (-3, -2, -1)$ for the **sinR** sequence. The **sin** dataset is made from the **sinL** and **sinR** datasets with a motion boundary

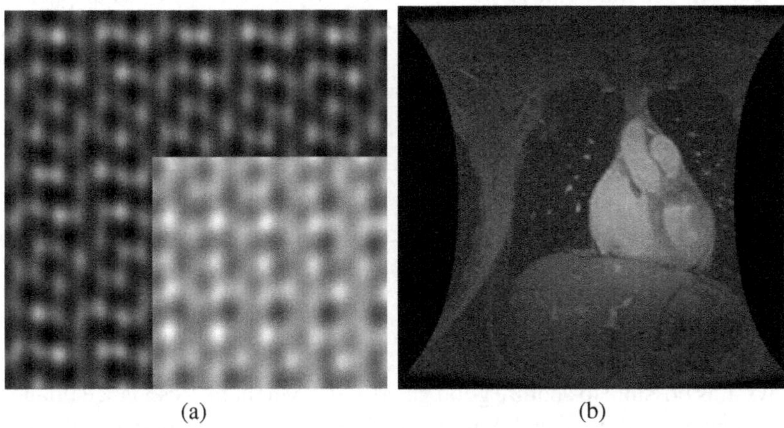

(a) (b)

Fig. 1. (a) The 10th slice of the **sin.9** sinusoid dataset and (b) the 40th slice of the **10phase.9** MRI dataset separating the two sinusoids. Initially, the upper left corner point of the boundary at (120,120,10) and this is displaced by $\boldsymbol{v}_R$ at each frame. Simple differentiation of Equation (4) gives all 1st and 2nd order derivatives for the 3 sequences. Figure 1a shows the 10th slice of the **sin.9** dataset while Figure 2 shows the correct flow.

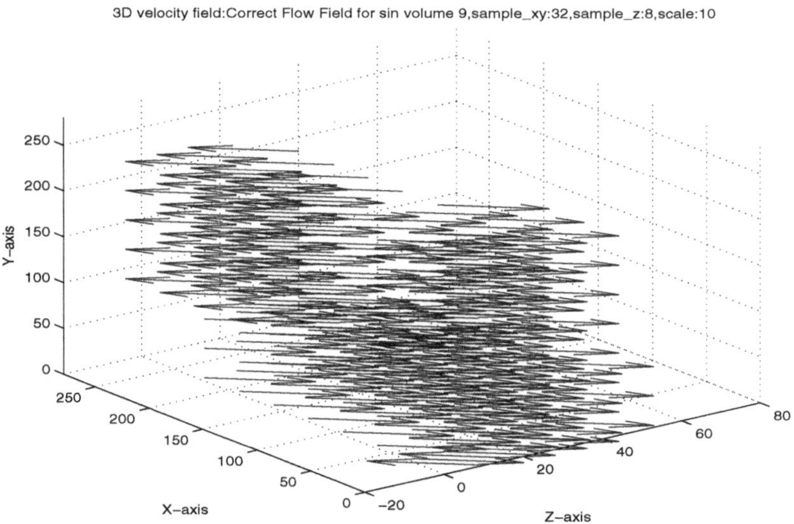

Fig. 2. The correct flow field of the 9th volume of **sin** with a downsampling rate 8 in the z dimension and 32 in the x and y dimensions and with a scale factor of 10

4 3D Quantitative Error Measurement

We use the average Fleet 4D angular error measurement [9] to measure the error between the correct 3D flow field and the computed 3D flow field. Velocity can be written as displacement per time unit as in $\boldsymbol{v} = (u, v, w)$ pixels/frame or as a space-time direction vector $(u, v, w, 1)$ in units of (pixel,pixel,pixel,frame). Let $\boldsymbol{v}_c = (u_c, v_c, w_c, 1)$ represent the correct velocity at the pixel (i, j, k) and $\boldsymbol{v}_e = (u_e, v_e, w_e, 1)$ represent the computed velocity at point (i, j, k). We can compute the 4D angular error as:

$$\psi_E = \arccos\left(\frac{(u_e, v_e, w_e, 1)}{\sqrt{u_e^2 + v_e^2 + w_e^2 + 1}} \cdot \frac{(u_c, v_c, w_c, 1)}{\sqrt{u_c^2 + v_c^2 + w_c^2 + 1}}\right). \tag{5}$$

Angular error has the advantages of encoding both magnitude and direction error as 1 number and preventing zero division problems.

5 Gated MRI Cardiac Datasets

Nowadays, it is possible to acquire good gated MRI (Magnetic Resonance Imagery) data of a human beating heart. The 3D motions of the heart can provide useful information for physician to detect heart disease. However, as we shall see, the measurement of 3D velocities of a beating heart is still challenging [8].

The gated MRI data we use was generated by the Robarts Research Institute at the University of Western Ontario [7]. Each set of this data contains 20 volumes of 3D data for one EGK synchronized heart beat. In our experiments, we examined the flow for one dataset, **10phase.9**, which has size $256 \times 256 \times 75$ and voxel intensities in the range $[0 - 4095]$. The MRI datasets is considerably more complex than our synthetic sinusoid datasets. A beating human heart is deformable, and its motion is discontinuous in space and time: different chambers in the heart are contracting/expanding at different times and the heart as a whole undergoes a twisting motion as it beats. The heart is also "textureless" in places, meaning good intensity derivatives are difficult to compute. The word "gated" refers to the way the data is collected: 1 or a few slices of each volume set are acquired at the same instance in a cardiac cycle. A patient lies in an MRI machine and holds his breath for approximately 42 second intervals to acquire each set of slices. This data acquisition method relies on the patient not moving or breathing during the the acquisition (this minimizes heart motion caused by a moving diaphragm or expanding/contracting lungs) [8]. Figure 1 shows the 40th slice of the **10phase.9** dataset.

6 Experimental Results

Table 1 show the comparison between a basic version of 3D Horn and Schunck [8] and a version that uses Brox et al.'s hierarchical pyramid and our implementation of 3D Brox

Table 1. The 3D angular errors and standard deviations for flow fields with 100% density computed by 3D Horn and Schunck ($\alpha = 100.0$, 100 iterations), hierarchical 3D Horn and Schunck ($\alpha = 100.0$, 10 levels of pyramid, 100 iterations), and 3D Brox et al.'s algorithm ($\alpha = 100.0, \gamma = 100.0$ and $\gamma = 0.0$, 10 levels in the pyramid, outer iterations=1, inner iterations=100, all with $3 \times 3 \times 3$ median prefiltering, for **sinL.9**, **sinR.9** and **sin.9**

sinL	AAE	STD
3D Horn and Schunck (Brox derivatives, $\alpha = 100.0$)	22.89°	9.08°
3D Horn and Schunck (Correct derivatives, $\alpha = 100.0$)	0.24°	0.25°
3D Horn and Schunck ($\alpha = 100.0$,10-levels pyramid)	10.09°	6.52°
3D Brox et al.($\alpha = 100.0, \gamma = 100.0$,10-levels pyramid)	2.06°	2.56°
3D Brox et al.($\alpha = 100.0, \gamma = 0.0$,10-levels pyramid)	0.65°	1.10°
sinR	AAE	STD
3D Horn and Schunck ($\alpha = 100.0$)	23.26°	12.56°
3D Horn and Schunck (Correct derivatives, $\alpha = 100.0$)	0.78°	0.41°
3D Horn and Schunck ($\alpha = 100.0$,10-levels pyramid)	9.90°	6.87°
3D Brox et al.($\alpha = 100.0, \gamma = 100.0$,10-levels pyramid)	1.99°	3.55°
3D Brox et al.($\alpha = 100.0, \gamma = 0.0$,10-levels pyramid)	0.75°	1.56°
sin	AAE	STD
3D Horn and Schunck ($\alpha = 100.0$)	32.32°	24.49°
3D Horn and Schunck (Correct derivatives, $\alpha = 100.0$)	12.78°	24.49°
3D Horn and Schunck ($\alpha = 100.0$,10-levels pyramid)	22.67°	29.00°
3D Brox et al.($\alpha = 100.0, \gamma = 100.0$,10-levels pyramid)	16.20°	29.96°
3D Brox et al.($\alpha = 100.0, \gamma = 0.0$,10-levels pyramid)	13.11°	28.71°

et al.'s algorithm. AAE is the abbreviation for Average Angular Error and STD is the abbreviation for Standard Deviation. Poor temporal differentiation (simple differences) is the main reason for the poor results for 3D Horn and Schunck. Error results computed using the correct derivatives are much more accurate than those computed using Brox et al.'s differentiation. For all three datasets, hierarchical 3D Horn and Schunck gives much better results than the original 3D Horn and Schunck, showing one of the benefit of using a pyramid. 3D Brox et al.'s results ($\alpha = 100$, $\gamma = 100$ and 10 levels of pyramid) surpasses them both. When the gradient parameter γ is turned off even better results can be obtained (see below). These results show that the worst results are for the **sin** dataset, which has significant error at the motion boundary. We sometimes use medium filtering in an attempt to remove outliers.

6.1 Inner and Outer Iterations and Convergence

Another experiment we performed with the correct derivatives was to see the effect of the inner and outer iteration on the result. The outer iterations update Ψ'_{data} and Ψ'_{smooth}, which are used in Equations (1) and (2) [6]. The inner iterations update the increments in velocities du, dv and dw. We can define the motion constraint term $eqID$, and gradient constraint terms, $eqIx$, $eqIy$ and $eqIz$, as:

$$eqID = (I_D + I_x du + I_y dv + I_z dw)^2, \quad (6)$$

$$eqIx = (I_{xD} + I_{xx} du + I_{xy} dv + I_{xz} dw)^2, \quad (7)$$

$$eqIy = (I_{yD} + I_{yx} du + I_{yy} dv + I_{yz} dw)^2, \quad (8)$$

$$eqIz = (I_{zD} + I_{zx} du + I_{zy} dv + I_{zz} dw)^2. \quad (9)$$

Then, the arguments to Ψ'_{Data} can be written as $\Psi'\,(eqID + \gamma(eqIx + eqIy + eqIz))$. As velocities get better, Equation (6) and Equations (7), (8) and (9) should become close to 0. Tables 1 shows the average change in Ψ'_{data} and Ψ'_{smooth}, the average values of $eqID$, $eqIx$, $eqIy$ and $eqIz$, the average angle error and standard deviation for 10 inner iterations for the 1st, the 5th and the 10th outer iteration of **sin** using Brox et al.'s spatial derivatives and temporal differences. We can see that as we perform more iterations, $eqID$, $eqIx$, $eqIy$ and $eqIz$ become smaller and smaller, which means the constraints are better satisfied. Brox et al. [3] used 10 inner and 10 outer iterations only for all their results. However, they never say why they used these number of iterations or report any investigation into the effect of this number of iterations on the flow accuracy.

The results for **sinL** and **sinR** converge quickly. By the 10th outer iteration, values of these equations are almost zeros, and the average angle error goes down to $0.65°$-$0.75°$, which is very close to the correct velocities (within roundoff error). Instead, we concentrate on the **sin** data, which is our worst case synthetic dataset.

Table 2 show the average change in Ψ'_{data} and Ψ'_{smooth}, the average values of $eqID$, $eqIx$, $eqIy$ and $eqIz$, the average angle error and standard deviation for each of the 10 inner iterations for the 1st, the 5th and the 10th outer iterations for the **sin** data using correct derivatives. [Using correct derivatives eliminates differentiation as a cause of poor performance.] In all cases, $eqID$, $eqIx$, $eqIy$ and $eqIz$ still decrease as more iteration are performed, although it is harder for them to approach 0 as there is now significant flow error at the motion boundary.

Table 2. Average differences in Ψ'_{data} and Ψ'_{smooth} and average values of $eqID$, $eqIx$, $eqIy$, $eqIz$, average angle error and standard deviation for 10 inner iterations of the 1st, 5th and 10th outer iterations for **sin.9** computed with Brox et al. derivatives ($\alpha = 100$, $\gamma = 100$, number of outer iteration=10, number of inner iteration=10, 1 pyramid level with no median filtering)

1st Outer Iteration							
Inner	Ψ'_{data} diff.	$\Psi'_{sm.}$ diff.	ave_eqID	ave_eqIx	ave_eqIy	ave_eqIz	AEE.±STD.
1	0.0005877	15.8113	170839.255	3722.804	2099.715	2897.699	$70.69° \pm 1.73°$
2	0.0000193	2.0166	159501.420	3455.617	1924.327	2623.482	$66.85° \pm 3.05°$
3	0.0000188	1.6025	148917.033	3230.226	1776.601	2400.909	$63.41° \pm 4.17°$
4	0.0000190	1.0764	139030.474	3031.516	1647.400	2211.502	$60.34° \pm 5.16°$
5	0.0000193	0.7289	129889.763	2853.311	1532.762	2047.496	$57.63° \pm 6.04°$
6	0.0000196	0.5160	121482.966	2691.516	1429.927	1903.577	$55.23° \pm 6.83°$
7	0.0000199	0.3808	113774.957	2543.410	1336.984	1776.070	$53.12° \pm 7.56°$
8	0.0000202	0.2914	106717.469	2407.037	1252.514	1662.261	$51.25° \pm 8.24°$
9	0.0000205	0.2304	100257.859	2280.929	1175.425	1560.079	$49.61° \pm 8.86°$
10	0.0000208	0.1866	94343.514	2163.938	1104.849	1467.893	$48.15° \pm 9.44°$
5th Outer Iteration							
Inner	Ψ'_{data} diff.	$\Psi'_{sm.}$ diff.	ave_eqID	ave_eqIx	ave_eqIy	ave_eqIz	AEE.±STD.
1	0.0131558	5.8824	2724.246	20.078	28.104	34.088	$19.30° \pm 20.05°$
2	0.0347743	0.2602	2234.735	14.483	22.206	25.176	$18.60° \pm 20.12°$
3	0.0059093	0.2068	2041.729	12.709	20.032	22.482	$18.00° \pm 20.19°$
4	0.0051870	0.2228	1905.180	11.597	18.541	20.735	$17.49° \pm 20.25°$
5	0.0045679	0.2327	1804.209	10.821	17.427	19.489	$17.03° \pm 20.30°$
6	0.0043418	0.2334	1726.878	10.246	16.556	18.546	$16.62° \pm 20.34°$
7	0.0042219	0.2287	1666.133	9.807	15.859	17.809	$16.25° \pm 20.37°$
8	0.0041733	0.2211	1617.340	9.465	15.294	17.218	$15.91° \pm 20.40°$
9	0.0041689	0.2119	1577.376	9.195	14.830	16.736	$15.61° \pm 20.43°$
10	0.0041954	0.2017	1544.075	8.977	14.445	16.335	$15.32° \pm 20.45°$
10th Outer Iteration							
Inner	Ψ'_{data} diff.	$\Psi'_{sm.}$ diff.	ave_eqID	ave_eqIx	ave_eqIy	ave_eqIz	AEE.±STD.
1	5.2872770	11.2821	684.577	5.147	7.791	9.594	$6.90° \pm 18.75°$
2	0.6563276	0.0441	667.738	5.040	7.599	9.421	$6.80° \pm 18.71°$
3	0.3178679	0.0422	658.751	5.011	7.546	9.373	$6.71° \pm 18.67°$
4	0.2265391	0.0396	651.884	4.991	7.507	9.332	$6.64° \pm 18.64°$
5	0.1905575	0.0362	646.484	4.975	7.475	9.300	$6.58° \pm 18.62°$
6	0.1744226	0.0330	642.100	4.963	7.449	9.274	$6.52° \pm 18.60°$
7	0.1672531	0.0301	638.465	4.952	7.427	9.252	$6.47° \pm 18.59°$
8	0.1642882	0.0275	635.401	4.943	7.408	9.233	$6.43° \pm 18.57°$
9	0.1626395	0.0252	632.786	4.935	7.392	9.216	$6.39° \pm 18.56°$
0	0.1600340	0.0232	630.531	4.928	7.377	9.201	$6.36° \pm 18.55°$

We used 1 outer iteration and 100 inner iterations for most of the results in this paper, because this gives much better results than using different sets of numbers of inner and outer iterations in our implementation. Again, we emphasize that Brox et al. [3] do not investigate this behaviour and do not (necessarily) iterate until convergence. One possible reason we don't get good results when using more outer iterations may be the poor quality of the temporal derivatives used. These derivatives are used to update

Ψ'_{data} and Ψ'_{smooth} in each outer iteration which, in turn, may propagate errors in further iterations.

6.2 Hierarchical Optical Flow

One way to understand the Brox et al. method is to see how fields change at different pyramid levels. Figures 3 and 4 show the start and end flow fields for **sin.9** for levels 10 and 1. At top of the pyramid (level 10), the size of the volume is the smallest and then both the volume size and the magnitude of the flow get larger and larger as it goes down the pyramid. Note the many erroneous flow vectors in the final flow field at the motion discontinuity at level 1. Since the correct velocities are known for the lowest pyramid image, we can compute the correct velocities at any level with a certain η value by resizing and rescaling. This allows quantitative evaluation at each level of the pyramid so we can see how the accuracy of the results evolve from level to level. Table 3 shows the angle error for the **sin.9** data at different levels. We can see that the angular error goes down to $13.56°$ at level 7, but then increase to $16.20°$ at level 0. This is most likely because we can't compute accurate velocities in the boundary area where the 2 sinusoids, **sinL** and **sinR**, meet. Warping in this area of the image will add error and these errors will accumulate as the pyramidal processing continues. When the refinement can't suppress the errors, the angular error increases.

We also exam the effect of 3D inverse warping. The spatial difference between the 1st and 2nd volume images should become smaller and smaller as we descend the pyramid, if the velocities we compute are becoming more and more accurate. Figures 5a to 5c shows the difference between the 15th slice of the 1st and 2nd volume images for the **sin.9** data at (a) the 10th, (b) the 9th and (c) the 1st pyramid level. White means no difference while black means large difference. We can observe obvious refinement between the top two levels, but for later levels, refinements between adjacent levels are very slight and so we can hardly tell if there is any changes in the image. Therefore,

Table 3. The 3D angular errors, standard deviation and density for the flow fields at all pyramid levels of the 9th volume of the **sin** data (10 levels of pyramid, $\alpha = 100.0$, $\gamma = 100.0$, number of outer iteration=1, number of outer iterations=1, number of inner iterations=100, size of median filter $(3 \times 3 \times 3)$

Pyramid level	AAE ± STD	Density(%)
10	$25.12° \pm 22.27°$	100.00
9	$20.12° \pm 24.19°$	100.00
8	$15.45° \pm 24.48°$	100.00
7	$13.56° \pm 25.26°$	100.00
6	$13.85° \pm 25.60°$	100.00
5	$14.95° \pm 26.86°$	100.00
4	$16.05° \pm 28.30°$	100.00
3	$16.89° \pm 29.79°$	100.00
2	$16.55° \pm 29.92°$	100.00
1	$16.20° \pm 29.96°$	100.00

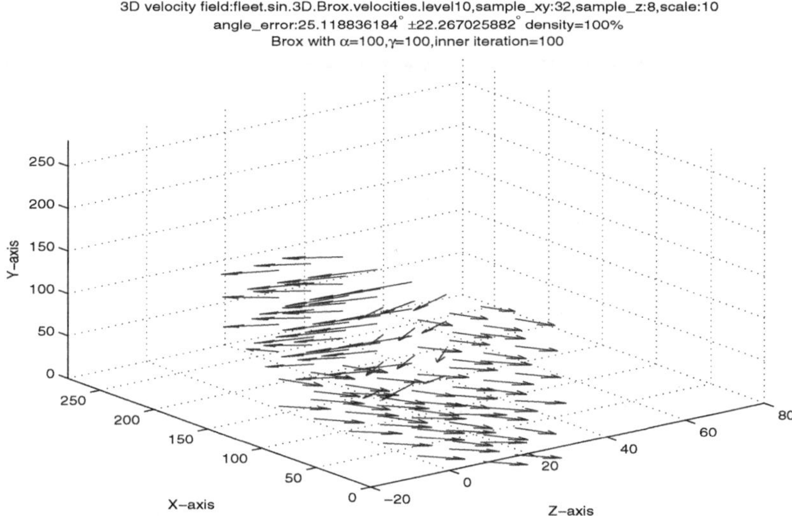

Fig. 3. The computed flow field at the 10th level for **sin.9** (with $\alpha = 100.0$, $\gamma = 100.0$, number of levels in the pyramid=10, outer iteration=1, inner iteration=100, $3 \times 3 \times 3$ median filtering), downsampling rate 32 in x and y dimensions and 8 in the z dimension and flow scaling by 10

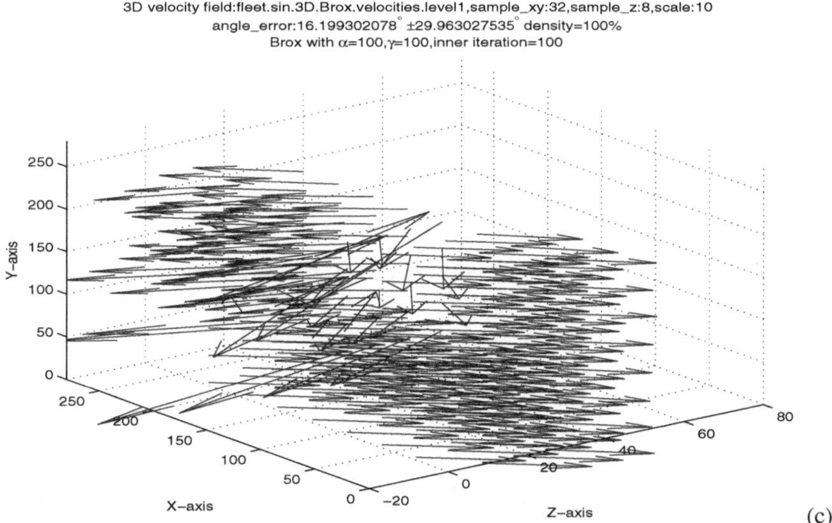

(c)

Fig. 4. The computed flow field at the 1st level for **sin.9** (with $\alpha = 100.0$, $\gamma = 100.0$, number of levels in the pyramid=10, outer iteration=1, inner iteration=100, $3 \times 3 \times 3$ median filtering), downsampling rate 32 in x and y dimensions and 8 in the z dimension and flow scaling by 10

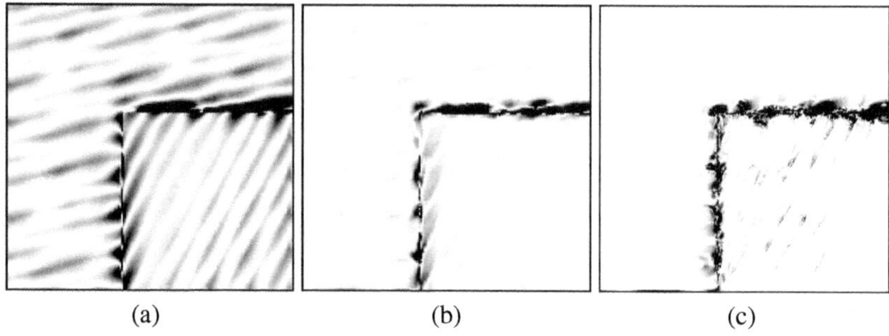

Fig. 5. The difference between the 15th slice of the left image and the right image for the 9th volume of **sin** data at (a) the 10th, (b) the 9th, and (c) the 1st pyramid levels

Table 4. Error analysis for parameter variation for **sin.9**

Parameter				Angular Error	
Pyramid levels	α	γ	filter size	AAE.	STD.
1	100	100	3	28.50°	25.08°
10	10	100	3	30.14°	32.89°
10	100	0	3	13.11°	28.71°
10	100	20	3	14.55°	28.27°
10	100	50	3	20.89°	33.53°
10	100	100	0	17.03°	30.40°
10	100	100	3	16.20°	29.69°
30	100	0	3	12.90°	28.83°

we just show the final difference image at the last level. We can see that at the motion boundary there is always a large difference because we can't actually compute the correct velocities here and therefore warping in this area will be wrong.

Lastly, we briefly discuss the effects of variation of the various parameters of the algorithm for the **sin** data in Table 4. We can conclude:

- The algorithm works better with more levels of pyramid. It never produce the best result if it is run at just one level. But it is also possible that more levels will cause the accumulation of errors introduced by warping and interpolation.
- Larger α means more smoothing which in most cases produces better results. But this may not be true when there are motion discontinuities in the input images.
- In all of these results, the algorithm seems to return more accurate results when $\gamma = 0$, which means no gradient constraint is used. It is quite likely that the reason we get better result when the gradient constraint is turned off is because of the inaccuracy of the I_{xD}, I_{yD} and I_{zD} values which use simple temporal differences.
- A median filter with larger size does remove more outliers and make the results more accurate.
- Computation costs are high. For example, with 1 outer iteration and 100 inner iterations or with 10 outer iterations and 10 inner iterations, about 2 hours of CPU

time (on a 2.8GHz PC running Linux with 8GB of RAM) was required to run our vectorized MatLab code).

7 Gated MRI Cardiac Datasets Result

We note that the different parts in heart are contracting/expanding at different rates and times and the heart is undergoing a twisting as a whole. Although the heart expansion seems to have been captured, the flow field for **10phase.9** is poor, with many outliers present and is not clinically useful.

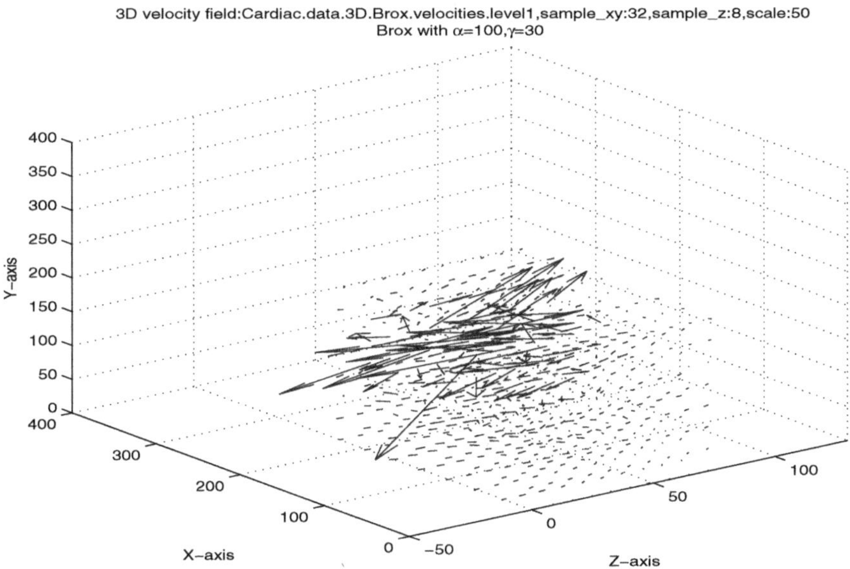

Fig. 6. The computed flow field for **10phase.9** (with $\alpha = 100$, $\gamma = 30$, number of level in the pyramid =10, outer iteration=1, inner iteration=100, $3 \times 3 \times 3$ median filtering, downsampling by 32 in the x and y dimensions and 8 in the z dimensions and a flow scale factor of 50

8 Conclusions and Future Work

Our 3D extension of Brox et al.'s optical flow algorithm produces very good flow for data sequences generated with continuous motion. The algorithm is not designed to handle motion discontinuities and performs poorly there. The flow field for the gated MRI cardiac dataset seems to capture some of essential motions of a beating heart but overall it is not good.

Future work includes finding out how to improve the temporal derivatives we compute and/or how to get the gradient constraint to work under poor temporal differentiation. We are investigating the parameterization of the 4D data using B-splines and a way to get good derivatives. We are currently building a functional model of the heart that will be able to predict the motion of individual parts of the heart over time. We would then add this as an additional constraint to our regularization. Lastly, we are investigating "Oriented Smoothness" in 3D as way to have motion discontinuities.

References

1. Horn, B., Schunck, B.: Determining optical flow. Artificial Intelligence 17(1-3), 185–203 (1981)
2. Lucas, B.D., Kanade, T.: An iterative image registration technique with an application to stereo vision. In: International Joint Conference on Artificial Intelligence, pp. 674–679 (1981)
3. Brox, T., Bruhn, A., Papenberg, N., Weickert, J.: High accuracy optical flow estimation based on a theory for warping. In: Proceedings of the 8th European Conference on Computer Vision, pp. 25–36 (2004)
4. Faisal, M., Barron, J.: High accuracy optical flow method based on a theory for warping: Implementation and qualitative/quantitative evaluation. In: Proceedings of the 4th International Conference on Image Analysis and Recognition, pp. 513–525 (2007)
5. Papenberg, N., Bruhn, A., Brox, T., Didas, S., Weickert, J.: Highly accurate optic flow computation with theoretically justified warping. International Journal of Computer Vision 67(2), 141–158 (2006)
6. Chen, W.: High accuracy optical flow method based on a theory for warping: 3d extension. Master's thesis, Dept. of Computer Science, The University of Western Ontario (2009)
7. Moore, J., Drangova, M., Wiergbicki, M., Barron, J., Peters, T.: A high resolution dynamic heart model. In: Ellis, R.E., Peters, T.M. (eds.) MICCAI 2003. LNCS, vol. 2878, pp. 549–555. Springer, Heidelberg (2003)
8. Barron, J.: Experience with 3d optical flow on gated mri cardiac datasets. In: Proceedings of the 1st Canadian Conference on Computer and Robot Vision, pp. 370–377 (2004)
9. Barron, J., Fleet, D., Beauchemin, S.: Performance of optical flow techniques. International Journal of Computer Vision 12(1), 43–77 (1994)

Improving Accuracy of Optical Flow of Heeger's Original Method on Biomedical Images

Vladimír Ulman

Centre for Biomedical Image Analysis
Masaryk University, Brno 602 00, Czech Republic
xulman@fi.muni.cz

Abstract. The paper is focused on the computation of optical flow from time-lapse 2D images acquired from fluorescence optical microscope. The Heeger's traditional established method based on spatio-temporal filtering is adopted and modified in order to solve issues that arose from this sort of image data. In particular, a scheme for effective and fast computations of complex Gabor convolutions is used. The filter tuning is changed to better support the detection of movement. The least squares fitting of the original method is also revised. A parametric study was conducted to assess optimal parameters. With optimal parameters, the proposed method showed lower average angular errors than the original. C++ implementation is available on the author's web pages.

1 Introduction

It is not unusual to see a method working beautifully on selected data while failing to give good results on other data, at least not in the field of image processing. This is probably the motivation that gives rise to new ideas, development of new or modifications of established methods. The same situation is observed in the field of optical flow computation where, only to give some illustration far from ambition to cite all papers in the field, new methods are being published, recently [3,7], and many derivates are appearing regularly [12,18,14,5]. In this paper, we aim to present some changes to the traditional method by Heeger [9,10] for computing optical flow using spatio-temporal filters. Our motivation are time-lapse 2D image sequences acquired from fluorescence optical microscope to estimate motion of whole cells and their individual inner structures.

Fluorescence optical microscopy focuses, among other targets, on live cell studies. The progress in staining of living cells together with advances in confocal microscopy devices has allowed detailed studies of the behaviour of intracellular components including structures inside the cell nucleus. To increase significance of a study, the typical number of investigated cells in one study varies from tens to hundreds. Trying to automate estimations of movements, for instance to correct for global movement of a cell before its subsequent analysis, in such image analyses has also become a challenge for computer vision methods [4,8].

Images of living cells are acquired periodically. Due to different nature of biological experiment and imaging limitations, such as bleaching of fluorescence in

A. Campilho and M. Kamel (Eds.): ICIAR 2010, Part I, LNCS 6111, pp. 263–273, 2010.

the specimen, the interval between two consecutive acquisitions varies in range from fractions of second up to tens of minutes. The interval is usually a compromise between quality of acquired images and rapidity of observed movement. Thus, we usually deal with an increased level of noise and displacements up to 10 pixels between two consecutive images. However, Hubený et al. reported recently good results achieved with PDE-based optical flow methods with multigrid numerical framework on similar data [11].

For example, time-lapse images of displacements of human HL-60 cell nucleus with moving HP1 protein domains exhibit both global movement of the nuclei with additional local movements of the domains. We artificially generated such image sequences to obtain reliable test data [17] for quantitative evaluations in this paper. Real images, unlike artificially generated images, do not provide dense correct information regarding its content. Without it, it is hard to judge on quality of any optical flow computation method, quantitatively.

We have decided to use a method based on spatio-temporal filters, namely the original Heeger's method [10]. The acknowledged survey by Barron et at. [1] showed that filtering-based methods tend to give good results on variety of different images. These methods are also typical for requiring increased temporal support, e.g. 7 subsequent images were used in Heeger's work. As a real movement of specimen in microscopy is usually not sampled ideally motion aliases may be observed when using only two images from a sequence. This is especially true for the noisy stained nuclei when not viewing it as a whole. This is referred to as the aperture problem in the literature. By considering more images we hoped to alleviate from this shortcoming. Moreover, by increasing the filters' support in the spatial domain we hoped to detect velocities larger than 1px/frame in an original full resolution without scaling input images. The recent results on fast IIR filtering of arbitrarily oriented anisotropic Gaussians [15] were hoped to support it as well as to allow for reasonable accuracy and computational times.

In the next section we will shortly describe the Heeger's method. The third section continues by explaining changes made to it. Results on artificially generated data and discussion will be given in the fourth section. The last section concludes the paper.

2 Heeger's Method

In this section we aim to give a closer look at the original method. We only outline the computation steps and give a bit more details on those where changes will be made in the next section. We do not want to discuss merits or drawbacks nor we want to explain all theory and reasoning that lead researches to their method. Also note that a sequence of 2D time-lapse images can be regarded as a 3D image in which all 2D images are stacked along the z-axis. We occasionally refer to such 3D images in the paper.

The first step of the original method is convolution with broad filter to remove local means or subtract Gaussian filtered copy of input image from the original input image. The Gaussian can be isotropic with large sigma to give very narrow peak in the Fourier domain. Both variants remove DC offset from input 3D image.

The next step is convolutions with three 3D Gabor filtering banks, four filters in each bank. All filters in a bank have the same Gaussian envelope and frequency band but different spatial-frequency tuning. If $[w_x, w_y, w_z]$ is the frequency tuning of a Gabor filter then $(w_x^2 + w_y^2)^{1/2}$ and w_z are constant within a bank. In particullar, there are always four differently spatially-tuned filters within a bank and three differently temporally-tuned banks. All Gaussians are separable along the Cartesian axes. The 2D illustration is shown in Figure 1.

In fact, quadrature pairs of filters, odd-phase and even-phase of identical orientation and bandwidth, is used in the original method. Convolutions with 3D *complex* Gabor filters can be used instead. What follows is computing sum of squared convolution results to compute Gabor energy. It returns the process back to the real domain.

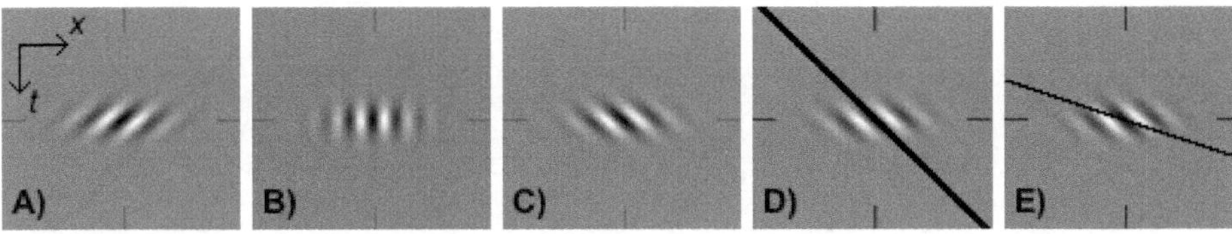

Fig. 1. An example with intentionally large sigmas of even-phase Gabor filters similar to those in the original method. A), B) and C) show x-t sections from 3D images of filters with the same fixed spatial-frequencies w_x, w_y and tuned to detect velocities $(-1, 0)$, $(0, 0)$ and $(1, 0)$, respectively. D) and E) show, in the x-t sections, relation of a patch moving with velocities $(1, 0)$ and $(3, 0)$, respectively, to Gabor filter tuned to $(1, 0)$.

A Gaussian filtering with small sigma then sets some maximum Fourier frequency present in the images. This allows for setting bounds to an integral that enumerate a Gabor filter response on an ideally random 2D image translating with given velocity (u, v). We will denote the results of Gaussian filtering the *measured* responses m_i and values of such integrals the *ideal* responses $R_i(u, v)$, $i = 1 \ldots 12$. There are twelve ideal responses for every combination of (u, v) in the original method.

Finally, the last step is to compare all twelve measured Gabor energies with all twelve-tuples of ideal responses, Figure 4. Heeger suggested to group the twelve-tuples by three, according to the heading of the spatial-frequency of their Gabor filters, and normalize each tripple independently of the others. The result is such (u, v) that minimizes a sum of squared differences between the measured and ideal twelve responses:

$$\sum_{i=1}^{12} \left[m_i - \bar{m}_{g(i)} \frac{R_i(u, v)}{\bar{R}_{g(i)}(u, v)} \right]^2, \tag{1}$$

$$\bar{R}_j(u, v) = \sum_{i, g(i)=j} R_i(u, v), \qquad \bar{m}_j = \sum_{i, g(i)=j} m_i \tag{2}$$

where $g(i) : \langle 1, 12\rangle \to \langle 1, 4\rangle$ is a function that groups responses as specified above. The method was originally presented with Gauss-Newton gradient-descent technique to locate, possibly, global minimum.

Most people start with Gaussian presmoothing of input image sequence before removing the DC offset. Thus, some may consider this a part of the method too.

3 Modifications to Heeger's Method

Basically, we have made only four changes to the original method in order to improve its results on our time-lapse fluorescence microscopy images.

The first change was in the setup of filtering banks. We increased the number of banks to nine. The temporal-frequency tunings matched velocities up to 4 pixels/frame, see Figure 2. Furthermore, the anisotropic Gaussian envelope was oriented according to the frequency tuning of each filter to closely wrap the motion-induced edge in time-lapse images.

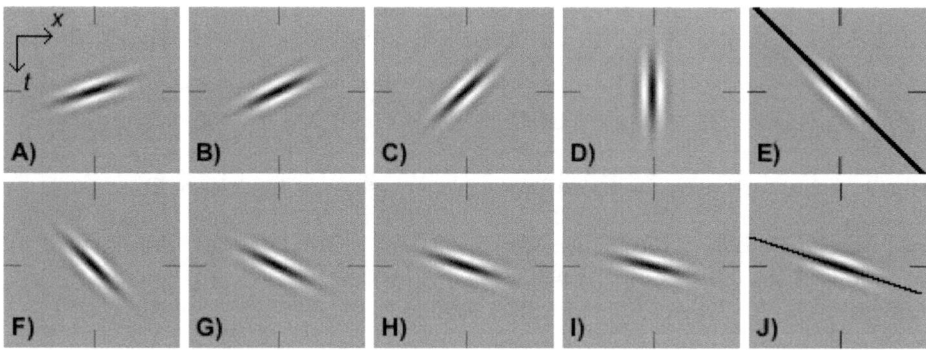

Fig. 2. An example with intentionally large sigmas of even-phase Gabor filters similar to those in the proposed method. A) to D) and F) to I) show x-t sections from 3D images of filters with the same fixed spatial-frequencies w_x, w_y and tuned to detect velocities from $(-3, 0)$ to $(4, 0)$, respectively. E) and J) show, in the x-t sections, relation of a patch moving with velocities $(1, 0)$ and $(3, 0)$ to Gabor filters tuned to $(1, 0)$ and $(3, 0)$, respectively.

The second change was in the normalization of measured and ideal responses. We wanted the least squares fitting to seek ideal responses with major peaks located over major peaks in measured responses, marginal peaks shouldn't have much influence. It is achieved by computing $m_i/\bar{m}_{g(i)}$ and $R_i(u, v)/\bar{R}_{g(i)}(u, v)$, both measured and ideal responses are then scaled linearly to interval $\langle 0, 1\rangle$ and the function $[0.5 - 0.5\cos(\pi x)]^2$ is applied but only on measured responses. This function slightly enhances x when above 0.77 and greatly lowers the smaller ones. This prepares responses m'_i and $R'_i(u, v)$ which are used to seek minimum of

$$\sum_{i=1}^{12} \left[m'_i(m'_i - R'_i(u, v))\right]^2 . \tag{3}$$

The third change was due to our observation that stationary pixels in the input image give rise to responses that form a ridge similar to the ideal response for (-0.5,0) in Figure 4. We employ the ridge detection which compares measured responses of stationary filters, on the ridge, with their two filters with the same spatial-frequencies and closest temporal-frequencies, at foots of both hillsides of the ridge. For each spatial tuning, two pairs of responses are always compared in this way. We count in how many pairs the stationary filter has equal or stronger response. If the number is greater than 1.5 times the number of spatial tunings used then a ridge is deemed present and velocity $(0, 0)$ is returned without any further search for other velocities. Similarly, if there is any measured response from a stationary filter greater than 0.6 or sum over all measured responses is close to zero then we also return with velocity $(0, 0)$.

The last change is that we always seek for correct (u, v) among 289 combinations spread on a grid $u, v \in \langle -4, +4 \rangle$ in steps of 0.5. The search is repeated once in the vicinity of global minimum with smaller step 0.1.

Returning to the convolutions with filtering banks, we have developed a scheme for convolution with arbitrarily-oriented anisotropic Gaussians previously [15]. A given 3D Gaussian is separated along six specifically oriented axes into six 1D Gaussians. We used IIR filtering [19] whose speed performance is not influenced by the magnitude of Gaussian's sigma. The scheme allowed for enhancement [16] such that convolutions with Gaussian or Gabor filtering banks can be computed nearly optimally according to [13,2] and with very low computational times. As a result, improved suppression of DC offset [2] could have been implemented as well.

Repetitive computing of ideal responses can be rather time demanding. We implemented software caches to alleviate this.

4 Results and Discussion

We compared three variants of the original Heeger's method for computations of optical flow: the original method as described in the section 2 with global search for optimal (u, v) instead of gradient descent, the extended original method, which is nearly the same as the proposed method from the section 3 with the difference that anisotropic Gaussian envelopes are (not oriented and hence) still separable only along Cartesian axes, and the proposed method with anisotropic oriented envelopes.

We generated two test sets of 2D+t time-lapse artificial images from cell microscopy [17]. Images in the first set showed only local movements of foreground, i.e. inner cell structures of interest. Images in the second set showed global movements of the whole cell with additional local foreground movements. There were 10 time-lapse images with different movements in each set.

Owing to the generation of images, we had correct flow fields at hand for evaluating quality by means of the average angular error [6]. The angular error is an angle between computed vector $(u, v, 1)$ and the expected vector $(u_{gt}, v_{gt}, 1)$:

$$\arccos\left(\frac{uu_{gt} + vv_{gt} + 1}{((u^2 + v^2 + 1)(u_{gt}^2 + v_{gt}^2 + 1))^{1/2}}\right). \tag{4}$$

We computed the *absolute* average angular error over a region of a whole cell and the *foreground* average angular error only over regions of foreground movement. The latter error is important indicator for tracking of cell structures of interest.

It must be noted that, unfortunately, the implementation of all three methods at the time of writing wasn't able to work with scaled versions of input image. It worked only with full resolution input images. After removing this limitation, we believe the proposed method has a potential to provide even better results on time-lapse 2D image sequences acquired from fluorescence optical microscope, especially of larger displacements. It managed already to detect movement of around 2px/frame, Figure 6C.

Table 1. The best Gabor parameters for image with only local movements

	5–1–4,8	4–1–3,8	5–1–4,8	4–1–3,8	3–2–2,8	4–3–2,8
original method	17°/60°	18°/59°				
extended original			6°/59°	6°/60°		
proposed method					9°/37°	9°/38°

Table 2. The best Gabor parameters for image with global and local movements

	5–4–4,8	5–4–1,7	5–4–4,8	5–4–3,7	5–4–4,8	5–3–4,8
original method	33°/28°	36°/24°				
extended original			28°/29°	33°/25°		
proposed method					23°/23°	25°/26°

We first conducted parametric studies on just one image from each set of test images to discover optimal setups of filtering banks. Tables 1 and 2 show setups that achieved the smallest *absolute/foreground* average angular errors for parameters σ_x–σ_y–σ_z, P where period $P = 100.0/[w_x^2 + w_y^2 + w_z^2]^{1/2}$. These parameters are constant for all filters. The frequency tuning of w_x, w_y and w_z depends on a filter's bank and filter position within a bank. Underlined setups were used in the rest of this section. The study was for sigmas from 1 to 5 and periods from 3 to 8.

We increased the number of filtering banks to be able to detect velocities of magnitudes up to 4px/frame. This is maximum, considering the best found setup, because the filters for velocities of 3px/frame and 4px/frame considerably overlap in the Fourier spectrum, see Figure 3. Increasing the Gaussian envelope in the spatio-temporal domain would reduce the overlap in the Fourier domain, thus increasing selectivity, but would also increase the spatio-temporal support of filters. This is not desirable due to weaker motion coherency observed between pairs of consecutive images, which is a consequence of longer intervals used during time-lapse image acquisition in optical microscopy.

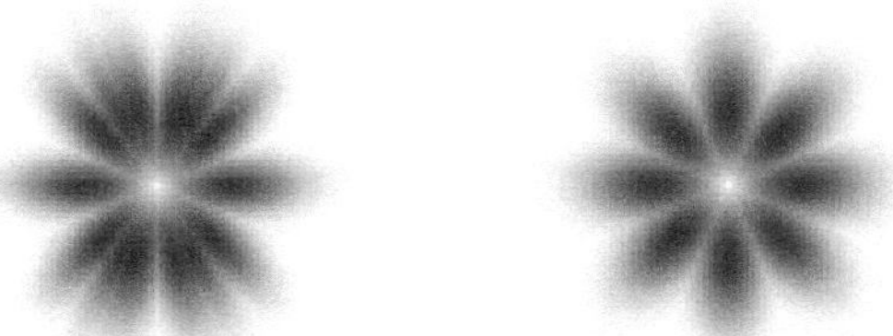

Fig. 3. Overlaid Fourier spectra of filters from the proposed method. The left-hand image shows w_x-w_t section of Fourier energy image of a filter with $w_y = 0$ from every filtering bank. The right-hand image shows w_x-w_y section of Fourier energy image of all filters with $w_t = 0$, i.e. the bank sensitive to stationary regions.

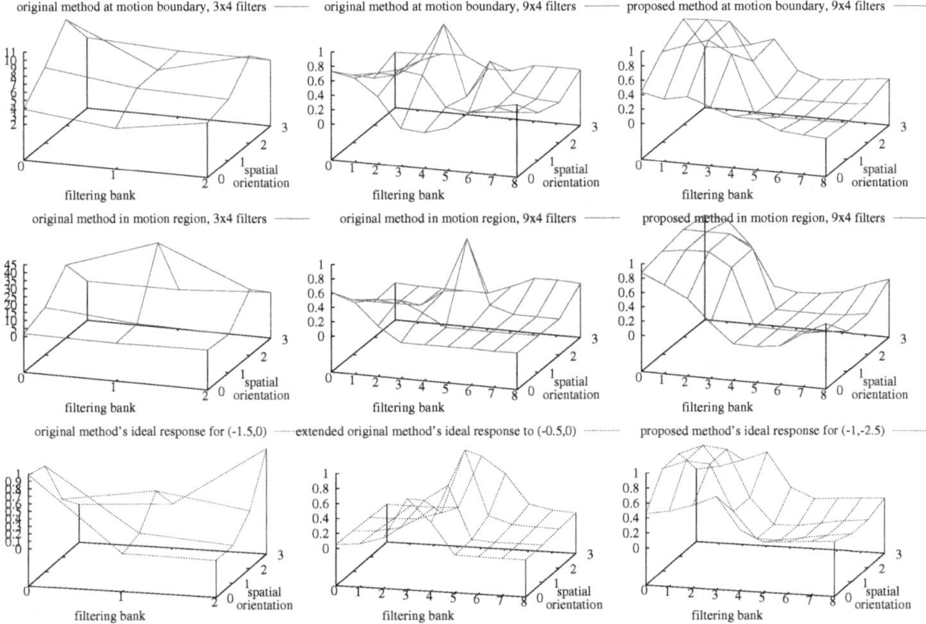

Fig. 4. Measured filtering responses for all filters in all banks for pixels denoted a and b in Figure 6. The upper row gives an example of responses at motion boundary, pixel a, while the middle row for pixel inside a region of motion $(0, -2)$, pixel b. The lower row shows ideal responses for velocities realizing global minima in graphs in lower row of Figure 5, i.e. best matching ideal responses inside the region of motion. From left to right, the original method with just 3 filtering banks and error given in equation (1), the original method after using 9 filtering banks and the new error (3) and the proposed method with oriented anisotropic filters and the new error.

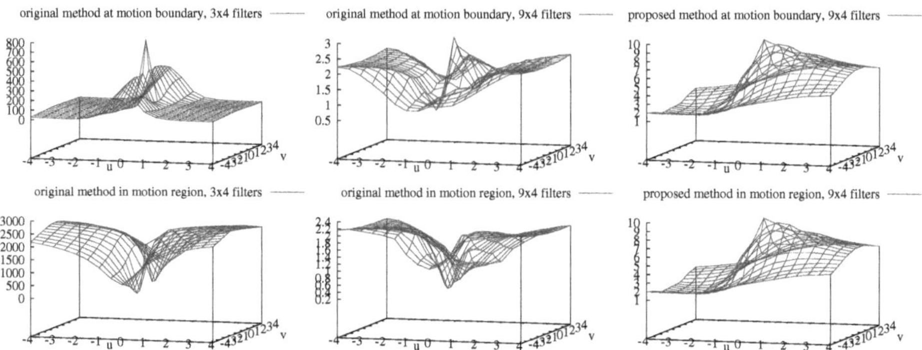

Fig. 5. Least squares evaluation on the 289 combinations of velocities (u, v), $u, v \in \langle -4, +4 \rangle$ in steps of 0.5 for pixels denoted a and b in Figure 6. The upper row gives an example of the fitting for pixel at motion boundary while the lower for pixel inside a region of motion. The velocity of motion was in both cases $(0, -2)$. From left to right, the original method with just 3 filtering banks and error given in equation (1), the original method after using 9 filtering banks and the new error (3) and the proposed method with oriented anisotropic filters and the new error.

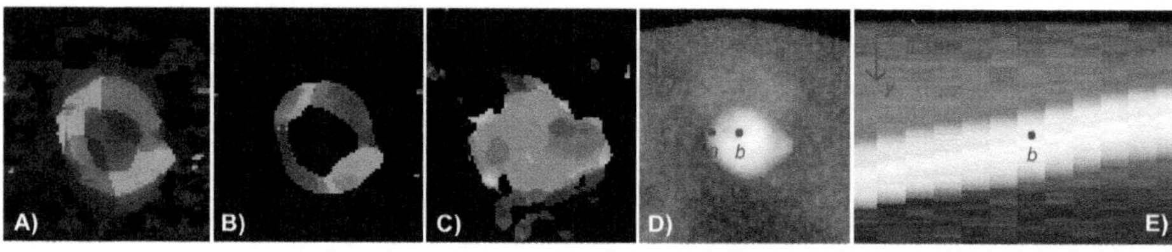

Fig. 6. Illustration of flow fields computed by the three variants of the method. A), B) and C) show flow field for the original method, the extended original and the proposed method, respectively, on input data shown in x-y and y-t sections in D) and E). Green colour denotes flow vectors with directions to up-left, yellow for direction up and red for directions up-right. The evaluation in Figure 5 was for pixels a and b.

The heart of Heeger's method is the matching of possible ideal responses to the measured responses. Figure 4 shows responses measured at some two pixels in a test image with only local movements. The original method has small number of filters resulting in poorer characterization of local texture in the test image. In other words, the characterization is too vague and no ideal response seems to fit perfectly while many seems to almost fit. The variant of original method with increased number of filters detected some significant patterns in the time-lapse image. The corresponding peaks were enhanced while minor peaks were suppressed by the new normalization procedure. Notice the weighted fitting, given in eq. (3), forced to select a set of ideal responses in which the strongest peak was obtained by the same filter as in the measured responses. As a result of this behaviour, errors dropped to the average of 6°, Table 1. Also notice that major features of measured responses computed with the proposed method in the

Table 3. Error rates on biomedical images

	absolute error, set1	foreground error, set1
original method	35.9°± 2.5°	36.9°± 8.6°
extended original	10.7°± 1.6°	62.9°± 5.0°
proposed method	13.8°± 1.6°	35.5°± 5.7°
	absolute error, set2	foreground error, set2
original method	37.4°± 1.4°	36.8°± 16.0°
extended original	34.0°± 2.4°	34.0°± 15.0°
proposed method	33.2°± 1.7°	32.4°± 14.6°

region and at the boundary of the region of a local motion were rather similar. The fitting should match the same or similar set of ideal responses, i.e. the same or similar velocities, in both situations. The process of fitting is illustrated in Figure 5.

Figure 6 presents visualization of flow fields computed by the all three methods as well as the input 3D image. The bright spot in the image is local foreground region of interest. It is moving straight up by 2 pixels per frame. The pictures D) and E) were enhanced for visualization purposes. We utilize a colour coding of flow vectors, direction is given by colour and vector's length is given by intensity, to inspect optical flow computation results easier. Only the proposed method managed to detect motion inside the foreground region without a need to resample the input image. The original method and its extension, which both are using Gabor filters separable only along Cartesian axes, presented a flow field with vectors in the direction of local gradient. This is typical for many optical flow computing methods when they start to suffer from the aperture problem. Owing to large Gaussian envelopes, the regions of local movement were overestimated in all cases.

Finally, we tested the methods for accuracy on the two test sets of images. Each method was tested with underlined setup according to Table 1 and 2, respectively. Table 3 shows error rates for the two test sets. The proposed method improved the error rates when compared to the original method. However, there seems to be left a great deal for future work and further improvements.

5 Conclusion

The paper was focused on the computation of optical flow from time-lapse 2D images acquired from fluorescence optical microscope. The aim was to obtain reasonably accurate flow fields, in reasonable computational times, that would eventually enable tracking of inner cell structures of interest simply from flow vectors.

The traditional established Heeger's method based on spatio-temporal filtering was adopted and modified in order to solve issues that arose from our image data. In particular, we used a scheme for effective and fast computations of complex Gabor convolutions. The filter tuning was changed to better support the detection of movement. We also revised the least squares fitting of the original method.

We conducted a parametric study to assess optimal parameters for the method to work the best. The parameters were used during the demonstrations in the section 4. With the tracking application in mind, we measured errors also only over some local regions of interests in order to see quality of flow field computed there. Generally, the proposed method shows lower average angular errors compared to the original method or similar errors but with smaller standard deviation. However, room is still left for further improvements.

The proposed method's C++ implementation is freely available at the URL http://cbia.fi.muni.cz/projects/optical-flow-for-live-cell-imaging.html. It can't scale input images, hence larger velocities can't be detected reliably. This will be the subject of our future research.

Acknowledgements. This work has been supported by the Ministry of Education of the Czech Republic (Grant No. MSM0021622419, LC535 and 2B06052).

References

1. Barron, J.L., Fleet, D.J., Beauchemin, S.S.: Performance of optical flow techniques. Int. J. Comput. Vision 12(1), 43–77 (1994)
2. Bernardino, A., Santos-Victor, J.: Fast iir isotropic 2-d complex gabor filters with boundary initialization. IEEE Transactions on Image Processing 15(11), 3338–3348 (2006)
3. Clocksin, W.: A new method for computing optical flow. In: BMVC 2000 (2000)
4. Eils, R., Athale, C.: Computational imaging in cell biology. The Journal of Cell Biology 161, 447–481 (2003)
5. Felsberg, M.: Optical flow estimation from monogenic phase. In: Jähne, B., Mester, R., Barth, E., Scharr, H. (eds.) IWCM 2004. LNCS, vol. 3417, pp. 1–13. Springer, Heidelberg (2007)
6. Fleet, D.J., Jepson, A.D.: Computation of component image velocity from local phase information. Int. J. Comput. Vision 5(1), 77–104 (1990)
7. Gautama, T., Hulle, M.V.: A phase-based approach to the estimation of the optical flow field using spatial filtering. IEEE Trans. Neural Networks 13(5), 1127–1136 (2002)
8. Gerlich, D., Mattes, J., Eils, R.: Quantitative motion analysis and visualization of cellular structures. Methods 29(1), 3–13 (2003)
9. Heeger, D.J.: Model for the extraction of image flow. J. Opt. Soc. Am. A 4(8), 1455–1471 (1987)
10. Heeger, D.J.: Optical flow using spatiotemporal filters. International Journal of Computer Vision 1(4), 279–302 (1988)
11. Hubený, J., Ulman, V., Matula, P.: Estimating large local motion in live-cell imaging using variational optical flow. In: VISAPP: Proc. of the Second International Conference on Computer Vision Theory and Applications, pp. 542–548. INSTICC - Institute for Systems and Technologies of Information, Control and Communication (2007)
12. Jähne, B.: Motion determination in space-time images. In: Faugeras, O. (ed.) ECCV 1990. LNCS, vol. 427, pp. 161–173. Springer, Heidelberg (1990)

13. Lampert, C., Wirjadi, O.: An optimal nonorthogonal separation of the anisotropic gaussian convolution filter. IEEE Transactions on Image Processing 15(11), 3501–3513 (2006)
14. Otero, A.: Robust optical flow estimation. In: ICIIS 1999: Proceedings of the 1999 International Conference on Information Intelligence and Systems, p. 370. IEEE Computer Society, Los Alamitos (1999)
15. Ulman, V.: Arbitrarily-oriented anisotropic 3d gaussian filtering computed with 1d convolutions without interpolation. In: ISCGAV 2008: Proc. of the 8th Conf. on Signal Processing, Computational Geometry and Artificial Vision, pp. 56–62. World Scientific and Engineering Academy and Society (WSEAS), Stevens Point (2008)
16. Ulman, V.: Filtering with anisotropic 3d gabor filter bank efficiently computed with 1d convolutions without interpolation. In: SPPRA 2010: Proc. of the 7th IASTED Int. Conf. on Signal Processing, Pattern Recognition, and Applications, pp. 33–42. ACTA Press (2010)
17. Ulman, V., Hubený, J.: Pseudo-real image sequence generator for optical flow computations. In: Scandinavian Conference on Image Analysis (June 2007)
18. Weber, J., Malik, J.: Robust computation of optical flow in a multi-scale differential framework. Int. J. Comput. Vision 14(1), 67–81 (1995)
19. Young, I., van Vliet, L., van Ginkel, M.: Recursive gabor filtering. Signal processing 50(11), 2798–2805 (2002)

Shape Reconstruction from Unorganized Set of Points

Yvan Maillot, Bruno Adam, and Mahmoud Melkemi

Faculty of Science and Technology
University of Haute Alsace
France, 68093 Mulhouse
{yvan.maillot,bruno.adam,mahmoud.melkemi}@uha.fr
http://www.lmia.uha.fr

Abstract. This paper deals with the problem of reconstructing shapes from an unorganized set of sample points (called S). First, we give an intuitive notion for gathering sample points in order to reconstruct a shape. Then, we introduce a variant of α-shape [1] which takes into account that the density of the sample points varies from place to place, depending on the required amount of details. The Locally-Density-Adaptive-α-hull (LDA-α-hull) is formally defined and some nice properties are proven. It generates a monotone family of hulls for α ranging from 0 to 1. Afterwards, from LDA-α-hull, we formally define the LDA-α-shape, describing the boundaries of the reconstructed shape, and the LDA-α-complex, describing the shape and its interior. Both describe a monotone family of subgraphs of the Delaunay triangulation of S (called $Del(S)$). That is, for α varying from 0 to 1, LDA-α-shape (resp. LDA-α-complex) goes from the convex hull of S (resp. $Del(S)$) to S. These definitions lead to a very simple and efficient algorithm to compute LDA-α-shape and LDA-α-complex in $O(n\ log(n))$. Finally, a few meaningful examples show how a shape is reconstructed and underline the stability of the reconstruction in a wide range of α even if the density of the sample points varies from place to place.

Keywords: Shape Reconstruction, Delaunay Filtration, α-shape, Local Density.

1 Introduction

Reconstruction of a sampled shape from a set of points is important in many domains such as computer vision, image analysis, clustering, or pattern recognition. The whole issue is the reconstruction of a shape only from its sample by respecting its topology.

Early works dealing with this problem did not define the reconstruction formally and their results was validated according to human perception. They proposed heuristics based on the Delaunay triangulation in order to capture the shape. These algorithms are quite difficult to implement, have numerous thresholds, and give results close to human perception for some sets of points. Among these works the Urquart's algorithm [2] makes use of the relative neighborhood graph while Ahuja and Tuceryan [3] propose a classification method mainly based on some geometrical properties of Voronoï regions.

Jarvis [4] was the first to consider the shape as a generalization of the convex hull. A couple of years later, Edelsbrunner, Kirkpatrick and Seidel [1] gave a formal definition of such a generalization, named α-shape. These concepts led to important progress

A. Campilho and M. Kamel (Eds.): ICIAR 2010, Part I, LNCS 6111, pp. 274–283, 2010.

in the domain of pattern recognition and gave rise to many works as $\mathscr{A}$-shape introduced by Melkemi [5]. More rigorous definitions of reconstruction and simpler algorithms without threshold were presented. The main step that followed concerns curve reconstruction in 2D or surface reconstruction in 3D with, for instance, methods with anisotropic-density-scaled-α-shapes of Techmann and Capps [6], or works of Dey and Kumar [7] or those of Amenta, Choi and Kolluri [8]. Amanta and Bern [9] introduced the notion of ε-sample and proved that their algorithm reconstructs a ε-sampled curve. Boissonnat and Oudot [10], Chazal and Lieutier [11], and Cohen-Steiner, Edelsbrunner and Harer [12], among others, were interested in insuring the reconstruction of a 3D surface, provided some sample conditions were fullfiled. Very recently, Dey et al. present in [13] an algorithm for the reconstruction of a surface with boundaries in 3D and guarantes that the output is isotopic to the sampled surface.

Our main interest is the reconstruction of shapes from a set of points, which are not just lying on the boundary of the shape but which cover its whole area. The issue is to find a subgraph of the Delaunay triangulation which is a good interpolation of the shape. The algorithms previously cited, using heuristics based on the Delaunay triangulation, are related to a significant number of thresholds to be adjusted and the interpretation of their results is often related to visual criteria. Some approaches consisting in reconstructing a region from slices can also be considered, in some way, close to ours. The interior of the region is also sampled, but with segments in $\mathbb{R}^2$ or with polygons in $\mathbb{R}^3$. Boissonnat and Geiger [14] reconstruct the region using the Delaunay complex of two consecutive slices, Barequet et al. suggest an interpolation algorithm that uses the medial axis of the overlay of two consecutive slices [15] and more recently, Boissonnat and Memari [16] generalized the algorithm proposed by Boissonnat and Geiger in 1993 for cutting planes whose positions and orientations may be arbitrary. Among the works concerning the problem of reconstructing a region, the α-shapes proposed by Edelsbrunner in 1981 and the $\mathscr{A}$-shapes proposed by Melkemi in 1999 are those that have inspired us most. The α-shape are very helpful with almost relatively uniform sampling but shows some limits with non uniform sampling. The Weighted α-shape [17], which extends this notion, aims at overcoming this problem. However, this solution does not completely solve irregular sampling problems because, as the influence zone associated with each point has a disc shape, the variation of density must be equal in all directions around it. Our works propose to provide a variant of α-shape which takes into account, in a very natural way, that the local density of the points can varie from place to place, depending on the required amount of details. This is a filtration of the Delaunay triangulation called *Locally-Density-Adaptive-α-shape* or shorter *LDA-α-shape* based on the local density of the points. LDA-α-shape is less generic than conformal-α-shape [18] since its main objective is to provide a precise and formal sense of what is the shape of a set of points according to α, even if these points are not uniformally distributed. And it leads to a simple and efficient algorithm with a local filtering condition which can adapt to the variations of the density of the set of points.

After presenting in section 2 different geometrical concepts on which our works are based, we give, in section 3, a definition of the reconstruction. Then, in section 4, we present the intuitive idea of the sampling conditions and of our reconstruction algorithm. In section 5, we give the LDA-α-empty disk definition and its properties. They allow to introduce the notions of LDA-α-hull, LDA-α-shape, and LDA-α-complex. In the last sections, we describe our algorithm and its complexity followed by some significant examples.

2 Related Geometrical Concept

In this section we briefly describe the background for our work.

Let S be a set of points of $\mathbb{R}^2$. We assume that points of S are in general position, that is, there are neither three points on a line nor four points on a circle. The general-assumption simplifies the forthcoming definitions, however the difficulties that arise without these assumptions could be treated by more elaborated definitions.

2.1 Delaunay Triangulation

The Delaunay Triangulation of a set of points S, introduced by Boris Delaunay in 1934, is the dual graph of the Voronoï Diagram of S. Such a triangulation can be defined with the following remarkable property: this is the only triangulation whose the circumcircle of each triangle does not contains any point of S in its interior. This property is generalized to the higher dimensions. In the following, the Delaunay triangulation of S is denoted $Del(S)$.

2.2 Maximal Disk

Definition 1. *We define a maximal disk of S as either, an open disk containing no point of S and whose boundary passes through three points of S, or, an open half plane containing no point of S and whose boundary passes through two points of S.*

In the first case, a maximal disk is the interior of the circumcircle of a Delaunay triangle and is also called a **finite maximal disk**. In the second case, a maximal disk is an open half plane containing no point of S and bounded by the straight line spanned by a convex hull edge. It is also called an **infinite maximal disks** and its radius is considered to be infinite.

It must be noticed that T.K. Dey et S. Goswami use in [19] a notion of "big Delaunay balls" close to our definition LDA-α-eliminated maximal balls.

3 Reconstruction

The issue of shape reconstruction is to produce an approximation of a shape from a degraded view, like the one from its set of sample points for instance. A problem is how to detect a well reconstructed shape. Albeit this seems quite subjective, it is possible to give an exact definition of the reconstruction. Indeed, the restricted Delaunay triangulation [20] is widely considered as a good approximation of the shape, as well topologically as geometrically if it is well sampled [21].

Let Σ be a shape sampled by S. A subgraph of $Del(S)$ reconstructs Σ if it is the Delaunay triangulation of S restricted by Σ.

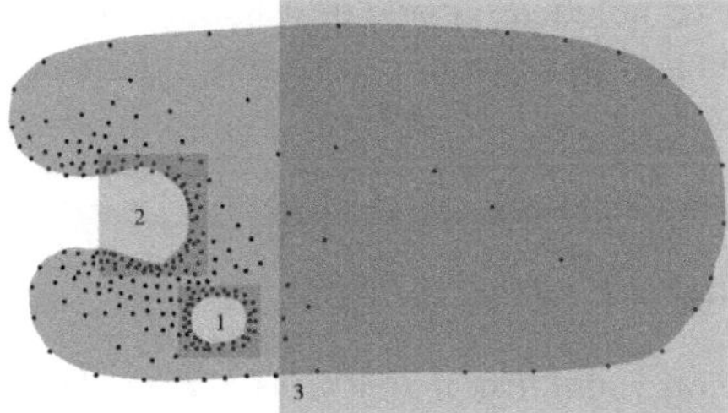

Fig. 1. Variation of the local density of the sample points

4 Sample Condition and Idea of Reconstruction

In order to reconstruct a region Σ, its set of sample points S has to carry enough information. Some assumptions about its quality have to be done. First, points of S are all supposed included in the interior of Σ. In other words, the sample is not noisy. Then, the repartition of the points and the local density of the sample is under consideration. The idea of a uniform sample points is avoided, i.e. such that its local density is approximatively the same everywhere. So, the sample is not regular but it is nevertheless under conditions. In particular, it has to be of strong density when a big amount of detail is required. For instance, the figure 1 shows a shape and its sample. The rectangular areas 1 and 2 have a relatively strong density close around holes because the difference between interior and exterior has to be explicitly shown. But, the rectangular area 3 does not need to be strongly sampled. Indeed, the local density of the sample can become less and less strong from left to right. However, in the rectangular area 3, the variation of the density need to be progressive to avoid the "apparition" of "nonexistent" holes.

To summarize, the assumptions done about a sample S in order to allow the reconstruction of Σ,

1. S is included in Σ.
2. Its local density is relatively dense close to an hole or an hollow. Relatively means that if we were able to measure a local density, the ratio between the local density around the hole and inside it might be smaller than a threshold, so called α.
3. On the other hand, its density variation inside the region must be progressive. Indeed, the ratio between the local density of two neighboring areas must be greater than α.

Our idea of reconstruction utilizes this concept. From the first assumption, points of S are inside Σ. It remains to find out the kind of areas between them. Which areas are inside the shape? Which areas are outside the shape? The problem comes down to determining how to efficiently obtain a significant measure of the local density of S for all points of S.

Correctly, by definition the maximal disks of S are empty. Thus, the larger they are, the sparser is the density at this place. Moreover, the ratio between two neighboring maximal disks gives a good interpretation of density variation between both places. The idea is to consider that a maximal disk relatively large compared with all its neighboring maximal disks means it is included outside the shape. Our reconstruction algorithm is based on this simple idea and its concept is formally explained in what follows.

5 Definitions

5.1 LDA-α-empty Disk

Definition 2. *let D be a maximal disk with radius r, P be the set of all points of S lying on the boundary of D and α a positive real number. The disk D is said to be LDA-α-empty if and only if, for all p of P, there exists a maximal disk with radius r', whose boundary passes through p, and such that $\frac{r'}{\alpha} \leq r$.*

Observation : all infinite maximal disks are LDA-α-empty, for all α since they are all infinitely larger than any finite maximal disk.

Property 1. *There exists an α' such that, for every $\alpha \leq \alpha'$, the maximal infinite disks are the LDA-α-empty disks.*

Proof: Let D be a finite maximal disk with radius r, let p be a point of S on the boundary of D, and D' be a maximal disk with radius r' whose boundary contains p. Let r_{max} be the radius of the largest finite maximal disk of S, r_{min} be the radius of the smallest one, and α such that $\frac{r_{min}}{r_{max}} > \alpha$. Since $\frac{r'}{\alpha} > r' / \frac{r_{min}}{r_{max}} = r' \frac{r_{max}}{r_{min}} \geq r_{min} \frac{r_{max}}{r_{min}} = r_{max} \geq r$, D is not LDA-α-empty.

Thus, we extend the range of α. It is a non-negative real number and we set that all infinite maximal disks and only infinite maximal disks are LDA-0-empty.

Property 2. *For every $\alpha \geq 1$, all maximal disks are LDA-α-empty.*

Proof: Let D be a maximal disk with radius r, p be a point of S on the boundary of D, D' be a maximal disk with radius r' whose boundary contains p, and α such that $\alpha \geq 1$. Now, choosing $D' = D$, we have $\frac{r'}{\alpha} = \frac{r}{\alpha} \leq r$ and thus D is LDA-α-empty.

From now on, α is ranging from 0 to 1.

5.2 LDA-α-hull

Definition 3. *The LDA-α-hull of S is the intersection of all complements of the closures of LDA-α-empty disks.*

Property 3. *The LDA-0-hull of S is the convex hull of S.*

Proof: From property 1, the infinite maximal disks are the LDA-0-empty disks. By definition, the LDA-0-hull is the intersection of all half-planes whose closures contain S and whose boundaries pass through two points of S. Hence, it is the convex hull of S.

Property 4. *The LDA-1-hull of S is S.*

Proof: The interiors of the circumcircles of the Delaunay triangles of S, i.e., the finite maximal disks cover the convex hull of S, but S since they are opens. The infinite maximal disks cover the complement of the convex hull of S, but S because S is included in the convex hull.

Property 5. *If $\alpha_1 \leq \alpha_2$ then the LDA-α_1-hull is a set containing the LDA-α_2-hull since an LDA-α_1-empty disk is also LDA-α_2-empty.*

Thus the LDA-α-hulls of S for α ranging from 0 to 1 form a discrete monotone family of hulls, from the convex hull of S to S itself.

5.3 LDA-α-shapes

Definition 4. *The points of S lying on the boundary of LDA-α-empty disks are LDA-α-extreme.*

From Property 5, we have:

Property 6. *For $\alpha_1 \leq \alpha_2$, the set of all LDA-α_1-extreme points is included in the set of all LDA-α_2-extreme points.*

Definition 5. *Two points p and p' of S are LDA-α-neighbors if p and p' are LDA-α-extreme, both lie on the same LDA-α-empty disk, and the open edge pp' is included in a unique LDA-α-empty disk.*

Now, LDA-α-shape can be defined similarly to α-shape [1].

Definition 6. *Given a set S and a real number α ranging from 0 to 1, the LDA-α-shape of S is the straight line graph whose vertices are the LDA-α-extreme points and whose edges connect the respective LDA-α-neighbors.*

It immediately follows from properties 3 and 4 that an LDA-0-shape of S is the boundary of the convex hull of S and that an LDA-1-shape of S is S.

An LDA-α-shape of S is obviously a sub-graph of $Del(S)$ because every edge connects two points which are on the same maximal disk.

An LDA-α-shape is a straight line graph. It describes the boundary of the reconstructed domain. It is sometime necessary to distinguish the interior from the exterior: interior faces are those which contain the sample. Another way to describe the domain considering interior and exterior is done by defining LDA-α-complex.

5.4 LDA-α-complex

Definition 7. *Given a set S and a real number α ranging from 0 to 1, the LDA-α-complex of S is the straight line graph whose edges are edges of $Del(S)$ included in at most one LDA-α-empty disk.*

By definition, the LDA-α-shape of S is included in the LDA-α-complex of S.

It immediately follows that an LDA-0-complex of S is the Delaunay triangulation of S and an LDA-1-complex of S is S.

Moreover, LDA-α-complexes of S describes a discrete monotone family of subgraphs of $Del(S)$ from $Del(S)$ to S, when α varies from 0 to 1.

6 Algorithm

Computing an LDA-α-shape and an LDA-α-complex of S can be done quite simply and in a efficient way since they are both subgraph of $Del(S)$ and may be obtained by filtering it in linear time.

For reason of efficiency, we consider the following observation deduced from the Definition 2:

We call $r_{min}(q)$, the radius of the smaller maximal disk whose boundary passes through q. If for each point p of P, $\frac{r_{min}(p)}{\alpha} \leq r$ then D is LDA-α-empty. So D is by definition LDA-α-empty.

The main algorithm has to compute if the circumdisk of a face f adjacent to an edge is LDA-α-empty or not. This is done in constant time in the following function:

Algorithm 1. Is the circumdisk of f LDA-α-empty ?

```
Let p, p′, p″ be the vertices of f
Let r be the radius of the circumcircle of f
return (r_min(p)/α ≤ r) ∧ (r_min(p′)/α ≤ r) ∧ (r_min(p″)/α ≤ r)
```

Algorithm 2. Computing of the LDA-α-shape and of the LDA-α-complex

```
Compute Del(S)
Associate each point p of S to the smaller radius r_min(p) of the circumcircles of the triangle
whose p is a vertex.
for all edges pp′ of Del(S) do
   {To reconstruct the LDA-α-complex}
   Let f and f′ be both faces incident to pp′
   if the circumdisk of f is not LDA-α-empty ∨ the circumdisk of f′ is not LDA-α-empty
   then
      pp′ is an edge of the LDA-α-complex
   end if
end for
for all edges pp′ of Del(S) do
   {To reconstruct the LDA-α-shape}
   Let f and f′ be both faces incident to pp′
   if the circumdisk of f is LDA-α-empty ⊕ the circumdisk of f′ is LDA-α-empty then
      pp′ is an edge of the LDA-α-shape
   end if
end for
```

The computation of the LDA-α-shape and the LDA-α-complex of S is done in time $O(nlog(n))$, where n is the number of points of S.

- The computation of $Del(S)$ is done in $O(nlog(n))$
- The computation of each $r_{min}(p)$ associated to each p is done in linear time because each circumcircle of a Delaunay triangle is processed once and only once for each vertex of a triangle.
- The test for each edge of $Del(S)$ is done in constant time.

Observation : For any new value of α, it is useless to compute $Del(S)$ again nor $r_{min}(p)$. This "piece" of computation has to be done once and for all α. It has to be done again every time S is modified.

7 Meaningful Examples

The algorithm presented in this paper has been implemented. For all experiments that we have done, it was always easy to find a set points and a value of α (or even a range of values) for which our algorithm was able to reconstruct the shape.

A few meaningful samples and their reconstruction are presented in what follows.

Figure 2 shows a shape whose bottleneck requires a sample with a relatively strong density at this place. On the over hand the density of the sample do not need to be so strong elsewhere.

Figure 3 shows a case of reconstruction with two connected components whose neighborhood implies a relatively strong sample.

Figure 4 shows a case of reconstruction when a component is nested in another. Interweaving may be as deep as wanted.

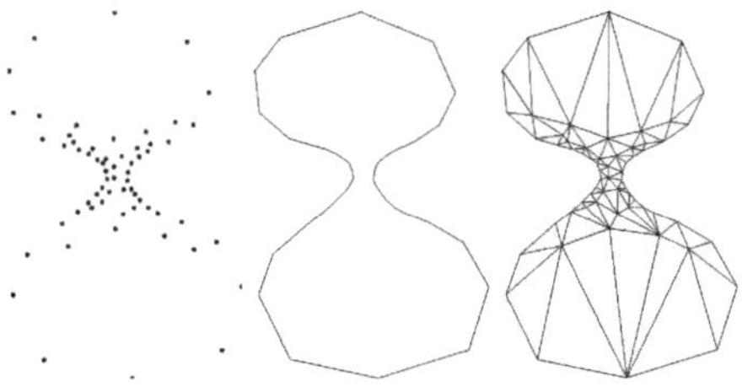

Fig. 2. The sample of a shape with a bottleneck, its reconstructed shape and complex for $\alpha = 0.5$

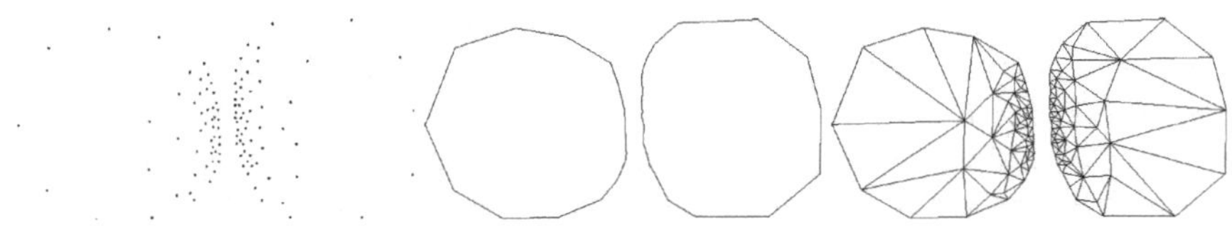

Fig. 3. The sample of two close different shapes, its reconstructed shape and complex for $\alpha = 0.5$

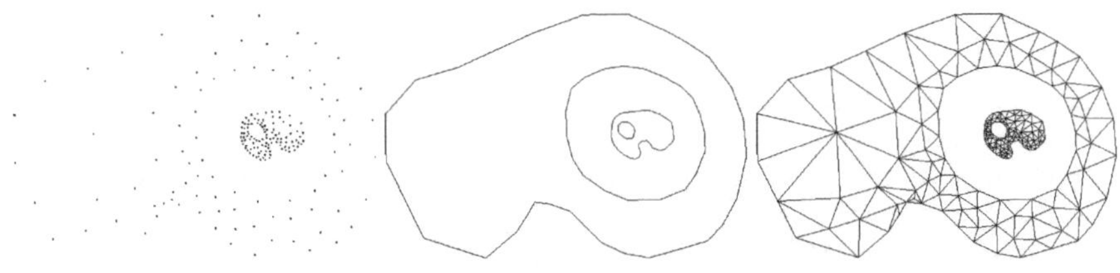

Fig. 4. The sample of two nested components, its reconstructed shape and complex for $\alpha = 0.5$

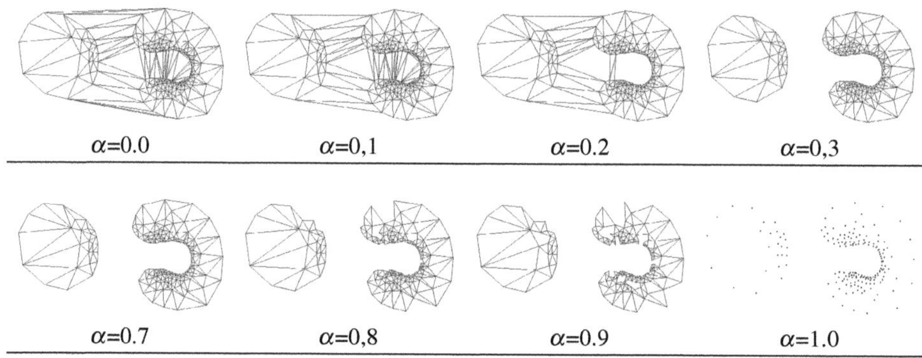

Fig. 5. A few LDA-α-complexes of the same sample according to different values of α

Figures 5 shows LDA-α-complexes of the same sample according to different values of α. The generated complex is $Del(S)$ for $\alpha = 0.0$. More and more edges are filtered while α is increasing. The sampled shape is reconstructed from $\alpha = 0.3$ to $\alpha = 0.7$. Next, edges are filtered until it remains only S. One may notice the outstanding stability of the reconstruction in a wide range of values of α: from 0.3 to 0.7.

8 Conclusion

We introduced the notion of LDA-α-hulls, LDA-α-shapes and LDA-α-complexes of a sample set of points of the plane. LDA-α-hull is a variant of α-hull [1], from which we took inspiration. Variation of α spawns a monotone family of hulls from the "crudest" to the "finest" interpretation of the sample. The main difference is that the variation of the density of the set of sample points is taken into consideration. A simple and efficient algorithm based on this concept was proposed. It was easily implemented and a few meaningful results have shown, on the one hand, some reconstructions from sets of sample points that included few points, on the other hand, the strong stability of the reconstruction because the shape is reconstructed in a wide range of α.

Our actual work is, first, to generalize it to higher dimensions and, second, to find out a distribution function of the sample such that the reconstruction of the sampled shape is guaranted for a range of values of α.

References

1. Edelsbrunner, H., Kirkpatrick, D., Seidel, R.: On the shape of a set of points in the plane. IEEE Transactions on Information Theory 29(4), 551–559 (1983)
2. Urquhart, R.: Graph theoretical clustering based on limited neighborhood sets. Pattern Recognition, 173–187 (1992)
3. Ahuja, N., Tuceryan, M.: Extraction of early perceptual structure in dot patterns: Integrating region, boundary, and component gestalt. Computer Vision, Graphics, and Image Processing 48(3), 304–356 (1989)

4. Jarvis, R.: Computing the shape hull of points in the plane. In: Pattern Recognition and Image Processing, pp. 231–241 (1977)
5. Melkemi, M.: A-shapes of a finite point set. In: SCG 1997: Proceedings of the Thirteenth Annual Symposium on Computational Geometry, pp. 367–369. ACM, New York (1997)
6. Teichmann, M., Capps, M.: Surface reconstruction with anisotropic density-scaled alpha shapes. In: Proceedings of IEEE Visualization, pp. 67–72. IEEE, Los Alamitos (1998)
7. Dey, T.K., Kumar, P.: A simple provable algorithm for curve reconstruction. In: SGP 2007: Proceedings of the Fifth Eurographics Symposium on Geometry Processing, pp. 893–894 (1999)
8. Amenta, N., Choi, S., Kolluri, R.K.: The power crust, unions of balls, and the medial axis transform. Computational Geometry: Theory and Applications 19, 127–153 (2001)
9. Amenta, N., Bern, M.: Surface reconstruction by voronoi filtering. Discrete and Computational Geometry 22, 481–504 (1999)
10. Boissonnat, J.D., Oudot, S.: Provably good surface sampling and approximation. In: SGP 2003: Proceedings of the 2003 Eurographics/ACM SIGGRAPH symposium on Geometry processing, pp. 9–18. Eurographics Association. Aire-la-Ville (2003)
11. Chazal, F., Lieutier, A.: Weak feature size and persistent homology: computing homology of solids in rn from noisy data samples. In: SCG 2005: Proceedings of the Twenty-First Annual Symposium on Computational Geometry, pp. 255–262. ACM, New York (2005)
12. Cohen-Steiner, D., Edelsbrunner, H., Harer, J.: Stability of persistence diagrams. In: SCG 2005: Proceedings of the Twenty-First Annual Symposium on Computational Geometry, pp. 263–271. ACM, New York (2005)
13. Dey, T.K., Li, K., Ramos, E.A., Wenger, R.: Isotopic reconstruction of surfaces with boundaries. Comput. Graph. Forum 28(5), 1371–1382 (2009)
14. Boissonnat, J.D., Geiger, B.: Three-dimensional reconstruction of complex shapes based on the delaunay triangulation. Research Report 1697, INRIA (1992)
15. Barequet, G., Goodrich, M.T., Levi-Steiner, A., Steiner, D.: Contour interpolation by straight skeletons. Graph. Models 66(4), 245–260 (2004)
16. Boissonnat, J.D., Memari, P.: Shape reconstruction from unorganized cross-sections. In: SGP 2007: Proceedings of the fifth Eurographics Symposium on Geometry Processing, pp. 89–98, Eurographics Association, Aire-la-Ville (2007)
17. Edelsbrunner, H.: Weighted alpha shapes. Technical report, Champaign, IL, USA (1992)
18. Cazals, F., Giesen, J., Pauly, M., Zomorodian, A.: The conformal alpha shape filtration. The Visual Computer 22(8), 531–540 (2006)
19. Dey, T.K., Goswami, S.: Provable surface reconstruction from noisy samples. Comput. Geom. Theory Appl. 35(1), 124–141 (2006)
20. Edelsbrunner, H., Shah, N.R.: Triangulating topological spaces. In: SCG 1994: Proceedings of the tenth annual symposium on Computational geometry, pp. 285–292. ACM, New York (1994)
21. Boissonnat, J.D., Oudot, S.: Provably good sampling and meshing of lipschitz surfaces. In: SCG 2006: Proceedings of the Twenty-Second Symposium on Computational Geometry, pp. 337–346. ACM, New York (2006)

Significantly Improving Scan-Based Shape Representations Using Rotational Key Feature Points

Yasser Ebrahim[1], Maher Ahmed[1],
Siu-Cheung Chau[1], and Wegdan Abdelsalam[2]

[1] Wilfrid Laurier University, Waterloo ON N2L 3C5, Canada
[2] University of Guelph, Guelph ON N1G 2W1, Canada

Abstract. In a previous paper we have presented the idea of representing the shape of a 2D object by scanning it following a Hilbert curve then performing wavelet smoothing and sampling. We also introduced the idea of using only a subset of the resulting signature for comparison purposes. We called that set the Key Feature Points (KFPs). In this paper we introduce the idea of taking the KFPs over a number of views of the original shape. The proposed improvement results in a significant increase in recognition rates when applied to the MPEG-7 and ETH-80 data sets when the Hilbert scan is used. Similar improvement is achieved when the raster scan is used.

1 Introduction

Shape representation is fundamental for many tasks, including object recognition, registration, and image retrieval. The exponential growth of digital multimedia data and the need to describe and identify the content of this information have made shape representation and description a very active research area. One user survey, regarding the cognitive aspects of image retrieval, indicates that users are more interested in retrieval by shape than by colour or texture [1]. Researchers are also more interested in the shape features due to the increased discrimination, easy handling, and the wide array of existing mathematical and geometrical models that can be applied to shape.

The current methods for shape representation can be categorized as either boundary-based (sometimes called contour-based) or region-based methods. For boundary-based methods, the shape's outer boundary is needed in computation, whereas in region-based ones, the inner shape pixels are utilized. Region-based methods reflect the global shape features, whereas the boundary-based ones exploit the local variations of a shape's outer boundary.

In a previous publication [2], we have introduced a new region-based shape representation technique that captures the shape features by scanning the object image with the Hilbert Curve (HC) producing a 1D version of the image which is smoothed and sampled to produce the image's Shape Feature Vector (SFV). The proposed technique runs in linear time and is invariant to translation, scaling, and stretching. Because in many cases the class which each object in a database

A. Campilho and M. Kamel (Eds.): ICIAR 2010, Part I, LNCS 6111, pp. 284–293, 2010.

belongs to is known (e.g., cars, planes) we have demonstrated how this knowledge can be applied to determine the most prominent features of the shape vectors of the objects within each class in order to maximize the representational power of these vectors. These prominent features are called Key Feature Points (KFPs).

In this paper we introduce the concept of Rotational KFPs (RKFPs). Instead of determining the KFPs based only on the SFVs of the class objects in their original view, we create SFVs for multiple views for each shape, concatenate them into a Cumulative SFV (CSFV), then use CSFVs to determine the RKFPs for the class. The different views are obtained by rotating the object a number of times. For each view, the SFV representing the shape is generated.

Section 2 describes the basics for shape representation using the Hilbert curve scan and KFPs. The proposed improvement is discussed in Section 3. Experimental results are presented in Section 4.

2 Shape Representation Using the HC and KFPs

A shape is distinguished from its background by its pixels intensity variation. Ebrahim et al. [2] suggest to capture this variation, which has the shape information embedded in it, by scanning the segmented out object image by using the HC. The intensity value for each visited pixel is saved in a vector, V. To smooth out noise while keeping the main shape features intact, the wavelet transform is applied to V, producing the vector WV which is then sampled to obtain the vector SWV which is normalized to produce the object's Shape Features Vector (SFV).

In many applications, the shapes in the database are grouped into classes (e.g., cars, airplanes, phones). In the same paper [2], the authors propose that in such applications, the knowledge about the class of each shape is utilized as leverage to improve the accuracy of the proposed approach. The idea is that shapes belonging to the same class share some features that make the class different from other classes. The SFV points that correspond to these features are called Key Feature Points (KFPs). At search time, only the SFV elements that correspond to the class' KFPs are used to compute the distance between the search object and each of the objects belonging to the class. Other feature points that have been deemed less important (i.e., are not KFPs) are ignored.

The KFPs for each class are determined first by calculating the standard deviation of each SFV element across the class. Next, and the elements with the lowest standard deviation are identified as the KFPs and their locations are saved.

When a query shape is to be searched for in the database, its SFV is compared to each SFV in the class and the closest distance of the two is recorded to be used in the nearest neighbour search. Note that only the elements that correspond to the KFPs of the class the database object belongs to are compared.

Figure 1 exhibits the 11 shape vectors for the "dude" class in the Kimia-99 dataset and the class' KFPs (at the top of the graph). As expected, the KFPs are located where the shape vectors differ the least.

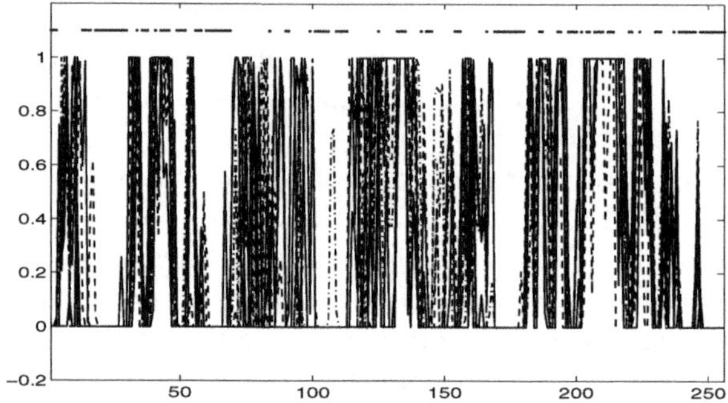

Fig. 1. Shape vectors of the 11 "dude" objects in Figure 2 and their associated KFPs (the dots at the top)

Fig. 2. The 11 "dude" objects in the Kimia-99 dataset

The use of KFPs improves the retrieval accuracy by maximizing the similarity of the objects within the same class. Figure 3 presents "dude-0" (top left) and "dude-8" (top right), along with their shape vectors, and the "dude" class KFPs. Correlating the full two SFVs yields a cross-correlation of 0.5, but when only the SFV points corresponding to the class KFPs are correlated, the cross-correlation becomes 0.88 better reflecting the similarity between the two objects (using SFV size of 512, rbio3.1 wavelet with approximation level 3, KFPs vector size is 20% of the SFV vector).

3 Proposed Improvement

In this section we propose a simple yet effective improvement over the technique described in Section 2. Instead of having an SFV that results from only one scan of the object, the CSFV will be the concatenation of the SFVs resulting from scanning the object at different views at different rotation angles. For example, the object can be scanned at the original position (0 rotation angle), rotated 120 degrees, and 240 degrees.

When scanning an object using the HC at each rotation angle, the HC intersects with the different pixels of the object in a different sequence. This results in a different representation of the shape of the object. Figure 4 shows an object with its three representations at angles 0, 120, and 240.

Figure 5 depicts three views of the dude0 object and the RKFPs superimposed on them in grey. From the figure, it is evident that the KFPs on each view capture different features of the shape. This increases the number of features

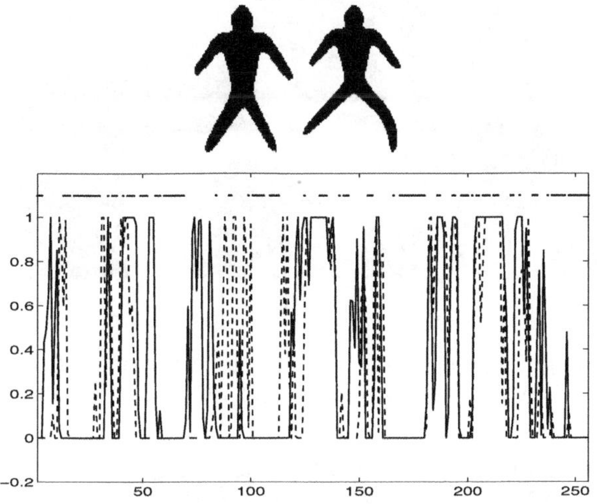

Fig. 3. Shape vectors of "dude-0" and "dude-8" in Kimia-99 and their associated KFPs

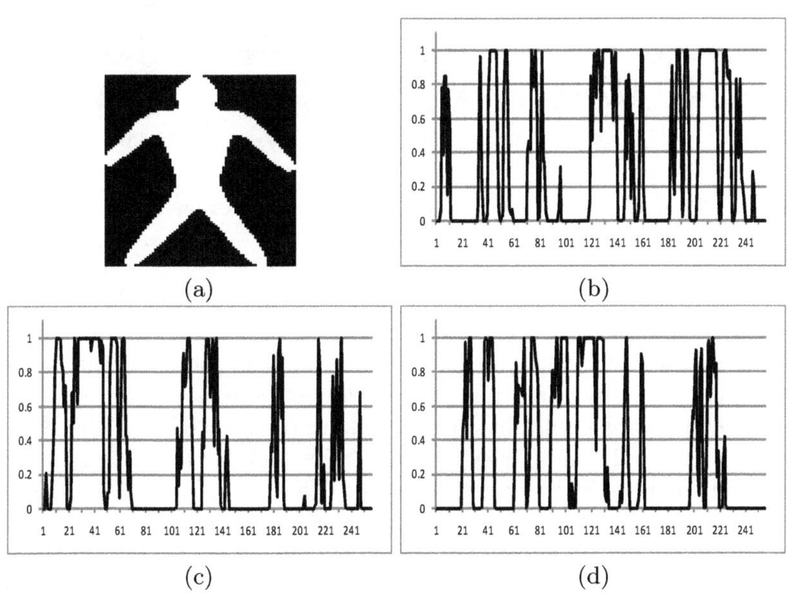

Fig. 4. (a)Dude0 from Kimia-99 and its SFVs at rotation angles (b) 0, (c) 120, and (d) 240 degrees

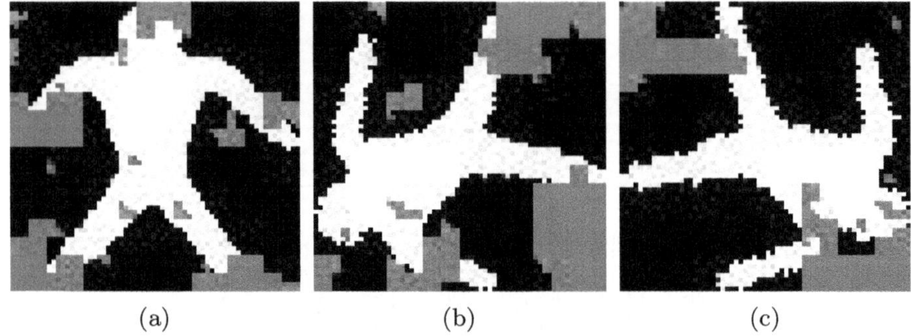

Fig. 5. RKFPs (in grey) superimposed on the three rotations of the dude0 object. (a) 0 degrees rotation, (b) 120 degrees rotation, (c) 240 degrees rotation.

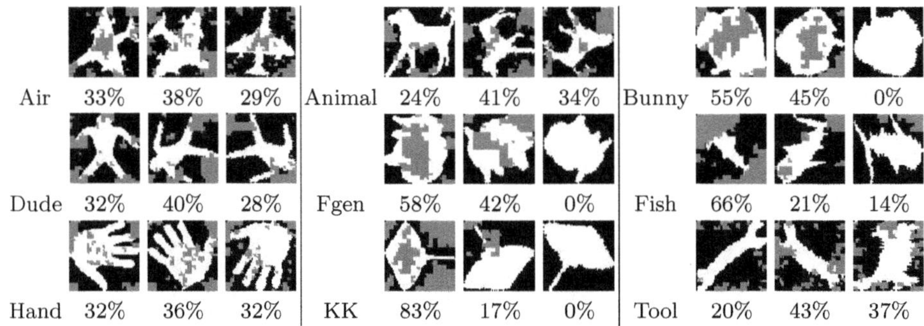

Fig. 6. RKFPs percentage in each of the 3 rotation angles for the Kimia-99 dataset

used for discrimination among different shapes. Using the RKFPs, the correlation between the two objects in Figure 3 goes from .88, when KFPs are used, to .93, when RKFPs are used, better reflecting the similarity between the two shapes.

Notice that the way the RKFPs are selected does not guarantee that each of the object rotations will get an equal share of the RKFPs. The RKFPs are selected based on the similarity of the CSFV elements throughout the class regardless of the view SFV that the element belongs to. Table 6 lists the RKFPs percentage in each of the 3 views for each class within the Kimia-99 dataset. It can be seen from the figure that some classes have a fairly even distribution of the RKFPs such as Air and Hand. Other classes have a high concentration of RKFPs in one of the object views such as KK and Fish.

4 Experimental Results

In this section, a number of experiments are described. The proposed technique is tested on the MPEG-7 and ETH-80 datasets. The first dataset contain a fair amount of affine (scaling, rotation, and translation) and visual (occlusion

Table 1. MPEG-7 core experiment (CE-Shape1) part B results of the proposed technique compared to some popular techniques in the literature

Algorithm	CSS [4]	DTW [5]	MCC [4]	Curve Edit [6]	Generative Model [7]	IDSC+DP [8]	HC+KFPs [2]	**HC+RKFPs 12 views**
Score	75.44%	77.76%	84.93%	78.17%	80.03%	85.4%	85.3%	**98%**
Complexity		$O(N^2)$	$O(N^3)$	$O(N^3)$		$O(N^3)$	$O(N)$	$O(NM)$ M=#of views

Table 2. ETH-80 results of the proposed technique compared to some popular techniques in the literature

	$D_x D_y$	Mag-Lap	PCA Masks	PCA grey	Cont. Greedy	Cont. DynProg	HC+KFPs	**HC+RKFPs** 12 views
Average	79.79%	82.23%	83.41%	82.99%	86.4%	86.4%	89.6%	**99.3%**

and articulation) transformations. The test datasets were not preprocessed in any way. The results are compared to those obtained when the object image is raster-scanned and with results of other techniques in the literature.

If it is not indicated otherwise, SFV size 512, wavelet approximation level 3, rbio3.1 wavelet, RKFPs vector size of 0.2 the CSFV size, and the Euclidean distance measure are used.

4.1 MPEG-7 Dataset

The MPEG-7 dataset consists of 70 classes of objects each of which has 20 different silhouette images (i.e. a total of 1400 silhouettes). The recognition rate is measured by the "Bulls eye" method [3]. Every image in the dataset is matched with all image and the 40 best matched candidates are determined. Within the 40 best matches, the objects belonging to the same class as the search object -except itself- are counted as correct hits.

Table 1 illustrates how the proposed approach fares against others in the literature. Note that for the HC+KFPs method the KFPs size is 25% of the SFV size while for the HC+RKFPs method the RKFPs size is 5% of the CSFV size. These settings gave the best results (please see below for more about this issue). All other variables are equal and are set to the default values mentioned above.

Figure 7 shows some observations made when the proposed improvement is applied to the MPEG-7 dataset. From (a), it is clear that the recognition rate tends to improve as the number of views increases. That is true for both the HC and raster scans (i.e., the algorithm in Section 2 is used but with two different scan types.) The raster scan seems to benefit more from the proposed improvement but is still inferior to the HC scan. From (b), it can be seen that the size of the RKFPs -that produces the best results- as a percentage of the size of the CSFV tends to decrease as the number of views increases. This helps to dampen the negative effect of increasing the number of views which results in larger

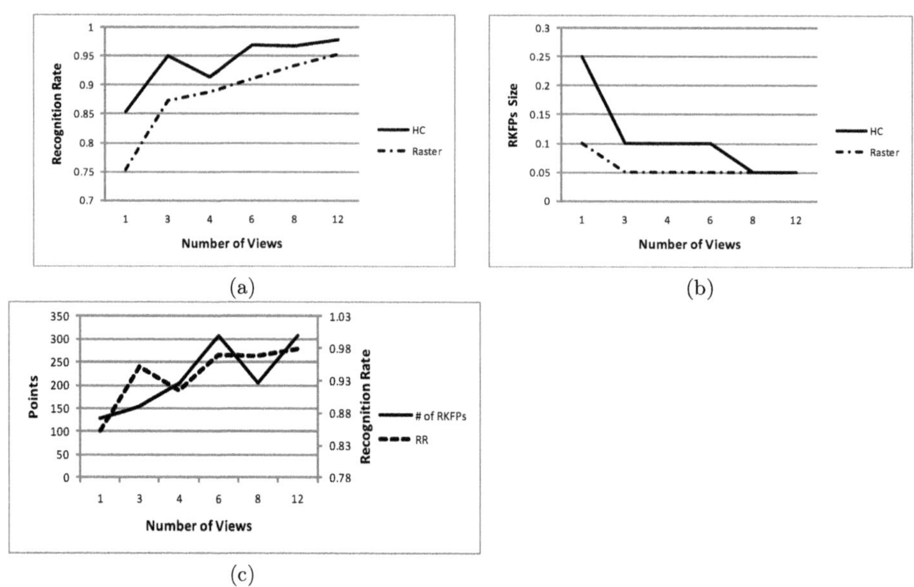

(a) (b) (c)

Fig. 7. MPEG-7 (a)HC and Raster scan Recognition Rate. (b) RKFPs size as a % of CSFV size. (c) RKFPs size as a # of points vs. Recognition Rate.

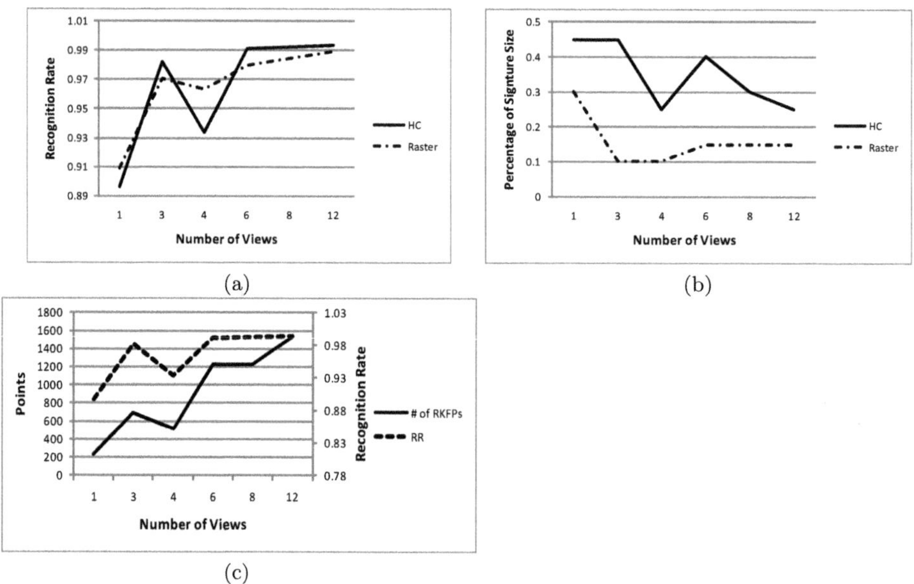

(a) (b) (c)

Fig. 8. ETH-80 silhouette (a)HC and Raster scan Recognition Rate. (b) RKFPs size as a % of RSVF size. (c) RKFPs size as a # of points vs. Recognition Rate.

CSFVs. The raster scan seems to require smaller number of RKFPs for lower numbers of views than the HC scan. The relationship between the recognition rate and the number of the RKFPs in points is depicted in (c). The figure shows a positive correlation between the two. Notice that even with the highest number of views tested, 12, the number of the RKFPs is still less than the SFV size of a single view of 512 points. This means that using the HC+RKFPs approach with its dramatically superior performance is still cheaper in terms of storage space than using the HC approach with no KFPs. Remember that we don't need to store the whole CSFV vector for each object. All we need is those elements of the CSFV that correspond to the RKFPs, discarding all the rest. This is where the savings are achieved.

4.2 ETH-80 Dataset

For this experiment, the ETH-80 dataset from ETH Zurich [9] is chosen. We use 1280 images of the dataset grouped into 8 classes with 10 objects each. Each object has 16 images obtained by *walking around* the object. The images are shot at an interval of 22 to 26 degrees. Each set of shots taken at a certain angle for each class of objects is considered a class on its own for RKFPs creation purposes. This means that the results obtained in this experiment not only show the success rate of identifying the right class of objects an object belongs to, but also that of identifying its right orientation. The experiment is performed on the silhouette version of the dataset.

Table 2 compares the results obtained by the HC+KFPs and HC+RKFPs techniques to that obtained by a set of techniques listed in [9]. Note that for the HC+KFPs method, the KFPs size is 45% of the SFV size, while for the HC+RKFPs method the RKFPs size is 25% of the CSFV size. These settings gave the best results. All other variables are equal and are set to the default values mentioned above.

Figure 8 shows the obtained results for the ETH-80 dataset in a similar way to that used with the MPEG-7 dataset above. From (a), it is evident that the recognition rate tends to improve as the number of views increases which is inline with the results obtained for the MPEG-7 dataset. That is true for both the HC and raster scans. Although both scans did comparably well, the HC scan performed better for higher number of views.

The trend found in Figure 7(b) is less evident in Figure 8(b). That is reflected in the significantly larger number of RKFPs at each number of views compared to that for the MPEG-7 dataset as can be seen in (c). Also from (b), the raster scan seems to require smaller number of RKFPs than the HC scan.

5 Conclusion

This paper describes a simple yet significant improvement over the the technique described in [2]. Instead of creating the shape signature based on one view of the object, it is created based on multiple views obtained by rotating the object a number of times. The result is the Cumulative Shape Features Vector (CSFV). The key feature points are determined over the entire CSFV resulting in a set of Rotational Key Feature Points (RKFPs). Because the different views of the object produce different SFVs, the CSFV contains a richer set of shape features than any one view SFV would have on its own. The RKFPs select the most prominent of those features for a superior representation of the shape class. As a result, the inter-correlation among the objects of each class is increased significantly. That extra discrimination power results in a dramatic increase in the recognition rate.

The results depicted in Section 4 raise a number of interesting issues. While the increase in the number of views tends to improve the recognition rate, when 4 and 8 views are used, the opposite tends to happen. This is more evident when the HC scan is used. Further investigation of how the number of views affects recognition rate may reveal interesting findings that may help in achieving higher recognition rates with fewer number of views.

Although the HC scan results in better recognition rate than the raster scan, the latter seems to need less RKFPs which translates into less storage requirements. Finding out why the raster scan requires less RKFPs may help in improving the HC scan in ways that lowers its storage requirements as well.

Experimental results show that the best results obtained for the ETH-80 dataset are associated with a much higher number of RKFPs than that of the MPEG-7 dataset. Further research into this phenomenon may lead to the determination of a set of dataset features that would help in estimating the number of RKFPs that would result in best results.

It would be interesting to see if the results obtained here would be affected significantly by changing the other parameters of the technique such as the wavelet used, the wavelet approximation level, and the SFV size. Testing on grey-scale objects is needed to see if the improvements achieved on the silhouette objects will carry over to grey objects.

References

1. Lambert, S., de Leau, E., Vuurpijl, L.: Using pen-based outlines for object-based annotation and image-based queries. In: Huijsmans, D.P., Smeulders, A.W.M. (eds.) VISUAL 1999. LNCS, vol. 1614, pp. 585–592. Springer, Heidelberg (1999)
2. Ebrahim, Y., Ahmed, M., Abdelsalam, W., Chau, S.-C.: Shape representation and description using the hilbert curve. Pattern Recognition Letters 30(4), 348–358 (2009)
3. Latechi, L.J., Lakamper, R., Eckhardt, U.: Shape descriptors for non-rigid shapes with a single closed contour. In: CVPR 2000: Proceedings of the 2005 IEEE Computer Society Conference on Computer Vision and Pattern Recognition (CVPR 2000), Washington, DC, USA. IEEE Computer Society, Los Alamitos (2000)
4. Adamek, T., O'Connor, N.E.: A multiscale representation method for non-rigid shapes with a single closed contour. IEEE Transactions on Circuits and Systems for Video Technology 14(5) (May 2004)

5. Adamek, T., O'Coonner, N.: Efficient contour-based shape representation and matching. In: The 5th ACM SIGMM International Workshop on Multimedia Information Retrieval (2003)
6. Sebastian, T.B., Klein, P.N., Kimia, B.B.: A survey of shape analysis techniques. IEEE Transactions on Pattern Analysis and Machine Intelligence 25(1), 116–125 (2003)
7. Tu, Z., Yuille, A.L.: Shape matching and recognition - using generative models and informative features. In: Pajdla, T., Matas, J(G.) (eds.) ECCV 2004. LNCS, vol. 3023, pp. 195–209. Springer, Heidelberg (2004)
8. Ling, H., Jacobs, D.W.: Using the inner-distance for classification of articulated shapes. In: CVPR 2005: Proceedings of the 2005 IEEE Computer Society Conference on Computer Vision and Pattern Recognition (CVPR 2005), Washington, DC, USA, vol. 2, pp. 719–726. IEEE Computer Society, Los Alamitos (2005)
9. Leibe, B., Schiele, B.: Analyzing appearance and contour based methods for object categorization. In: International Conference on Computer Vision and Pattern Recognition (CVPR 2003), Madison, Wisconsin, June 2003, pp. 409–415 (2003)

An Experimental Comparison of Seven Shape Descriptors in the General Shape Analysis Problem

Dariusz Frejlichowski

West Pomeranian University of Technology, Szczecin,
Faculty of Computer Science and Information Technology,
Zolnierska 49, 71-210, Szczecin, Poland
dfrejlichowski@wi.zut.edu.pl

Abstract. The general shape analysis is a problem similar to the recognition or retrieval of shapes. The most important difference is that the processed object does not have to belong to a base class, but usually is only similar to the template representing the class. The most general information about a shape is here concluded, i.e. how round, elliptical, triangular, etc. it is. Such a problem can occur in applications with few general base classes, e.g. in pre-classification or the assignment of stamps extracted from an image to few classes in order to find the fraudulent stamp images (mainly governmental, official ones). In the paper seven shape descriptors were explored using the template matching approach. In order to select the best approach their performance was compared with results provided by almost two hundred humans and collected using appropriate inquiry forms.

1 Introduction

The idea of image recognition can be realized in many various ways. One of them is based on identification of objects placed in a digital image. In that case one can use some features that are supposed to appropriately represent an object. Usually color, texture and shape are taken for that purpose. In many cases the last one is especially useful (however, lately the idea of combining completely different features becomes more tempting and popular, [1]). The recognition of shapes can be realized through the *template matching* approach. Roughly speaking, in this method an object under identification is matched with the base objects (*templates*). Obviously the matching of objects itself is insufficient, since in real situations they tend to be strongly deformed. In case of shapes not only the affine transformations (e.g. rotation, scaling and translation), but also noise and occlusion have to be considered ([2]). Therefore the so-called *shape descriptors* are used in order to represent a shape invariantly to particular deformations.

However, there is a class of applications, where the problem can not be considered as the traditional recognition. In that case the processed shape does not have to belong to a base class and usually it does not. The problem is depicted

A. Campilho and M. Kamel (Eds.): ICIAR 2010, Part I, LNCS 6111, pp. 294–305, 2010.

in Fig. 1. An object can be considered as similar to the three templates, yet it definitely is not one of them. It can be assigned to one of the very general shape classes, e.g. circle, triangle, star. It means that we are not interested in the exact identification of an object, but we are trying to find one or several general shapes, which are the most similar to the object being processed.

Fig. 1. Illustration of the problem — which general shape is the most similar?

The problem of general shape analysis can be utilized in various applications. Here, three examples will be briefly described. The first one is searching for probable false documents stored on a hard drive in a digital form ([3]). In this problem we can identify a general template — type of seal (e.g. official, public, business, institutional) instead of performing the exact process of recognition. This is based on the assumption that particular types of seal have an expected shape, e.g. official ones are round, whereas medical ones are rectangular (in Poland). The second example is the process of initial classification when working with large databases. In order to speed up the whole process the object is firstly matched with small number of general classes. Later it can be recognized more precisely within the preliminarily selected general class. In fact this process can be performed several times, with the classes becoming more detailed at each subsequent iteration. The third example is the possibility of using voice commands (e.g. 'find round red objects') for shape retrieval in large multimedia databases.

In the paper seven shape descriptors are used to indicate the general shapes. Their selection is not random. Firstly two so-called *simple shape descriptors* were taken, namely *Roundness* ([4]) and one of the *Feret measures* ([5]) — the *X/Y ratio.* Also, five more sophisticated shape descriptors were applied: *Moment Invariants* ([6]), *2D Fourier Descriptors* ([7]), *Point Distance Histogram* ([8]), *UNL* ([9]) and *UNL-Fourier* ([10]). Two of them (*PDH* and *UNL-F*) are invariant to rotation, the other three — are not. The *2D FD* is known for its ability of generalization ([7]). On the other hand, the *PDH*, thanks to the combination of polar coordinates with histogram, can emphasize the small differences between objects. Finally, the *UNL* and *UNL-F* have been successfully used in shape recognition, and are invariant (especially *UNL-F*) to many shape deformations.

The shape under analysis can be represented in two different ways. The first one is the outline of an object and the second is the whole region covered by it ([11]). Each of the algorithms works with one of the mentioned representations. In the paper the contour was explored using *PDH*, *UNL*, *UNL-F*, *Roundness* and the *X/Y Feret ratio.* The region was a subject for *MI* and *2D FD.*

The rest of the paper is organized as follows. The second section describes precisely the shape descriptors used in the problem of general shape analysis.

The third one provides experimental results achieved using them. The fourth section presents the results provided by humans and the comparison between them and the artificial algorithms. The last section concludes the paper and provides some ideas for future work with the problem.

2 The Description of Selected Algorithms

Seven various shape descriptors were taken for the experiments with the general shape analysis. The methods were selected deliberately in order to consider various ways of performing the analysis by humans. Some people take into account only the simplest features, e.g. curvature; others analyze an object on a higher level. However, the most important difference is the invariance to rotation. Some people during general analysis of a shape assume that it can not be rotated and that strongly influences the results.

In this section all the algorithms used will be described in detail. The order of their presentation is based on the ascending level of difficulty and sophistication.

2.1 Simple Shape Descriptors

Two simple shape descriptors were chosen. Those fast methods for measuring a shape are less popular nowadays thanks to the increasing computational power of computers. However, the generality of the simple methods can be an advantage in the presented problem. The first one, *X/ Y Feret shape measure*, can be computed using the formula:

$$F_{xy} = \frac{x_{max} - x_{min}}{y_{max} - y_{min}}, \tag{1}$$

where:
x_{min}, x_{max} — minimal and maximal horizontal coordinates of a contour shape,
y_{min}, y_{max} — minimal and maximal vertical coordinates of a contour shape.

The *Roundness* (a measure of the sharpness of a shape) was a second simple descriptor used in the experiment. This measure is based on two other shape features: the area (A) and the perimeter (P), and can be formulated as ([4]):

$$R = \frac{4\pi A}{P^2}. \tag{2}$$

2.2 Moment Invariants

The *moments* are commonly used in image representation. Hu introduced the name of *Moment Invariants* in 1962 ([12]). For shapes this representation uses only two values of image function $f(x, y)$ — 1 for a pixel belonging to an object and 0 for background pixels — instead of 256 levels as for a grayscale image. The *general geometrical moments* are given by the formula ([6]):

$$M_{pq} = \int_{-\infty}^{\infty} x^p y^q f(x, y) dx dy, \tag{3}$$

where: $p, q = 0, 1, ..., \infty$.

For the discrete values in the image the above can be written as ([13]):

$$m_{pq} = \sum_x \sum_y x^p y^q f(x,y). \tag{4}$$

We calculate the *centre of gravity* of an object ([13]):

$$x_c = \frac{m_{10}}{m_{00}} \qquad y_c = \frac{m_{01}}{m_{00}}. \tag{5}$$

Then we calculate the *central moments* ([13]):

$$\mu_{pq} = \sum_x \sum_y (x - x_c)^p (y - y_c)^q f(x,y). \tag{6}$$

The next step is the calculation of *normalized central moments* ([13]):

$$\eta_{pq} = \frac{\mu_{pq}}{\mu_{00}^{\frac{p+q+2}{2}}}. \tag{7}$$

Finally we can derive the *Moment Invariants* (*MI*). In practice usually the first seven MI are used ([14]):

$$\begin{aligned}
\varphi_1 &= \eta_{20} + \eta_{02} \\
\varphi_2 &= (\eta_{20} + \eta_{02})^2 + 4\eta_{11}^2 \\
\varphi_3 &= (\eta_{30} - 3\eta_{12})^2 + (3\eta_{21} - \eta_{03})^2 \\
\varphi_4 &= (\eta_{30} - \eta_{12})^2 + (\eta_{21} - \eta_{03})^2 \\
\varphi_5 &= (\eta_{30} - 3\eta_{12})(\eta_{30} + \eta_{12})[(\eta_{30} + \eta_{12})^2 - 3(\eta_{03} + \eta_{21})^2] \\
&\quad + (3\eta_{21} - \eta_{03})(\eta_{03} + \eta_{21})[3(\eta_{30} + \eta_{12})^2 - (\eta_{03} + \eta_{21})^2] \\
\varphi_6 &= (\eta_{20} - \eta_{02})[(\eta_{30} + \eta_{12})^2 - (\eta_{21} + \eta_{03})^2] + 4\eta_{11}(\eta_{30} + \eta_{12})(\eta_{03} + \eta_{21}) \\
\varphi_7 &= (3\eta_{21} - \eta_{03})(\eta_{30} + \eta_{12})[(\eta_{30} + \eta_{12})^2 - 3(\eta_{03} + \eta_{21})^2] \\
&\quad - (\eta_{30} - 3\eta_{12})(\eta_{03} + \eta_{21})[3(\eta_{30} + \eta_{12})^2 - (\eta_{03} + \eta_{21})^2].
\end{aligned} \tag{8}$$

The received shape representation is very compact — it is constituted by a vector of only seven values.

2.3 Fourier Descriptors

The *Fourier transform* is widely used in pattern recognition. In case of shapes usually its one-dimensional version is applied to the contour representation (e.g. [15]). However, in the literature another approach is also present. The so-called *2D Fourier Descriptors* are applied to a region (e.g. [16]).

In the experiments the *2D FD* were utilized. They can be derived using the *2D Fourier transform*, where only the absolute spectrum is used ([7]):

$$C(k,l) = \frac{1}{HW} \left| \sum_{h=1}^{H} \sum_{w=1}^{W} P(h,w) \cdot e^{\left(-i\frac{2\pi}{H}(k-1)(h-1)\right)} \cdot e^{\left(-i\frac{2\pi}{W}(l-1)(w-1)\right)} \right|, \tag{9}$$

where:
H, W — height and width of the image in pixels,
k — sampling rate in vertical direction ($k \geq 1$ and $k \leq H$),
l — sampling rate in horizontal direction ($l \geq 1$ and $l \leq W$),
$C(k,l)$ — value of the coefficient of *discrete Fourier transform* in the coefficient matrix in k row and l column,
$P(h,w)$ — value in the image plane with coordinates h, w.

2.4 UNL and UNL-Fourier Shape Descriptors

The *UNL* (*Universidade Nova de Lisboa*) descriptor is based on the transform of the same name ([9]). It uses complex representation of Cartesian coordinates for points and parametric curves in discrete manner ([9]):

$$z(t) = (x_1 + t(x_2 - x_1)) + j(y_1 + t(y_2 - y_1)), \qquad t \in (0,1), \tag{10}$$

where $z_1 = x_1 + jy_1$ and $z_2 = x_2 + jy_2$ (complex numbers) and z_i denotes point with coordinates x_i, y_i. The *centroid* O is now calculated ([9]):

$$O = (O_x, O_y) = (\frac{1}{n}\sum_{i=1}^{n} x_i, \frac{1}{n}\sum_{i=1}^{n} y_i), \tag{11}$$

and the maximal Euclidean distance between points and centroid is found ([10]):

$$M = \max_i\{\|z_i(t) - O\|\}, \qquad \forall i = 1...n, \qquad t \in (0,1). \tag{12}$$

Now coordinates are transformed ([9]):

$$U(z(t)) = R(t) + j \times \theta(t) = \frac{\|z(t) - O\|}{M} + j \times atan(\frac{y(t) - O_y}{x(t) - O_x}). \tag{13}$$

The discrete version is formulated as follows ([9]):

$$\begin{aligned} U(z(t)) = &\tfrac{\|(x_1+t(x_2-x_1)-O_x)+j(y_1+t(y_2-y_1)-O_y)\|}{M} \\ &+j \times atan(\tfrac{y_1+t(y_2-y_1)-O_y}{x_1+t(x_2-x_1)-O_x}). \end{aligned} \tag{14}$$

The parameter i is discretized in the interval [0,1] with significantly small steps ([10]). Derived coordinates are put into a matrix, in which the row corresponds to the distance from centroid, and the column — to the angle. The obtained matrix is 128 × 128 pixels size.

Since after the *UNL-transform* we obtain 2-dimensional binary image again, the author of the approach proposed using the *2D Fourier transform* as the next step ([9]). That gave one of the best descriptors in shape recognition called shortly the *UNL-F* and achieved using the *UNL-Fourier transform*.

2.5 Point Distance Histogram

The *Point Distance Histogram* ([8]) is an algorithm combining the advantages of histogram with the transformation of contour points into polar coordinates. Firstly the mentioned coordinates are derived (with $O = (O_x, O_y)$ as the origin of the transform) and put into two vectors Θ^i for angles and P^i for radii ([8]):

$$\rho_i = \sqrt{(x_i - O_x)^2 + (y_i - O_y)^2}, \qquad \theta_i = atan\left(\frac{y_i - O_y}{x_i - O_x}\right). \tag{15}$$

The resultant values are converted into nearest integers ([8]):

$$\theta_i = \begin{cases} \lfloor \theta_i \rfloor, & if\ \theta_i - \lfloor \theta_i \rfloor < 0.5 \\ \lceil \theta_i \rceil, & if\ \theta_i - \lfloor \theta_i \rfloor \geq 0.5 \end{cases}. \tag{16}$$

The next step is the rearrangement of the elements in Θ^i and P^i according to increasing values in Θ^i. This way we achieve the vectors Θ^j, P^j. For equal elements in Θ^j only the one with the highest corresponding value P^j is selected. That gives a vector with at most 360 elements, one for each integer angle. For further work only the vector of radii is taken — P^k, where $k = 1, 2, ..., m$ and m is the number of elements in P^k ($m \leq 360$). Now, the normalization of elements in vector P^k is performed ([8]):

$$M = \max_k \{\rho_k\}, \qquad \rho_k = \frac{\rho_k}{M}, \tag{17}$$

The elements in P^k are assigned to r bins in histogram (ρ_k to l_k,[8]):

$$l_k = \begin{cases} r, & if\ \rho_k = 1 \\ \lfloor r\rho_k \rfloor, & if\ \rho_k \neq 1 \end{cases}. \tag{18}$$

3 Conditions and Results of the Experiments

The problem was explored using objects from [17]. The database included 10 templates (the general shapes) and 40 tested objects (see Fig. 2).

For each explored algorithm the idea of a test was simple. A test object was represented using a shape descriptor, and so were all the general shapes (templates). Basing on the typical template matching approach, the description of a test shape was matched using *Euclidean distance* with all the descriptions of the templates. The three smallest dissimilarity values indicated the general shapes closest to a test shape according to the algorithm explored. Pictorial representations of the results will be provided in consecutive figures.

The results for *Roundness* presented in Fig. 3 can not be considered ideal. However, in some cases they seem to be correct. For example, the rhombus (object no. 2) is firstly connected to the square and secondly to the rectangle, which is a very good result. However, the third indicated general shape — the disc — completely does not fit to the previous ones. Similarly, the triangle was

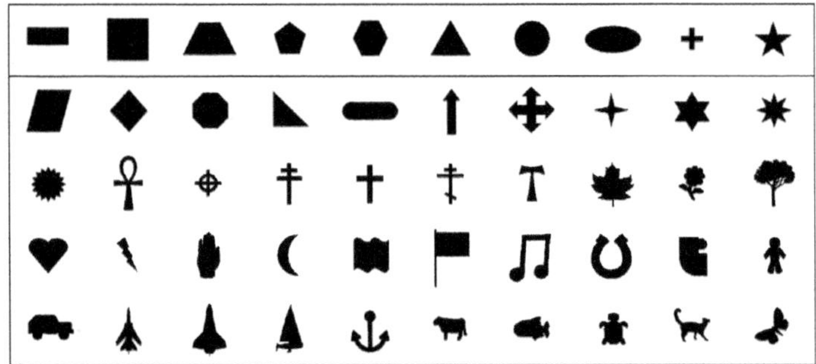

Fig. 2. The division of shapes into 10 templates and 40 test objects

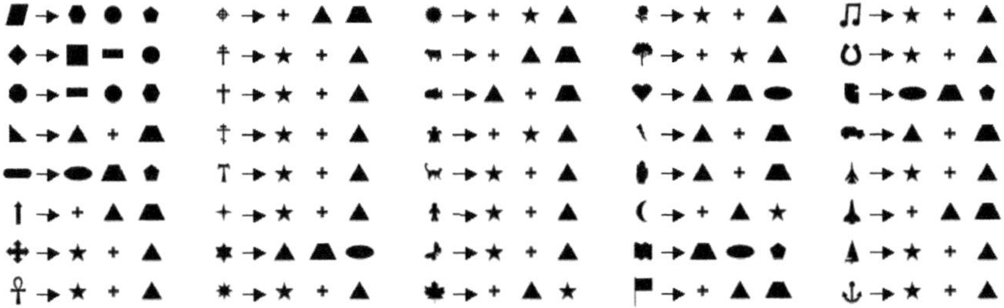

Fig. 3. Results of the experiment on general shape analysis using Roundness

properly matched with a general triangular shape and a rectangle with rounded corners was connected with the ellipse. In other cases the star was selected very often. That is clearly seen for various versions of crosses. The results for more complicated objects are less convincing. Usually the star, triangle or cross were indicated.

As one might expect, the results achieved using such a simple approach (*X/Y Feret*) are not satisfactory (see Fig. 4). Only in a few cases the results can be considered correct. The rounded rectangle is the first example. Definitely the first two selected general shapes (the square and the ellipse) are similar to the test object. Similarly, the results achieved for the car are correct in all three cases — the ellipse, the trapezoid and the rectangle. In few other cases the first result is also acceptable. That concerns for example the heart that is similar to the triangle and the flag close to the square.

The results provided using *Moment Invariants* (see Fig. 5) are usually very promising for the first indicated general shape. This time the rectangle 'became more popular'. In fact, e.g. for the crosses, the human, the cat, the car, etc. one can agree that when rotated they are very similar to the rectangle.

The results for *Fourier Descriptors* (see Fig. 6) are very interesting. For the first five simple test shapes they are very proper. The results achieved for group

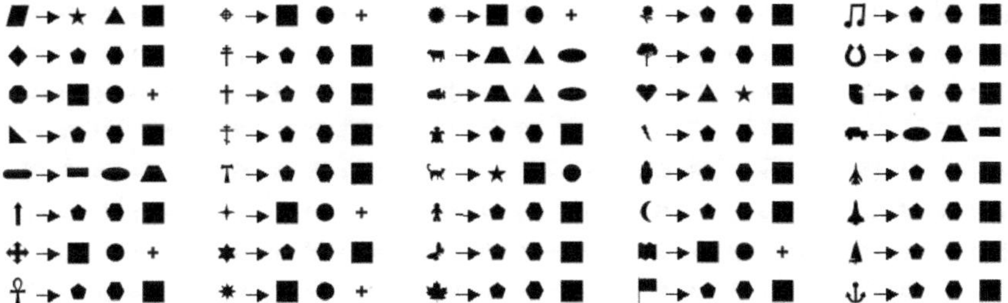

Fig. 4. Results of the experiment on general shape analysis using X/Y Feret ratio

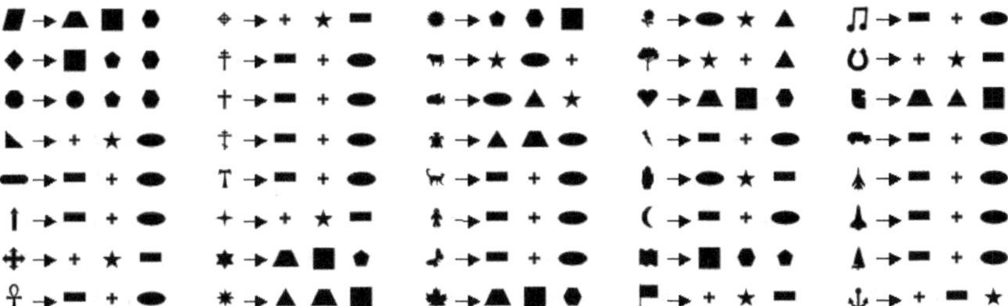

Fig. 5. Results of the experiment on general shape analysis using Moment Invariants

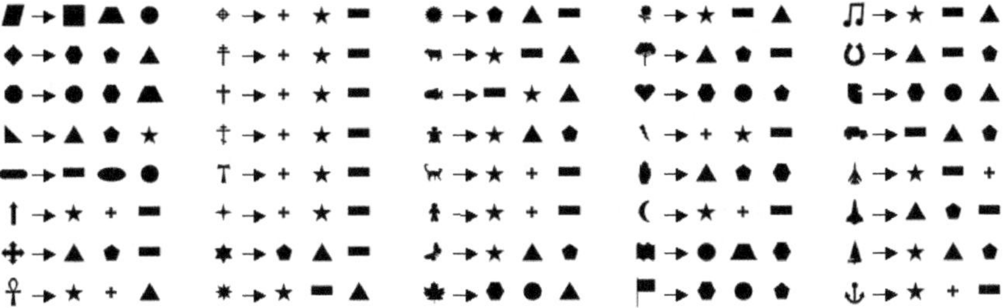

Fig. 6. Results of the experiment on general shape analysis using Fourier Descriptors

of crosses is also correct. For other objects the indication of the star as the most similar general shape is clearly visible yet not always correct.

The *UNL* shape descriptor (Fig. 7) completely failed in our problem. Only few results can be considered proper. Usually, the square and the disc were taken at the first place. Sometimes the pentagon and the hexagon were selected. The lack of other templates in results, e.g. the star and the cross, is noticeable.

The results achieved using the *UNL-Fourier* descriptor (Fig. 8) are not convincing. This method is very effective in the traditional shape recognition problem. However in the general shape analysis it often fails. For example, the results achieved for the parallelogram and crosses are irrational. On the other hand, the

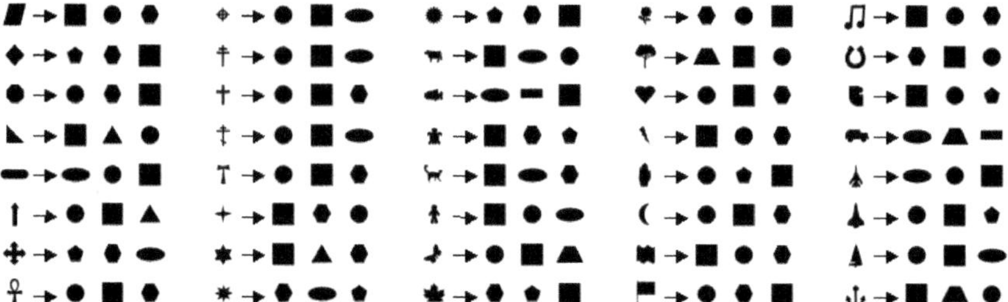

Fig. 7. Results of the experiment on general shape analysis using UNL descriptor

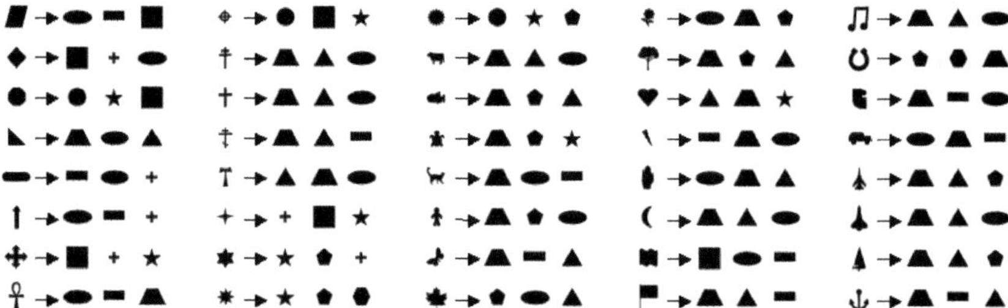

Fig. 8. Results of the experiment on general shape analysis using UNL-F descriptor

performance of the descriptor for the rounded rectangle, stars and some more complicated test objects is definitely acceptable.

In some cases the results of the *PDH* approach (Fig. 9) are acceptable. For example, the first indicated general shape is correct for the first four test objects as well as for the human, the hand and the car. In general this approach gave slightly worse results than expected. However, in comparison with other explored descriptors it seems to be appropriate for the problem under consideration.

4 The Results Provided by Humans

The results provided and briefly described in the previous section can not be adequately judged, since we can only guess if a method is working well or not. There is no independent measure for the problem of general shape analysis. Therefore, in order to estimate the behavior of explored algorithms an inquiry form was filled by 187 persons (124 men, 63 women, aged from 9 to 62) in. This inquiry was conducted in order to investigate the manner in which humans perform the task of general shape analysis. This can serve as a benchmark, an ideal result. Now, we only have to investigate which of the explored artificial algorithms is the most similar to it and in what degree. The results of general shape analysis performed by humans are depicted in Fig. 10. The analysis of the inquiry forms could be performed in various ways. Here, the most popular result at the particular place was selected. That gave the most common general shapes indicated by humans for particular test objects.

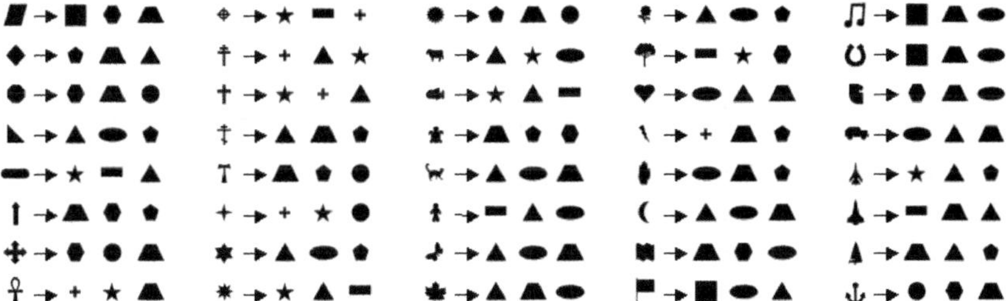

Fig. 9. Results of the experiment on general shape analysis using PDH descriptor

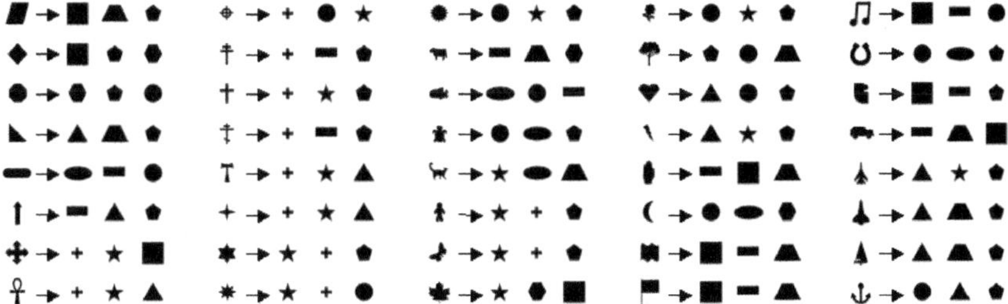

Fig. 10. Results of the general shape analysis test performed by humans, a benchmark for the artificial methods

Table 1. The comparison of the general shape analysis performed by humans and artificial algorithms - the percentage of convergence between a shape descriptor and benchmark human results

Shape descriptor	1st indication	2nd indication	3rd indication
1. FD	35%	22.5%	17.5%
2. PDH	25%	15%	27.5%
3. Roundness	25%	12.5%	17.5%
4. MI	20%	17.5%	5%
5. UNL-F	17.5%	10%	12.5%
6. UNL	15%	5%	12.5%
7. X/Y Feret	10%	7.5%	2.5%

The degree of similarity between shape descriptors and results provided by humans is presented in Table 1. Each time the percentage of proper indications compared to benchmark human results is presented, separately for the three firstly selected templates. As it can bee seen, the *Fourier Descriptors* work most similarly to the benchmark human statistics. The result is much higher than for other six explored methods. That concerns the first (35%) and the second (22.5%) selected template. Only for the third selection another descriptor works better, namely — *PDH.* It is accordant with the human results in 27.5 %.

5 Conclusions and Future Plans

The paper described experimental results on usage of shape descriptors in the general shape analysis. The problem is similar to recognition or retrieval, but here we assume that the processed shape does not have to belong to any of the template classes, which include the few most general shapes — triangle, square, disc, etc. Therefore, it can be considered as the determination how triangular, square, round, etc. is a tested shape. The general shape analysis, as presented in the paper, can be useful in many applications. The first one is the preliminary classification of shapes, when we firstly assign an object to a major class and subsequently we increase the level of details in the identification. Another example is the shape retrieval based on the similarity of an object to few the most general shapes. It can be combined for example with the usage of voice commands. The third example is the analysis of seals when searching for probable false documents stored on a hard drive in a digital form.

During the experiments on the problem seven shape descriptors were explored. In order to measure their performance, a special inquiry form was developed. It was similar to the performed tests and it was filled in by almost two hundred persons. The selection of the best method was based on a very simple criterion. The artificial method with the results most similar to the ones provided by humans was treated as the best. As it turned out the *FD* were the best among the tested approaches. The second place went to the *PDH*, which was also rather successful. However, the achieved numerical results can not be treated as satisfactory enough. On the other hand, the results provided by humans are in many cases ambiguous as well. Nevertheless, the results achieved by the best shape descriptor among tested are worse than expected (35%), therefore there is still necessity of exploring some other algorithms in the problem. This is the first conclusion related to the future work. The second important issue is the a different way of constructing the benchmark. In the paper the simplest approach was utilized. Plainly, the most popular result in the inquiry forms was treated as the proper one. However, in some cases the differences between the most popular indication and the second one were very small. This can be taken into consideration in the future improved method of comparing the artificial results with the human benchmark. Finally, the experiments on some practical examples will be performed to illustrate the capabilities of the best approaches. The first problem to explore is the identification of document seals by means of the general shape analysis.

References

1. Forczmanski, P., Frejlichowski, D.: Strategies of Shape and Color Fusions for Content Based Image Retrieval. Advances in Soft Computing 45, 3–10 (2007)
2. Frejlichowski, D.: An Algorithm for Binary Contour Objects Representation and Recognition. In: Campilho, A.C., Kamel, M.S. (eds.) ICIAR 2008. LNCS, vol. 5112, pp. 537–546. Springer, Heidelberg (2008)

3. Forczmanski, P.: Stamp Detection in Scanned Documents. In: 8th Int. Conf. on Comp. Science - Research and Applications (IBIZA 2008), Poland, Kazimierz Dolny (2009)
4. Nafe, R., Schlote, W.: Methods for Shape Analysis of two-dimensional closed Contours — A biologically important, but widely neglected Field in Histopathology. Electronic Journal of Pathology and Histology 8(2) (2002)
5. Whang, S.S., Kim, K., Hess, W.M.: Variation of silica bodies in leaf epidermal long cells within and among seventeen species of Oryza (Poaceae). American Journal of Botany 85, 461–466 (1998)
6. Rothe, I., Süsse, H., Voss, K.: The method of normalization to determine invariants. IEEE Trans. on Pattern Anal. and Mach. Int. 18, 366–375 (1996)
7. Kukharev, G.: Digital Image Processing and Analysis. SUT Press (1998) (in Polish)
8. Frejlichowski, D.: The Point Distance Histogram for Analysis of Erythrocyte Shapes. Polish Journal of Environmental Studies 16(5b), 261–264 (2007)
9. Rauber, T.W., Steiger-Garcao, A.S.: 2-D form descriptors based on a normalized parametric polar transform (UNL transform). In: Proc. of MVA 1992 IAPR Workshop on Machine Vision Applications (1992)
10. Rauber, T.W.: Two-dimensional shape description. Technical Report: GR UNINOVA-RT-10-94, Universidade Nova de Lisboa (1994)
11. Zhang, D., Lu, G.: Review of shape representation and description techniques. Pattern Recognition 37(1), 1–19 (2004)
12. Flusser, J., Suk, T.: Affine moment invariants: A new tool for character recognition. Pattern Recognition Letters 15(4), 433–436 (1994)
13. Hupkens, T.M., de Clippeleir, J.: Noise and intensity invariant moments. Pattern Recognition Letters 16(4), 371–376 (1995)
14. Liu, C.-B., Ahuja, N.: Vision Based Fire Detection. In: Proc. of 17th Int. Conf. on Pattern Recognition (ICPR 2004), UK, Cambridge (2004)
15. Osowski, S., Nghia, D.: Fourier and wavelet descriptors for shape recognition using neural network — a comparative study. Pattern Recognition 35(9), 1949–1957 (2002)
16. Yadav, R.B., Nishchal, N.K., Gupta, A.K., Rastogi, V.K.: Retrieval and classification of shape-based objects using Fourier, generic Fourier, and wavelet-Fourier descriptors technique: A comparative study. Optics and Lasers in Engineering 45, 695–708 (2007)
17. Frejlichowski, D.: General Shape Analysis Using Fourier Shape Descriptors. In: Swiatek, J., et al. (eds.) Information Systems Architecture and Technology – System Analysis in Decision Aided Problems, pp. 143–154 (2009)

Generic Initialization for Motion Capture from 3D Shape

Benjamin Raynal, Michel Couprie, and Vincent Nozick

Université Paris-Est
Laboratoire d'Informatique Gaspard Monge, Equipe A3SI

Abstract. Real time and markerless motion capture is an active research area, due to applications in human-computer interactions, for example. A large part of the existing markerless motion capture methods require an initialization step, consisting in finding the initial position of the different limbs of the subject. In this paper, we propose a new method for interactive time initialization step, only based on morphological and topological information and which can be easily adapted to any kind of model (full human body or only hand, animals, for example).

Keywords: motion capture, initialization, markerless, skeleton, tree matching.

1 Introduction

Motion capture without marker is a highly active research area, as shown by Moeslund and al. [1]: between 2000 and 2006, more than 350 papers on this topic have been published. Markerless motion capture approaches can be classified in two categories: those which detect the pose of the subject independently at each frame, and those which start by an initialization step in order to find the initial pose of the subject, then use tracking to find the pose in the following frames. In most of these methods, the initialization step uses an a priori model, which can be of several kinds, describing different information: kinematic skeleton, shape, color priors. Most of real-time methods for full body pose estimation purpose [2,3,4] use both color priors and shape fitting for the initialization. The method proposed in [5], close to interactive time (about one iteration per second), uses kinematic skeleton fitting.

In a context of generic motion capture, where the subject can be a full body, the hand, or the the upper part of the body for example, some of this information cannot be retained, as the color information (hands have homogeneous color). Furthermore, the shape of the subject can differ from person to person. In addition, from our point of view, the a priori model in generic motion capture must be as simple as possible.

Our goal is to propose a motion capture initialization method which has the following properties:

interactive runtime: our method must be fast enough to be usable in online context (more than one iteration per second).

A. Campilho and M. Kamel (Eds.): ICIAR 2010, Part I, LNCS 6111, pp. 306–315, 2010.

markerless: our method does not require any kind of marker.
generic: our method has to be compatible with any subject.

Our method is based on 3D shape obtained by visual hull reconstruction. In our method, we use a voxel representation of the visual hull, in opposition to the polygonal representation. This choice is guided by the fact that we use some discrete treatments which are easy and fast to perform in a voxel grid. As we use a small number of cameras, the 3D shape can contain some deformities, as variations of limb thickness, noisy surface, or ghost limbs (i.e. parts which not exist in the subject). Thus only the global topology of the shape and the length of the different limbs are well preserved.

The method starts by extracting this information by skeletonizing the shape (Sec.2). Then a data tree representation of the skeleton is extracted (Sec.3), containing all the information we need: positions of ending and intersection points on vertices, and distances between them on edges.

Then, according to the edge information, we proceed to a matching between an a priori model and the data tree (Sec.4). Finally, we discriminate similar limbs if necessary, using "between" constraints of the model (Sec.5). Figure 1 show the complete pipeline of our method.

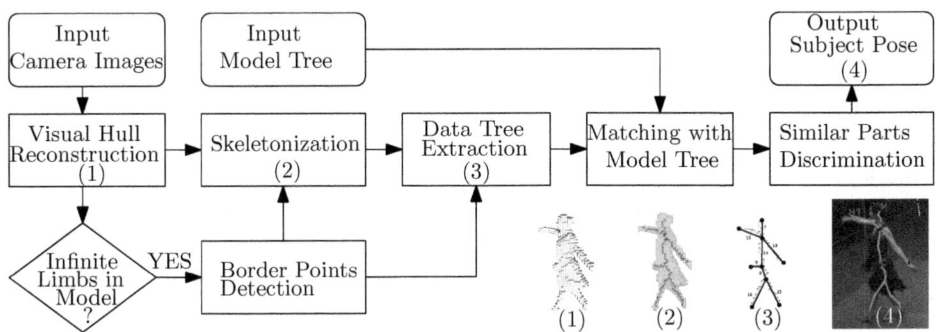

Fig. 1. Pipeline of our method

1.1 A Priori Model Definition

Our a priori model is very simple. It consists of a tree, where vertices represent the different parts of the subject, and the edges contain information on distance between the different parts. Two kinds of models can be considered (see Fig.2 for examples):

- the *incomplete models*, which are part of a biggest shape, as in the case of hand pose estimation. In this case, a part of the shape intersects the border of the 3D acquisition space, and we represent this part in the model as an *infinite limb*.
- the *complete models*, for subjects fully contained in the 3D acquisition space, as in the case of full human body motion capture, for example.

In addition of the tree representation, two kinds of constraints can be added:

"Between" constraints specify the position of a part between two others, in order to discriminate similar limbs. In the case of hand model, we require e.g. that the index is between the thumb and the middle finger. More details about "between" constraints are given in Sec.5.

"Coordinate" constraints require a particular spatial position of a part in regard of the spatial position of one of its neighbor. In the case of full body model for example, we require that the head is above the torso. Constraints of this kind improve the matching robustness.

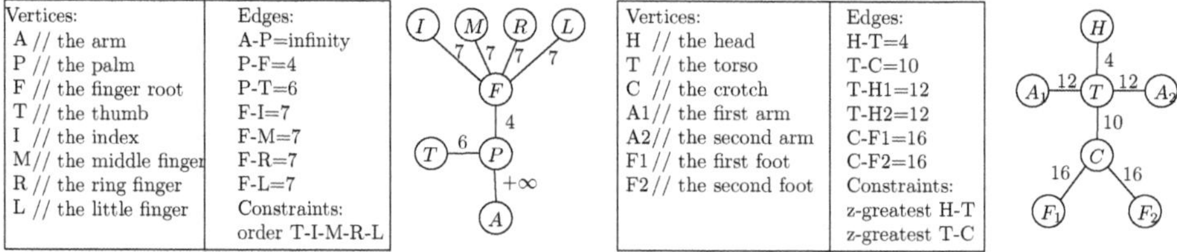

Vertices:
A // the arm
P // the palm
F // the finger root
T // the thumb
I // the index
M// the middle finger
R // the ring finger
L // the little finger
Edges:
A-P=infinity
P-F=4
P-T=6
F-I=7
F-M=7
F-R=7
F-L=7
Constraints:
order T-I-M-R-L
Vertices:
H // the head
T // the torso
C // the crotch
A1// the first arm
A2// the second arm
F1 // the first foot
F2 // the second foot
Edges:
H-T=4
T-C=10
T-H1=12
T-H2=12
C-F1=16
C-F2=16
Constraints:
z-greatest H-T
z-greatest T-C

Fig. 2. From left to right: description of the hand model, model tree for the hand, description of the full body model, and model tree for the full body

2 Skeletonization

Topology-preserving operators, like homotopic skeletonization, are used to transform an object while leaving unchanged its topological characteristics. In discrete grids ($\mathbb{Z}^n$, with $n = 2, 3$), such a transformation can be defined thanks to the notion of simple point [6,7,8]: intuitively, a point of an object (*i.e.*, a subset of $\mathbb{Z}^n$) is called simple if it can be deleted from this object without changing its topological characteristics.

The most "natural" way to thin an object consists of removing some of its border points in parallel, in a symmetrical manner. By repeating such a procedure until stability, one can obtain a well-centered "skeleton" of the original object. However, parallel deletion of simple points does not, in general, guarantee topology preservation. In fact, such a guarantee is not obvious to obtain, even for the 2D case. To check whether a point is simple or not, it is sufficient to examine its $3 \times 3 \times 3$ neighborhood (3×3 in 2D), but such a local criterion does not allow to check whether a simple point may be safely removed together with other ones.

In [9], G. Bertrand introduces a general framework for the study of parallel thinning in any dimension. The most fundamental result proved in [9] is that, if a subset Y of an object X contains the so-called critical kernel of X, then Y has the same topological characteristics as X. In [10], several new parallel algorithms to compute curvilinear skeletons are proposed, in which topological and geometrical conditions are clearly separated, unlike in many previous works. The topological soundness of these algorithms is proved thanks to the aforementioned property of critical kernels. Furthermore, these algorithms may be expressed by the way of masks and are relatively simple to implement.

The skeletonization algorithm that we use in this study is named ACK^3 in [10]. We choose this algorithm for its computational speed, the possibility of parallelization, and for the quality of the resulting skeleton (very low amount of noise, and guaranty that branch thickness is always of one voxel (asymmetrical skeleton). Since the complete presentation of this algorithm is beyond the scope of this paper, we give here a sketch of its main lines.

Let us describe one step of algorithm ACK^3. Let X be the current object. The set S of all simple points of X is computed, as well as the set I of all $1D$ isthmuses of X (points of which the removal would break locally X into several components). Then a subset Y of X is computed, that verifies the following conditions: i) Y is a superset of $X \setminus S$, ii) Y contains the critical kernel of X, and iii) Y contains I. If $Y = X$ then the algorithm stops, otherwise X is set to Y and the algorithm continues.

In the case of incomplete models, we have to take into consideration the part of the shape which is on the border of grid. Preliminary to the iterative process, the set B of the border points contained in X is computed. For each connected component of B, we compute the centroid, which will be preserved during the skeletonization process. Figure 3 show iterations results, for both kinds of model.

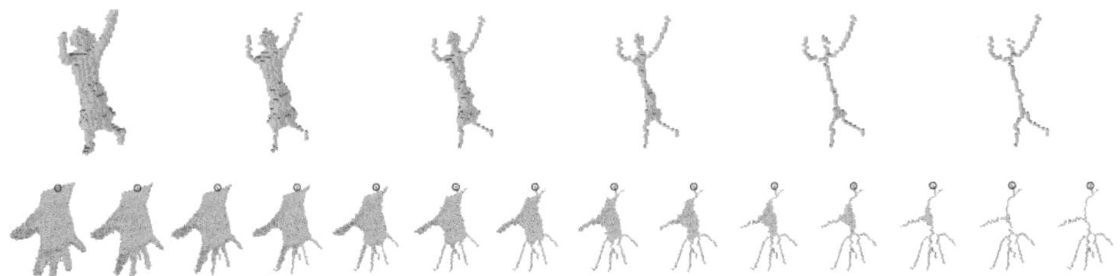

Fig. 3. From left to right, results of successive iterations of skeletonization. Top: human model. Bottom: hand model. The circled points represent the border points used to constraint the skeletonization.

3 Extraction of Data Tree Representation

We extract the data tree representation from the skeleton obtained in the previous step. The skeleton points can be classified into three classes, in regard of their number of neighbors included in the skeleton: ending points (exactly one neighbor included in the skeleton), linking points (exactly two neighbors), and intersection points (strictly more than two neighbors).

The points of interest are the ending points and the intersection points. For each of them, we create a vertex in the data tree. If several intersection points are neighbors, we merge their associated vertices. Intersection points and ending points are connected by sequences of linking points. For each sequence, we create an edge between vertices associated to its extremities, weighted by the length of the sequence, incremented by one if an extremity is an ending point. See Fig.4 for an example with a 2D skeleton.

In the case of an incomplete model, the skeleton is tied to contain at least one border point. It implies that at least one intersection point is a border point. We consider that all the edges having a bounding vertex associated with a border point have infinite weight. See Fig.4 for an example with a 2D skeleton.

A skeleton can contain cycles, for example if a sequence of linking points has its two extremities in the same point. In this case, the extracted graph is not a tree, and we stop the pipeline for the current frame.

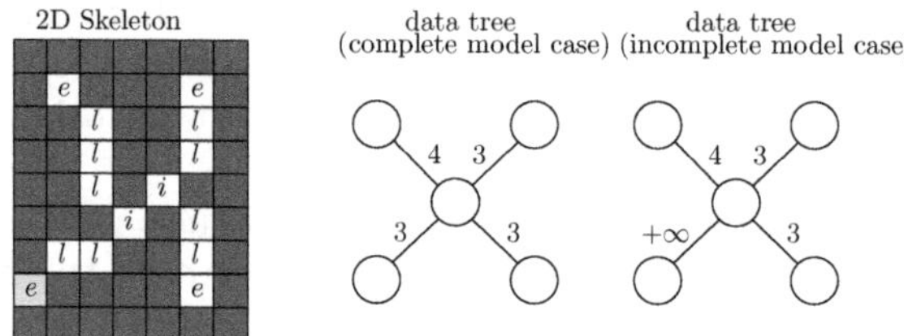

Fig. 4. Example of a data tree extraction from a 2D skeleton. On the left, the 2D skeleton, with pixels labeled by i,e and l representing respectively intersection, ending, and linking points. The light gray pixel represents a border point. In the middle, the data tree extracted from the skeleton, with the edge weights represented by numbers. On the right, the data tree extracted from the skeleton, in the case of an incomplete model.

4 Matching with a Priori Model

The data tree can be affected by different kinds of noise, which must be taken into consideration during the matching: due to the irregularities of the shape surface, skeleton branches without important topological signification can appear. These branches are not difficult to remove, but the problem is that it generates new vertices in the tree. These vertices, after the removing of branches, uselessly split an edge (and its weight) into two parts, making difficult a good matching. The second kind of noise is due to the skeletonization: a cluster of vertices linked by weakly weighted edges in the data tree can correspond to a vertex with more than three neighbors in the model tree. In order to match the data tree with the model tree, we use the optimal homeomorphic alignment method [11], especially designed to be robust in regard of these kinds of noise.

4.1 Preliminary Definitions and Notations

In order to present the homeomorphic alignment, some definitions and notations are necessary. We denote a weighted tree as a triplet $T = (V, E, \omega)$, V is a finite set (called the *vertex* set), A is a subset of $V \times V$ (called the *edge* set), and ω is a mapping from A to $\mathbb{R}^+$, corresponding to the weights.

An homeomorphic alignment is based on edit operations: *deletion* consists of removing an edge in the tree, *resizement* consists of changing the weight of an edge, and *merging* consists of replacing two edges (a, b) and (b, c), where b has exactly two neighbors, by an unique edge (a, c) weighted by $\omega((a, b)) + \omega((b, c))$, and removing b from V. The *merging kernel* of a tree T is obtained by applying iteratively all possible mergings on T.

The *cost* of an operation is equal to the variation of weights in the tree before and after the application of the involved operation. Then, the cost of a deletion is equal to the deleted edge weight and is denoted by $\gamma(w, 0)$, where w is the weight of the deleted edge. In the same way, the cost of a resizement $\gamma(w, w')$ is equal to the difference between the former weight w and the new weight w' of the resized edge. A merging has a null cost, since the total weight of the tree is preserved by the operation.

In order to match trees with infinite weights, we have to take the convention that $\gamma(+\infty, +\infty) = 0$.

Two weighted trees $T = (V_T, E_T, \omega_T)$ and $T' = (V'_T, E'_T, \omega'_T)$ are said to be *isomorphic* if there exists a bijection $f : V_T \rightarrow V'_T$, such as for any pair $(x, y) \in V_T \times V_T$, $(x, y) \in E_T$ if and only if $(f(x), f(y)) \in E'_G$.

Two weighted graphs $T = (V_T, E_T, \omega_T)$ and $T' = (V'_T, E'_T, \omega'_T)$ are *homeomorphic* if there exists an isomorphism between the merging kernel of T and the merging kernel of T'.

4.2 Homeomorphic Alignment Definition

Let $T_1 = (V_1, E_1, \omega_1)$ and $T_2 = (V_2, E_2, \omega_2)$ be two weighted trees. Let $T'_1 = (V'_1, E'_1, \omega'_1)$ and $T'_2 = (V'_2, E'_2, \omega'_2)$ be weighted graphs obtained by deleting edges in T_1 and T_2, such that there exists an homeomorphism between T'_1 and T'_2 (not necessarily unique). Let $T''_1 = (V''_1, E''_1, \omega''_1)$ and $T''_2 = (V''_2, E''_2, \omega''_2)$ be the merging kernel of T'_1 and T'_2, respectively. By definition, there exists an isomorphism $\mathcal{I}$ between T''_1 and T''_2. The set of all couples of arcs $\mathcal{H} = \{(e, e'); e \in E''_1, e' \in E''_2, e' = \mathcal{I}(e)\}$ is called an *homeomorphic alignment* of T_1 with T_2 (see figure 5).

The *cost* $C_{\mathcal{H}}$ of $\mathcal{H}$ is the sum of the costs of all operations used to homeomorphically align T_1 and T_2: the deletion of edges in T_1 and T_2, to obtain T'_1 and T'_2 respectively, and the resizement for each edge $e_1 \in E''_1$ to the weight of $\mathcal{H}(e_1)$. An homeomorphic alignment with minimal cost is said to be optimal. The cost of an optimal homeomorphic alignment is called the homeomorphic alignment distance.

The "coordinate" constraints of the model are applied during the optimal homeomorphic alignment: the cost of the resizement of a model edge (v_M, v'_M), weighted by w_M, to the weight w_D of a data edge (v_D, v'_D) is equal to $+\infty$ if there is a constraint C associated to (v_M, v'_M), and the 3D point associated to v_D is in a spatial position relative to the one associated to v'_D which is not compatible with C.

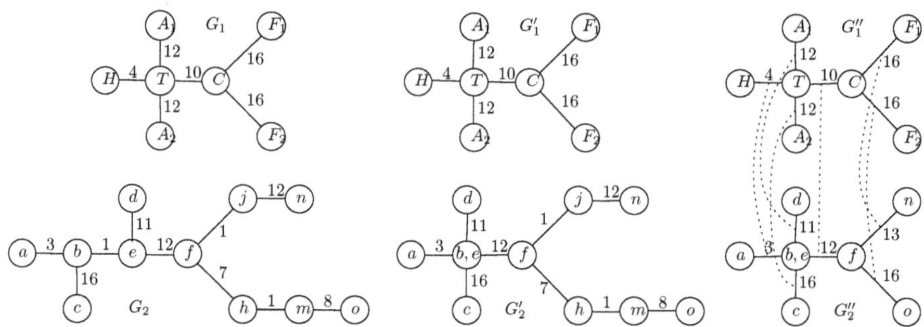

Fig. 5. Left: two trees G_1 and G_2. Middle: G'_1 and G'_2, obtained respectively from G_1 and G_2 by deletion of edges, and which are homeomorphic. Right: G''_1 and G''_2, which are respectively the merging kernels of G'_1 and G'_2, and which are isomorphic. Dotted lines represent an optimal homeomorphic alignment.

For the purpose of scale invariance, we start our method by normalizing the weights of the two trees, so that the sum of all non-infinite weights in a tree is equal to 1. Then, we compute the homeomorphic alignment distance between the trees, using the algorithm described in [11]. If the distance is greater than a given threshold T_d, which is defined by the user, we assume that the data tree is not enough similar to the model tree to provide a good matching, and we stop the pipeline for this frame. Otherwise, we use the optimal homeomorphic alignment to match the labels associated to the model vertices, with the 3D positions associated to the data vertices.

The number of non-aborted matchings, and their quality, obviously depend on the choice of the threshold value T_d : as both trees are normalized, $T_d = 2$ means that none matching will be aborted (both trees can be deleted), but the resulting matching can be null, or poor. In the other hand, a lower value yields a lower amount of non-aborted matchings, but with better probability of good matching.

5 Discrimination of Similar Limbs Using Model "Between" Constraints

In case of similar limbs, the matching with the model tree can generate multiple solutions. In the case of our model of hand, as the descriptions of index, middle finger, ring finger and little finger are identical, the matching will give a set of four possible positions for each finger. To solve this problem, we introduce the "between" constraints. It consists of constraining the position of some limb to lie between two other ones. A usual ternary relation "between" definition is based on collinearity [12] : a point is said to be between two other points P_1 and P_2 if it belongs to the segment P_1P_2. However, this definition is too restrictive. On the other hand, we could say that a point is between P_1 and P_2 if its projection on the line (P_1P_2) is between P_1 and P_2. In this case, the problem is that there exist triplets (P_1, P_2, P_3) such that each point is between the two others (for example, the three vertices of an equilateral triangle).

We propose a definition which is not too restrictive, and which gives at most one possibility of "betweenness" for three points P_1, P_2, P_3 : consider $\mathcal{B}$ the unique ball with diameter P_1P_2 which contains P_1 and P_2. The point P_3 is between P_1 and P_2 iff $P_3 \in \mathcal{B}$. An other formulation is that P_3 is between two points P_1 and P_2 iff the angle between $\overrightarrow{P_3P_1}$ and $\overrightarrow{P_3P_2}$ is greater than $\frac{\pi}{2}$. Then, if there exists a "between" constraint on model vertices, defined by a sequence $m_0, ..., m_n$, the corresponding data vertices $d_0, ..., d_n$, with associated positions $P_0, ..., P_n$ must be chosen such as, for each $i \in [1, n-1]$, $\overrightarrow{P_iP_{i-1}} \cdot \overrightarrow{P_iP_{i+1}} < 0$.

6 Implementation and Results

Our method has been tested on a computer with a processor Intel(R) Core(TM) 2 Quad Q8200 at 2.33 GHz, a GPU Nvidia(R) Geforce(TM) 9800 GT and 3 Go of RAM. Our implementation is implemented in C++, and the visual hull is computed on GPU, using GLSL. For the tests, we use two data sets: a set for hand pose estimation, produced by our team, and the dancer set produced by the INRIA Perception Group [1], for full body body pose estimation. We have tested our method for different voxel grid resolutions, in order to estimate the computation speed, the number of matchings and their quality.

Figure 6 shows the results of speed measurement. Since image data must be loaded from a data base instead of being captured online, shape acquisition cost is overestimated. It can be observed that our implementation reaches interactive time. However, our program is still a prototype, and can be widely optimized, e.g. by parallelizing the skeletonization step. The difference of initialization speed for both models can be explained by taking into consideration the time complexities of the two costly steps: the skeletonization time complexity is in $O(S_G + S_S \times T_S)$, where S_G and S_S are respectively the size of the voxel grid and the size of the shape(in voxels), and T_S the thickness of the shape. The tree matching time complexity is in $O(S^2 * (D \times 2^{3\times D} + S^2 \times D))$, where S and D are respectively the maximal size and the maximal degree of the both trees. The values of S_S, T_S, S and D being higher for the hand model, the initialization speed for this model is slower than for the other one.

Figure 7 shows some results of pose initialization. The accuracy of our method obviously depends on the size of the voxel grid. It is also the case for the proportion of non-aborted matchings, and for their probability to give robust matching, as shown in Table 1. The reasons of high rate of aborted matching are different for the two sets. In the case of the dancer set, it is due to the fact that the pose of the subject does not always allow the initialization : an arm can be too close to the torso, cycles can appear, or other cases of the same kind. In the case of the hand set, even if the hand always has a good pose for initialization (spread fingers), the positioning of the four cameras is not efficient enough to provide a good shape reconstruction.

[1] http://4drepository.inrialpes.fr/

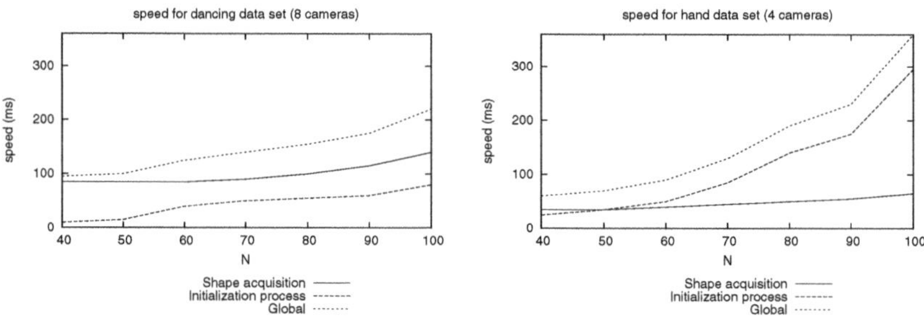

Fig. 6. Speed results for two kinds of model (Left: human body model; Right: hand model) for different sizes of grid. (The complete grid size is N^3 voxels).

Table 1. Matching count. Left: dancer data set. Right: hand data set. A matching is considered as a false positive (FP) if at least one part is not in a correct position.

grid side	threshold T_d	matching	FP
40	0.4	52%	16.3%
40	2.0	87.5%	15.4%
100	0.4	58%	11.2%
100	2.0	71.5%	9%

grid side	threshold T_d	matching	FP
40	0.4	40.6%	40%
40	2.0	75.6%	49.8%
100	0.4	18.1%	0.0%
100	2.0	61.5%	0.0%

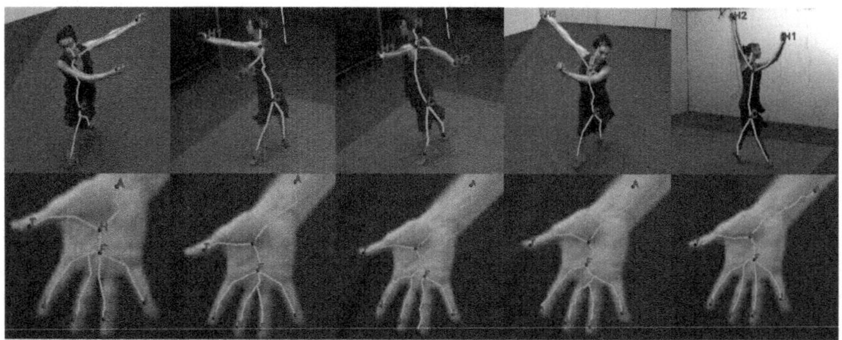

Fig. 7. Some examples of initial pose estimation results. In green, the skeleton, in blue the matched points, and in red, the corresponding labels.

7 Conclusion

In this paper, we have presented a new method for generic pose initialization, for markerless motion capture purpose. The performances of our method allow the initialization in interactive time, for an online usage. Our future works will focus on the detection of 2-degree joints, as elbows or shoulders, which can be detected by the presented method, and on the optimization of our prototype, in the aim to reach real time initialization.

References

1. Moeslund, T., Hilton, A., Krüger, V.: A survey of advances in vision-based human motion capture and analysis. Computer vision and image understanding 104(2-3), 90–126 (2006)
2. Alisi, T., Del Bimbo, A., Pucci, F., Valli, A.: Motion capture based on color error maps in a distributed collaborative environment. In: Proceedings of the Pattern Recognition, 17th International Conference on ICPR 2004, vol. 4, pp. 953–956. IEEE Computer Society, Washington (2004)
3. Michoud, B., Guillou, E., Briceno, H., Bouakaz, S.: Real-Time Marker-free Motion Capture from multiple cameras. In: IEEE 11th International Conference on Computer Vision, pp. 1–7 (2007)
4. Colombo, C., Del Bimbo, A., Valli, A.: A real-time full body tracking and humanoid animation system. Parallel Computing 34(12), 718–726 (2008)
5. Menier, C., Boyer, E., Raffin, B.: 3d skeleton-based body pose recovery. In: Proceedings of the 3rd International Symposium on 3D Data Processing, Visualization and Transmission, pp. 389–396. IEEE Computer Society, Los Alamitos (2006)
6. Kong, T.Y., Rosenfeld, A.: Digital topology: Introduction and survey. Computer Vision, Graphics, and Image Processing 48(3), 357–393 (1989)
7. Bertrand, G., Malandain, G.: A new characterization of three-dimensional simple points. Pattern Recognition Letters 15(2), 169–175 (1994)
8. Couprie, M., Bertrand, G.: New characterizations of simple points in 2D, 3D and 4D discrete spaces. IEEE Transactions on Pattern Analysis and Machine Intelligence 31(4), 637–648 (2009)
9. Bertrand, G.: On critical kernels. Comptes Rendus de l'Académie des Sciences, Série Mathématiques 1(345), 363–367 (2007)
10. Bertrand, G., Couprie, M.: Three-dimensional parallel thinning algorithms based on critical kernels. Technical report (2010)
11. Raynal, B., Couprie, M., Biri, V.: Homeomorphic Alignment of Edge-Weighted Trees. In: Proceedings of the 7th IAPR-TC-15 International Workshop on Graph-Based Representations in Pattern Recognition, pp. 134–143. Springer, Heidelberg (2009)
12. Greenberg, M.: Euclidean and non-Euclidean geometries: Development and history. WH Freeman, New York (1993)

Topology Preserving 3D Thinning Algorithms Using Four and Eight Subfields

Gábor Németh, Péter Kardos, and Kálmán Palágyi

Department of Image Processing and Computer Graphics,
University of Szeged, Hungary
{gnemeth,pkardos,palagyi}@inf.u-szeged.hu

Abstract. Thinning is a frequently applied technique for extracting skeleton-like shape features (i.e., centerline, medial surface, and topological kernel) from volumetric binary images. Subfield-based thinning algorithms partition the image into some subsets which are alternatively activated, and some points in the active subfield are deleted. This paper presents a set of new 3D parallel subfield-based thinning algorithms that use four and eight subfields. The three major contributions of this paper are: 1) The deletion rules of the presented algorithms are derived from some sufficient conditions for topology preservation. 2) A novel thinning scheme is proposed that uses iteration-level endpoint checking. 3) Various characterizations of endpoints yield different algorithms.

Keywords: 3D image analysis, Shape representation, Feature extraction, Thinning algorithms, Topology preservation.

1 Introduction

Skeleton-like shape features (i.e., centerline, medial surface, and topological kernel [1]) extracted from volumetric binary images play an important role in numerous applications of image processing, pattern recognition, and visualization, such as topological analysis [2], measurement [12], surface generation [4], shape matching [15], or automatic navigation [16].

Parallel thinning algorithms [3] are capable of extracting skeleton-like shape descriptors in a topology preserving way [5]. They use parallel reduction operations: some points having value of "1" in a binary image that satisfy certain topological and geometric constraints are deleted (i.e., changed some "1" points to "0" ones) simultaneously, and an iteration step is repeated until stability is achieved. Thinning algorithms use operators that delete some points which are not endpoints, since preserving endpoints provides important geometrical information relative to the shape of the objects.

Thinning has a major advantage over other skeletonization methods: it is capable of extracting all the three kinds of skeleton-like shape features: surface-thinning algorithms extract *medial surfaces* by preserving *surface-endpoints*, curve-thinning algorithms produce *centerlines* by preserving *curve-endpoints*,

A. Campilho and M. Kamel (Eds.): ICIAR 2010, Part I, LNCS 6111, pp. 316–325, 2010.

and *topological kernels* (i.e., minimal structures which are topologically equivalent to the original objects) can be generated if no endpoint criteria are considered during the thinning process. Medial surfaces are generally extracted from general shapes, tubular structures can be represented by their centerlines, and topological kernels are fairly useful in topological description. Note that thinning is sensitive to coarse object boundaries, hence it is to be coupled with an efficient pruning method [14].

One type of parallel thinning algorithms is the *subfield-based* technique [3]. In existing subfield-based 3D thinning algorithms, the digital space denoted by $\mathbb{Z}^3$ is partitioned into two [7,8,11], four [9], and eight [1] subfields which are alternatively activated (see Fig. 1b-d). At a given iteration step of a k-subfield algorithm, k successive parallel reductions associated to the k subfields are performed. In each parallel reduction, some border points in the active subfield can be designated to be deleted.

In [11], we proposed some 2-subfield 3D thinning algorithms that are based on Ma's sufficient conditions for parallel reduction operators [6]. In this paper, we introduce a set of 4- and 8-subfield 3D thinning algorithms that satisfy some sufficient criteria for such operators. In addition, a new thinning scheme with iteration-level endpoint checking is suggested.

2 Basic Notions and Results

Let p be a point in the 3D digital space $\mathbb{Z}^3$. Let us denote $N_j(p)$ (for $j = 6, 18, 26$) the set of points that are j-*adjacent* to point p (see Fig. 1a).

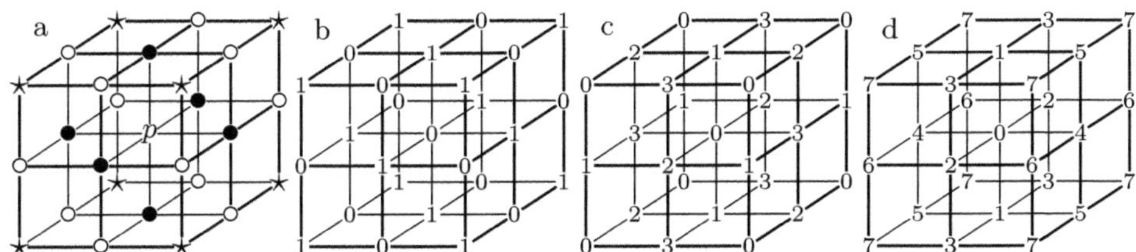

Fig. 1. Frequently used adjacencies in $\mathbb{Z}^3$ (a). The set $N_6(p)$ contains point p and the 6 points marked "•". The set $N_{18}(p)$ contains $N_6(p)$ and the 12 points marked "◦". The set $N_{26}(p)$ contains $N_{18}(p)$ and the 8 points marked "⋆".
The divisions of $\mathbb{Z}^3$ into 2 (b), 4 (c), and 8 (d) subfields. If partitioning into k subfields is considered, then points marked "i" are in the subfield $SF_k(i)$ ($k = 2, 4, 8$, $i = 0, 1, \ldots, k-1$).

The *3D binary* (26, 6) *digital picture* $\mathcal{P}$ is a quadruple $\mathcal{P} = (\mathbb{Z}^3, 26, 6, B)$ [5], where each element of $\mathbb{Z}^3$ is called a *point* of $\mathcal{P}$, each point in $B \subseteq \mathbb{Z}^3$ has a value of "1", each point in $\mathbb{Z}^3 \backslash B$ has a value of "0". 26-connectivity (i.e., the reflexive and transitive closure of the 26-adjacency relation) is considered for "1" points forming the objects, and 6-connectivity (i.e., the reflexive and transitive closure of the 6-adjacency) is considered for "0" points [5] (see Fig. 1a).

A "1" point is called a *border point* in a $(26, 6)$ picture if it is 6-adjacent to at least one "0" point. A "1" point is called an *interior point* if it is not a border point.

A *parallel reduction operator* changes a set of "1" points to "0" ones (which is referred to as deletion). A 3D parallel reduction operator does *not* preserve topology if any object (i.e., maximal 26-connected component of "1" points) is split or is completely deleted, any cavity (i.e., maximal 6-connected component of "0" points) is merged with another cavity, a new cavity is created, or a hole (that donuts have) is eliminated or created.

A "1" point is called a *simple* point if its deletion does not alter the topology of the image [5]. Note that simplicity of point p in $(26, 6)$ images is a local property that can be decided by investigating the set $N_{26}(p)$ [5].

Parallel reduction operators delete a set of "1" points. Hence we need to consider what is meant by topology preservation when a number of "1" points are deleted simultaneously. First we define the concept of simple sets.

Definition 1. [6] *The set of "1" points $D = \{d_1, \ldots, d_k\} \subset B$ in a picture $\mathcal{P} = (\mathbb{Z}^3, 26, 6, B)$ is called a* simple set *if D can be arranged in a sequence $\langle d_{i_1}, \ldots, d_{i_k} \rangle$ in which point d_{i_1} is simple in $\mathcal{P}$, and each point d_{i_j} is simple in $(\mathbb{Z}^3, 26, 6, B \backslash \{d_{i_1}, \ldots, d_{i_{j-1}}\})$, for $j = 2, \ldots, k$. (By definition, let the empty set be simple.)*

The following theorem provides *sufficient conditions* for 3D parallel reduction operators to preserve topology.

Theorem 1. [6] *A 3D parallel reduction operation* preserves topology *for $(26, 6)$ pictures if all of the following conditions hold:*

1. *Only simple points can be deleted.*
2. *If a set of two, three, or four mutually 18-adjacent "1" points are deleted, then it is a simple set.*
3. *No object formed by mutually 26-adjacent points can be deleted completely.*

The three kinds of partitionings of $\mathbb{Z}^3$ into two, four, and eight subfields are illustrated in Fig. 1b-d. Without loss of generality, we can assume that $(0, 0, 0) \in SF_k(0)$ $(k = 2, 4, 8)$. We can state the following properties:

Proposition 1. *For the 4-subfield case (see Fig. 1c), two points p and $q \in N_{26}(p)$ are in the same subfield if $q \in N_{26}(p) \backslash N_{18}(p)$.*

Proposition 2. *For the 8-subfield case (see Fig. 1d), two points p and $q \in N_{26}(p)$ are* not *in the same subfield.*

As consequences of Propositions 1 and 2 if a 3D parallel reduction operation may delete some points from the subfield $SF_k(i)$ ($k = 4, 8,\ i = 0, 1, \ldots, k-1$), then we get the following simplified versions of Theorem 1:

Theorem 2. [9] *A 4-subfield 3D parallel reduction operation* preserves topology *for* (26, 6) *pictures if both of the following conditions hold:*

1. *Only simple points can be deleted.*
2. *No object formed by two 26-adjacent, but not 18-adjacent points can be deleted completely.*

Theorem 3. [9] *An 8-subfield 3D parallel reduction operation* preserves topology *for* (26, 6) *pictures if only simple points can be deleted.*

3 Existing Thinning Algorithms Using Four and Eight Subfields

Each existing k-subfield ($k = 4, 8$, see Fig. 1c-d) 3D thinning algorithm can be sketched by the following program:

Input: picture $(\mathbb{Z}^3, 26, 6, X)$
Output: picture $(\mathbb{Z}^3, 26, 6, Y)$

$Y = X$
repeat
  // one iteration step
  for $i = 0$ **to** $k - 1$ **do**
    // subfield $SF_k(i)$ *is activated*
    $D(i) = \{\ p \mid p \text{ is "deletable" in } Y \cap SF_k(i)\ \}$
    $Y = Y \setminus D(i)$
until $\bigcup_{i=0}^{k-1} D(i) = \emptyset$

Ma, Wan, and Lee [9] proposed the following two 4-subfield 3D thinning algorithms:

- **SF-4-C-MWL**: 4-subfield curve-thinning algorithm,
- **SF-4-S-MWL**: 4-subfield surface-thinning algorithm.

Deletable points of both algorithms are defined by three-color matching templates.

The three existing 8-subfield thinning algorithms proposed by Bertrand and Aktouf [1] are:

- **SF-8-C-BA**: 8-subfield curve-thinning algorithm,
- **SF-8-S-BA**: 8-subfield surface-thinning algorithm,
- **SF-8-K-BA**: 8-subfield algorithm for extracting topological kernels.

They are based on Theorem 3 (i.e., sufficient conditions for topology preservation in eight subfields) and two types of endpoint characterizations: one for surface-endpoints and one for curve-endpoints. Their algorithm for extracting topological kernels does not consider any endpoint criteria.

4 The New Subfield-Based Thinning Algorithms

We propose a set of new thinning algorithms using four and eight subfields. Their deletable points are derived directly from Theorems 2 and 3 (i.e., sufficient conditions for topology preservation for four and eight subfields, respectively). The proposed algorithms using the same kind of partitioning differ from each other just in the considered endpoint characterizations. In order to reduce the noise sensitivity and the number of skeletal points (without overshrinking the objects), we introduce a new subfield-based thinning scheme. It takes the endpoints into consideration at the beginning of iteration steps, instead of preserving them in each parallel reduction as it is accustomed in existing subfield-based thinning algorithms.

Let us consider an arbitrary characterization of endpoints that is called as type $\mathcal{E}$. Our algorithm denoted by **SF-k-$\mathcal{E}$** uses k subfields ($k = 4, 8$) and endpoints of type $\mathcal{E}$. It is outlined as follows:

Algorithm SF-k-$\mathcal{E}$.

Input: picture $(\mathbb{Z}^3, 26, 6, X)$
Output: picture $(\mathbb{Z}^3, 26, 6, Y)$
$Y = X$
repeat
$E = \{ p \mid p$ is a border point, but not an endpoint of type $\mathcal{E}$ in $Y \}$
for $i = 0$ **to** $k - 1$ **do**
$D(i) = \{ q \mid q$ is an *SF-k-deletable* point in $E \cap SF_k(i) \}$
$Y = Y \setminus D(i)$
until $\bigcup_{i=0}^{k-1} D(i) = \emptyset$

We are to lay down *SF-k-deletable* points ($k = 4, 8$):

Definition 2. *A "1" point in a* $(26, 6)$ *picture* $\mathcal{P}$ *is* self-SF-4-deletable *if it is simple in* $\mathcal{P}$ *(see Condition 1 of Theorem 2).*

Definition 3. *A "1" point* p *in a* $(26, 6)$ *picture* $\mathcal{P}$ *is* SF-4-deletable *if* p *is self-SF-4-deletable (see Definition 2), and it does not come first in the lexicographic ordering of any object* $\mathcal{O}$ *of two self-SF-4-deletable points, where* $\mathcal{O} = \{p, q\}$ *and* $q \in N_{26}(p) \setminus N_{18}(p)$ *(see Condition 2 of Theorem 2).*

It can be readily seen that simultaneous deletion of SF-4-deletable points satisfies both conditions of Theorem 2, hence it preserves the topology.

Let us define SF-8-deletable points:

Definition 4. *A "1" point in a* (26, 6) *picture* $\mathcal{P}$ *is* SF-8-deletable *if it is simple in* $\mathcal{P}$.

It is easy to see that Definition 4 was derived directly from Theorem 3, hence simultaneous deletion of SF-8-deletable points is a topology preserving reduction.

We can state that all of our algorithms **SF**-k-$\mathcal{E}$ ($k = 4, 8$) with any endpoint characterizations are topologically correct.

At the end of this section, we illustrate the usefulness of the new subfield-based thinning scheme (i.e., just border points in the input picture of the iteration step may be deleted). Figure 2 compares the conventional and the proposed methods using four subfields.

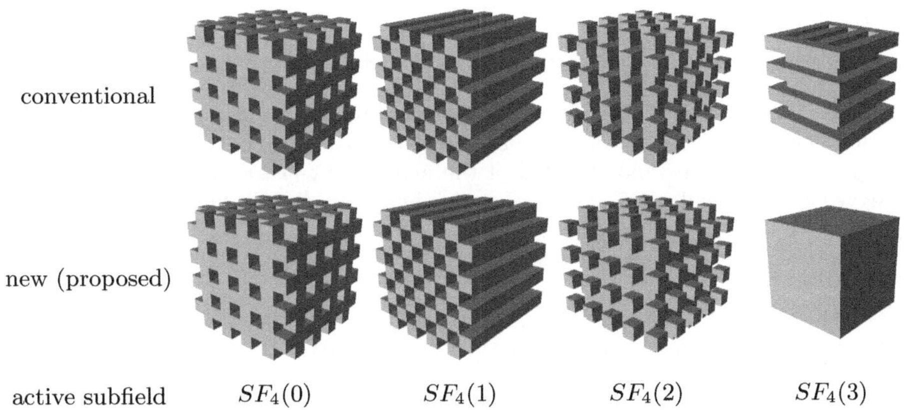

Fig. 2. One iteration step of the conventional 4-subfield thinning process (see Section 3) and the proposed thinning scheme. For simplicity, no endpoints are preserved when the 9×9×9 cube is thinned. The conventional thinning may delete some points that are interior ones in the picture at the beginning of the iteration step, hence some objects may not be reduced uniformly. It may create numerous unwanted endpoints (according to some endpoint characterizations) that blocks the rest of the thinning process.

5 Examples of the Subfield-Based Thinning Algorithms

In Section 4, we defined the deletable points of the proposed algorithms that follow our new thinning scheme using iteration-level endpoint checking. Various characterizations of endpoints yield different algorithms. Here, we define six types of endpoints that determine twelve new algorithms.

Definition 5. *Any "1" point in a* (26, 6) *picture is* not *an endpoint of type* **K**. *(If no endpoints are preserved, then we get topological kernels.)*

Definition 6. *A "1" point* p *in picture* $(\mathbb{Z}^3, 26, 6, B)$ *is a curve-endpoint of type* **C1** *if* $(N_{26}(p)\setminus\{p\}) \cap B = \{q\}$ *(i.e.,* p *is 26-adjacent to exactly one "1" point* q*).*

Definition 7. *A "1" point p in picture* $(\mathbb{Z}^3, 26, 6, B)$ *is a curve-endpoint of type* **C2** *if* $(N_{26}(p) \setminus \{p\}) \cap B = \{q\}$ *and*

- $(N_{26}(q) \setminus \{q\}) \cap B = \{p\}$ *or*
- $(N_{26}(q) \setminus \{q\}) \cap B = \{p, r\}$.

Note that the characterization of C2 curve-endpoints is inspired by the concept of "twig voxel" that was introduced by Ma, Wan, and Chang [8].

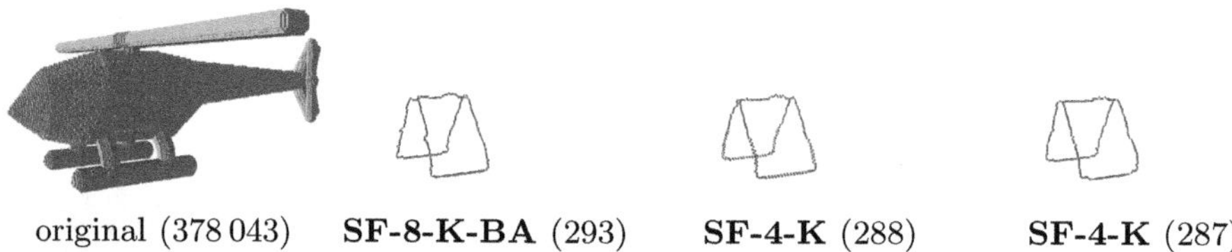

original (378 043) **SF-8-K-BA** (293) **SF-4-K** (288) **SF-4-K** (287)

Fig. 3. A $304 \times 96 \times 261$ image of a helicopter and its topological kernels produced by the three algorithms under comparison. The original image contains just one object, there is no cavity in it, and the skid of the helicopter consists of two holes that are preserved in the topological kernels (i.e., minimal structures which are topologically equivalent to the original helicopter).

Definition 8. *A "1" point p in picture* $(\mathbb{Z}^3, 26, 6, B)$ *is a surface-endpoint of type* **S1** *if there is no interior point in the set* $N_6(p) \cap B$.

Definition 9. *A "1" point p in picture* $(\mathbb{Z}^3, 26, 6, B)$ *is a surface-endpoint of type* **S2** *if there is no interior point in the set* $N_{18}(p) \cap B$.

Definition 10. *A "1" point p in picture* $(\mathbb{Z}^3, 26, 6, B)$ *is a surface-endpoint of type* **S3** *if there is no interior point in the set* $N_{26}(p) \cap B$.

Note that these three characterizations of surface-endpoints are hidden in the algorithms proposed by Manzanera et al. [10].

In experiments our twelve new algorithms based on four and eight subfields and using the endpoints according to Definitions 5-10 were tested on objects of different shapes. Here we present some illustrative examples below (Figures 3-7). Numbers in parentheses mean the count of "1" points. Skeleton-like shape features produced by the proposed twelve algorithms are compared with the results of 8-subfield algorithms **SF-8-K-BA**, **SF-8-C-BA**, and **SF-8-S-BA** proposed by Bertrand and Aktouf [1]. Unfortunately, we could not make credible implementations of the two existing 4-subfield algorithms **SF-4-C-MWL** and **SF-4-S-MWL** proposed by Ma, Wan, and Lee [9].

Note that our algorithms are not time consuming and it is easy to implement them on conventional sequential computers by adapting the efficient implementation method presented in [13]. Skeleton-like features can be extracted from large 3D shapes within one second by the proposed algorithms on a usual PC.

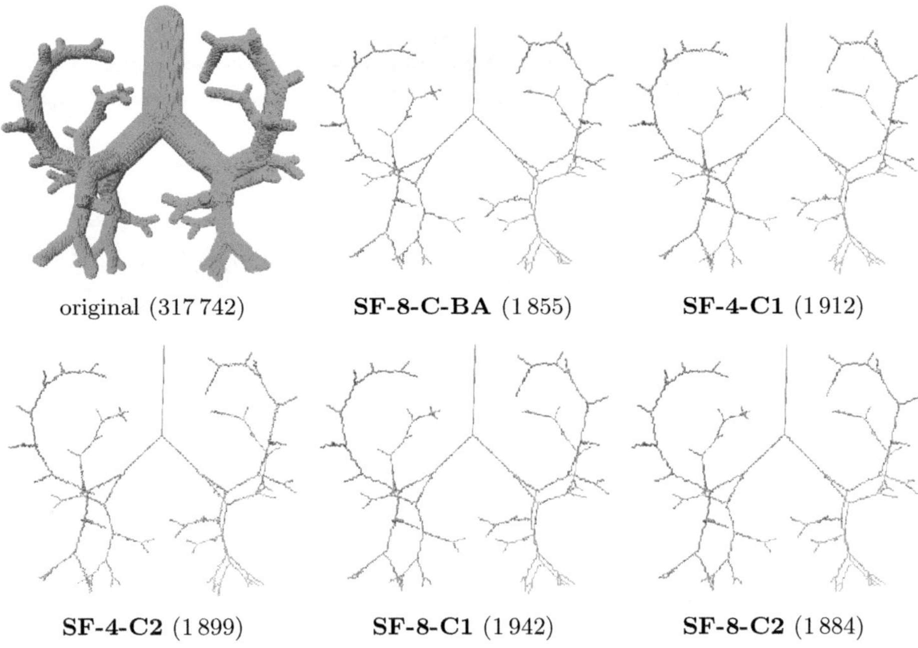

Fig. 4. A $300 \times 300 \times 300$ image of a tubular structure and its centerlines produced by the five algorithms under comparison

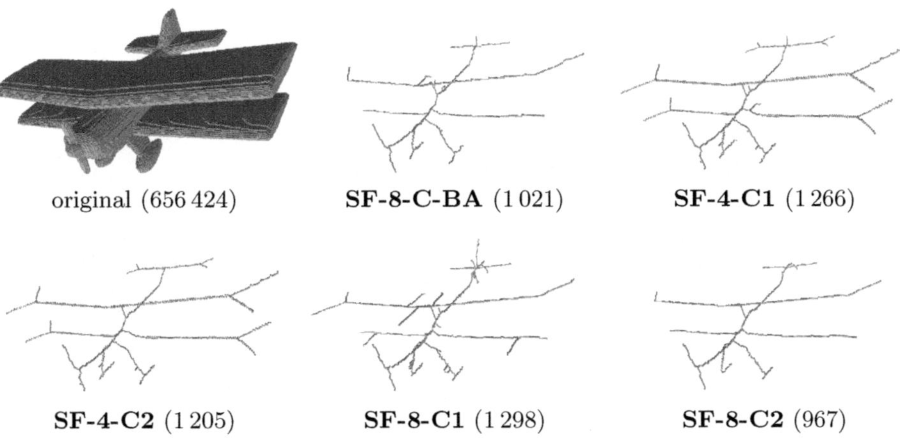

Fig. 5. A $217 \times 304 \times 98$ image of an airplane and its centerlines produced by the five algorithms under comparison

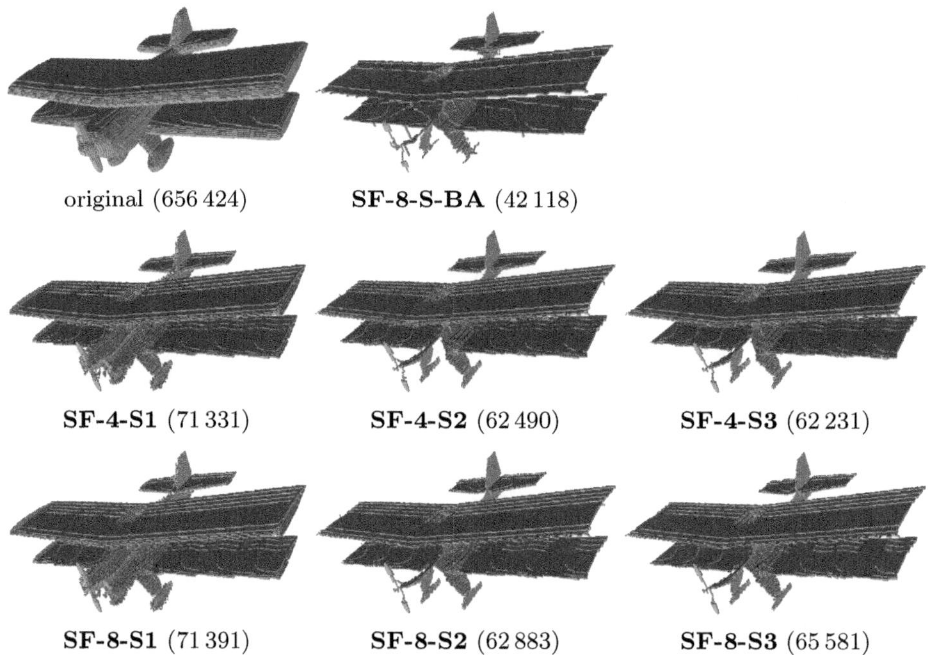

Fig. 6. A $217 \times 304 \times 98$ image of an airplane and its medial surfaces produced by the seven algorithms under comparison

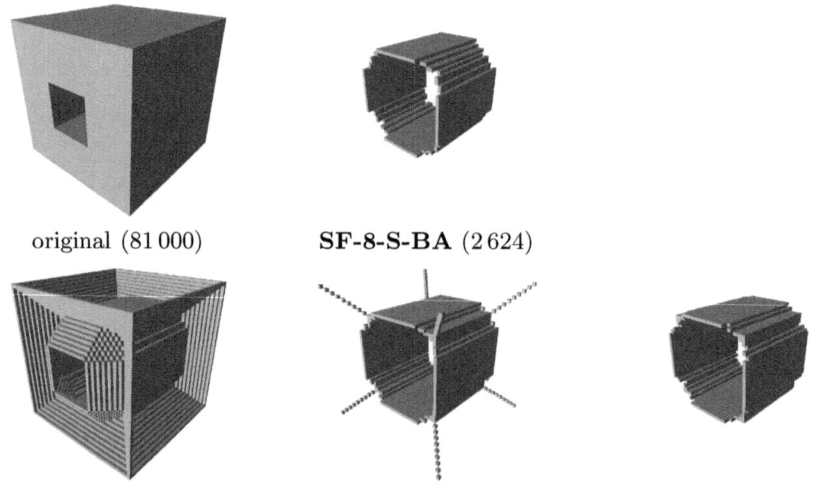

Fig. 7. The 3D image of a $45 \times 45 \times 45$ cube with a hole and its medial surfaces produced by the seven algorithms under comparison. Interestingly, the same medial surfaces are extracted from this special object by the corresponding 4-subfield and 8-subfield algorithms

Acknowledgements. This research was supported by the TÁMOP-4.2.2/08/1/2008-0008 program of the Hungarian National Development Agency, the NKTH-OTKA-CNK80370 Grant, and the National Office for Research and Technology.

References

1. Bertrand, G., Aktouf, Z.: A 3D thinning algorithm using subfields. In: SPIE Proc. of Conf. on Vision Geometry, pp. 113–124 (1994)
2. Gomberg, B.R., Saha, P.K., Song, H.K., Hwang, S.N., Wehrli, F.W.: Topological analysis of trabecular bone MR images. IEEE Transactions on Medical Imaging 19, 166–174 (2000)
3. Hall, R.W.: Parallel connectivity-preserving thinning algorithms. In: Kong, T.Y., Rosenfeld, A. (eds.) Topological algorithms for digital image processing, pp. 145–179. Elsevier Science, Amsterdam (1996)
4. Itoh, T., Yamaguchi, Y., Koyamada, K.: Fast isosurface generation using the volume thinning algorithm. IEEE Transactions on Visualization and Computer Graphics 7, 32–46 (2001)
5. Kong, T.Y., Rosenfeld, A.: Digital topology: Introduction and survey. Computer Vision, Graphics, and Image Processing 48, 357–393 (1989)
6. Ma, C.M.: On topology preservation in 3D thinning. CVGIP: Image Understanding 59, 328–339 (1994)
7. Ma, C.M., Wan, S.Y.: A medial-surface oriented 3-d two-subfield thinning algorithm. Pattern Recognition Letters 22, 1439–1446 (2001)
8. Ma, C.M., Wan, S.Y., Chang, H.K.: Extracting medial curves on 3D images. Pattern Recognition Letters 23, 895–904 (2002)
9. Ma, C.M., Wan, S.Y., Lee, J.D.: Three-dimensional topology preserving reduction on the 4-subfields. IEEE Transaction on Pattern Analysis and Machine Intelligence 24, 1594–1605 (2002)
10. Manzanera, A., Bernard, T.M., Prêteux, F., Longuet, B.: Medial faces from a concuise 3D thinning algorithm. In: Proc. 7th IEEE Int. Conf. on Computer Vision, pp. 337–343 (1999)
11. Németh, G., Kardos, P., Palágyi, K.: Topology preserving 2-subfield 3D thinning algorithms. In: Proc. 7th IASTED Int. Conf. Signal Processing, Pattern Recognition and Applications, pp. 310–316 (2009)
12. Palágyi, K., Tschirren, J., Hoffman, E.A., Sonka, M.: Quantitative analysis of pulmonary airway tree structures. Computers in Biology and Medicine 36, 974–996 (2006)
13. Palágyi, K., Németh, G.: Fully parallel 3D thinning algorithms based on sufficient conditions for topology preservation. In: Brlek, S., Reutenauer, C., Provençal, X. (eds.) DGCI 2009. LNCS, vol. 5810, pp. 481–492. Springer, Heidelberg (2009)
14. Shaked, D., Bruckstein, A.: Pruning medial axes. Computer Vision and Image Understanding 69, 156–169 (1998)
15. Sundar, H., Silver, D., Gagvani, N., Dickinson, S.: Skeleton based shape matching and retrieval. In: Proc. Int. Conf. Shape Modeling and Applications, pp. 130–139 (2003)
16. Wan, M., Liang, Z., Ke, Q., Hong, L., Bitter, I., Kaufman, A.: Automatic centerline extraction for virtual colonoscopy. IEEE Transactions on Medical Imaging 21, 1450–1460 (2002)

Robust Approaches to 3D Object Secret Sharing

Esam Elsheh and A. Ben Hamza

Concordia Institute for Information Systems Engineering
Concordia University, Montréal, QC, Canada

Abstract. Inspired by the successful application of image secret sharing schemes in multimedia protection, we present in this paper two secret sharing approaches for 3D models using Blakely and Thien & Lin schemes. We show that encoding 3D models using lossless data compression algorithms prior to secret sharing helps reduce share sizes and remove redundancies and patterns that possibly ease cryptanalysis. The proposed approaches provide a higher tolerance against data corruption/loss than existing 3D protection mechanisms, such as encryption. Experimental results are provided to demonstrate the secrecy and safety of the proposed schemes. The feasibility of the proposed algorithms is demonstrated on various 3D models.

Keywords: Secret sharing; Lossless data compression; 3D graphics.

1 Introduction

The ongoing developments in computer technologies and the rapid increase in internet users have led to the increasing usage of network-based data transmission. In numerous applications, such as military documents and sensitive business data, this information must be kept secret and safe. Recently, 2D images and 3D models are considered as important as any other text sensitive information. As a result, several 2D image-protection techniques, such as data encryption in [1, 2] and steganography in [3, 4], have been proposed to increase the security of secret images. One common disadvantage of the traditional protection techniques, such as encryption, is their policy of centralized storage, in that an entire protected model is usually maintained in a single information storage. If an intruder detects security vulnerability in the information storage in which the protected model resides, then s/he may attempt to decipher the secret model inside, or simply damage the entire information storage. Hence, the secret sharing is a defense mechanism to protect the secret that does not suffer from these problems. It works by splitting the secret into n shares that are transmitted and stored separately. One can then reconstruct the original secret if at least a preset number t ($1 \leqslant t \leqslant n$) of these n shares are obtained. However, knowledge of less than t shares is insufficient for revealing the secret. The idea of secret sharing was introduced independently in [5] and [6]. These schemes are based on the use of Lagrange interpolation polynomial and the intersection of affine hyperplanes, respectively. Since then, several studies have investigated the different

A. Campilho and M. Kamel (Eds.): ICIAR 2010, Part I, LNCS 6111, pp. 326–335, 2010.

implementations of the (t, n)-threshold scheme by mainly concentrating on the communication of keys in cipher systems. Most of these schemes are based on different mathematical primitives, such as matrix theory and prime numbers [7]. These protocols are specifically designed for text and numeric data. Due to the main distinctive nature of multimedia, in the sense that they have a large amount of data and the difference between two neighboring values is typically very small, it is considerably difficult to apply directly traditional secret sharing schemes to digital images or 3D objects.

On the other hand, visual secret sharing scheme, which is based on the human visual system, is an important cryptographic technique for secret sharing of 2D images [18]. This method is a visual variant of the ordinary secret sharing schemes. The secret image is divided into n shares (transparencies) such that if any t transparencies are overlaid together, then the image becomes visible. However, if less than t transparencies are overlaid together, then nothing can be seen. Such a scheme is constructed by viewing the secret image as a set of black and white dots, and by handling each dot separately. The scheme is shown to be perfectly secure and easily implementable without any cryptographic computation. A further development allows each transparency to be a valid image instead of noisy dots in order to hide the fact that the secret sharing is taking place, for example an image of a landscape or an image of a human. Over the last decade, various construction methods based on visual cryptography have been proposed [8,9,10,11,12]. Also, other protocols for digital images sharing have been designed based on vector quantization [13], Shamir-based schemes [14,15,16], sharing circle [7,12], and cellular automata [1,2]. Just a few of these schemes generate shares that have smaller sizes than the original image. The method proposed in [15] generates shares of $1/t$ the size of the secret image, whereas the method proposed in [14], that involves using Huffman coding generates shared images 40% smaller than that of the approach in [15].

Recently, a flurry of research efforts have been carried out to design secure and efficient approaches for 2D image protection. However, 3D models have received less attention due to the fact that 2D image algorithms do not generally extend to 3D models. Besides, the rapid development in computer and information technology has increased the use of 3D models in various application domains, including manufacturing industries, entertainment and even in the military. Thus, the need for protection techniques to keep these 3D models secret and safe is of paramount importance. Inspired by the successful application of image secret sharing schemes, we propose in this paper two secret sharing approaches for 3D models using Blakely scheme [6] and Thien & Lin [15]. We then show that encoding the 3D models using Huffman coding [19] or ZLIB [20] prior to secret sharing reduces the shares sizes significantly.

The rest of the paper is organized as follows. In Section 2, we briefly review some traditional and image secret sharing schemes. In Section 3, we propose two secret sharing approaches for 3D models, and describe their algorithmic steps. In Section 4, we provide experimental results to validate the effectiveness of the proposed secret sharing schemes on various 3D models. Finally, we conclude in Section 5.

2 Previous Work

The simplest secret splitting method is the $(2,2)$-scheme [7], where the secret K is split into two shares X and Y. Neither X nor Y independently provide any information about the secret.

Let $K = k_1, \ldots, k_n$ be a binary string of length n, called the secret.

1. for $1 \leq i \leq n$, let $x_i \in \mathbb{F}_2$ be chosen at random.
2. for $1 \leq i \leq n$, let $y_i = x_i + k_i \mod 2$.
3. then $X = x_1, \ldots, x_n$ and $Y = y_1, \ldots, y_n$ are two shares of the secret K.
4. To recover the secret, K is computed as $K = X + Y \mod 2$.

The majority of existing secret sharing schemes are generalized within the so-called (t,n)-threshold framework [7]. This framework confidentially splits the content of a secret message into n shares in a way that requires the presence of at least t shares for the secret message reconstruction. If $t = n$, then all the shares are required in the (n,n)-threshold scheme to recover the secret. Conversely, the lost of any of the produced shares results in inaccessible secret messages. Therefore, apart from the simplest $(2,2)$-schemes that are commonly used as a private key cryptosystem solution, the general (t,n)-threshold schemes with $t < n$ are often the point of interest due to their ability to recover the secret message even if several shares are lost. In this case, any possible combinations of t shares can be used to recover the secret message. Since protection against cryptoanalytic attacks, including brute force enumeration, should remain unchanged regardless of how many shares are available until the threshold t is reached, the use of $(t-1)$ shares should not reveal any valid information about the secret compared to that obtained by only one share.

2.1 Blakley (t,n)-Threshold Scheme [6]

The Blakley scheme uses hyperplane geometry to solve the secret sharing problem. The secret is a point in a t-dimensional space. The n shares are constructed such that each share is defined as an affine hyperplane that passes through the secret point. An affine hyperplane can be described by a linear equation of the following form $a_1x_1 + \ldots + a_tx_t = b$. The intersection point is obtained by finding the intersection of any t of these hyperplanes. The secret can be any of the coordinates of the intersection point or any function of the coordinates.
For $(3,n)$-threshold scheme:

1. Choose a prime number P larger than the point coordinates.
2. Given a secret point (x_0, y_0, z_0), n shares are generated as follows:
   For each share:
   2.1. Choose $a, b \in \mathbb{F}_P$ independently at random.
   2.2. Let
   $$c = z_0 - ax_0 - by_0 \mod P \tag{1}$$
   where $z = ax + by + c$ is the equation of a hyperplane.

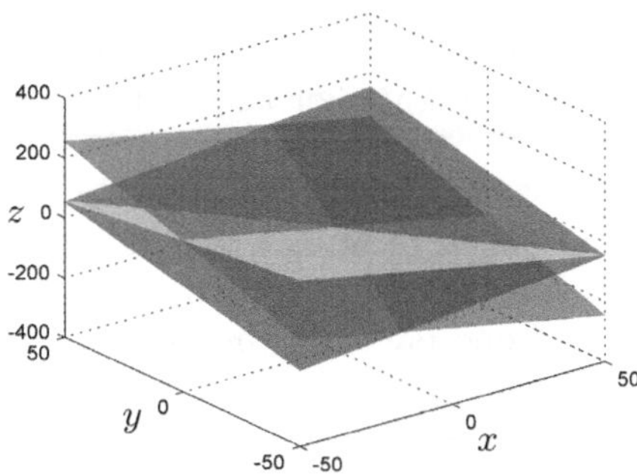

Fig. 1. The secret point is the intersection point between the three planes

Given any t hyperplanes, the secret point is the intersection point of these t hyperplanes. Fig. 1 illustrates how the three hyperplanes intersect in only one point (secret point).

The traditional Shamir and Blakley (t, n)-threshold schemes produce shares with same size as the original secret. In multimedia, a secret can be an image or 3D model. Typically, these files have a large amount of data. Thus, applying these traditional schemes may be inefficient in terms of the storage space. Consequently, several image secret sharing approaches have been proposed to reduce the shares sizes. The method proposed in [15] produces shares of $1/t$ the size of the secret image, whereas the Huffman coding-based scheme introduced in [14] generates shared images 40% smaller than the method in [15].

2.2 Thien & Lin (t, n)-Threshold Scheme [15]

Thien & Lin proposed a (t, n)-threshold-based approach using Shamir scheme [5] for grayscale images to generate image shares. Suppose we want to divide an image S into n image shares $(S_1, \ldots, S_n)$, and the secret image S cannot be revealed without t or more image shares. The essential idea is to use a polynomial of degree $(t-1)$ to construct n image shares, by letting the t coefficients be the gray values of t pixels. The main difference is that these coefficients are randomly chosen in Shamir scheme. To this end, we first divide an image into several sections. Each section has t pixels of the image, and for each section j, the following $t-1$ degree polynomial is defined

$$q_j(x) = \left(a_0 + a_1 x + \ldots + a_{t-1} x^{t-1}\right) \bmod P, \tag{2}$$

where the coefficients $a_0, a_1, \ldots, a_{t-1}$ are the values of the t pixels of the section. Then $q_j(1), q_j(2), ..., q_j(n)$ are computed. These n values of the section are distributed to the n participants to assign them sequentially to their n image shares. Since for each given section (of t entries) of the secret image, each image share receives only one value of the generated shares, the size of each image share is $1/t$ of the secret image. This method reduces the size of image shares to become $1/t$ of the size of the secret image. We should note here that since the gray values are between 0 and 255, the value of P was set to 251, which is the greatest prime number less than 255. For this method to be valid, all the pixel values greater than 250 must be rounded down to 250. Obviously, there will be some loss in terms of pixel values during the reconstruction of the secret image. Thus, Thien & Lin modified their technique to offer a lossless image secret sharing method. It should be noted here that applying Thien & Lin's scheme directly to the image shares can outline partially the original secret image. Therefore, some sort of initial permutation is needed before employing the scheme. Fig. 2 illustrates Thien & Lin's image secret sharing scheme, where $t = 2$ and $n = 4$.

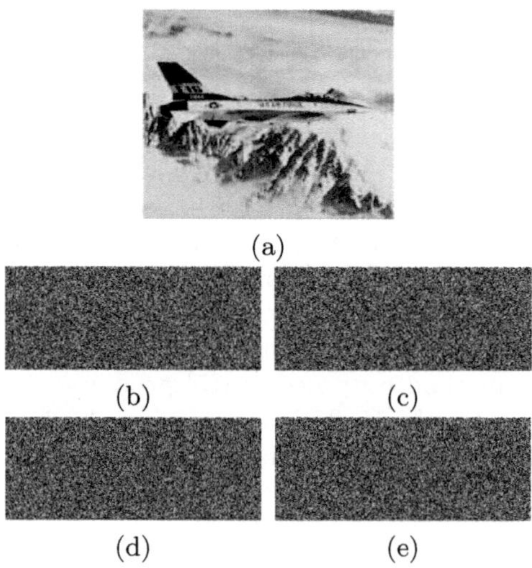

Fig. 2. Thien & Lin's secret sharing process for a Jet plane: (a) original image 512×512; (b)-(e) the four share images after the original image is permuted, each of size 1/2 of the original image size

3 Proposed 3D Secret Sharing Schemes

In computer graphics and geometric-aided design, a 3D triangle mesh $\mathbb{M}$ may be defined as $\mathbb{M} = (\mathcal{V}, \mathcal{T})$ where $\mathcal{V} = \{\boldsymbol{v}_1, \ldots, \boldsymbol{v}_m\}$ is the set of vertices and $\mathcal{T} = \{\boldsymbol{t}_1, ..., \boldsymbol{t}_\ell\}$ is the set of triangles (faces). In matrix form, the sets $\mathcal{V}$ and $\mathcal{T}$ may be written as follows:

$$\mathcal{V} = \begin{pmatrix} \boldsymbol{v}_1 \\ \boldsymbol{v}_2 \\ \vdots \\ \boldsymbol{v}_m \end{pmatrix} = \begin{pmatrix} v_{1_x} & v_{1_y} & v_{1_z} \\ v_{2_x} & v_{2_y} & v_{2_z} \\ \vdots & \vdots & \vdots \\ v_{m_x} & v_{m_y} & v_{m_z} \end{pmatrix}, \qquad \mathcal{T} = \begin{pmatrix} \boldsymbol{t}_1 \\ \boldsymbol{t}_2 \\ \vdots \\ \boldsymbol{t}_\ell \end{pmatrix} = \begin{pmatrix} t_{1_i} & t_{1_j} & t_{1_k} \\ t_{2_i} & t_{2_j} & t_{2_k} \\ \vdots & \vdots & \vdots \\ t_{\ell_i} & t_{\ell_j} & t_{\ell_k} \end{pmatrix}$$

3.1 3D Secret Sharing Using Blakley Scheme

Our proposed approach is motivated by Blakley secret sharing scheme. The main idea is to split every vertex in the vertices matrix $\mathcal{V}$ and every face in the faces matrix $\mathcal{T}$ into n share hyperplanes, where $n > 3$. Each share hyperplane is represented by an equation $z = ax + by + c$. The main algorithmic steps of the proposed scheme are shown in Table 1.

Table 1. Algorithmic steps of the proposed approach

For the faces matrix $\mathcal{T}$:

1. We choose $P_{\mathcal{T}}$ as the next prime number larger than m.
2. For each i-th share, $1 \leq i \leq n$:

   (i) Select two random numbers independently $a_i, b_i \in \mathbb{F}_{P_{\mathcal{T}}}$.

   (ii) Find c_i using Eq.(1), where x, y, z are the values of the face and a_i, b_i, c_i are the coefficients of the hyperplane equation $z = a_i x + b_i y + c_i$.

   (iii) Distribute the hyperplane equations (coefficients) to all n participants.
3. Repeat step (2) for all ℓ faces in $\mathcal{T}$ matrix.

In the recovery phase, the original values of the faces and vertices coordinates are the intersection points between any three or more hyperplane equations. To be able to draw the shares, all the calculations are performed in a prime field $\mathbb{F}_{P_{\mathcal{T}}}$. This $\mathbb{F}_{P_{\mathcal{T}}}$ is essential in splitting the faces matrix $\mathcal{T}$ to ensure the coefficients of the shared hyperplanes are within the range of the number of vertices, i.e. less than m. Moreover, since all the values in $\mathcal{T}$ are integers, $\mathbb{F}_{P_{\mathcal{T}}}$ is necessary to avoid the prediction of the range of the faces original values from the shared-values. On the other hand, the modular operation in splitting the vertices matrix is not crucial. In this case, the scheme is still secure since the vertices coordinates (v_x, v_y, v_z) and the random coefficients (a_i, b_i) are floating negative

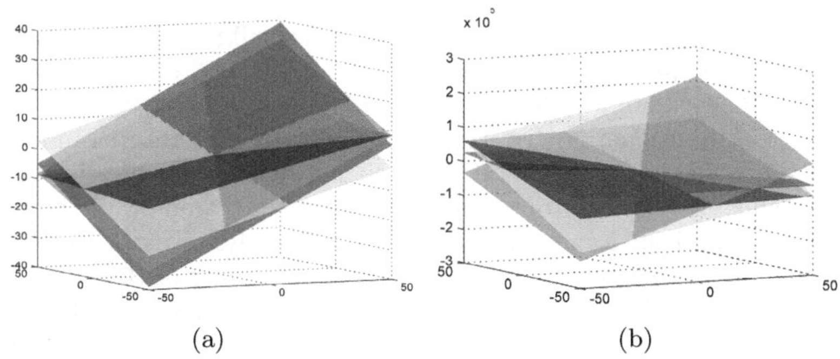

Fig. 3. Four planes generated by Blakley secret sharing scheme of F15 model for (a) vertex $\boldsymbol{v}_1$, and (b) face $\boldsymbol{t}_1$

numbers, and sometimes integers. For this reason, the values of the shares will not correlate with the original values of the vertices matrix. Thus, we apply the same algorithmic steps to split the vertices matrix $\mathcal{V}$, with the exception of using the prime field. Knowing the numbers of the vertices and faces of the share, the adversary can guess the share corresponds to which 3D model. To resist this statistical attack, we duplicate the last vertex (resp. last face) by a random numbers prior to finding $P_{\mathcal{T}}$. Fig. 3 shows how Blakley scheme split the vertex $\boldsymbol{v}_1$ and face $\boldsymbol{t}_1$ of the 3D F15 fighter jet into four share hyperplanes.

3.2 3D Secret Sharing Using Thien & Lin Scheme

Blakley's scheme produces shares of the same size as the original (secret) 3D model. However, Thien & Lin scheme produces shares 1/3 the size of the secret model. In the latter scheme, we divide each vertex (resp. each face) into n shares, where each share is $m \times 1$ array (resp. $\ell \times 1$ array), as shown in Fig. 4.

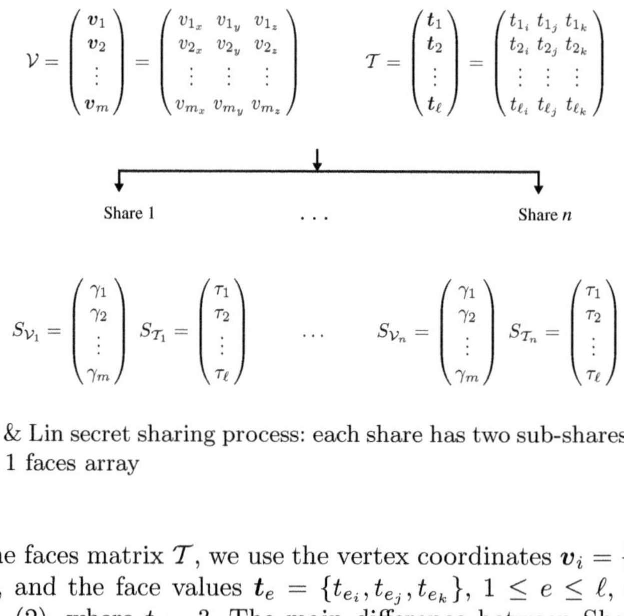

$$\mathcal{V} = \begin{pmatrix} \boldsymbol{v}_1 \\ \boldsymbol{v}_2 \\ \vdots \\ \boldsymbol{v}_m \end{pmatrix} = \begin{pmatrix} v_{1_x} & v_{1_y} & v_{1_z} \\ v_{2_x} & v_{2_y} & v_{2_z} \\ \vdots & \vdots & \vdots \\ v_{m_x} & v_{m_y} & v_{m_z} \end{pmatrix} \qquad \mathcal{T} = \begin{pmatrix} \boldsymbol{t}_1 \\ \boldsymbol{t}_2 \\ \vdots \\ \boldsymbol{t}_\ell \end{pmatrix} = \begin{pmatrix} t_{1_i} & t_{1_j} & t_{1_k} \\ t_{2_i} & t_{2_j} & t_{2_k} \\ \vdots & \vdots & \vdots \\ t_{\ell_i} & t_{\ell_j} & t_{\ell_k} \end{pmatrix}$$

Share 1 . . . Share n

$$S_{\mathcal{V}_1} = \begin{pmatrix} \gamma_1 \\ \gamma_2 \\ \vdots \\ \gamma_m \end{pmatrix} \quad S_{\mathcal{T}_1} = \begin{pmatrix} \tau_1 \\ \tau_2 \\ \vdots \\ \tau_\ell \end{pmatrix} \quad \cdots \quad S_{\mathcal{V}_n} = \begin{pmatrix} \gamma_1 \\ \gamma_2 \\ \vdots \\ \gamma_m \end{pmatrix} \quad S_{\mathcal{T}_n} = \begin{pmatrix} \tau_1 \\ \tau_2 \\ \vdots \\ \tau_\ell \end{pmatrix}$$

Fig. 4. Thian & Lin secret sharing process: each share has two sub-shares $m \times 1$ vertices array and $\ell \times 1$ faces array

To split the faces matrix $\mathcal{T}$, we use the vertex coordinates $\boldsymbol{v}_i = \{v_{i_x}, v_{i_y}, v_{i_z}\}$, $1 \leq i \leq m$, and the face values $\boldsymbol{t}_e = \{t_{e_i}, t_{e_j}, t_{e_k}\}$, $1 \leq e \leq \ell$, as the coefficients to Eq. (2), where $t = 3$. The main difference between Shamir's scheme and Thien & Lin scheme is that the coefficients are not taken randomly. An important issue in the implementation of secret sharing schemes is the size of the shares, since the security of a system lessens as the amount of the information that must be kept secret increases. Unfortunately, in most secret sharing schemes the size of the shares cannot be less than the size of the secret. Therefore, to reduce the share size, we compress the 3D models using lossless data compression methods such as Huffman coding [19] or ZLIB [20] before applying the secret sharing schemes. Besides, compression prior to secret sharing helps remove redundancies and patterns that might facilitate cryptanalysis. ZLIB is a

Table 2. Compression results of 3D objects using Huffman coding and ZLIB algorithm

3D Model	Uncompressed	Compressed using Huffman Coding		Compressed using ZLIB	
F15	176.55 KB	150.56 KB	85.2 %	98.38 KB	55.7 %
Tank	317.96 KB	271.15 KB	85.2 %	143.90 KB	45.2 %
Engine	81.32 KB	66.88 KB	82.2 %	23.12 KB	28.4 %

lossless data compression library that uses a compression algorithm called Deflate. This lossless compression algorithm uses a combination of LZ77 algorithm and Huffman coding, and provides good compression on a wide variety of data with minimal use of system resources. Table 2 displays the sizes of different 3D models before and after compression using Huffman coding and ZLIB.

4 Experimental Results

We applied the (3, 4)-Blakley scheme on two 3D models: F15 fighter jet and tank. The 3D F15 model consists of 5401 vertices and 9665 faces, whereas the 3D tank model consists of 8659 vertices and 18474 faces. All the share models have the same number of vertices and faces. If the number of vertices is not a prime number, then we may duplicate the last vertex until we reach the next prime number greater than the number of vertices. For F15 and tank models, we used the prime numbers $P_T = 5407$ and $P_T = 8663$, respectively. From

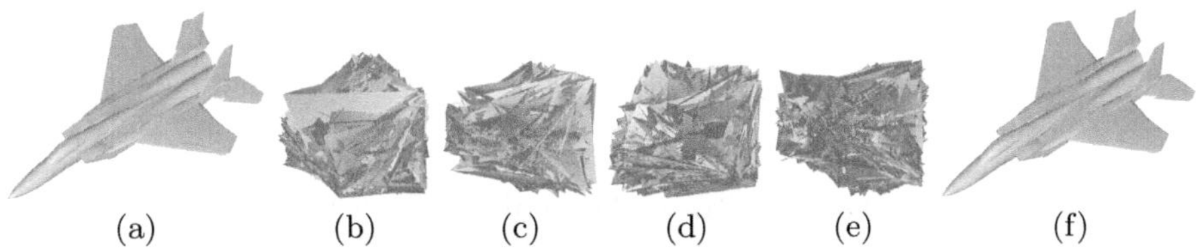

Fig. 5. (3, 4)-Blakley secret sharing process for the 3D F15 model: (a) original model; (b)-(e) the four split shares; (f) reconstructed model using any 3 shares

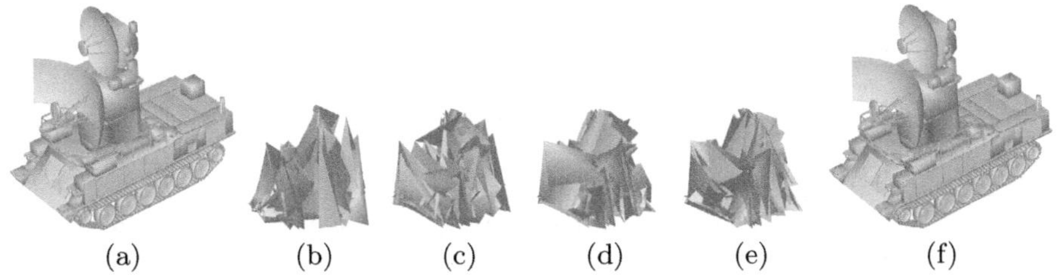

Fig. 6. (3, 4)-Blakley secret sharing process for the 3D tank model: (a) original model; (b)-(e) the four split shares; (f) reconstructed model using any 3 shares

Table 3. Comparison between the sizes of the shares generated by Blakely and Thien & Lin schemes using Huffman coding and ZLIB

3D Model	Blakely Scheme Share Size			Thien & Lin Scheme Share Size		
	Uncompressed	Huffman	ZLIB	Uncompressed	Huffman	ZLIB
F15	176.55 KB	150.56 KB	98.38 KB	58.85 KB	50.18 KB	32.79 KB
Tank	317.96 KB	271.15 KB	143.90 KB	105.98 KB	90.38 KB	47.96 KB
Engine	81.32 KB	66.88 KB	23.12 KB	27.1 KB	22.29 KB	7.7 KB

Fig. 5 and Fig. 6, it is clear that the four generated shares of both models are unrecognizable, indicating that the secret property is satisfied. Combining any 3 shares from Fig. 5(b)-(e) (resp. Fig. 6(b)-(e)), we can reconstruct the original secret model as shown in Fig. 5(f) (resp. Fig. 6(f)). Therefore, the lost of one share will not prevent recovery of the model. These results are in fact consistent with numerous 3D models used for experimentation. To further increase the security by minimizing the share sizes and also reduce the overhead calculations of the sharing process, we compress the 3D models before applying the secret sharing schemes. Table 3 shows the comparison results between the sizes of the shares generated by Blakely and Thien & Lin schemes using Huffman coding [19] and ZLIB [20].

5 Conclusions

A geometric framework for 3D secret sharing is proposed in this paper. The proposed algorithms were motivated by Blakley and Thien & Lin secret sharing schemes. To increase the security of the schemes by decreasing the amount of the information that must be kept secret, we used two lossless data compression algorithms Huffman Coding and ZLIB prior to splitting the 3D models. The experimental results on several 3D models indicate the feasibility of the proposed approaches.

References

1. Cheng, H., Xiaobo, L.: Partial encryption of compressed images and videos. IEEE Trans. Signal Process. 48(8), 2439–2451 (2000)
2. Bourbakis, N., Dollas, A.: Scan-based compression-encryption hiding for video on demand. IEEE Multimedia Mag. 10, 79–87 (2003)
3. Marvel, L.M., Boncelet, C.G., Retter, C.T.: Spread spectrum image steganography. IEEE Trans. Image Process. 8(8), 1075–1083 (1999)
4. Petitcolas, F.A.P., Anderson, R.J., Kuhn, M.G.: Information hiding-a survey. Proc. IEEE 87(7), 1062–1078 (1999)
5. Shamir, A.: How to share a secret. ACM Comm. 22(11), 612–613 (1979)
6. Blakley, G.R.: Safeguarding cryptography keys. In: Proc. of the AFIPS 1979 National Computer Conference, vol. 48, pp. 313–317 (1979)

7. Menezes, A., van Oorschot, P., Vanstone, S.: Handbook of Applied Cryptography. CRC Press, Boca Raton (1997)
8. Chen, Y., Chan, Y., Huang, C., Tsai, M., Chu, Y.: A multiple-level visual secretsharing scheme without image size expansion. Inform. Sciences 177, 4696–4710 (2007)
9. Cimato, S., De Prisco, R., De Santis, A.: Colored visual cryptography without color darkening. Theor. Comput. Sci. 374, 261–276 (2007)
10. Lin, S., Lin, J.: VCPSS: A two-in-one two-decoding-options image sharing method combining visual cryptography (VC) and polynomial-style sharing (PSS) approaches. Pattern Recogn. 40, 3652–3666 (2007)
11. Shyu, S.: Efficient visual secret sharing scheme for color images. Pattern Recogn. 39, 866–880 (2006)
12. Tsai, D., Chen, T., Horng, G.: A cheating prevention scheme for binary visual cryptography with homogeneous secret images. Pattern Recogn. 40, 2356–2366 (2007)
13. Chang, C., Hwang, R.: Sharing secret images using shadow codebooks. Inform. Sciences 111, 335–345 (1998)
14. Wang, R.Z., Su, C.H.: Secret image sharing with smaller shadow images. Pattern Recognition Letters 27(6), 551–555 (2006)
15. Thien, C.C., Lin, J.C.: Secret image sharing. Computers and Graphics 26(1), 765–770 (2002)
16. Wu, Y., Thien, L., Lin, J.: Sharing and hiding secret images with size constraint. Pattern Recogn. 37, 1377–1385 (2004)
17. Feng, J., Wu, H., Tsai, C., Chu, Y.: A new multi-secret image sharing scheme using Lagrange's interpolation. Journal Syst. Software 76, 327–329 (2005)
18. Noar, N., Shamir, A.: Visual cryptography. In: De Santis, A. (ed.) EUROCRYPT 1994. LNCS, vol. 950, pp. 1–12. Springer, Heidelberg (1995)
19. Storer, J.A.: Data Compression: Methods and Theory. Computer Science Press, Rockville (1988)
20. ZLIB Specification, `http://www.zlib.org`

A Fast PDE Algorithm Using Adaptive Matching Scan Order for Real-Time Video Coding

Jong-Nam Kim[1] and Tae-Kyung Ryu[2]

[1] Dept. of IT Convergence and Application Engineering
Pukyong National University, Busan, 608-737 Korea
[2] Dept. of Visual Contents, Dongseo University, San69-1, Jurye2-Dong, Sasang-Gu, Busan 617-716, Korea
jongnam@pknu.ac.kr, tkryu@gdsu.dongseo.ac.kr

Abstract. In this paper, we propose a new fast full search block matching algorithm which significantly reduces unnecessary computations without degradation of prediction quality. The proposed algorithm identifies computational matching order from initial matching errors. According to the computational order obtained, matching errors are calculated based on partial distortion elimination (PDE) method. The proposed algorithm reduces significantly computational cost for calculating block matching errors compared with the conventional lossless motion estimation algorithm without degrading prediction quality. The proposed algorithm will be useful for realizing fast real-time video coding applications, such as H.264 video coding, that require large amount of computations for motion estimation.

1 Introduction

In video coding, full search (FS) algorithm based on block matching finds optimal motion vectors which minimize the matching differences between reference blocks and candidate blocks in search area. FS algorithm has been widely used in video coding applications because of its simple and easy hardware implementation. However, high computational cost of the FS algorithm with very large search area has been considered as a serious problem for realizing fast real-time video coding.

Several fast motion estimation algorithms have been studied in recent years in order to reduce the computational cost required. These algorithms can be classified into two main groups. One group of algorithms is based on lossy motion estimation technique with degradation of prediction quality compared with the conventional FS algorithm. The other group of algorithms is based on lossless estimation technique that does not degrade the prediction quality. The lossy group of algorithms includes unimodal error surface assumption algorithm, multi-resolution algorithm, variable search range algorithm with spatial/temporal correlation of the motion vectors, half-stop algorithm using threshold of matching distortion, and others [1]. Lossless group of algorithms includes successive elimination algorithm (SEA), modified SEAs [2]-[8], fast algorithm using a fast 2-D filtering method [9], massive projection algorithm, [10], partial distortion elimination (PDE) algorithm and modified PDE algorithms [11]-[13].

A. Campilho and M. Kamel (Eds.): ICIAR 2010, Part I, LNCS 6111, pp. 336–343, 2010.

The PDE algorithm as a lossless motion estimation technique has been known as a very efficient algorithm in the sense that it could reduce unnecessary computations required for matching error calculations. To further reduce unnecessary computational cost in calculating matching errors, Kim et al. proposed fast PDE algorithms based on adaptive matching scan [11]-[12]. Obtaining adaptive matching scan order however requires additional computational overhead.

In this paper, we propose a new lossless fast full search motion estimation algorithm by reducing unnecessary computations. To achieve this, we divide the matching blocks into sub-square blocks and determine the computational order for the sub-blocks from initial calculation of matching errors. Instead of using conventional top-to-bottom matching ordering, we calculate the matching errors adaptively according to the pre-determined computational order. The proposed algorithm requires very little additional computational overhead for checking computational order, which makes our algorithm efficient enough to be used with other fast algorithms such as SEA or Multilevel SEA (MSEA). The proposed novel algorithm reduces large amount of computational cost for block matching error calculations compared with the conventional PDE algorithm without any loss of prediction quality.

2 Conventional Fast Full Search Algorithms

SEA algorithm is well known lossless FS motion estimation algorithm. It removes impossible candidate motion vectors by using the sum of the current block, the sum of the candidate blocks and the minimum sum of absolute difference (SAD) [2]. At first, the algorithm computes sum of the rows or the columns in the reference and candidate blocks. After calculating initial matching error of the search origin in the search area, the algorithm removes impossible candidate motion vectors by the Eq. (1). In the Eq. (1), R means the norm of the reference block in the current frame and $C(x,y)$ represents the summation of the norms of the candidate blocks with the motion vector (x,y) in the previous frame. SAD_{min} means the sum of absolute differences as a distortion measure at that checking time. By the Eq. (1), useless computations required for impossible candidates can be eliminated without any degradation of predicted images. If the summation of the candidate blocks is satisfied with the Eq. (1), the candidate blocks are calculated for matching errors with SAD, otherwise the candidate blocks are removed and next candidate blocks will be checked.

$$R - SAD_{\min} \leq C(x, y) \leq R + SAD_{\min} \tag{1}$$

A few modified algorithms based on SEA have been reported. The performance of the various modified SEA algorithms depends on the way how to calculate the initial matching errors. Oliveira et al. [3] proposed the modified algorithm with less initial matching distortion from the adjacent motion vectors. Lu and others [4] could reduce impossible candidate vectors further by using hierarchical structure of Minkowski's inequality with pyramid of 5 levels. Coban and others [5] used the concept of the Eq. (1) to determine motion vector with optimized rate-distortion. They extended the Eq. (1) by adding the weighted rate term to avoid unnecessary computations. Wang et al. [6] used the Eq. (1) by adding PDE, square root, and square term in order to reduce computational cost. Meanwhile, Gao et al. [7] proposed an algorithm for reducing impossible candidate vectors by using tight boundary levels shown in Eq. (2) and Eq. (3).

$$SSAD_L = \sum_{u=0}^{2^L-1} \sum_{v=0}^{2^L-1} | R_L^{(u,v)}(i,j) - C_L^{(u,v)}(i+x, j+y) | \tag{2}$$

$$SSAD_0 \le SSAD_1 \le \cdots \le SSAD_l \le \cdots \le SSAD_L \le SAD_{\min} \tag{3}$$

Another algorithm to reduce the computational cost is the PDE approach [11]-[13]. The algorithm uses the partial sum of matching distortion to eliminate impossible candidates before completing calculation of matching distortion in a matching block. That is, if an intermediate sum of matching error is larger than the minimum value of matching error at that time, the remaining computations for matching errors is abandoned. The *kth* partial SAD can be expressed by the Eq. (4),

$$\sum_{i=1}^{k} \sum_{j=1}^{N} \left| f_{t+1}(i,j) - f_t(i+x, j+y) \right| \quad k = 1,2,...N \tag{4}$$

where N represents matching block size. The term, $f_{t+1}(i,j)$, means image intensity at the position (i,j) of the $(t+1)th$ frame. The variables x and y are the pixel coordinate of a candidate vector. If the partial sum of matching distortion exceeds the current minimum matching error at k, then we can abandon the remaining calculation of matching error ($k+1$ to Nth rows) by assuring that the checking point is an impossible candidate for the optimal motion vector. Kim et al. [11] calculated block matching errors to reduce unnecessary calculations with the four-directional scan order based on the gradient magnitude of images instead of the conventional top-to-bottom matching scan order. Block matching errors are calculated to further reduce unnecessary computations with adaptive matching scan [12]. While these approaches could reduce unnecessary computations for getting block matching errors, they need additional computations to determine the matching scan order.

3 Proposed Algorithm

Modified PDE algorithms, such as spiral search algorithm and cascaded algorithms, have used adjacent motion vectors [11]-[12]. Ability to reject impossible candidate vectors in the PDE algorithm depends on the search strategy, which makes minimum matching errors can be detected faster. Because PDE algorithm with spiral search rejects impossible candidate vectors faster than simple PDE, we employ the spiral search in the proposed matching scan algorithm. The relationship between matching error and image gradient of the matching block can be summarized by Taylor series expansion [11]-[12]. Let the image intensity at the position (x,y) of the $(t+1)th$ frame be $\{f_{t+1}(p), p=(x,y)\}$, and the motion vector of the position p be $mv=(mvx, mvy)$. We can describe the relationship between the reference frame and the candidate frame as shown in the Eq. (5).

$$f_{t+1}(p)=f_t(p+mv) \tag{5}$$

$$
\begin{aligned}
d_{t+1}(p) &= \left|f_{t+1}(p) - f_t(p+cmv)\right| \\
&= \left|f_t(p+mv) - f_t(p+cmv)\right| \\
&\approx \left|\tfrac{\partial f_t(p+mv)}{\partial x}(cmvx - mvx) + \tfrac{\partial f_t(p+mv)}{\partial y}(cmvy - mvy)\right| \\
&\approx \left|\tfrac{\partial f_{t+1}(p)}{\partial x}(cmvx - mvx)\right| + \left|\tfrac{\partial f_{t+1}(p)}{\partial y}(cmvy - mvy)\right|
\end{aligned} \tag{6}
$$

By using modified form of the Taylor series expansion, we can express the relationship between the matching distortion and the gradient magnitude of the reference block as shown in the Eq. (6). Here, *cmv=(cmvx, cmvy)* represents candidate motion vector corresponding to the matching distortion. From the Eq. (6), we can find out an important fact that the matching distortion at pixel *p* is proportional to the gradient magnitude of the reference block in the current frame, which corresponds to the complexity of the image data. By localizing the image complexity well, we can further reduce unnecessary computations. In general, image complexity is well localized in a block rather than the whole span of an image. In this paper, we calculate the matching error using 4x4 square sub-blocks instead of the conventional 1x16 row vector to measure the complexity of the matching block.

Efficiently identifying complex square sub-block needs to be performed at early stage of the algorithm. In the previous algorithms, a complex square sub-block was found by calculating gradient magnitude in the reference matching block, where the additional computation for calculating the gradient magnitude can be avoided with large candidates. The previous algorithms [11]-[12] can increase computational load more than the original PDE algorithm when both of them are cascaded to other fast algorithms such as SEA [2]. Instead of calculating the gradient magnitude of the matching blocks, we find complex sub-blocks from initial SAD computation at center point of search area. At first, we calculate block matching error by the unit of square sub-block, and then accumulate the sum of sub-blocks. According to the accumulated sum of sub-blocks, we determine the matching order for the following candidates. The idea further reduces unnecessary computations as explained below. Our proposed algorithm uses the matching order for sub-blocks in all candidates of the search range. The Eq. (7) shows our modified PDE algorithm which employs sub-blocks and the matching order of the sub-blocks from initial computation of SAD.

$$
\sum_{s}^{k}\sum_{i=1}^{N/s}\sum_{j=1}^{N/s}\left|f_{t+1}(i,j) - f_t(i+x, j+y)\right| \tag{7}
$$

$$
k = 1,2,\ldots N/s * N/s, \quad s \in matching\_order[]
$$

In the Eq. (7), *matching_order[]* is obtained from the initial partial SAD value for each block. We put the pixel number of *N/s*N/s* square sub-block as *N*. The matching order of *N* candidate sub-blocks is calculated for every reference block of the search range. Thus, we increase the probability of scan order that have larger matching earlier. The computational cost required for sorting *N/s*N/s* sub-block is small and negligible compared to the overall computational cost of the block matching algorithm.

If we cascade the SEA algorithm to our proposed algorithm, we can further remove unnecessary computation. In the MSEA algorithm, block size is 16x16 and sub-block size for adaptive matching scan is 4x4. If we employ the multilevel SEA instead of the original SEA, we can further reduce computations.

4 Experimental Results

To compare the performance of the proposed algorithm with the conventional algorithms, we use 100 frames of 'foreman', 'car phone', 'trevor', 'clair', 'akio' and 'grand mother' image sequences. In these sequences, 'foreman', and 'car phone' have higher motion variance than the other image sequences. 'clair', 'akio' and 'grand mother' are rather inactive sequences compared with the first two sequences. 'trevor' sequence has intermediate level of motion variance. Matching block size is 16x16 pixels and the search window is ± 7 pixels. Image format is QCIF(176×144) for each sequence and only forward prediction is used.

The experimental results shown in Figures 2 and 3 and Tables 1 and 2 are presented in terms of average number of checking rows with reference to that of full search without any fast operation. Table 3 shows the peak-to-peak-signal-to-noise ratio (PSNR) performance of the proposed algorithm. All the algorithms employed spiral search scheme to make use of the distribution of motion vectors.

Figure 1 and Figure 2 show the reduced computation of average checking rows using 4x4 square sub-blocks based on PDE algorithm. The adaptive matching scan algorithm significantly reduces unnecessary calculations compared with the conventional sequential scan algorithm. We apply the proposed algorithm to SEA [2] and MSEA [7] to further demonstrate the performance. From the experimental results, we can see that the proposed matching algorithm can reduce unnecessary computations efficiently for PDE itself and cascaded algorithms with SEA and MSEA. Note that the computational reduction is different among three algorithms. In PDE, all candidates in search area are involved in calculating partial matching distortion. However fewer candidates are involved in calculating distortion by SEA or MSEA because many candidates are filtered out by the Eq. (1) and the Eq. (3). MSEA has smaller candidates in calculating partial matching distortion than SEA because of tighter boundary.

Table 1 and Table 2 summarize average numbers of checking rows computed for various algorithms in all sequences of 30Hz and 10Hz, respectively. The average number of checking rows of the conventional full search algorithm without any fast operation is 16.

The importance and efficiency of spiral scan in PDE algorithm was shown in [11]-[12]. From the Table 1, we can see that the computational reduction ratios from the proposed adaptive matching scan combined with SEA and MSEA are 37% and 16% compared with the conventional sequential matching. The results of PSNR are all the same for all algorithms as shown in Table 3 because they are all lossless algorithms.

With the experimental results, we can conclude that our adaptive matching scan algorithm can reduce computational cost significantly without any degradation of prediction quality and any additional computational cost for obtaining matching order.

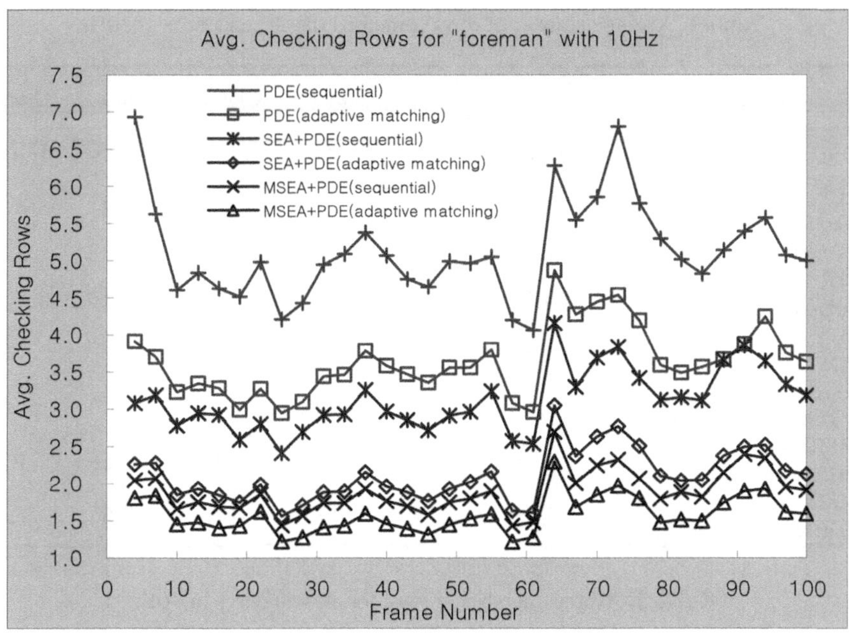

Fig. 1. Average number of rows computed for "foreman" sequence of 10Hz

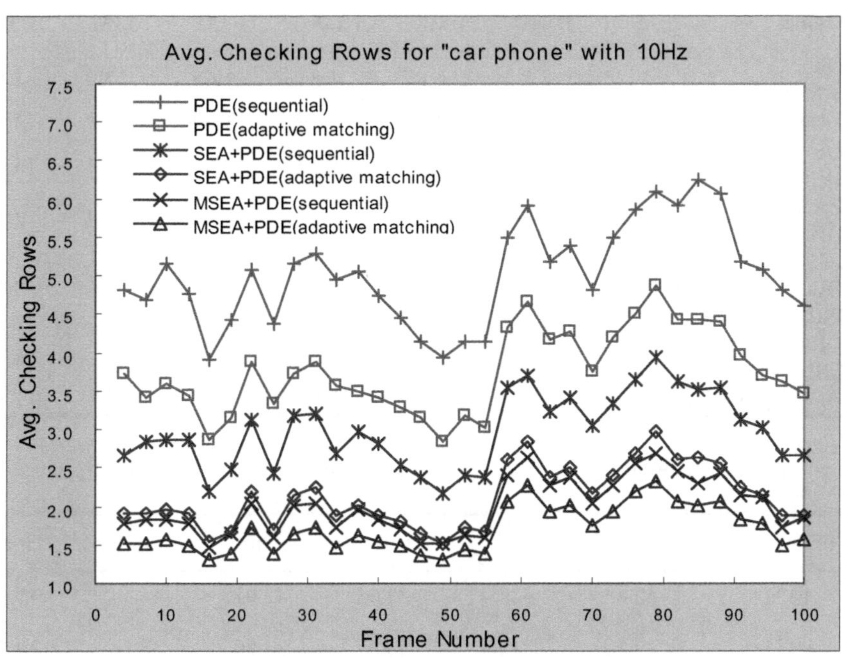

Fig. 2. Average number of rows computed for "carphone" sequence of 10Hz

Table 1. Average number of rows computed for all sequences of 30Hz

Algorithms	Fore-man	Car phone	Trevor	Claire	Akio	Grand
Original FS	16.00	16.00	16.00	16.00	16.00	16.00
PDE (sequential)	4.05	4.23	3.18	3.94	1.71	4.18
PDE Hadamard [13]	2.84	3.21	2.46	3.21	1.33	3.55
PDE (adaptive matching)	2.86	3.16	2.43	3.26	1.36	3.45
SEA + PDE (sequential)	1.90	2.03	1.29	1.31	0.62	1.99
SEA + PDE(adaptive matching)	1.55	1.72	1.12	1.24	0.56	1.85
MSEA + PDE(sequential)	1.32	1.56	0.91	1.05	0.58	1.77
MSEA + PDE(adaptive matching)	1.19	1.43	0.85	1.03	0.57	1.71

Table 2. Average number of rows for all sequences of 10Hz

Algorithms	Fore-man	Car phone	Trevor	Claire	Akio	Grand
Original FS	16.00	16.00	16.00	16.00	16.00	16.00
PDE (sequential)	4.81	4.87	4.51	4.45	2.09	4.66
PDE Hadamard [13]	3.65	3.76	3.49	3.64	1.58	3.78
PDE (adaptive matching)	3.64	3.75	3.53	3.54	1.54	3.80
SEA + PDE (sequential)	2.50	2.48	2.17	1.57	0.83	2.30
SEA + PDE(adaptive matching)	2.10	2.12	1.84	1.44	0.69	2.10
MSEA + PDE(sequential)	1.75	1.88	1.46	1.21	0.68	1.92
MSEA + PDE(adaptive matching)	1.58	1.71	1.32	1.16	0.63	1.83

Table 3. Average PSNR of all sequences for the frame rates 30Hz and 10 Hz

Frame rate	Foreman	Car phone	Trevor	Claire	Akio	Grand
30 Hz	32.85	34.04	34.05	42.97	44.14	43.44
10 Hz	29.50	31.54	28.63	37.50	38.66	39.01

5 Conclusions

In this paper, we propose a new block matching algorithm by sorting square sub-blocks according to the initial matching distortion. The proposed algorithm reduces unnecessary computations for motion estimation while keeping the same prediction quality compared with the conventional full search algorithm. Unlike the conventional fast PDE algorithms, the proposed algorithm does not require additional computations to identify matching order. The proposed algorithm can be efficiently cascaded to other fast algorithms such as MSEA or SEA. The proposed algorithm will be particularly useful for realizing fast real-time video coding applications, such as MPEG-4 advanced video coding, that require large amount of computations.

Acknowledgement

This work was supported from Advanced Technology Project by SMBA and RIS Project by KOTEF.

References

1. Dufaus, F., Moscheni, F.: Motion estimation techniques for digital TV: A review and a new contribution. IEEE Proceeding 83, 858–876 (1995)
2. Li, W., Salari, E.: Successive elimination algorithm for motion estimation. IEEE Trans. Image Processing 4, 105–107 (1995)
3. de Oliveira, G., Alcaim, A.: On fast motion compensation algorithms for video coding. In: Proc. PCS, pp. 467–472 (1997)
4. Lu, J., Wu, K., Lin, J.: Fast full search in motion estimation by hierarchical use of Minkowski's inequality (HUMI). Pattern Recog. 31, 945–952 (1998)
5. Coban, M., Mersereau, R.: A fast exhaustive search algorithm for rate-constrained motion estimation. IEEE Trans. Image Processing 7, 769–773 (1998)
6. Wang, H., Mersereau, R.: Fast algorithms for the estimation of motion vectors. IEEE Trans. Image Processing 8, 435–438 (1999)
7. Gao, X., Duanmu, C., Zou, C.: A multilevel successive elimination algorithm for block matching motion estimation. IEEE Trans. Image Processing 9, 501–504 (2000)
8. Ahn, T., Moon, Y., Kim, J.: Fast full-search motion estimation based on Multilevel Successive Elimination Algorithm. IEEE Trans. Circuits System for Video Technology 14, 1265–1269 (2004)
9. Naito, Y., Miyazaki, T., Kuroda, I.: A fast full-search motion estimation method for programmable processors with a multiply-accumulator. In: Proc. ICASSP, pp. 3221–3224 (1996)
10. Do, V., Yun, K.: A low-power VLSI Architecture for full-search block-matching motion estimation. IEEE Trans. Circuits System for Video Technology 8, 393–398 (1998)
11. Kim, J.: Adaptive matching scan algorithm based on gradient magnitude for fast full search in motion estimation. IEEE Trans. Consumer Electronics 45, 762–772 (1999)
12. Kim, J., Byun, S., Kim, Y., Ahn, B.: Fast Full Search Motion Estimation Algorithm Using Early Detection of Impossible Candidate Vectors. IEEE Trans. Signal Processing 50, 2355–2365 (2002)
13. Jin, S., Lee, H.: Fast partial distortion elimination algorithm based on hadamard probability model. IEE Electron. Letters 44(1), 17–19 (2008)

New Non Predictive Wavelet Based Video Coder: Performances Analysis

Tarek Ouni, Walid Ayedi, and Mohamed Abid

National Engineering School of Sfax, Road Sokkra, Km 3 Sfax, 3078 Tunisia
tarek.ouni@gmail.com, walid.ayedi@gmail.com,
Mohamed.abid@enis.rnu.tn

Abstract. A non-predictive video coding is a new branch of emerging research area in video coding, where the motion estimation/compensation or prediction step in the temporal domain is omitted. One direction was to look for the exploitation of temporal decomposition of video frames. The proposed method consists on 3D to 2D transformation of the temporal frames that allows exploring the temporal redundancy of the video using 2D wavelet transforms and avoiding the computationally demanding motion compensation step. Although the many advantages presented by the proposed coder, some annoying artifacts still exist. In this paper, we will explore the performances of the proposed method and try to better show what it actually offers to users. The paper presents also the extensions chosen in order to reduce the perceived artifacts and increase the perceptual as well as objective (PSNR) decoded video quality, which is actually competitive with state-of-the-art video coder algorithms, especially when low computational demands of the proposed approach are taken into account.

Keywords: video coding, temporal decomposition, wavelet, correlation.

1 Introduction

Video compression has generated a lot of discussion and increasing attention from the research in recent years.

Among many proposed methods, motion compensated coding has taken the most attention and taken its place in many standards. These include Mpeg, H.26L, etc. Such encoders exploit inter-frame correlation in order to further improve its compression. However, the main challenge of these methods lies on the motion estimation process which is known to be computationally intensive. Besides, its real time implementation is difficult and costly [1],[2]. Nevertheless, new applications such as sensor networks and portable video devices necessitate a low processing capability for the compression, which makes the encoding complexity a big burden. To deal with this problem, motion-based video coding standard MPEG was primarily developed for stored video applications, where the encoding process is typically carried out off-line on powerful computers. With the explosive growth of video devices ranging from hand-held digital cameras to low-power video sensors, a new class of multimedia devices is required which includes the following architectural requirements: Low

A. Campilho and M. Kamel (Eds.): ICIAR 2010, Part I, LNCS 6111, pp. 344–353, 2010.

power, less- complexity encoding and real time constraint. Therefore, there have been extensive research efforts in video coding in order to give response to the new requirements of video applications different than those targeted by conventional coding schemes in the past years [1]. A non-predictive video coding is a new branch of emerging research area in video coding, where the motion estimation/compensation or prediction step in the temporal domain is omitted.

In [3]-[4], authors exploit 3D transforms in order to exploit temporal redundancy. Coder based on 3D transform produces video compression ratio which is comparable to some motion estimation based coding one but with lower processing complexity [5]. However, 3D transform based video compression methods process temporal and spatial redundancies in the 3D video signal in the same way. This can reduce the efficiency of these methods as pixel's values variations in spatial or temporal dimensions are not uniform and hence, temporal and spatial redundancies have not the same pertinence. It is known that the temporal redundancies are more relevant than spatial one [2], hence its practical importance. It is more beneficial to utilize the proposed method rather than the 3D based methods, because it is able to achieve higher compression by more exploiting the redundancies in the temporal domain.

This method consists on 3D to 2D transformation of the video frames; it will then explore the temporal redundancy of the video using 2D transforms and avoids the computationally demanding motion compensation step. In particular, the used method projects temporal redundancy of each group of pictures into spatial domain and combines it with spatial redundancy in one representation with high spatial correlation [6]. Then, the new representation will be compressed as still image using wavelet transform based coder (JPEG2000). Actually, the proposed approach presents many advantages. It exploits objectively temporal and spatial redundancy. It omits the temporal prediction step and transforms a 3D processing into 2D one while reducing considerably the complexity processing. Furthermore, it inherits the JPEG 2000 proprieties such as scalability ROI and error resilience.

In this paper, we focus on the analysis of experimental results, solutions and extensions proposed to remove some annoying artifacts of the presented method. Experimental results will show the efficiency of the proposed method at an expense of some annoying artifacts.

The rest of paper is organized as follows. In section 2, we review the basics of the used approach. In section 3, we present some experimental results and explore the method performances and limitations. Section 4 presents the extensions chosen to further improve the compression ratio. Finally, conclusions are drawn in Section 5.

2 Description of the Used Approach

Actually, the used method relies on the following assumption: high frequency data is more difficult to compress compared to low frequency one.

The main idea of the proposed method is to make some geometric transformation of the 3D data in order to make one representation with very high correlation, and consequently without high frequencies data. We will play on the disposition of pixel's data in the video cube.

2.1 Hypothesis

The video stream contains more temporal redundancies than spatial ones [2]. This assumption will be the basis of the proposed method where we will try to put pixels - which have a very high temporal correlation - in spatial adjacency. Thus, video data will be presented with high correlated form which exploits both temporal and spatial redundancies in video signal with appropriate portion that put in priority the temporal redundancy exploitation.

2.2 Accordion Representation

The input of our encoder is the so called video cube (GoF), which is made up of a number of frames. This cube will be decomposed into temporal frames which will be gathered into one 2D representation. Temporal frames are formed by gathering the video cube pixels which have the same column index.

These frames will be projected on 2D representation (further called "IACC" frame) while reversing the direction of odd frames, i.e. the odd temporal frames will be turned over horizontally in order to more exploit the spatial correlation of the video cube frames extremities. In this way, Accordion representation also minimizes the distances between the pixels spatially correlated in the source. This representation transforms temporal correlation of the 3D original video source into a high spatial correlation in the 2D representation ("IACC") [6]. Figure 1 illustrates the principle of this representation.

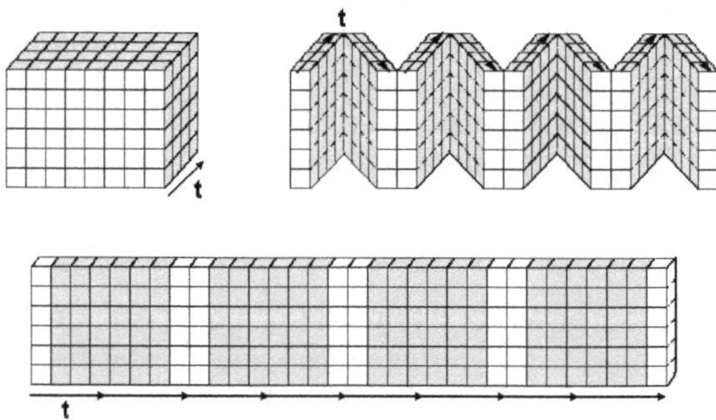

Fig. 1. Accordion representation [6]

In the following, we will present the diagram of coding based on the Accordion representation further called ACC-JPEG2000.

2.3 ACC-JPEG2000 Coding Scheme

The proposed ACC-JPEG2000 coding scheme follows the following steps:

- The decomposition of video sequence into groups of frames (GOF).
- Accordion representation of the GOF.

- DWT transform.
- Quantization of obtained coefficients (Q).
- Arithmetic coding of obtained coefficients.

Figure 2 presents ACC-JPEG2000 coding scheme.

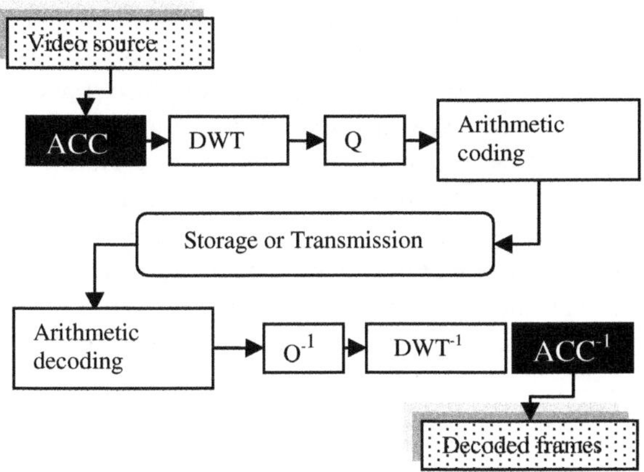

Fig. 2. ACC-JPEG2000 coding scheme

The video encoder takes a video sequence and passes it to a frame buffer. The buffer dispatches a group of frames at a time to Accordion process before being sending to DWT blocks. Each of the DWT blocks performs a 2-D discrete wavelet transform using JPEG2000 wavelet filters coefficients. In this system, we use five levels of wavelet decomposition which is sufficient for CIF sequences.

3 Experiments

In the following, we summarize the experimental results with some analysis and comments.

3.1 PSNR Evaluation

In these experiments, we use the XVID MPEG-4 video coder including P frames. The GOF number of frames relative to ACC-JPEG 2000 is fixed to 8 frames.

The experiments prove the efficiency of ACC-JPEG2000 on slow and uniform motion sequences. In these sequences, temporal redundancy is **relevant**; the spatial representation “IACC” performs a pertinent correlation. In this case, it is expected that the proposed method proves its efficiency. However, the method shows a remarkable sensitivity to very fast motion video sequences.

Among the studied sequences, we have got worst compression performance with “tennis” sequence. Tennis sequence contains very fast motion with fast complex background changes. The generated spatial representation still presents some high frequencies. The ACC-JPEG2000 efficiency decreases with the apparition of transparency effect due to background change, measured PSNR is relatively low with an alternate character. In fact, such results are expected as ACC-JPEG2000 eliminates "IACC" frame's high frequency data which actually contains the high temporal frequency produced by the fast motion. Foreman sequence contains fast non-uniform motion which is caused by the camera as well as the man's face movement. So measured PSNR is relatively low and visual quality suffers from some blur effect, especially in face detail which represents the high resolution data. Hall monitor sequence seems to involve less motion compared to the Foreman sequence; the motion takes place only in a very concentrated area. Due to the little amount of motion taking place on the overall image, we observed that our method get better results. Miss America is a low motion sequence. The motion is confined to the person's lips and head. Since motion is low, temporal redundancy is high and it is expected that ACC- JPEG2000 becomes efficient.

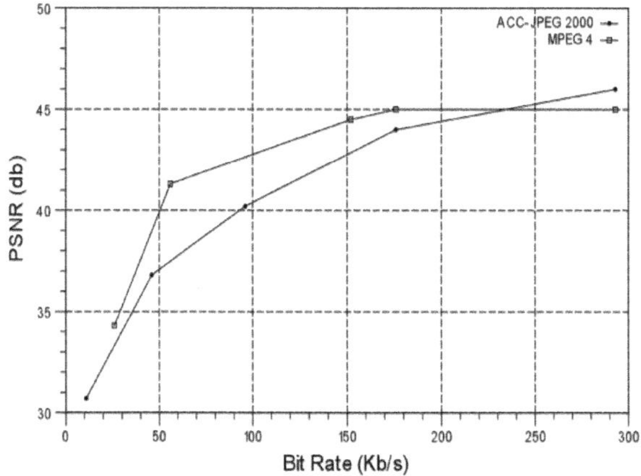

Fig. 3. PSNR evaluation: Miss America (QCIF, 25Hz)

Figure 3 shows results of PSNR based comparative study between ACC-JPEG2000 and MPEG 4 relative to miss america sequence. Up to 230 kb/s, the proposed coder outperforms the mpeg 4, the relative PSNR continue to increase until lossless level. Otherwise, MPEG 4 can not rich less than 22 kbits/s, but ACC-JPEG2000 can go less than 10 kbits/s. we can state that the proposed coder is highly scalable.

Table 1 shows results of PSNR based comparative study between ACC-JPEG2000 and MPEG 4 coder.

Table 1. PSNR EVALUATION

	Bit Rate (Kbits/s)	ACC-JPEG 2000	MPEG 4
Water fall	*100*	29	30.1
(CIF, 25Hz)	*1000*	35	36.6
Bus	*200*	21.3	25.5
(CIF, 25Hz)	*1000*	26.6	32
Tennis	*100*	25	29
(CIF, 25Hz)	*1000*	34	43
Mobile	*100*	22	23
(QCIF, 25Hz)	*1000*	36	35
Hall monitor (CIF, 25Hz)	*50*	28.3	29.1
	1000	41.2	39.1
Miss America (QCIF, 25Hz)	*25*	33.8	34.2
	300	46.7	44.9

Moreover, the PSNR Curve relative to the ACC-JPEG2000 coding is in continuous alternation from one frame to another unlike MPEG PSNR which is almost stable as it is shown in figure 4.

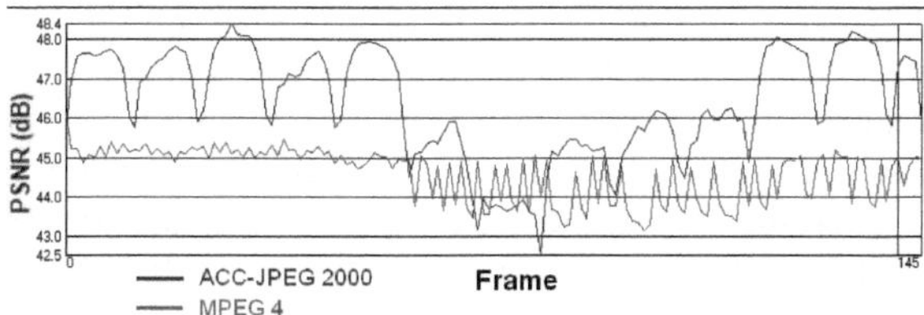

Fig. 4. PSNR evaluation: Miss America (QCIF, 25Hz)

In one hand, ACC-JPEG2000 affects the quality of some frames of a GOP, but on the other hand, it provides relevant quality frames in the same GOP, while MPEG produces frames practically of the same quality. In video compression, such feature could be useful for video surveillance field; Generally, we just need some good quality frames in a GOP to identify the objects (i. e. person recognition) rather than medium quality for all the frames. The example of the "hall monitor" in table 1proves that the video surveillance is one of the best application field to the proposed coder.

Differently to MPEG codec's, ACC-JPEG2000 can reach very low bit rates. In high bit rates it provides a relevant quality (until lossless level).

3.2 Visual Evaluation

Some artifacts existing in DWT based compression methods (MJPEG 2000) and in motion estimation based method (MPEG) such as spatial distortions generated through the massive elimination of the high spatial frequencies (tiling artefact and blocking artefact) as it is presented in figure 6, does not exist in the proposed method as shown in figure 5. It's actually replaced by some less annoying blur artifact.

Blocking artefact in MPEG 4
Blur artefact in ACC-JPEG 2000

Fig. 5. Perceptual evaluation: Tennis (CIF, 25Hz)

In the proposed method, the DWT is exploited in both spatial and temporal domain. Actually, temporal and spatial redundancy is projected on spatial domain forming the IACC representation. The application of the DWT on IACC allows the transformation from the spatial domain to the frequency domain.

After quantification process, we will eliminate the high spatial frequencies of "IACC" frame which actually include the high temporal frequencies of the 3D signal source. As temporal redundancy is more exploited than spatial one, a strong quantification will not seriously affect the quality of image but will rather affect the fluidity of the video. Spatial high frequency is mainly made of fast pixel's values change from one frame to another. Once some of the coefficients have been quantized (set to zero) the signal is smoothed out. Thus some fast changes over time is somewhat distorted. As a consequence, the PSNR significantly decreases in very fast motion sequences leading for some annoying blur artifact.

However, some sudden pixels change will be eliminated. This will offer a useful functionality such as the noise removal. Indeed, the very high temporal frequency (sudden change of a pixels value over time) is generally interpreted as a noise.

As it is shown in figure 6, some other artifacts appear in video sequences containing cuts: transparency [7]:

Fig. 6. Transparency artefact: Tennis (CIF, 25Hz)

The input data stream is divided into n frames (in our case n=8) as shown in Fig. 7. These groups of n frames are completely independent to each other. The problem appears when one group contains several types of video sequences. In consequence, particular frames compound images from different video sequences.

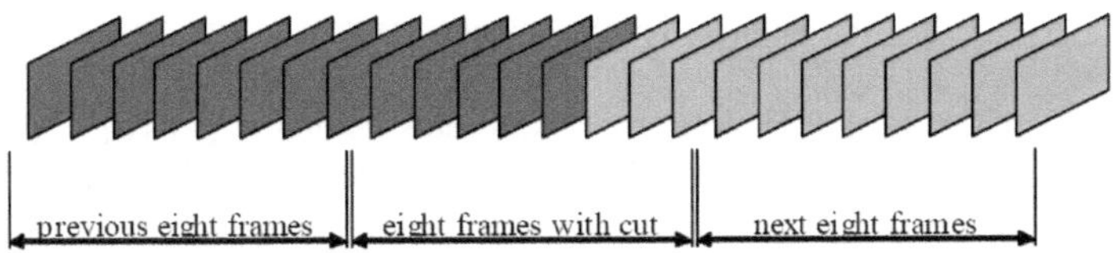

Fig. 7. Video sequence with cut

There are many solutions for this known issue in the prior art [7][8]. However their integration are not well adapted to our coder, and it increase the coder complexity.

4 Performance Improvement

Our current work is directed towards finding solutions to treat certain weaknesses shown by our method. First, the proposed method exhibits significant boundary effects at GOF boundaries. The PSNR drops every N frames, leading to annoying jittering artifacts in video playback. This well known [12] issue can be resolved by extending some temporal filtering indefinitely in time [13][14]. Some spatial filter can be useful when applied on IACC frame. We are currently testing this approach.

Second, the proposed method lose its efficiency in very fast motion sequences especially fast moving objects details are often lost in video playback. Iin this case, we are trying to exploit the ROI propriety of the JPEG 2000. Moreover the extension presented below should clearly decrease this weakness effect.

Another annoying artifact is the transparency; we don't look for some post processing but rather we look for eliminate the cause of this anomaly. Thus, proposed solution is to work with a dynamic strategy in the construction of the GOF, the number of frames will not be previously fixed, but rather will vary according to the semantics of the video in order to avoid cuts in the video inputs GOFs. For this reason an additional inter frame change detection module will be integrated (figure 8).

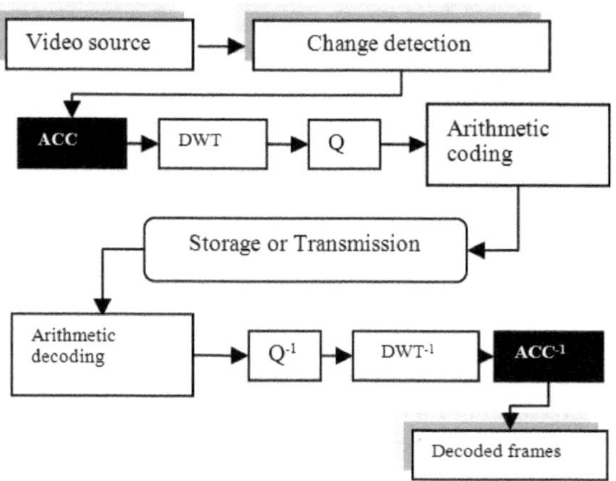

Fig. 8. Integration of the change detection module

There are many existing techniques related to cuts detection [9][10][11], in our case, we don't only look for cuts detection, but also local frames change due to fast moving objects, that's why we proceeded with local comparison with threshold based method. This module is responsible for detecting significant and fast inter-frames changes. This module allows removing transparency artifact by avoiding cuts in inputs GOFs. It also contributes in the improvement of the video quality by reducing the number of frames in the GOF in fast video sequence. The figure 9 shows the disappearance of the transparency artefact after the integration of the change detection module.

Fig. 9. Transparency removal: Tennis (CIF, 25Hz)

5 Conclusion

In this paper, we tended to explore a new non predictive wavelet based video coder; many experiments were conducted in order to prove the method performances and point out its limits. Taking into account its operating simplicity in one hand, and its competitive performances in other hand, we can state that this approach can be useful in large application domains, especially, in embedded systems and video surveillance applications. There are various directions for future investigations. First of all, we will try to combine the Accordion representation with other image coding techniques. Another direction could be to explore others possibilities of video representation in order to look for one more correlated one.

References

1. Molino, A., Vacca, F.: Low complexity video codec for mobile video conferencing. In: EUSIPCO, Vienna, Austria, pp. 665–668 (2004)
2. Gokturk, S.B., Aaron, A.M.: Applying 3d methods to video for compression. In: Digital Video Processing (EE392J) Projects Winter Quarter (2002)
3. Servais, M.P.: Video compression using the three dimensional discrete cosine transform. In: Proc. COMSIG, pp. 27–32 (1997)
4. Burg, R.A.: 3d-dct real-time video compression system for low complexity singlechip vlsi implementation. In: Mobile Multimedia Conf, MoMuC (2000)
5. Koivusaari, J.J., Takala, J.H.: Simplified three-dimensional discrete cosine transform based video codec. In: Proc. SPIE-IS&T EI Symposium, San Jose, CA (January 2005)
6. Ouni, T., Ayedi, W.: New low complexity DCT based video compression method. In: International Conference on Telecommunication, Morocco 2009 (2009)
7. Fryza, T.: Compression of Video Signals by 3D-DCT Transform. Diploma Thesis. Institute of Radio Electronics, FEKT Brno University of Technology, Czech Republic (2002)
8. Fryza, T.: Improving Quality of Video Signals Encoded by 3D DCT Transform. In: 48th International Symposium ELMAR 2006 focused on Multimedia Signal Processing and Communications, pp. 89–93 (elmar 2006)
9. Brunelli, R., Mich, O., Modena, C.M.: A survey on the automatic indexing of video data. Journal of Visual Communication and Image Representation 10(2), 78–112 (1999)
10. Idris, F., Panchanathan, S.: Review of image and video indexing techniques. Journal of Visual, Communication and Image Representation 8(2), 146–166 (1997)
11. Ahanger, G., Little, T.D.C.: A survey of technologies for parsing and indexing digital video. Journal of Visual Communication and Image Representation 7(1), 28–43 (1996)
12. Xing, Q., Yan, X.: Tiling artifact reduction for JPEG2000 image at low bit-rate. In: ICME 2004, Taipei, Taiwan (2004)
13. Marusic, B., Skocir, P.: Video post-processing with adaptive 3-D filters for wavelet ringing artifact removal. IEICE Transactions on Information and Systems 88(5), 1031–1040 (2005)
14. Liang, J., Tu, C.: Optimal block boundary pre/post-filtering for wavelet-based image and video compression. In: ICIP 2004, Singapore (2004)

3D Texton Spaces for Color-Texture Retrieval

Susana Alvarez[1], Anna Salvatella[2], Maria Vanrell[2], and Xavier Otazu[2]

[1] Universitat Rovira i Virgili, Department of Computer Science & Mathematics, Campus Sescelades, Avinguda dels Països Catalans, 26, 43007 Tarragona, Spain
susana.alvarez@urv.es
[2] Computer Vision Center,
Edifici O, Campus UAB, 08193 Bellaterra, Barcelona, Spain
{anna.salvatella,maria.vanrell,xavier.otazu}@uab.es

Abstract. Color and texture are visual cues of different nature, their integration in an useful visual descriptor is not an easy problem. One way to combine both features is to compute spatial texture descriptors independently on each color channel. Another way is to do the integration at the descriptor level. In this case the problem of normalizing both cues arises. In this paper we solve the latest problem by fusing color and texture through distances in texton spaces. Textons are the attributes of image blobs and they are responsible for texture discrimination as defined in Julesz's Texton theory. We describe them in two low-dimensional and uniform spaces, namely, shape and color. The dissimilarity between color texture images is computed by combining the distances in these two spaces. Following this approach, we propose our TCD descriptor which outperforms current state of art methods in the two different approaches mentioned above, early combination with LBP and late combination with MPEG-7. This is done on an image retrieval experiment over a highly diverse texture dataset from Corel.

Keywords: color-texture descriptors, retrieval, Corel dataset.

1 Introduction

In the literature there are several works dealing with color and texture in different applications, however the integration of both features is still an open problem [6]. The different nature of these two visual cues has been studied from different points of view. While texture is essentially a spatial property, color has usually been studied as a property of a point.

Computational approaches have proposed several algorithms to integrate these features. Some approaches [2,3] process texture and color separately, using different descriptors, they combine both descriptions at a similarity measure level afterwards. This means that for every visual cue a dissimilarity measure is obtained, each one in a different space and then they are combined to obtain a final similarity that needs to be scaled in order to be comparable. Other approaches [6,11,12] use the same descriptor over the three components on the chosen color space. The final descriptor is composed by the concatenation of the three feature vectors obtained separately from each color channel.

A. Campilho and M. Kamel (Eds.): ICIAR 2010, Part I, LNCS 6111, pp. 354–363, 2010.

In this paper we propose a perceptual approach to combine color and texture in order to define a compact color-texture descriptor. Our combination is based on two low-dimensional spaces that describe color textures through the texton concept. Here we use the original definition of texton given by Julesz in his Texton Theory [4]. Textons are defined as the attributes of image blobs. The differences of their first order statistics are the responsibles for texture discrimination. We use two different spaces, one to represent shape textons and a second one to represent color textons. In this way we obtain a combination of cues directly from the attributes of the blobs.

The paper is organized as follows: in section 2 we review the perceptual considerations justifying the attribute spaces and describe the computational method to obtain the important blobs of an image. In section 3 we describe the two texton spaces where the descriptor, *Texture Component Descriptor* (TCD), we propose is derived from the fusion of similarities computed in each of these spaces. Section 4 contains the experiment that evaluates our approach, showing that our descriptor achieves better performance than current descriptors in retrieval. We compare our TCD with MPEG-7 and LBP descriptors in standard Corel datasets. In the last section we sum up our proposal of a perceptual integration of color and texture descriptors.

2 Texture and Blobs

Texton theory [4] was originally introduced as the basis for the first steps in texture perception. This theory states that preattentive vision directs attentive vision to the location where differences in density of textons occur, ignoring positional relationships between textons and defines the concept of textons as the attributes of elongated blobs, terminators and crossings. From several psychophysical experiments they conclude that preattentive texture discrimination is achieved by differences in first-order statistics of textons, which are defined as line-segments, blobs, crossings or terminators; and their attributes, width, length, orientation and color.

Inspired by this idea, we consider hereby that a texture can be defined as a set of blobs, but we will not consider terminators or crossings, since it is not clear whether they would be necessary for natural images. Therefore, we propose a texture descriptor based on the attributes of the image blobs or *perceptual blobs*.

Thus, following the assumption that a texture can be described by their blobs then we are also assuming that a texture is provided by the existence of groups of similar blobs. This is the basis of the repetitiveness nature of texture images. Some examples of this proposal can be seen in Fig. 1. In image (a) a striped texture is described by two different types of blobs: blue elongated blobs and grey elongated blobs. In the same figure, texture (b) can be described in terms of 6 different types of blobs, which are blue, green and orange, of different sizes and shapes. The groups of blobs sharing similar features (size, orientation and color) are called *texture components* (TC). This description of a textured image in terms of the attributes of blobs or textons is the basis of our descriptor.

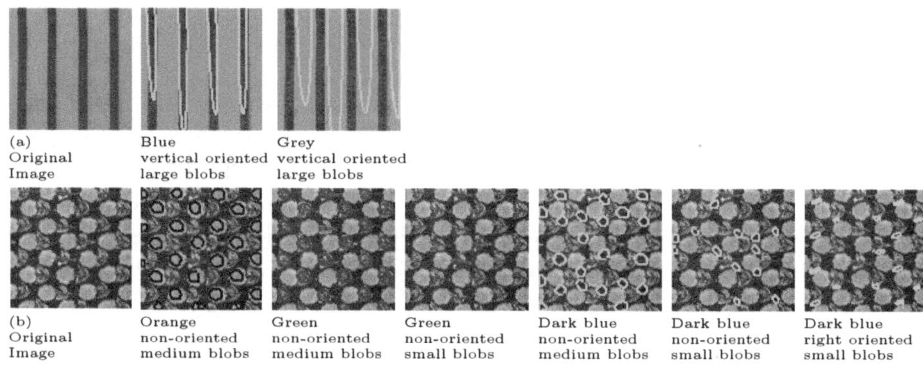

Fig. 1. Texture components and their description

2.1 Blob Detection

To obtain the attributes of the image blobs we use the differential operators in the scale-space representation proposed in [5]. We use the normalised differential Laplacian of Gaussian operator to detect the blobs of the image ($\nabla^2_{norm} L_\sigma$). This operator also allows us to obtain the scale and the location of the blobs. The aspect-ratio and orientation of non-isotropic blobs are obtained from the eigenvectors and eigenvalues of the windowed second moment matrix [5].

Since blob information emerge from both intensity and chromaticity variations, this procedure is applied to each component in the opponent color space in order to obtain the blobs of a color image. Previously, all the components were normalized to be invariant to intensity changes and then a perceptual filtering was carried out. This perceptual filtering is performed with a winner-take-all mechanism that selects the blobs of higher response of $\nabla^2_{norm} L_\sigma$ from those that overlap in different channels. This last step provides us with a list of *perceptual blobs* and their attributes, that we refer as Blob Components (BC), which are given in matrix form as:

$$\mathbf{B} = [\mathbf{B_{sha}}\mathbf{B_{col}}] \tag{1}$$

where $\mathbf{B}$ is formed by joining two matrices: $\mathbf{B_{sha}}$ that contains blob shape attributes and $\mathbf{B_{col}}$ contains blob color attributes. These matrices can be defined as:

$$\mathbf{B_{sha}} = [\mathbf{WL\Theta}], \ \mathbf{B_{col}} = [\overline{\mathbf{I}}\,\overline{\mathbf{RG}}\,\overline{\mathbf{BY}}] \tag{2}$$

where $\mathbf{W^T} = [w_1 \dots w_n]$, $\mathbf{L^T} = [l_1 \dots l_n]$, $\mathbf{\Theta^T} = [\theta_1 \dots \theta_n]$ being (w_j, l_j, θ_j) shape attributes of the j-th blob (width, length and the orientation respectively), and $\overline{\mathbf{I}}^\mathbf{T} = [\bar{i}_1 \dots \bar{i}_n]$, $\overline{\mathbf{RG}}^\mathbf{T} = [\overline{rg}_1 \dots \overline{rg}_n]$, $\overline{\mathbf{BY}}^\mathbf{T} = [\overline{by}_1 \dots \overline{by}_n]$ being $(\bar{i}_j, \overline{rg}_j, \overline{by}_j)$ color attributes of the j-th blob (median of the intensity and chromaticities of the pixels forming the winner blob respectively).

3 Texton Spaces

At this point we have to deal with the problem of the different nature of the attributes we have computed for the Blob Components that is given by $\mathbf{B}$. We will use two different texton spaces to represent the two sets of attributes, $\mathbf{B_{sha}}$ and $\mathbf{B_{col}}$. The first one is the shape texton space and the second one is the color texton space. Both need to be perceptual spaces since the fusion of color and texture is done through the Euclidean distances in these two spaces separately.

The uniform space used to represent shape is a three dimensional cylindrical space where two axes represent the shape of the blob (aspect-ratio and area) and the third axis represents its orientation. The space we have used is shown in Fig. 2.(a). This perceptual shape space is obtained by performing a non linear transformation U,

$$\begin{array}{rccc} U: & \mathbb{R}^3 & \rightarrow & \mathbb{R}^3 \\ & (w,l,\theta) & \rightarrow & (r,z,\phi) \end{array} \tag{3}$$

where $r = \log(ar), z = \log(A)$ and $\phi = 2\theta$, being ar the blob aspect ratio ($ar = w/l$), A its area ($area = w \cdot l$) and θ its orientation.

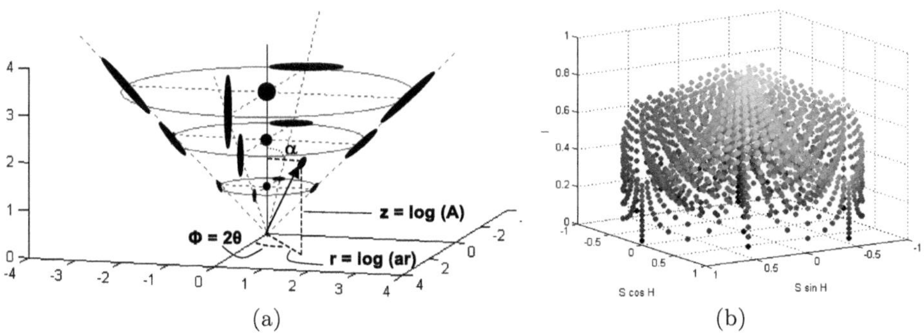

Fig. 2. (a) Shape Texton Space in cylindrical coordinates. (b) Color Texton space (HSI).

To represent the color attributes of blobs we use the HSI color space corresponding to the transform given in [1]. This space is shown in Fig. 2.(b). Although this color space is not perceptually uniform, our choice is based on the fact that is close to an uniform space when we need to represent non-calibrated color.

Following our initial assumption that a texture is provided by the existence of groups of similar blobs (*Texture Components*), in the next section we propose a color-texture descriptor based on clustering blob attributes in these two texton spaces.

3.1 Texture Component Descriptor (TCD)

Considering the properties of the texton spaces we can state that similar blobs are placed on different unidimensional varieties such as lines, rings or arcs. To group blobs of similar shapes and colors we use a clustering method that groups data with these points distributions and, at the same time, makes it possible to combine spaces with different characteristics, specifically color and shape. The clustering algorithm which has these properties is the Normalized Cut (N-cut) [9], that obtains the clusters by partitioning a graph. In the graph the nodes are the points of the feature space and the edges between the nodes have a weight equal to the similarity between nodes. To determine the similarity between nodes we need to define a distance. Since the shape space has been designed to be uniform and the HSI color space is almost uniform, it is reasonable to use the Euclidean distance.

The N-Cut clustering algorithm can be defined as

$$NCUT([U(\mathbf{B_{sha}}), HSI(\mathbf{B_{col}})], \boldsymbol{\Omega}) = \{\hat{\mathbf{B}}^{\mathbf{1}}, \hat{\mathbf{B}}^{\mathbf{2}}, \ldots, \hat{\mathbf{B}}^{\mathbf{k}}\} \tag{4}$$

where, $\boldsymbol{\Omega}$ is the weight matrix, and its elements define the similarity between two nodes through the calculation of the distance in each one of the texton spaces (shape and color) in an independent way. These weights are defined as,

$$\omega_{pq} = e^{-\frac{\|U(\mathbf{B_{sha}})_p - U(\mathbf{B_{sha}})_q\|_2^2}{\sigma_{sha}^2}} \cdot e^{-\frac{\|HSI(\mathbf{B_{col}})_p - HSI(\mathbf{B_{col}})_q\|_2^2}{\sigma_{col}^2}} \tag{5}$$

This weight represents the similarity between blob p and blob q that depends on the similarity of its shape features and the similarity of its color features. $U(\mathbf{B_{sha}})_p$ and $HSI(\mathbf{B_{col}})_p$ are the p-th row of the matrices $U(\mathbf{B_{sha}})$ and $HSI(\mathbf{B_{col}})$ respectively. As in [9], σ_{sha} and σ_{col} are defined as a percentage

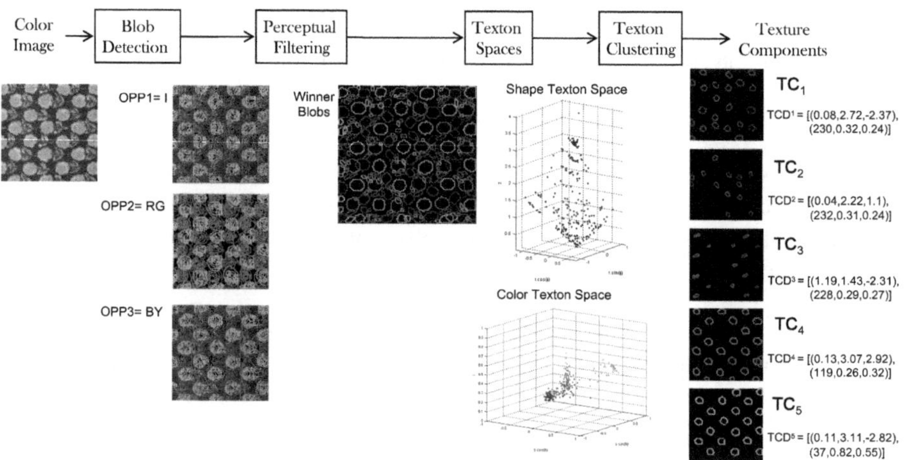

Fig. 3. Stages of TCD Computation

of the total range of each feature distance function, the first one in the shape space and the second one in the color space.

The result of the clustering obtained by the N-cut algorithm is represented by $\hat{\mathbf{B}}^{\mathbf{i}}$, $\forall i = 1, ..., k$ (where k is the total number of clusters). The prototype of each cluster i becomes our *Texture Component Descriptor* (TCD^i). This is computed by estimating the median of all the blob attributes in the i cluster, $[\hat{\mathbf{B}}^{\mathbf{i}}_{\mathbf{sha}}\hat{\mathbf{B}}^{\mathbf{i}}_{\mathbf{col}}]$. This give a 6-dimensional description for each cluster or Texture Component (TC):

$$TCD^i = (r^i, z^i, \phi^i, h^i, s^i, i^i) \quad (6)$$

In this way the descriptors of an image are the shape (3D) and color attributes (3D) of its TC. In figure 3 we show the over all scheme to obtain the TCD.

4 Experiment

This experiment evaluates the performance of our TCD descriptor in an image retrieval application. In order to compute the similarity between two textures we need to define an adequate measure which considers that the TCD of images can have different number of texture components. For a given image, the number of texture components in its TCD depends on the complexity of the texture content. A metric presenting this property is the Earth Mover's Distance [8]. In our case this distance adapts perfectly because our feature spaces are bounded independently of the image content. Shape space has the limits of blob attributes and color space is bounded by the maximum luminance. Therefore we define the ground distance between two TCD and the weighting parameters as

$$d(TCD^i, TCD^j) = \alpha \cdot d_{shape}(TCD^i, TCD^j) + \beta \cdot d_{color}(TCD^i, TCD^j) \quad (7)$$

where d_{shape} and d_{colour} are Euclidean distances in the shape space and color space, respectively. The shape space has been built taking into account perceptual considerations allowing it to be considered as a uniform space, therefore distances are correctly estimated. This is not the case in the HSI color space that is not real uniform space, therefore the distances are not accurate. The parameters α and β are the weights of these two distances.

To perform this experiment we have used three different datasets, these are Texture images from the Corel stock photography collection[1]: Textures (137000), Various Textures I (593000) and Textures II (404000). In the experiment we refer to them as *Corel*, *Corel1* and *Corel2* respectively. Each Corel group has 100 textures (768 x 512 pixels) and every texture is divided into 6 subimages, then the total number of images is $6x100 = 600$ for each Corel dataset. In figure 4 we show some textures of the three Corel datasets.

We use the Recall measure [10] to evaluate the performance of the retrieval and the precision-recall curves. The results have been computed by using all the images in each dataset as query images. In the ideal case of the retrieval, the top 6 retrieved images would be from the same original texture.

[1] Corel data are distributed through http://www.emsps.com/photocd/corelcds.htm

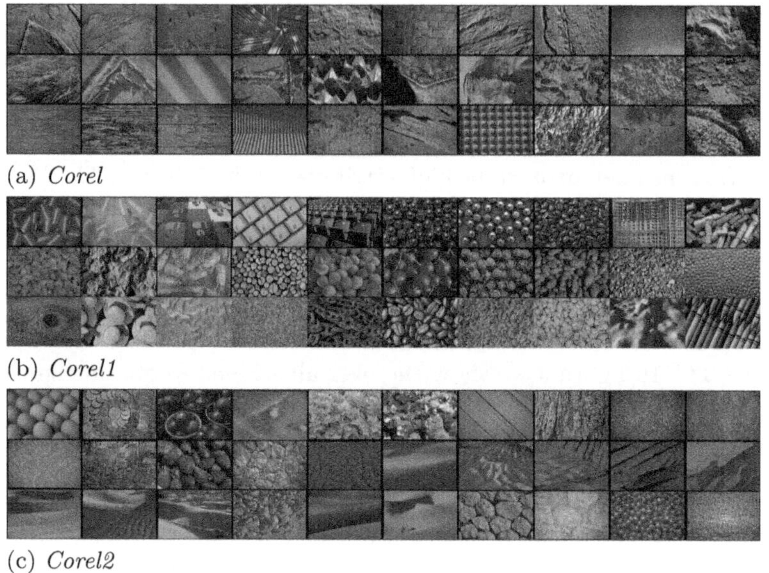

(a) *Corel*

(b) *Corel1*

(c) *Corel2*

Fig. 4. Corel datasets

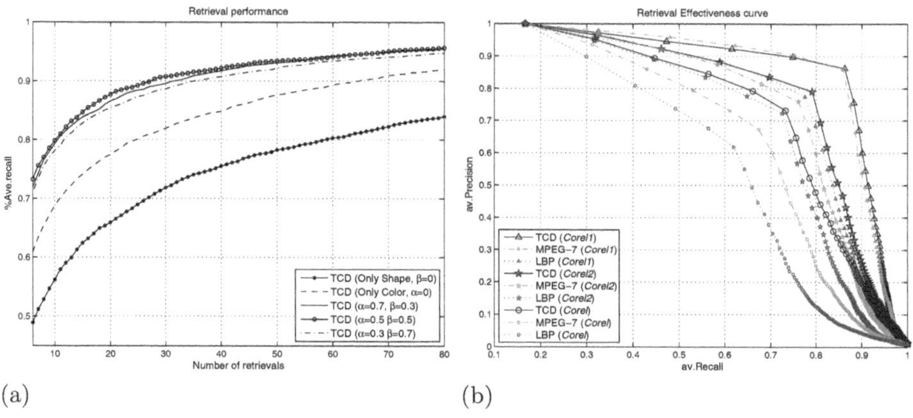

(a) (b)

Fig. 5. (a) Retrieval performance of TCD with different weights on the *Corel* dataset. (b) Precision-Recall curves of TCD, MPEG-7 and LBP descriptors for different datasets.

We find that using similar weights in the combination of shape and color descriptors to compute the distance (α, β in equation 7) do not have a relevant influence on the average recall measure. This is because color and texture information are already integrated at the blob level, before building the descriptor *TCD*. This fact is illustrated in Fig.5.(a) for *Corel* datasets. Best results in all

Table 1. Average Recall Rates

Descriptor	*Corel*	*Corel1*	*Corel2*
TCD	**73.25%**	**86.25%**	**79.11%**
MPEG-7 (SCD+HTD)	67.33%	85.94%	76.11%
$LBP_{8,1}RGB$	61.89%	77.53%	72.5%
TCD(Only Color)	60.89%	78.56%	69.25%
TCD(Only Shape)	48.92%	49.33%	49.33%
MPEG-7 (SCD)	48.5%	64.56%	61.58%
MPEG-7 (HTD)	55.56%	74.22%	63.64%

datasets are obtained when both color and shape are combined (when using only color or shape the average rate substantially decreases).

For comparing purposes, in table 1 we show the Recall rates for the 3 datasets using our *TCD* and two different descriptors that combine color and texture in different ways. These two descriptors are the standard MPEG-7 descriptors [7] (HTD and SCD as they are combined in [2]) and the color extension of the LBP descriptor proposed in [6]. The computed Average Retrieval rate shows how our *TCD* overcomes both the $LBP_{8,1}RGB$ descriptor and the MPEG-7 descriptors for the three Corel datasets. The LBP parameters we have chosen are those that produce the best results over the Corel datasets. In Fig.5.(b) there are the precision-recall curves that confirms the previous results. The last four rows of Table 1 show the retrieval rates using either color or texture for *TCD* and MPEG-7 descriptors respectively, showing the contribution of each separate feature on the discrimination experiment.

The best results of the *TCD* are achieved with *Corel1* dataset because it has more homogeneous textures than the other Corel datasets. That is, any subimage of the given texture preserves the same appearance of the texture. The *TCD* is a good descriptor to model the repetitive properties of textures.

5 Conclusions

This paper proposes a perceptual integration of color and texture in an unified descriptor. To this end, we propose a computation procedure to implement the original definition of texton given in the Julesz's perceptual theory [4]. It is done by using two spaces to represent shape and color attributes of the image blobs. Both spaces show two important properties, they are low-dimensional and have perceptual transformations over the axes in order to easily derive similarities from distances.

Although blobs are initially computed separately in the channels of an opponent color space, they are fused with a winner-take-all mechanism over different spatially coincident responses. The shape attributes of these winner blobs (aspect-ratio, area and orientation) that we call *perceptual blobs* are uniformly transformed in the shape space and their median color is represented in a perceptual HSI space. Similarities in these two spaces are combined a posteriori to obtain a final similarity between blobs which is the input of a clustering algorithm. Clusters of blobs are coping with the inherent repetitive property of the image texture. Therefore, the fusion of texture and color is done at the level of their attributes independently of their spatial location.

By combining previous spaces we propose a high level color-texture descriptor, the Texture Component Descriptor (TCD), that arises from the decomposition of the image in its *textural components*, which are the clusters of the blob attributes. Each cluster is defined by a 6-dimensional vector and our TCD will be a list of these vectors, depending on the inherent complexity of the texture. To sum up, the TCD is compact, low-dimensional and it inherits the semantic derived from the blob attributes.

In order to test the efficiency of the proposed descriptor we have performed a retrieval experiment on a highly diverse dataset of Corel Texture images. We compared our descriptor with a late combination of two MPEG-7 descriptors [7] (HTD and SCD) and an early combinations with the LBP RGB [6] descriptor in a retrieval experiment. Our descriptor overcomes both in the three Corel datasets of textures analysed.

Acknowledgments

This work has been partially supported by projects TIN2007-64577 and Consolider-Ingenio 2010 CDS2007-35100018 of Spanish MEC (Ministry of Science).

References

1. Carron, T., Lambert, P.: Color Edge Detector Using Jointly Hue, Saturation and Intensity. Int. Conference on Image Processing 3, 977–981 (1994)
2. Dorairaj, R., Namuduri, K.R.: Compact combination of MPEG-7 color and texture descriptors for image retrieval. In: Conference on Signals, Systems and Computers. Conference Record of the Thirty-Eighth Asilomar, vol. 1, pp. 387–391 (2004)
3. Flickner, M., Sawhney, H., Niblack, W., Ashley, J., Huang, Q., Dom, B., Gorkani, M., Hafner, M.J., Lee, D., Petkovic, D., Steele, D., Yanker, P.: Query by Image and Video Content: The QBIC System. Computer 28(9), 23–32 (1995)
4. Julesz, B., Bergen, J.R.: Textons, the fundamental elements in preattentive vision and perception of textures. Bell. Syst., Tech. Journal 62(6), 1619–1645 (1983)
5. Lindeberg, T.: Scale-space theory in computer vision. Kluwer Academic Publishers, Dordrecht (1994)
6. Maënpää, T., Pietikäinen, M.: Classification with color and texture: jointly or separately? Pattern Recognition 37, 1629–1640 (2004)
7. Manjunath, B.S., Salembier, P., Sikora, T.: Introduction to MPEG-7. John Wiley & Sons, Chichester (2003)

8. Rubner, Y., Tomasi, C., Leonidas, J.G.: The earth mover's distance as a metric for image retrieval. Int. Journal of Computer Vision 40(2), 99–121 (2000)
9. Shi, J., Malik, J.: Normalized cuts and image segmentation. IEEE Trans. on PAMI 22(8), 888–905 (2000)
10. Smith, J.R.: Image Retrieval Evaluation. In: Proc. IEEE Workshop on Content - Based Access of Image and Video Libraries, pp. 112–113 (1998)
11. Yu, H., Li, M., Zhang, H.J., Feng, J.: Color Texture Moments for Content-Based Image Retrieval. In: International Conference on Image Processing, pp. 24–28 (2003)
12. Zhong, Y., Jain, A.: Object localization using color, texture and shape. Pattern Recognition 33(4), 671–684 (2000)

A Geometric Data Structure Applicable to Image Mining and Retrieval

T. Iwaszko, M. Melkemi, and L. Idoumghar

Université de Haute Alsace,
LMIA-MAGE, 4 Rue des Frères Lumière,
68093 Mulhouse, France
{thomas.iwaszko,mahmoud.melkemi,lhassane.idoumghar}@uha.fr

Abstract. Due to improvements in image acquisition and storage technology, terabyte-sized databases of images are nowadays common. This abundance of data leads us to two basic problems: how to exploit images (image mining)? Or how to make it accessible to human beings (image retrieval)? The specificity of image mining/retrieval among other similar topics (object recognition, machine vision, computer vision, etc.) is precisely that their techniques operate on the whole collection of images, not a single one. Under these circumstances, it is obvious that the time complexity of related algorithms plays an important role. In this paper, we suggest a novel general approach applicable to image mining and retrieval, using only compact geometric structures which can be pre-computed from a database.

1 Introduction

In recent years, the amount of "non-standard" or multimedia data (in contrast with standard alphanumeric data) has greatly increased. Terabyte-sized databases of images are now available for various purposes: medicine, astronomy, physics, etc. but also digital photography: monitoring, online photo albums or entertainment.

In general, the problem of extracting implicit relevant information has already been studied for decades by researchers from *data mining*. However, and as described in [1], data mining techniques are not sufficient or fully appropriate for *image* databases. Singularities of information in images make necessary the design specific techniques and tools. These are being developed in the young area of *image mining* [2,3].

The other way to deal with such large collections of images is to make them easily accessible to human beings. To tackle this problem of *image retrieval*, one must provide a user interface to make the collection browsable, and a relevant method for specifying search queries. In addition, the image collection has to be ordered (*indexing*) in such way the system can quickly compute database matches with user queries. Finally, the user should be able to give feedback on the relevance of the results so the searching engine can possibly improve its

A. Campilho and M. Kamel (Eds.): ICIAR 2010, Part I, LNCS 6111, pp. 364–373, 2010.

performance aftewards. Content-Based Image Retrieval systems [4,5] i.e. CBIR systems are the realization of these ideas.

Image mining and image retrieval share the fact they both operate on whole collections of images, in contrast to fields of *object recognition, machine vision, computer vision, etc.* which analyse a single image, try to recognize a single scene. Consequently, the time complexity of image mining/retrieval-related algorithms must be taken into account, as well as the size of intermediate representation.

Until now, many indexing techniques have been reported in the literature [6,7]. Thanks to an indexing schema, it is possible to filter the complete list of elements in the database, in order to reduce the actual number of considered images.

In this paper, we suggest a general approach for working with image collections, which can be seen as an alternative or a complement to indexing. It offers the possibility to reduce the dataset by converting images to sets of points. This compact representation along with a pre-computation step might speed-up detection of spatial patterns in image mining or retrieval techniques. We present a geometric data structure, variant of the Voronoi diagram, for recording in advance locations of empty shapes (i.e. spatial patterns) and thus saving time on later treatments.

The next section deals with the "feature extraction step" for converting raw image data to geometric data. Section 3 introduces the notations used throughout the paper. A particular shape representation is presented in section 4. Then, in section 5, we define the new geometric structure based on shapes. We outline computation and present some results in section 6. Finally, we conclude by describing future work and other possible applications in the last section.

2 Image Analysis and Computational Geometry

Recently in image analysis, some research has been done in order to detect so-called *interest points* in images. Interest points are sometimes called Spatial Interest Pixels, [8]. Intuitively, an interest point corresponds to a pixel that has stronger interest in strength than most of pixels in an image. Interest point detection is often a particular form of edge/corner detection, but it can also concern search methods in the color space [9]. For a list and evaluation of interest point detectors, see [10]. This method constitutes a fast pre-processing step and allows to work on more compact representations. A remarquable advantage is that it can be combined easily with other tools of image analysis (histograms for instance, as demonstrated in [11]) or computer science.

In connection with point sets, computational geometry (or CG) is a field devoted to the study of algorithms which can be stated in terms of geometry. The algorithms and data structures (e.g. the Voronoi decomposition) of CG are designed for efficiency and have found numerous applications in various fields of computer science, in particular: image processing [12], analysis, indexing [9], retrieval [4].

In this paper, we present a novel data structure, based on points and geometric shapes, suitable for image mining/retrieval tasks. This structure works with specified models of shape, but these models could be learned on labelled images as well.

Let us first consider the abstract problem and the structure in general before showing its application to the field of Image Analysis. Let S be a set of points in the plane. Being given a plane geometric shape, (that is, an open bounded region of $\mathbb{R}^2$) is it possible to translate and rescale the shape in such way it has at least one point of S on its boundary, while remaining empty? More generally what is the set of solutions to the problem? i.e. how to locate all the free spaces for fitting a particular shape into the set of points? The shape representation and geometric structure introduced in the following bring an answer to this question. The structure can be seen as a variant of the Voronoi diagram.

3 Notations and Basic Terminology

In this work, we use the following notations:

- pq: the euclidean distance between p and q
- $[pq]$: the segment of extremities p and q
- $\boldsymbol{b}(x,r)$: the *open disc* of radius r centered on x. Its boundary is a *circle*, we note it $\partial\boldsymbol{b}(x,r)$. Mathematically:

$$\boldsymbol{b}(x,r) = \{y \in \mathbb{R} \mid yx < r\}$$
$$\partial\boldsymbol{b}(x,r) = \{y \in \mathbb{R} \mid yx = r\}$$

- $R(p)$: the Voronoi region of a point p of S, S being a given set of points. Mathematically, we have:

$$R(p) = \{x \in \mathbb{R}^2 \mid px \leq qx, \forall q \in S\}$$

- $L = (l_1, \ldots, l_n)$ denotes a n-tuple (i.e. a sequence) of objects, where l_1 is the first one, l_2 the second one, etc.

Moreover, we introduce the following terms:

- Given a set of points S, an open bounded region A is said to be *empty* if and only if $A \cap S = \emptyset$.
- We call *weighted point* a pair constituted by a point and a real positive number, formally: $w = (p,r)$ where $p \in \mathbb{R}^2, r \in \mathbb{R}$ and $r > 0$.

4 Shape Representation

For convenience, in this section we shall abuse language slightly: Given a tuple T of weighted points, *a disc of* T refers to an open disc $b(c,r)$ where (c,r) is an object of T.

Definition 1 requires two preliminary concepts introduced below. Given $T = ((c_1,r_1),(c_2,r_2),\ldots,(c_m,r_m))$ a m-tuple of weighted points, it is accepted that:

- Two discs $b(c_i, r_i)$, $b(c_j, r_j)$ of T are said to be **adjacent discs** iff: the smallest has its center on the boundary of the biggest, i.e.

$$\begin{cases} c_j \in \partial b(c_i, r_i) \text{ if } r_j \leq r_i \\ c_i \in \partial b(c_j, r_j) \text{ otherwise} \end{cases}$$

- Let $V = \{c_1, \ldots, c_m\}$ be the set of all the points listed in T. Let E be the set of segments $[c_i c_j]$ such that the discs $b(c_i, r_i)$ and $b(c_j, r_j)$ of T are adjacent discs. The resulting straight-line graph (V, E) is **the adjacency graph** of T. An example of an adjacency graph is shown on Fig. 1.

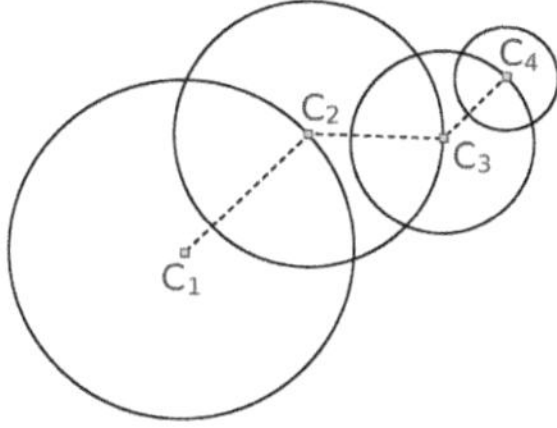

Fig. 1. Representation of a shape-parameter list and its adjacency graph (dashed line)

Definition 1. *A* ***shape-parameter list*** $\mathcal{C} = \big((c_1, r_1), (c_2, r_2), \ldots, (c_m, r_m)\big)$ *is an m-tuple of weighted points which satisfies the three conditions:*

1. *The two first discs of $\mathcal{C}$ are adjacent discs*
2. *The adjacency graph of $\mathcal{C}$ is connected*
3. *Let $\mathcal{C}'$ be the list of the k first discs of $\mathcal{C}$, where $1 < k < m$. The adjacency graph of $\mathcal{C}'$ is also connected.*

Definition 2. *Given a shape-parameter list $\mathcal{C} = ((c_1, r_1), \ldots, (c_m, r_m))$, we define the* ***shape-model*** $\mathfrak{p}_m(\mathcal{C})$ *as being the open bounded region obtained by the union of all the discs of $\mathcal{C}$:*

$$\mathfrak{p}_m(\mathcal{C}) = \bigcup_{i=1}^{m} b(c_i, r_i)$$

The centers of the two first discs of $\mathcal{C}$ (i.e. the points c_1, c_2) are called ***reference points*** *of the shape-model $\mathfrak{p}_m(\mathcal{C})$.*

An example of a simple shape-model is shown on Fig. 2a (its shape-parameter list is represented on Fig. 2b).

As we will see in section 6, that representation allows us to create complex shape-models and even good approximations of real objects silhouettes. Some examples are given on Fig. 3.

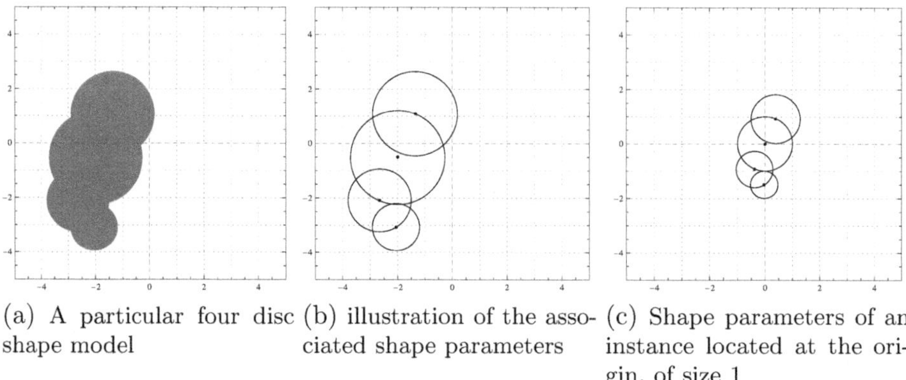

(a) A particular four disc shape model (b) illustration of the associated shape parameters (c) Shape parameters of an instance located at the origin, of size 1

Fig. 2. Concepts for shape representation

Fig. 3. Sophisticated shape-models, approximating real-world 2D pictures

Definition 3. *Given a shape-model $\mathfrak{p}_m(\mathcal{C})$, we call* ***instance of*** $\mathfrak{p}_m(\mathcal{C})$ ***located at*** x ***and of size*** λ *the region of the plane defined by:*

$$\mathfrak{p}_m(\mathcal{C}, x, \lambda) = \bigcup_{i=1}^{m} b(c'_i, r'_i)$$

Where c'_i, r'_i are given by:

- $c'_1 = x$, *and* $r'_1 = \lambda$
- $c'_i = x + \alpha(c_i - c_1)$, *and* $r'_i = \alpha r_i$ *for* $2 \leq i \leq m$
- α *is the rescaling factor:* $\alpha = \frac{\lambda}{r_1}$, *and* λ *the resulting size.*

For short, in the following we shall use the term ***instance*** *for refering to: instance of a shape-model located at a certain point and of a new size.*

An illustration of this concept is shown on Fig. 2c.

Interpretation: Given a shape-model $M_1 = \mathfrak{p}_m(\mathcal{C})$, its instance is a new shape-model $M_2 = \mathfrak{p}_m(\mathcal{C}, x, \lambda)$, which has x and $x+\alpha(c_2-c_1)$ for reference points while being similar to M_1 (similar: there exists an affine transformation that takes M_2 to M_1).

Interpretation of the Calculi: The coordinates c'_i and numbers r'_i are defined in order to perform an affine transformation which combines: translation and homothety (no rotation). This transformation is fully parameterized by x, λ.

5 Regions of Expanded Empty Shape-Models

We have defined shape-models precisely. Thanks to this preliminary work, given a shape-model, new instances can be computed. An instance is parameterized by a point and a real number, and the original shape parameters are known. Thus all the geometric information (boundary, disc overlap, etc.) is computable. We can determine wether a particular instance is empty or not, has a point on its boundary or not, etc.

Definition 4. *Given S a set of points of the plane, and $\mathfrak{p}_m(\mathcal{C})$ a m-disc shape-model, we call* ***region of expanded empty shape-model*** *associated to $p \in S$ (*REES *of $p \in S$ for short) the region defined such that:*

$$R_{\mathcal{C}}(p) = \left\{x \in \mathbb{R}^2 \mid \mathfrak{p}_m(\mathcal{C}, x, px) \cap S = \emptyset\right\}$$

Fig. 4 illustrates this definition with few simple shape-models and associated regions.

Intuitively, $R_{\mathcal{C}}(p)$ represents the locations $x \in \mathbb{R}$ where the shape-model $\mathfrak{p}(\mathcal{C})$ can be translated (ie. its first center becomes x) *and expanded until it has p on its boundary*, while remaining empty of S.

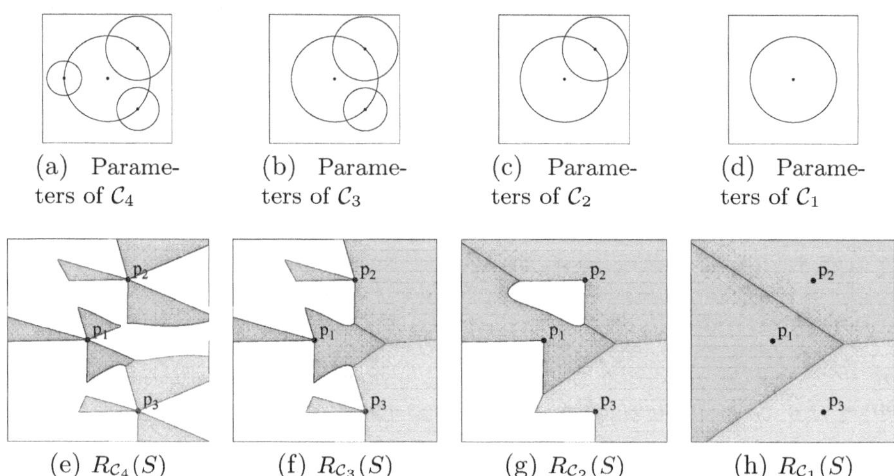

(a) Parameters of $\mathcal{C}_4$ (b) Parameters of $\mathcal{C}_3$ (c) Parameters of $\mathcal{C}_2$ (d) Parameters of $\mathcal{C}_1$

(e) $R_{\mathcal{C}_4}(S)$ (f) $R_{\mathcal{C}_3}(S)$ (g) $R_{\mathcal{C}_2}(S)$ (h) $R_{\mathcal{C}_1}(S)$

Fig. 4. Trivial shape-models and associated regions for a 3-point set. The regions are the set of points where the shape-model can be translated then rescaled until it has a site on its boundary, while remaining empty.

6 Practical Results

Shape-models as presented in this paper can be build from 2d silhouettes as shown on Fig. 5. Actually, that disc-based shape representation has already been introduced in previous work. Despite of the fact the representation has been slightly modified and reformulated since then, the construction of approximations from silhouette remains identical. For details on this process, see [13].

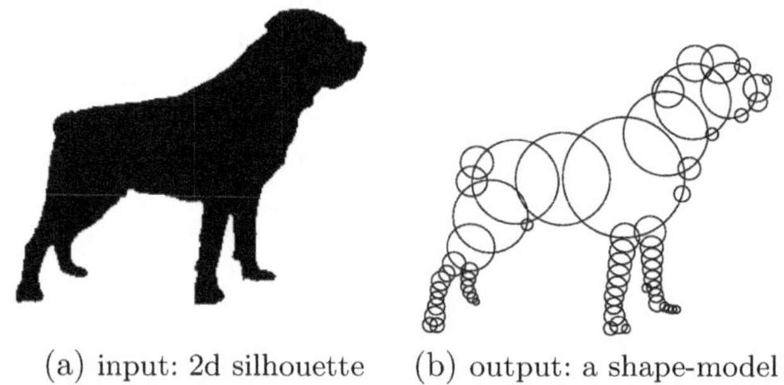

(a) input: 2d silhouette (b) output: a shape-model

Fig. 5. Building process of a shape-model for approximating a given silhouette as well as possible (notably using the so-called medial axis or topological skeleton)

In the section 2, several methods for interest points detection, along with the possibility to used it as a pre-processing and combine it with other tools, were mentionned.

Accordingly, we can use our geometric structure with images. After choosing a suitable method for interest points detection ('suitable" would be application-specific), a whole image collection used for data mining/retrieval can be converted in advance into a collection of point-sets.

Firstly we have implemented the well-known Harris detector [14] and ran the algorithm on grayscale images in order to find mostly spatial pixels of interest. Some results are shown on Fig. 6. Note that by pre-computing point sets for each image in a database, it is possible to save space and the structure proposed previously can still work on these points sets for detecting spatial pattern (empty of points).

Fig. 6. Computation of pixels of interest using the Harris detector on grayscale images, in order to convert images to simple point sets

Having computed both point sets and shape-models from silhouettes, the computation of REES can be made. The definition mentionned in previous sections being equivalent to a system of inequalities (to test the emptiness of the shape is to test the distances between centers of discs constituting the shape-model and points/sites), the calculation boils down to approximating each region using linear algebra and algebraic methods, like the resultants. Thus each elements of the system is simply considered as the part of a two variable polynomial.

The result of such regions computation is shown on Fig. 7.

For application to CBIR systems, geometric models could be specified in advance and the user would select one and indicate its approximate location in the picture he is searching. We have chosen this scenario but other possibilities are offered by the geometric structure. Two facts are worth of interest with this approach:

- The computation of regions is still a part of the pre-computation (before any user query)
- The classic matching step is replaced by a simple point in polygon algorithm (REES being approximated by polygons).

Therefore good performances are expected but this cannot be strictly demonstrated yet.

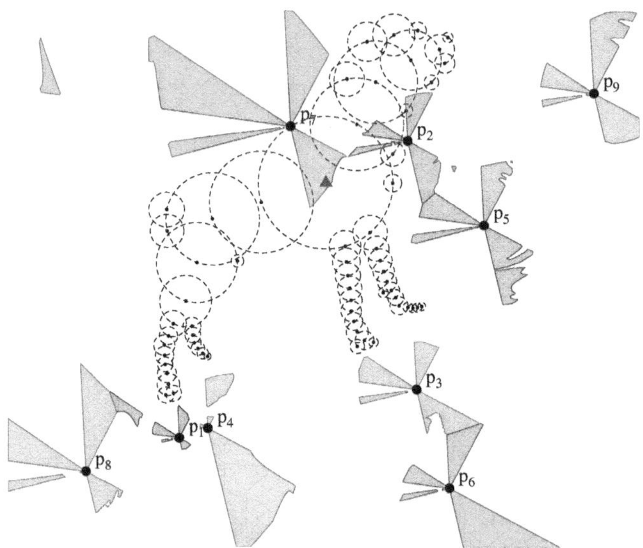

Fig. 7. REES and an empty instance. The triangle represents the location of the instance's reference point. According to the definition, it is know the whole shape-model is empty of points iff its reference point is located inside a region $R_C(p)$.

Similarly, at this point of our work it was not possible to make comparisons with other CBIR methods as a whole framework integrating all steps is required (image to point sets conversion, shape-models construction and computation of regions). However in this section we have presented meaningful results already obtained for each separate step.

7 Conclusion

This paper presented a theoretical structure and explained in what way it could be used for image mining and retrieval. The actual computation of this structure relies on algebraic calculus. Indeed, the introduced REES can be decomposed into simpler regions (just like the Voronoi diagram can be decomposed in halfplanes). Each simpler subregion can then be expressed with an inequation. All in all, the computation boils down to algebraic system of inequations solving.

Currently we use the computational software *Mathematica* for this task. Our implementation produced the illustations presented in this paper and scales up well, up to hundreds of discs and points.

If fully described and developed, this new structure might have numerous applications, like the classic Voronoi diagram, because problems involving proximity informations are general, and found in many of areas of science. New shape-based algorithms for image mining and retrieval could be designed. The presented structure could also be worth of interest for robotics path planning problems.

As future work, we aim at: achieving the REES computation and coding in a stand-alone application. It would let us studying precise time complexity for both region construction and point query. The most interesting future prospect is the setting of a whole CBIR framework (using together all the steps presented in the previous section) in order to test the structure in practice and notably compare it to existing techniques.

References

1. Hsu, W., Lee, M.L., Zhang, J.: Image mining: Trends and developments. J. Intell. Inf. Syst. 19(1), 7–23 (2002)
2. Zhang, J., Hsu, W., Lee, M.-L.: An information-driven framework for image mining. In: Mayr, H.C., Lazanský, J., Quirchmayr, G., Vogel, P. (eds.) DEXA 2001. LNCS, vol. 2113, pp. 232–242. Springer, Heidelberg (2001)
3. Burl, M.C., Fowlkes, C., Roden, J.: Mining for image content. In: Systemics, Cybernetics, and Informatics/Information Systems: Analysis and Synthesis (1999)
4. Veltkamp, R.C., Tanase, M., Sent, D.: Features in content-based image retrieval systems: a survey. In: State-of-the-Art in CBIR, pp. 97–124. Kluwer, Dordrecht (2001)
5. Datta, R., Li, J., Wang, J.Z.: Content-based image retrieval: approaches and trends of the new age. In: MIR 2005: Proceedings of the 7th ACM SIGMM International Workshop on Multimedia Information Retrieval, pp. 253–262. ACM, New York (2005)
6. Chiueh, T.-c.: Content-based image indexing. In: VLDB 1994: Proceedings of the 20th International Conference on Very Large Data Bases, pp. 582–593. Morgan Kaufmann Publishers Inc., San Francisco (1994)
7. Doermann, D.S.: The indexing and retrieval of document images: A survey. CVIU 70(3), 287–298 (1998)
8. Li, Q., Ye, J., Kambhamettu, C.: Spatial interest pixels (SIPs): useful low-level features of visual media data. Multimedia Tools Appl. 30(1), 89–108 (2006)
9. Tao, Y., Grosky, W.I.: Spatial color indexing: A novel approach for content-based image retrieval. In: ICMCS 1999, pp. 530–535. IEEE Computer Society, Los Alamitos (1999)
10. Schmid, C., Mohr, R., Bauckhage, C.: Evaluation of interest point detectors. Int. J. Comput. Vision 37(2), 151–172 (2000)
11. Meng, F.-j., Guo, B.-l., Guo, L.: Image retrieval based on 2d histogram of interest points. In: IAS 2009: Proceedings of the 2009 Fifth International Conference on Information Assurance and Security, Washington, DC, USA, pp. 250–253. IEEE Computer Society, Los Alamitos (2009)
12. Du, Q., Gunzburger, M., Ju, L., Wang, X.: Centroidal voronoi tesselation algorithms for image compression, segmentation, and multichannel restoration. J. Math. Imaging Vis. 24(2), 177–194 (2006)
13. Idoumghar, L., Melkemi, M.: Pattern retrieval from a cloud of points using geometric concepts. In: Kamel, M.S., Campilho, A. (eds.) ICIAR 2007. LNCS, vol. 4633, pp. 460–468. Springer, Heidelberg (2007)
14. Harris, C., Stephens, M.: A combined corner and edge detection. In: Proceedings of the 4th Alvey Vision Conference, pp. 147–151 (1988)

SIA: Semantic Image Annotation Using Ontologies and Image Content Analysis

Pyrros Koletsis and Euripides G.M. Petrakis

Department of Electronic and Computer Engineering
Technical University of Crete (TUC)
Chania, Greece
{pierce,petrakis}@intelligence.tuc.gr

Abstract. We introduce SIA, a framework for annotating images automatically using ontologies. An ontology is constructed holding characteristics from multiple information sources including text descriptions and low-level image features. Image annotation is implemented as a retrieval process by comparing an input (query) image with representative images of all classes. Handling uncertainty in class descriptions is a distinctive feature of SIA. Average Retrieval Rank (AVR) is applied to compute the likelihood of the input image to belong to each one of the ontology classes. Evaluation results of the method are realized using images of 30 dog breeds collected from the Web. The results demonstrated that almost 89% of the test images are correctly annotated (i.e., the method identified their class correctly).

1 Introduction

Image annotation is the process of assigning a class or description to an unknown image. The goal of automatic image annotation in particular is to produce coherent image descriptions which are as good as human authored annotations. This will not only permit faster and better understanding of the contents of image collections but also, can be viewed as a tool for enhancing the performance of image retrievals by content.

In large image collections and the Web [1] images are typically indexed or retrieved by keywords or text descriptions which are automatically extracted or assigned to them manually by human experts. This approach has been adopted by general purpose image search engines such as Google Image Search[1] as well as by systems providing specific services to users ranging from simple photo sharing in the spirit of Flickr[2] to unauthorized use of images and licensing in the spirit of Corbis[3].

Image annotations are compact consisting of a few meaningful words of phrases summarizing image contents. Human-based image annotation can lead

[1] http://images.google.com
[2] http://www.flickr.com
[3] http://www.corbisimages.com

A. Campilho and M. Kamel (Eds.): ICIAR 2010, Part I, LNCS 6111, pp. 374–383, 2010.

to more comprehensive image descriptions and allow for more effective Web browsing and retrieval. However, the effectiveness of annotations provided by humans for general purpose retrievals is questionable due to the specificity and subjectivity of image content interpretations. Also, image annotation by humans is slow and costly and therefore does not scale-up easily for the entire range of image types and for large data collections such the Web. A popular approach relates to extracting image annotations from text. This approach is particularly useful in applications where images co-exist with text. For example, images on the Web are described by surrounding text or attributes associated with images in `html` tags (e.g., filename, caption, alternate text etc.). Google Image Search is an example system of this category.

Overcoming problems related to uncertainty and scalability calls for automatic image annotation methods [2]. Automatic annotation is based on feature extraction and on associating low-level features (such as histograms, color, texture measurements, shape properties etc.) with semantic meanings (concepts) in an ontology [3,4,5]. Automatic annotation can be fast and cheap however, general purpose image analysis approaches for extracting meaningful and reliable descriptions for all image types are not yet available. An additional problem relates to imprecise mapping of image features to high level concepts, referred to as the "semantic gap" problem. To handle issues relating to domain dependence, diversity of image content and achieve high quality results, automatic image annotation methods need to be geared towards specific image types.

Recent examples of image annotation methods include work by Schreiber et.al. [4] who introduced a photo annotation ontology providing the description template for image annotation along with a domain specific ontology for animal images. Their solution is not fully automatic, it is in fact a tool for assisting manual annotation and aims primarily at alleviating the burden of human annotators. Park et. al. [5] use MPEG-7 visual descriptors in conjunction with domain ontologies. Annotation in this case is based on semantic inference rules. Along the same lines, Mezaris et.al. [6] focus on object ontologies (i.e., ontologies defined for image regions or objects). Visual features of segmented regions are mapped to human-readable descriptor values (e.g., "small", "black" etc.). Lacking semantics, the above derived descriptors can't be easily associated with high-level ontology concepts. Also, the performance of the method is constraint by the performance of image segmentation.

SIA (Semantic Image Annotation) is motivated by these ideas and handles most of these issues. To deal with domain dependence of image feature extraction we choose the problem of annotating images of dog breeds as a case study for the evaluation of the proposed methodology. High-level concept descriptions together with low-level information are efficiently stored in an ontology model for animals (dog breeds). This ontology denotes concept descriptions, natural language (text) descriptions, possible associations between classes and associations between image classes and class properties (e.g., part-of, functional associations). Descriptions in terms of low-level color and texture features are also assigned to each image class. These class descriptions are not fixed but are augmented with features pertaining to virtually any image variant of each particular class. Image annotation is implemented as a retrieval process. Average Retrieval Rank (AVR) [7] is used to compute the likelihood of the query image to belong to an ontology class. Evaluation results of the method are realized on images of 30 dog breeds collected from the Web. The results demonstrated that almost 89% of the test images are annotated correctly.

The method is discussed in detail in Sec. 2. The discussion includes SIA resources and processes in detail, namely the ontology, image analysis, image similarity and image annotation. Evaluation results are presented in Sec. 3 and the work is concluded in Sec. 4.

2 Proposed Method

SIA is a complete prototype system for image annotation. Given a query image as input, SIA computes its description consisting of a class name and the description of this class. This description may be augmented by class (ontology) properties depicting its shape, size, color, texture (e.g., "has long hair", "small size" etc.). The system consists of several modules. The most important of them are discussed in the following.

2.1 Ontology

The image ontology has two main components namely, the class hierarchy of the image domain and the descriptions hierarchy [5]. Various associations between concepts or features between the two parts are also defined:

Class Hierarchy: The class hierarchy of the image domain is generated based on the respective nouns hierarchy of Wordnet[4]. In this work, a class hierarchy for dog breeds is constructed (e.g., dog, working group, Alsatian). The leaf classes in the hierarchy represent the different semantic categories of the ontology (i.e., the dog breeds). Also a leaf class (i.e., a dog breed) may be represented by several image instances for handling variations in scaling and posing. For example, in SIA leaf class "Labrador" has 6 instances.

Descriptions hierarchy: Descriptions are distinguished into high-level and low-level descriptions. High-level descriptions are further divided into concept descriptions (corresponding to the "glosses" of Wordnet categories) and visual text descriptions (high-level narrative information). The later, are actually descriptions that humans would give to images and are further specialized based on animal shape and size properties (i.e., "small", "medium" and "big") respectively. The low-level descriptions hierarchy represents features extracted by 7 image descriptors (see Sec. 2.3). Because an image class is represented by more than one image instances (6 in this work), each class is represented by a set of 7 features for each image instance. An association between image instances and low-level features is also defined denoting the existence of such features (e.g., "hasColorLayout", "hasCEDD"). Fig. 1 illustrates part of the SIA ontology (not all classes and class properties are shown).

[4] http://wordnet.princeton.edu

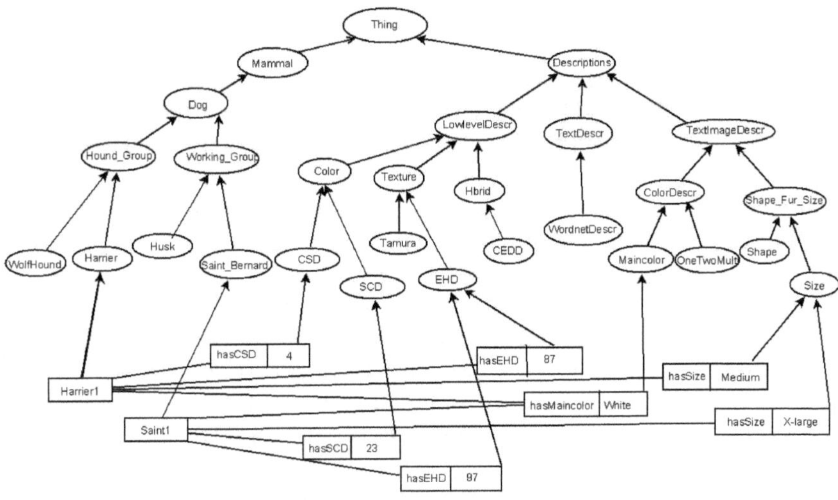

Fig. 1. Part of the SIA ontology

2.2 ROI Selection

The input image may contain several regions from which some may be more relevant to the application than others. In this work, dog's head is chosen as the most representative part of a dog image for further analysis. This task is implemented by manual Region of Interest (ROI) placement (the user drags a rectangle around a region) followed by background substraction by applying GrabCut [8] and noise reduction. Fig. 2 illustrates an original image and its corresponding ROI.

Fig. 2. Original image and Region Of Interest (ROI)

2.3 Image Feature Extraction

Automatic image annotation requires that content descriptions be extracted from images and used to represent image content. The focus of this work is not on novel image feature extraction but on showing how to enhance the accuracy of automatic annotation for a given and well established set of features. Images of dog breeds are mainly characterized by the spatial distribution of color intensities. This information is mostly captured by the following 7 descriptors (the first 4 descriptors are included in MPEG-7 [7]). The implementations are from LIRE [9].

Scalable Color Descriptor (SCD): This is a 256-bin color histogram in the HSV color space encoded by a Haar transformation. Histogram values are mapped to a 4-bit representation, giving higher significance to the small values with higher probability. The matching function is the L_1 metric.

Color Structure Descriptor (CSD): A color histogram in the HMMD color space that captures both color content and information about the structure of this content (position of color). First, a non-uniform quantification is applied on the HMMD color space resulting to an 256-bin histogram. Then, a 8x8 pixel structure element is applied on the image for counting the CSD bins for colors found in the respective location. Its purpose is to avoid the loss of structure information as in typical histograms. The matching function is the L_1 metric.

Color Layout Descriptor (CLD): Captures the spatial layout of the representative colors in an image. The image is partitioned into 8x8 blocks. For each block, representative colors are selected and expressed in YCbCr color space. DCT (Discrete Cosine Transform) is applied on each one of the three components (Y, Cb and Cr). The resulting DCT coefficients are zigzag-scanned and the first few coefficients are non-linearly quantized to form the descriptor. The default matching function is a weighted sum of squared differences between the corresponding descriptor components (Y, Cb and Cr).

Edge Histograms Descriptor (EHD): Represents the spatial distribution of edges in an image. A gray-intensity image is divided in 4×4 regions. A 5-bin histogram is computed to each region. These 5 bins correspond to the 5 edge types: vertical, horizontal, 45°diagonal, 135°diagonal, and isotropic. The final histogram contains a total of 80 bins (16 regions times 5 bins each). The matching function is the L_1 metric.

Color and Edge Directivity Descriptor (CEDD): A hybrid feature combing color and texture information in one histogram with 144 bins. The histogram is a result of a fuzzy system providing information about color in the HSV color space, and a second fuzzy system providing information about 5 types of edges in the same spirit as EHD. Matching is based on the Tanimoto coefficient.

Fuzzy Color and Texture Histogram (FCTH): Similar to CEDD but despite CEDD it applies texture information extraction and results in a histogram with 192 bins. Matching is based on the Tanimoto coefficient.

Tamura Descriptor: This is a vector of 6 features representing texture (coarseness, contrast, directionality, line-likeness, regularity, roughness). The matching function is the L_1 metric.

2.4 Image Retrieval

Given a query image, the problem of image annotation is transformed into an image retrieval one. The input image is compared with the representative images of each class. The SIA ontology holds information for 30 classes (dog breeds) and each class is represented by 6 instances. Therefore, the query is compared with

180 images. The output consists of the same 180 images ordered by similarity with the query. Image similarity between any two images A and B is computed as a weighted sum of differences on all features:

$$D(A,B) = \sum_{i=1}^{7} w_i d_i(A,B), \tag{1}$$

where i indexes features from 1 through 7, $d_i(A,B)$ is the distance between the two images for feature i and w_i represents the relative importance of feature i. All distances $d_i(A,B)$ are normalized in $[0,1]$ by Gaussian normalization

$$d_i(A,B) = \frac{1}{2}\left(1 + \frac{d_i(A,B) - \mu}{3\sigma}\right), \tag{2}$$

where μ is the mean value computed over all $d_i(A,B)$ and σ is the standard deviation. The advantage of Gaussian normalization is that the presence of a few large or small values does not bias the importance of a feature in computing the similarity.

Notice that not all features are equally important. Instead of manually selecting weights this is left to machine learning to decide algorithmically. Appropriate weights for all features are computed by a decision tree: The training set consists of 1,415 image pairs collected from the Web (559 pairs of similar images and 856 pairs of dissimilar images). For each image pair a 6-dimensional vector is formed. The attributes of this vector are computed as the Gaussian normalized feature distances. The decision tree accepts pairs of images and classifies them into similar or not (i.e., a yes/no answer). The decision tree was pruned with confidence value 0.1 and achieved 80.15% classification accuracy. The evaluation method is stratified cross validation. Appropriate weights are computed from the decision tree as follows:

$$w_i = \sum_{nodes\ of\ feature\ i} \frac{maxdepth + 1 - depth(feature_i)}{\sum_{j=1}^{all\ nodes} maxdepth + 1 - depth(node_j)}, \tag{3}$$

where i indexes features from 1 through 7, j indexes tree nodes ($node_j$ is the j-th node of the decision tree), $depth(feature_i)$ is the depth of feature i and $maxdepth$ is the maximum depth of the decision tree. The summation is taken over all nodes of feature i (there may exist more than nodes for feature i in the tree). This formula suggests that the higher a feature is in the decision tree and the more frequently it appears, the higher its weight will be.

2.5 Image Annotation

The input image is compared with the 180 ontology images (30 classes with 6 instances each) by applying Eq. 1. The answer is sorted by decreasing similarity. The class description of the input image can be computed by any of the following methods:

Best Match: Selects the class of the most similar instance.

Max Occurrence: Selects the class that has the maximum number of instances in the first n answers (in this work n is set to 15). If more than one classes have the same number of instances within the first n answers then Best Match is applied.

Average Retrieval Rank (AVR) [7]: Selects the description of the class with the higher AVR (Best Match is applied if more than one). Assuming that there are $NG(q)$ images similar to the input image q (ground truth) in the top n answers and $rank(i)$ is the rank of the i-th ground truth image in the results list, AVR is computed as:

$$AVR(q) = \sum_{i=1}^{NG(q)} \frac{rank(i)}{NG(q)} \tag{4}$$

2.6 Semantic Web - MPEG-7 Interoperability

SIA outputs annotation results in OWL[5], the description language of the Semantic Web. MPEG-7[6] provides a rich set of standardized tools to describe multimedia content and is often the preferred data format for accessing image and video content and descriptions (meta-data). To ensure interoperabiltiy between OWL and MPEG-7 applications, as a last (optional) step, SIA incorporates a two-way transformation between the two formats: Tsinaraki et. al. [10] proved that OWL ontologies can be transformed to MPEG-7 abstract semantic entity hierarchies. Fig. 3 illustrates that SIA image annotations can be described in either format and also shows the correspondences between the two representations. MPEG-7 annotations depict not only the class hierarchy that an image belongs to but

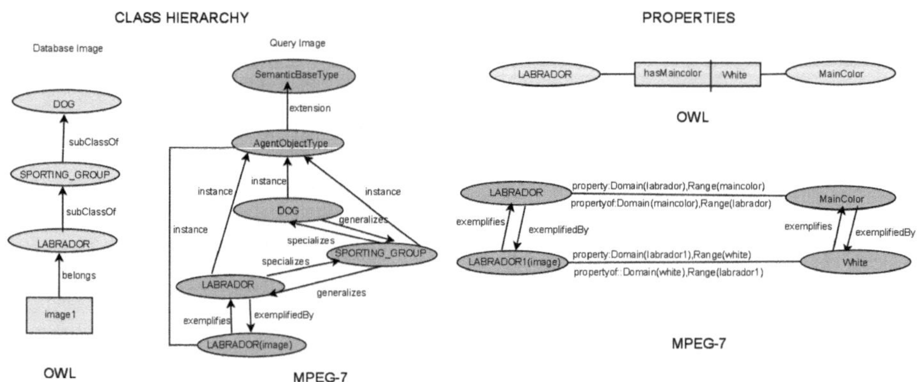

Fig. 3. Mapping OWL to MPEG-7

[5] http://www.w3.org/TR/owl-features

[6] http://www.chiariglione.org/mpeg/standards/mpeg-7/mpeg-7.htm

also, high level information obtained from object properties thus making the annotation the richest possible.

3 Experimental Evaluation

We conducted two different experiments. The purpose of the first experiment is to demonstrate that retrievals using the combination of descriptors in Eq. 1 indeed performs betters than any descriptor alone. Fig. 4 illustrates precision and recall values for retrievals using Eq. 1 and retrieval using each one of the 7 descriptors in Sec. 2.3. Each method is represented by a precision-recall curve. For the evaluations, 30 test images are used as queries and each one retrieves the best 15 answers (the precision/recall plot of each method contains exactly 15 points). The k-th (for $k = 1, \ldots 15$) point represents the average (over 30 queries) precision-recall for answer sets with the best k answers. A method is better than another if it achieves better precision and recall. Obviously, retrievals by Eq. 1 outperform retrievals by any individual descriptor alone. In addition, Eq. 1 with weights computed by machine learning achieves at least 15% better precision and 25% better recall than retrieval with equal weights.

The purpose of the second experiment is to demonstrate the annotation efficiency of SIA. All the 30 query images of the previous experiment are given as input to SIA. Table 1 illustrates the accuracy of the Best Match, Max. Occurrence and AVR methods of Sec. 2.5. All measurements are average over 30

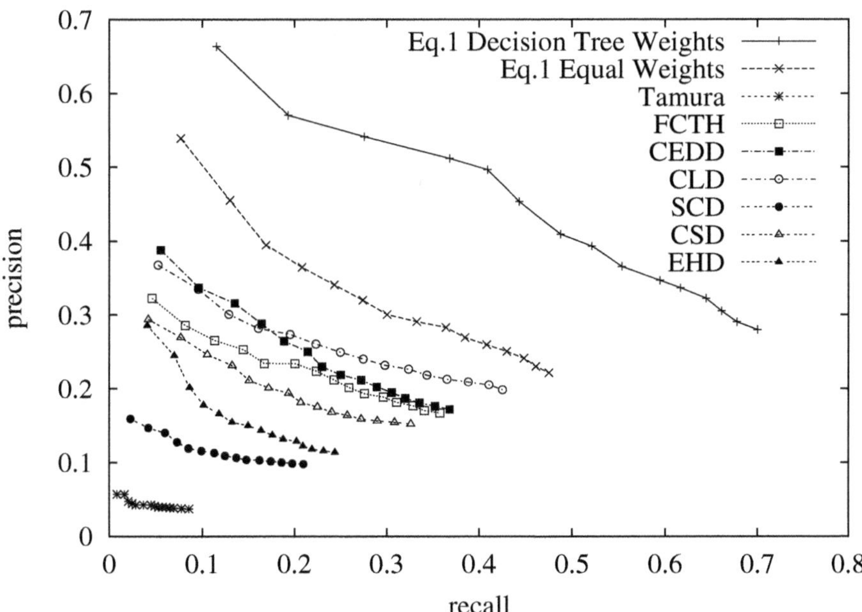

Fig. 4. Average precision and recall for retrievals in SIA

test images. The image ranked first has always higher probability of providing the correct annotation. There are cases where the correct annotation is provided by the image ranked second or third. AVR outperforms all other methods: The image ranked first in correctly annotated in 63% of the images tested. Overall, the correct annotation is provided by any of the top 3 ranked images in 89% of the images tested.

Table 1. Annotation results corresponding to Best Match, Max. occurrence and AVR

Annotation Result	**Best Match**	**Max. Occurrence**	**AVR**
Ranked 1st	53%	60%	63%
Ranked 2nd	10%	12%	20%
Ranked 3rd	7%	10%	6%

Fig. 5 illustrates the annotation for the image of a Collie (shown on the left). The images on its right are the 10 top ranked images by AVR (most of them are Collies).

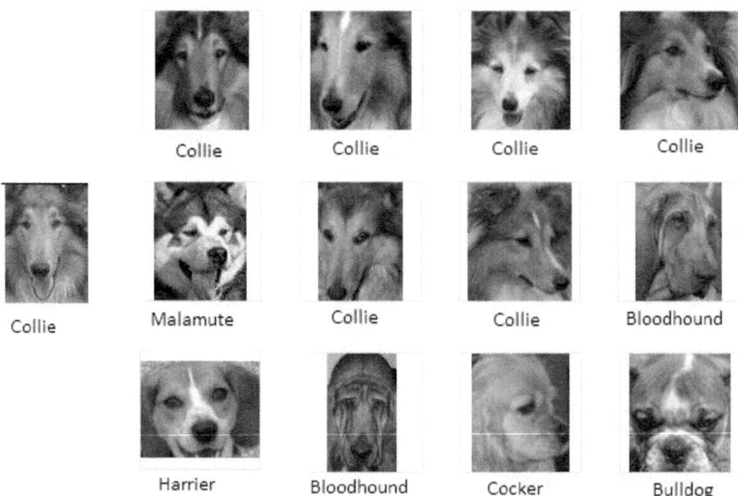

Fig. 5. Examples of annotation for test image "Collie"

4 Conclusion

We introduce SIA, a framework for annotating images automatically using information from ontologies and image analysis. Handling uncertainty in class descriptions is a distinctive feature of SIA and is achieved by combining information from multiple information sources for representing ontology classes. The results indicate that it is possible for the method to approximate algorithmically the human notion of image description reaching up to 89% accuracy (89% of the test images are correctly annotated). Extending SIA for handling more image categories and incorporating more elaborate image analysis (e.g., for handling different poses of animal heads) and classification methods are promising issues for further research.

Acknowledgements

We are grateful to Savvas A. Chatzichristofis of the Dept. of Electrical and Computer Engineering at the Democritus University of Thrace, Greece, for valuable contributions into this work.

References

1. Kherfi, M., Ziou, D., Bernardi, A.: Image Retrieval from the World Wide Web: Issues, Techniques, and Systems. ACM Computing Surveys 36(1), 35–67 (2004)
2. Hanbury, A.: A Aurvey of Methods for Image Annotation. Journal of Visual Languages and Computing 19(5), 617–627 (2008)
3. Jeon, J., Lavrenko, V., Manmatha, R.: Automatic Image Annotation and Retrieval using Cross-Media Relevance Models. In: Proc. of ACM SIGIR 2003, Toronto, CA, pp. 119–126 (July 2003)
4. Schreiber, A., Dubbeldam, B., Wielemaker, J., Wielinga, B.: Ontology-Based Photo Annotation. IEEE Intelligent Systems 16(3), 66–74 (2001)
5. Park, K.W., Jeong, J.W., Lee, D.H.: OLYBIA: Ontology-Based Automatic Image Annotation System Using Semantic Inference Rules. In: Kotagiri, R., Radha Krishna, P., Mohania, M., Nantajeewarawat, E. (eds.) DASFAA 2007. LNCS, vol. 4443, pp. 485–496. Springer, Heidelberg (2007)
6. Mezaris, V., Kompatsiaris, J.: MStrintzis: Region-Based Image Retrieval using an Object Ontology and Relevance Feedback. EURASIP Journal on Applied Signal Processing 2004(1), 886–901 (2004)
7. Manjunath, B., Ohm, J., Vasudevan, V., Yamada, A.: Color and Texture Descriptors. IEEE Trans. on Circuits and Systems for Video Technology 11(1), 703–715 (2001)
8. Rother, C., Kolmogorov, V., Blake, A.: GrabCut: Interactive Foreground Extraction using Iterated Graph Cuts. ACM Transactions on Graphics (TOG) 23(3), 309–314 (2004)
9. Lux, M., Chatzichristofis, S.: LIRE: Lucene Image Retrieval: An Extensible Java CBIR Library. In: Proc. of the 16th ACM Intern. Conf. on Multimedia (MM 2008), Vancuver, CA, pp. 1085–1088 (November 2008)
10. Tsinaraki, C., Polydoros, P., Christodoulakis, S.: Interoperability Support between MPEG-7/21 and OWL in DS-MIRF. IEEE Transactions on Knowledge and Data Engineering 19(2), 219–232 (2007)

Using the Fisher-Rao Metric to Compute Facial Similarity

Simone Ceolin* and Edwin R. Hancock **

Computer Vision and Pattern Recognition Group,
Department of Computer Science, University of York
York YO10-5DD, UK
{sceolin,erh}@cs.york.ac.uk

Abstract. In this paper we show how the Fisher-Rao metric can be used to compute the similarity of fields of surface normals, under the assumption of a von-Mises Fisher (vMF) distribution. We use the similarity measure to analyse differences in facial shape due to gender and expression. Finally, we show the results achieved using BU-3DFEDB and Max Planck datasets.

Keywords: Surface normals, Density model, Fisher-Rao metric.

1 Introduction

Over the past decade there has been a considerable growth in interest in the statistical theory of shape [4],[16]. This field of study has been the result of a synthesis of ideas from a number of different areas including statistics, computer vision, pattern recognition and machine learning, and the realization that the areas share a considerable common ground [2]. One recent and powerful development in this area has been to explore the use of techniques motivated by information theory, and in particular to use the Fisher-Rao metric to measure the similarities of statistical shape models and construct shape-spaces. In the literature Maybank [9] shows how to use Fisher information for line detection, Mio et al. [11] apply non-parametric Fisher-Rao metrics for image segmentation and Peter [14] has presented a unified framework for shape representation and deformation.

In this paper we are particularly interested in the use of these ideas to represent variations in facial shape, and to determine the modes of variation due to factors such as gender and expression. The reported work is motivated in part by the fact that faces have multiple shape properties, which can be used to categorize them according to different levels of specificity. Examples include gender, ethnicity, age, expression, identity, attractiveness and distinctiveness [20].

* Supported by the Programme Alban, the European Union Programme of High Level Scholarships for Latin America, scholarship no. E06D101062BR.

** This worked was supported by EU FET Project SIMBAD and the author was supported by Royal Society Wolfson Research Merit Award Holder.

A. Campilho and M. Kamel (Eds.): ICIAR 2010, Part I, LNCS 6111, pp. 384–393, 2010.

In particular, we are interested in how such shape variations manifest themselves in terms of changes in the field of surface normals. The reason for this is that we aim to fit statistical models of shape to 2D facial images, and from these images recover information concerning 3D shape. One natural way of doing this that captures features of the human vision system is to employ shape from shading to recover surface shape from variations in brightness. Here it is more natural to represent the facial surface using fields of surface normals rather than surface height information, since the former are more directly linked to the physical process of light reflectance.

As a result surface normal models are more suitable for the purposes of fitting to image data. However, due to their non-Cartesian nature the statistical modeling of variations in surface normal direction is more difficult than that for landmark positions. Fields of surface normals can be viewed as distributions of points residing on a unit sphere and may be specified in terms of the elevation and azimuth angles. It is natural to parameterise such statistical variations in direction sing the von-Mises Fisher (vMF) distribution, which is specified in terms of a mean surface normal direction and a concentration parameter. Our goal in this paper is to explore how to use the vMF distribution for shape representation, and in particular to recognise variations in facial shape due to expression and gender difference.

Working in the surface normal domain, we show how to use the vMF distribution to represent unstructured surface normal data without landmarks. To measure the similarity between two fields of surface normals parameterized using the vMF distribution we make use of the Fisher-Rao metric. In this way facial similarity is measured by the geodesic distance between the shapes on a statistical manifold.

The remainder of the paper is organized as follows. Section 2 describes how the Fisher-Rao metric can be used to measure the similarity of facial needle-maps. Section 3 discusses how multidimensional scaling can be used to embed faces into a low-dimensional pattern space based on the Fisher-Rao metric. Section 4 provides some experiments on gender discrimination and facial expression analysis using the BU-3DFEDB database and Max Planck dataset. Finally, Section 5 offers some conclusions and suggests some directions for future research.

2 Geodesic Distances between Fields of Surface Normals using the Fisher-Rao Metric

The aim in this paper is to explore whether the Fisher-Rao metric can be used to measure different facets of facial shape estimated from fields of surface normals using the von-Mises Fisher (vMF) distribution. In particular we aim to characterise the shape changes due to differences in gender and due to different facial expressions. We make use of the vMF distribution since we are dealing with surface normal data over the sphere $\Re^2$.

2.1 The von-Mises Fisher Distribution (vMF)

We choose to work with the von-Mises Fisher distribution because it is the natural probability distribution for high-dimensional directional data. The space of vMF distributions forms a differentiable manifold, which can be considered to be embedded in a higher dimensional space [17]. The embedding space induces a metric on the manifold that allows for an intrinsic way to measure distances on the manifold. A Riemannian manifold is a smooth manifold supplied with a Riemannian metric [5],[12].

The vMF distribution for multivariate directional data is

$$f_p(\underline{x}, \underline{\mu}, \kappa) = \frac{\kappa^{\frac{p}{2}-1}}{(2\pi)^{\frac{p}{2}} I_{\frac{p}{2}-1}(\kappa)} \exp(\kappa \underline{\mu}^T \underline{x}) \tag{1}$$

where $\underline{x}$ is a p dimensional vector residing on the hyper-sphere S^{p-1} submersed in $\Re^p$, $\underline{\mu}$ is the mean direction on the hyper-sphere and κ is the concentration parameter and $I_l(\kappa)$ is the modified Bessel function of the first kind of order l. The concentration parameter κ, quantifies how tightly the distribution function is distributed around the mean direction $\underline{\mu}$, and plays a role analogous to variance. The distribution is unimodal and rotationally symmetric around the direction $\underline{\mu}$. Finally, the distribution is uniform over the hyper-sphere for $\kappa = 0$.

The maximum likelihood estimators for the two parameters are obtained as follows. Suppose we have m samples of $\underline{x}$, i.e. $\underline{x}_1, ...\underline{x}_m$. The estimator of the mean direction is given by

$$\underline{\mu} = \frac{\sum_{i=1}^{m} \underline{x}_i}{||\sum_{i=1}^{m} \underline{x}_i||}$$

There is no closed form estimator of concentration parameter $\hat{\kappa}$. Instead it is the solution of the transcendental equation

$$\frac{I_{\frac{p}{2}}(\hat{\kappa})}{I_{\frac{p}{2}-1}(\hat{\kappa})} = \frac{1}{m} ||\sum_{i=1}^{m} \underline{x}_i||$$

In practice we solve this equation using the Newton-Raphson method [3]. It is worth noting that Jupp and Mardia [7] have developed some non-iterative approximations which apply under small and large values of κ.

For p=3, the distribution is referred to as the vMF distribution. In the next subsection, we use the Fisher-Rao Riemannian metric to compute the geodesic between vMF distributions.

2.2 Fisher Information Matrix

The Fisher information matrix is a Riemannian metric which can be defined on a smooth statistical manifold, i.e. a smooth manifold whose points are probability measures defined on a common probability space [8],[13],[11].

Let I = [0,1] and p: I $\times$ $\Re^k \rightarrow \Re^+$, (x,$\underline{\theta}$) $\mapsto$ p(x;$\underline{\theta}$), a k-dimensional family of positive probability density functions parameterized by the vector of parameters $\underline{\theta} = (\theta_1, ..., \theta_k)^T \in \Re^k$. In classical information geometry the Riemannian structure of the parameter space $\Re^k$ defined by the Fisher information matrix with elements

$$g_{ij}(\underline{\theta}) = \int p(\underline{x}|\underline{\theta}) \frac{\partial}{\partial \theta_i} \log p(\underline{x}|\underline{\theta}) \frac{\partial}{\partial \theta_j} \log p(\underline{x}|\underline{\theta}) d\underline{x} \tag{2}$$

The notation ∂_{θ_i} is used for the partial derivative with respect to the component θ_i of $\underline{\theta}$, where $\underline{\theta}$ is a vector of parameters associated with the density p. The Fisher-Rao metric tensor (2) is an intrinsic measure, allowing us to analyze a finite, k-dimensional statistical manifold M without considering how M resides in an R^{2k+1} space. In our case, we have 4 parameters and $\underline{\theta} = (.\kappa, \mu_1, \mu_2, \mu_3)^T$. where $\mu = (\mu_1, \mu_2, \mu_3)^T$ density parameter vector

In practice we divide each field of surface normals into windows whose size is determined by the overall image size. In our experiments, the window size is 4x4. This provides sufficient statistics to make stable estimates of the mean direction and concentration parameter.

For simplicity, we concatenate the components of the mean surface normal $\underline{\mu}$ and write $\underline{\theta} = (\kappa, \underline{\mu}^T)^T$. We perform vector-differentiation with respect to μ to simplify our calculations.

We commence by computing

$$g_{\kappa,\kappa} = \int f_p(\underline{x}, \kappa, \underline{\mu}) \frac{\partial}{\partial \kappa} \log f_p(\underline{x}, \kappa, \underline{\mu}) \frac{\partial}{\partial \kappa} \log f_p(\underline{x}, \kappa, \underline{\mu}) d\underline{x} \tag{3}$$

Substituting for the vMF distribution, we have

$$g_{\kappa,\kappa} = \frac{1}{\kappa^2}\left(\kappa^2 <\cos\alpha_\mu>^2 + 2\frac{(\frac{p}{2}-1)I_{\frac{p}{2}-1}(\kappa) - \frac{\kappa}{2}\left(I_{\frac{p}{2}-2}(\kappa) + I_{\frac{p}{2}(\kappa)}\right)}{I_{\frac{p}{2}-1}(\kappa)}\kappa <\cos\alpha_\mu>\right)$$
$$+ \left(\frac{(\frac{p}{2}-1)I_{\frac{p}{2}-1}(\kappa) - \frac{\kappa}{2}\left(I_{\frac{p}{2}-2}(\kappa) + I_{\frac{p}{2}(\kappa)}\right)^2}{I_{\frac{p}{2}-1}(\kappa)}\right) \tag{4}$$

where $\cos\alpha_\mu = \underline{\mu}^T\underline{x}$. Using the change of variables $y = \kappa\underline{\mu}^T\underline{x}$, we have $<\cos\alpha_\mu> = \int y e^y dy = e^y(y-1)$. With the substitution, we have

$$g_{\kappa,\kappa} = \frac{1}{\kappa^2}\left(\kappa^2 (e^{\kappa\underline{\mu}^T\underline{x}}((\kappa\underline{\mu}^T\underline{x}) - 1)^2) + 2\frac{(\frac{p}{2}-1)I_{\frac{p}{2}-1}(\kappa) - \frac{\kappa}{2}\left(I_{\frac{p}{2}-2}(\kappa) + I_{\frac{p}{2}(\kappa)}\right)}{I_{\frac{p}{2}-1}(\kappa)}\right.$$
$$\left. \times\ \kappa(e^{\kappa\underline{\mu}^T\underline{x}}((\kappa\underline{\mu}^T\underline{x}) - 1))\right) + \left(\frac{(\frac{p}{2}-1)I_{\frac{p}{2}-1}(\kappa) - \frac{\kappa}{2}\left(I_{\frac{p}{2}-2}(\kappa) + I_{\frac{p}{2}(\kappa)}\right)^2}{I_{\frac{p}{2}-1}(\kappa)}\right) \tag{5}$$

In the above we can set $p = 3$ since we are dealing with a vMF distribution over a 2D field of surface normals.

Next, we compute

$$g_{\kappa,\underline{\mu}} = \int f_p(\underline{x}, \kappa, \underline{\mu}) \frac{\partial}{\partial \kappa} \log f_p(\underline{x}, \kappa, \underline{\mu}) \frac{\partial}{\partial \underline{\mu}} \log f_p(\underline{x}, \kappa, \underline{\mu}) d\underline{x} \tag{6}$$

Again, substituting for the vMF distribution, and making use of the rules of vector differentiation we have

$$g_{\kappa,\underline{\mu}} = \int \frac{(2\pi)^{\frac{p}{2}} I_{\frac{p}{2}-1}(\kappa)}{\kappa^{\frac{p}{2}-1}} e^{-\kappa\underline{\mu}^t\underline{x}} \Big[\frac{\partial}{\partial\kappa}\Big(\frac{\kappa^{\frac{p}{2}-1}}{(2\pi)^{\frac{p}{2}} I_{\frac{p}{2}-1}(\kappa)} e^{\kappa\underline{\mu}^t\underline{x}}\Big)\Big]\Big[\frac{\partial}{\partial\mu}\Big(\frac{\kappa^{\frac{p}{2}-1}}{(2\pi)^{\frac{p}{2}} I_{\frac{p}{2}-1}(\kappa)} e^{\kappa\underline{\mu}^t\underline{x}}\Big)\Big] d\underline{x} \tag{7}$$

On simplification this becomes

$$g_{\kappa,\underline{\mu}} = \left[\frac{2(\frac{p}{2}-1) I_{\frac{p}{2}-1}(\kappa) - \kappa\big(I_{\frac{p}{2}-2}(\kappa) - I_{\frac{p}{2}}(\kappa)\big)}{2(I_{\frac{p}{2}-1}(\kappa))^2}\right]\underline{\mu} \tag{8}$$

which is a 3-vector and concatenates the derivatives with respect to each component of $\underline{\mu}$.

Finally, we compute

$$g_{\underline{\mu},\underline{\mu}} = \int f_p(\underline{x},\kappa,\underline{\mu}) \frac{\partial}{\partial\underline{\mu}} \log f_p(\underline{x},\kappa,\underline{\mu}) \frac{\partial}{\partial\underline{\mu}} \log f_p(\underline{x},\kappa,\underline{\mu}) d\underline{x} \tag{9}$$

Substituting for the vMF distribution, we have

$$g_{\underline{\mu},\underline{\mu}} = \int \frac{(2\pi)^{\frac{p}{2}} I_{\frac{p}{2}-1}(\kappa)}{\kappa^{\frac{p}{2}-1}} e^{-\kappa\underline{\mu}^t\underline{x}} \Big(\frac{\kappa^{\frac{p}{2}-1}}{(2\pi)^{\frac{p}{2}} I_{\frac{p}{2}-1}(\kappa)}\Big)^2 \Big\{\frac{\partial}{\partial\underline{\mu}}(e^{\kappa\underline{\mu}^t\underline{x}})\Big\}\Big\{\frac{\partial}{\partial\underline{\mu}}(e^{\kappa\underline{\mu}^t\underline{x}})\Big\}^T d\underline{x} \tag{10}$$

On simplification

$$g_{\underline{\mu},\underline{\mu}} = \frac{\kappa^{\frac{p}{2}-1}}{(2\pi)^{\frac{p}{2}} I_{\frac{p}{2}-1}(\kappa)} \kappa^2 \underline{\mu}\underline{\mu}^T \tag{11}$$

which is a 3x3 matrix.

We make use of the Fisher-Rao metric to compute the geodesic distance between the two parametric densities. Consider two corresponding 4x4 image regions for which the estimated parameter vectors are $\theta^{a_k} = (\kappa_{a_k}, \underline{\mu}_{a_k})^T$ and $\theta^{b_k} = (\kappa_{b_k}, \underline{\mu}_{b_k})^T$. Let $\hat{\kappa} = \frac{1}{2}(\kappa_{a_k} + \kappa_{b_k})$ and $\hat{\underline{\mu}} = \frac{1}{2}(\mu_{a_k} + \mu_{b_k})$. For small changes in parameters the geodesic distance between parameter vectors is

$$\begin{aligned} ds^2_{a_k,b_k} &= g_{\hat{\kappa},\hat{\kappa}}(\kappa_{a_k} - \kappa_{b_k})^2 + 2(g_{\hat{\kappa},\hat{\underline{\mu}}})^T(\kappa_{a_k} - \kappa_{b_k})(\underline{\mu}_{a_k} - \underline{\mu}_{b_k}) \\ &+ (\underline{\mu}_{a_k} - \underline{\mu}_{b_k})^T g_{\hat{\underline{\mu}},\hat{\underline{\mu}}}(\underline{\mu}_{a_k} - \underline{\mu}_{b_k}) \end{aligned} \tag{12}$$

To compute the total facial dissimilarity, we sum the geodesic distances over all 4x4 non-overlapping image blocks. The total dissimilarity is given by $D^2_{a,b} = \sum_k ds^2_{a_k,b_k}$.

3 Embedding Techniques

To visualise the distribution of geodesic distances we use a number of manifold embedding techniques to embed the facial shapes into a two-dimensional pattern space. The method studied is multi-dimensional scaling (MDS) [6].

MDS is a family of methods that maps measurements of similarity or dissimilarity among pairs of feature items, into distances between feature points with given coordinates in a low-dimensional space. The first step is to compute the squared distance matrix $DS = [D^2_{a,b)}]a, b = 1, ..., n$. This matrix is subjected to the eigendecomposition $DS = \Phi_D \Lambda \Phi_D^T$ where Λ_D is the diagonal eigenvalue matrix with the eigenvalues ordered in decreasing size along the leading diagonal. The embedding co-ordinate matrix $Y_D = \sqrt{\Lambda_D} \Phi_D^T$ has the vectors of embedding co-ordinates of the n data-points as columns.

4 Experimental Results

Our experiments are concerned with assessing shape variation in fields of surface normals due to both facial expression and gender difference. In the case of facial expression we aim to explore the changes in facial shape due to subjects pulling seven different expressions namely, happiness, sadness, surprise, fear, anger, disgust and neutral. We also aim to explore if the techniques outlined in this paper can be used to distinguish the gender of different subjects.

The procedure adopted is as follows. We commence with a range of image faces captured using a Cyberware 3030 full-head scanner. From the range of images, we estimate fields of surface normals by computing the derivatives of the height data, and projecting these onto a fronto-parallel plane. We refer to the fields of surface normals obtained as facial needle maps. We align the needle maps obtained from the different range of images to give the maximum overlap (correlation). Each field of surface normals is tessellated into non-overlapping 4x4 blocks. For each pair of block we estimate the mean surface normal direction and the concentration parameter. For each pair of facial needle maps be compute the Fisher-Rao metric on a block-by-block basis, and then compute the dissimilarity by summing over the blocks. For the set of n faces under consideration we construct a $n \times n$ dissimilarity matrix. We then apply embedding technique (MDS) to the dissimilarity matrix to obtain embedding co-ordinates for the n faces.

We assess the quality of the resulting low-dimensional data representation by evaluating to what extent the local structure of the data is retained. The evaluation is performed by measuring the generalization error of a 1-nearest neighbour (1-NN) classifier that is trained on the low-dimensional data representation. Here an object is simply assigned to the class of its nearest neighbour [1],[18].

In addition, we use the Rand Index to assess the degree of agreement between two partitions of the same set of objects. Based on extensive empirical comparison of several such measures, (Milligan and Coooper, 1986) recommended the Rand Index as the measure of agreement even when comparing partitions having different numbers of clusters [10],[15],[19].

4.1 Gender Discrimination

Our first experiment concerns gender discrimination. The data used here consists of 200 facial needle-maps extracted from range images in the Max Planck

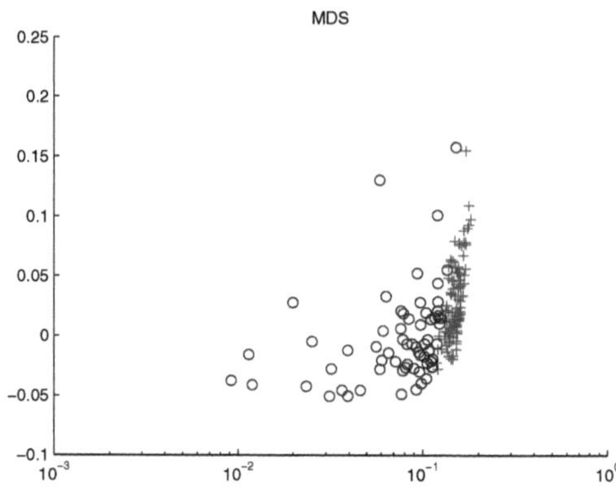

Fig. 1. Gender difference

dataset[1]. There are 100 females and 100 males, annotated with ground truth gender information.

Figure 1, shows the MDS embedding of the pattern of distances into a 2-dimensional space. The blue markers are used to denote male subjects, and the red ones female subjects. We can draw the following conclusions from this plot. Firstly, turning our attention to the embeddings, using the Fisher-Rao metric the distribution of male and female markers are concentrated differently. In particular the female markers are more densely concentrated. This would suggest that probabilistic separation may be feasible, and the unambiguous male subjects are separated from the female ones. Secondly, by it is worth noting that attempting to discriminate male and females faces on the basis of shape alone is a difficult task, and human observers make numerous additional cues such as hair-style.

In Table 1, we observe that the performance from the 1-NN classifier gives us the best result.

Table 1. Generalization errors of 1-NN and Rand Index classifier trained

	Rand Index	1-NN
Gender	0.1450	0.0455

4.2 Face Expressions

The second experiment explores the ability of the Fisher-Rao metric to distinguish the same face when presented in a different expression. We use 6 different sets of data from the BU-3DFEDB database[2]. Again, we work with surface normals estimated from range-images.

[1] The Database is available at http://vdb.kyb.tuebingen.mpg.de/

[2] The database is available at http://www-users.cs.york.ac.uk/ nep/tomh/3DFaceDatabase.html

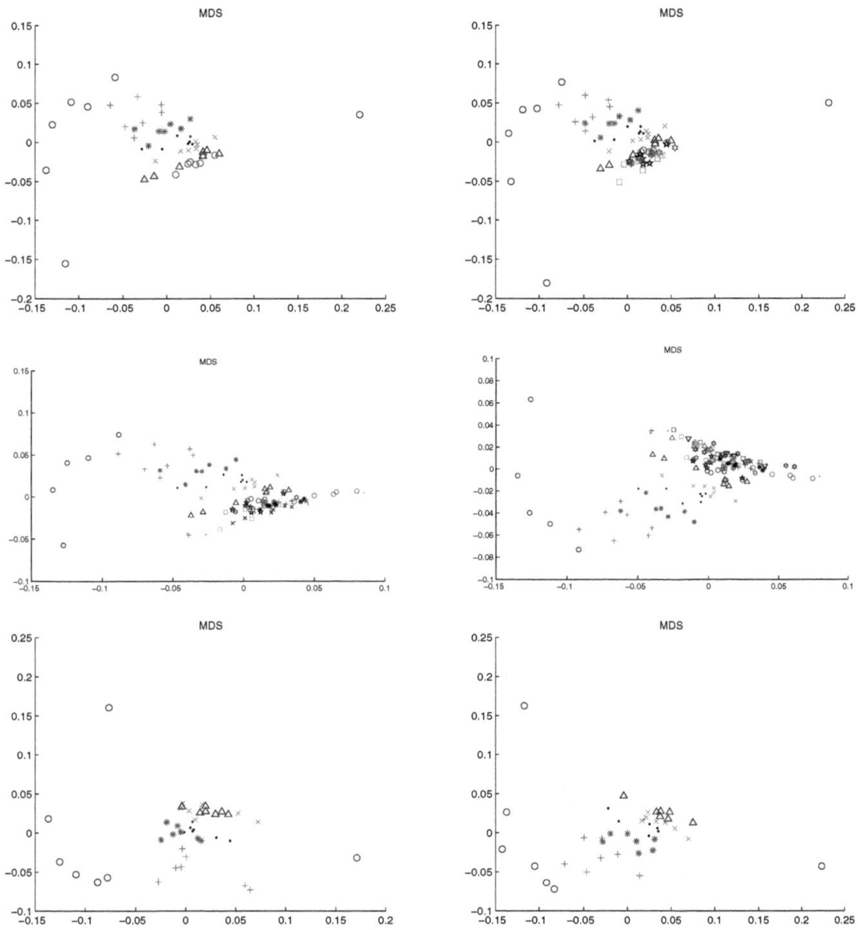

Fig. 2. In the first row, the left-hand embedding consists of 7 different faces each of which appears in 7 different facial expressions for each embedding and the right-hand embedding consists of 10 different faces each of which appears in 7 different facial expressions for each embedding. In the second row, the left-hand embedding consists of 15 different faces each of which appears in 7 different facial expressions for each embedding and the right-hand embedding consists of 20 different faces each of which appears in 7 different facial expressions for each embedding. In the third row, the left-hand embedding consists f 6 different faces each of which appears in 7 different facial expressions for each embedding, where there are only male faces and the right-hand embedding consists of 6 different faces each of which appears in 7 different facial expressions for each embedding, where there are only female faces.

Table 2. Generalization errors of 1-NN and Rand Index classifier trained

Face Expressions/DataBase	Rand Index	1-NN
7 Faces 7 Expressions	0.2041	0.1633
10 Faces 7 Expressions	0.2571	0.2857
15 Faces 7 Expressions	0.3163	0.3429
20 Faces 7 Expressions	0.4662	0.4429
6 Faces 7 Expressions (Male Expressions)	0.1190	0.1905
6 Faces 7 Expressions (Female Expressions)	0.1667	0.2619

The results of our analysis are shown in Fig. 2. In our embedding visualizations (MDS) we show the different expressions for the same subject with the same symbol. We have 6 different databases. The first four databases contain male and female subjects. The first database contains 7 faces with 7 expressions, the second database has 10 faces with 7 expressions, the third contains 15 faces with 7 expressions and the fourth contains 20 faces with 7 expressions. The final two databases, respectively,contain 7 expressions for 6 males and 6 females.

We observe that the performance from the MDS embedding and the classifier trained are similar (see Fig. 2 and Table 1). For a small number of faces in the data-set, achieves good separation of different faces under varying expression. However, as the number of faces increases the overlap becomes significant. Also, for data sets with the same gender the results achieved a better performance.

Globally, the difference between the results of the experiments using the 1-NN and the Rand Index classifier trained are: for database with mixture of male and female the best results, generally, is using 1-NN and for database where we have only one subject, female or male, the best classifier trained is Rand Index.

5 Final Remarks

In this paper we are able to show a notion of distance, using Fisher-Rao metric, between fields of surface normals on a shape manifold. The immediate next step is to construct individual shape-spaces for each class of object. Another line of investigation will be to revisit the problem of computing geodesic distance between needle-maps, in a way that explicitly accounts for the shape of manifold on which they reside.

References

1. Bremmer, D., Demaine, E., Erickson, J., Lacono, J., Langerman, S., Morin, P., Toussaint, G.: Output-sensitive algorithms for computing nearest-neighbor decision boundaries (2005)
2. Cootes, T.F., Taylor, C.J., Cooper, D.H., Graham, J.: Active shape models, their training and application. Computer Vision and Image Understanding 61, 38–59 (1995)
3. Deuflhard, P.: Newton methods for nonlinear problems. affine invariance and adaptive algoritms (2004)

4. Dryden, I.L., Mardia, K.V.: Statistical Shape Analysis. Wiley Series in Probability and Statistics (1998)
5. Jurgen, J.: Riemannian Geometry and Geometry Analysis, 5th edn. Springer, Heidelberg (2008), ISBN 978-3540773405
6. Le, H., Small, C.G.: Multidimensional scaling of simplex shapes. Pattern Recognition 32(9), 1601–1613 (1999)
7. Mardia, K.V., Jupp, P.E.: Directional Statistics. John Wiley and Sons Ltd., Chichester (2000)
8. Maybank, S.J.: Detection of image structures using the fisher information and the rao metric. IEEE Transactions on Pattern Analysis and Machine Intelligence 26(2) (2004)
9. Maybank, S.J.: The fisher-rao metric for projective transformations of the line 63, 191–206 (2005)
10. Milligan, G.W., Cooper, M.C.: A study of the comparability of external criteria for hierarchical cluster analysis (1986)
11. Mio, W., Badlyang, D., Liu, X.: A computational approach to fisher information geometry with applications to image analysis. In: Rangarajan, A., Vemuri, B.C., Yuille, A.L. (eds.) EMMCVPR 2005. LNCS, vol. 3757, pp. 18–33. Springer, Heidelberg (2005)
12. Pennec, X.: Probabilities and statistics of riemannian manifolds: A geometric approach. Research Report RR-5093, INRIA (2004)
13. Peter, A., Rangarajan, A.: A new closed-form information metric for shape analysis. In: Larsen, R., Nielsen, M., Sporring, J. (eds.) MICCAI 2006. LNCS, vol. 4190, pp. 249–256. Springer, Heidelberg (2006)
14. Peter, A., Rangarajan, A.: Shape analysis using the fisher-rao riemannian metric: unifying shape representation and deformation. In: 3rd IEEE International Symposium on Biomedical Imaging: Nano to Macro, pp. 1164–1167 (2006)
15. Rand, M.W.: Objective criteria for the evaluation of clustering methods (1971)
16. Small, C.G.: The Statistical Theory of Shape. Springer, Heidelberg (1996)
17. Stephens, M.A.: Use of the von mises distribution to analyse continuous proportions (1982)
18. Toussaint, G.T.: Geometric proximity graphs for improving nearest neighbor methods in instance-based learning and data mining (2005)
19. Yeung, K.Y., Ruzzo, L.: Principal component analysis for clustering gene expression data (2001)
20. Zhang, J., Zhang, X., Krim, H., Walter, G.G.: Object recognition and recognition in shape spaces. Pattern Recognition 36, 1143–1154 (2003)

Adaptation of SIFT Features for Robust Face Recognition

Janez Križaj, Vitomir Štruc, and Nikola Pavešić

Faculty of Electrical Engineering, University of Ljubljana,
Tržaška 25, SI-1000 Ljubljana, Slovenia
{janez.krizaj,vitomir.struc,nikola.pavesic}@fe.uni-lj.si

Abstract. *The Scale Invariant Feature Transform* (SIFT) is an algorithm used to detect and describe scale-, translation- and rotation-invariant local features in images. The original SIFT algorithm has been successfully applied in general object detection and recognition tasks, panorama stitching and others. One of its more recent uses also includes face recognition, where it was shown to deliver encouraging results. SIFT-based face recognition techniques found in the literature rely heavily on the so-called keypoint detector, which locates interest points in the given image that are ultimately used to compute the SIFT descriptors. While these descriptors are known to be among others (partially) invariant to illumination changes, the keypoint detector is not. Since varying illumination is one of the main issues affecting the performance of face recognition systems, the keypoint detector represents the main source of errors in face recognition systems relying on SIFT features. To overcome the presented shortcoming of SIFT-based methods, we present in this paper a novel face recognition technique that computes the SIFT descriptors at predefined (fixed) locations learned during the training stage. By doing so, it eliminates the need for keypoint detection on the test images and renders our approach more robust to illumination changes than related approaches from the literature. Experiments, performed on the Extended Yale B face database, show that the proposed technique compares favorably with several popular techniques from the literature in terms of performance.

Keywords: SIFT, keypoint detector, SIFT descriptor, face recognition.

1 Introduction

Face recognition is extensively used in a wide range of commercial and law enforcement applications. Over the past years many algorithms have been proposed for facial recognition systems. These algorithms include two basic aspects: holistic, e.g. PCA (Principal Component Analysis [1]) and LDA (Linear Discriminant Analysis [2]), and feature-based, e.g., Gabor- and Scale Invariant Feature Transform-based (or SIFT-based) methods [3], [4]. Holistic approaches use the entire face region for the task of feature extraction and, therefore, avoid difficulties in the detection of specific facial landmarks. Feature-based approaches,

A. Campilho and M. Kamel (Eds.): ICIAR 2010, Part I, LNCS 6111, pp. 394–404, 2010.

on the other hand, extract local features from specific feature points of the face. Generally, holistic approaches obtain better results on images captured in controlled conditions, while feature-based approaches exhibit robustness to variations caused by expression or pose changes.

One of the more recent additions to the group of feature-based face recognition techniques is the Scale Invariant Feature Transform (SIFT) proposed by Lowe in [4]. The SIFT technique and its corresponding SIFT features have many properties that make them suitable for matching different images of an object or a scene. The features are invariant to image scaling and rotation, (partial) occlusion and to a certain extent also to changes in illumination and 3D camera view point. The SIFT technique works by first detecting a number of interest points (called keypoints) in the given image and then computing local image descriptors at these keypoints. When performing recognition (or classification), each keypoint descriptor from the given image is matched independently against all descriptors extracted from the training images, and based on the outcome of the matching procedure, the image is assigned to a class featured in the database.

Event though the SIFT technique represent one of the state-of-the-art approaches to object detection/recognition, it has some deficiencies when applied to the problem of face recognition. Compared to general objects, there are less structures with high contrast or high-edge responses in facial images. Since keypoints along edges and low-contrast keypoints are removed by the original SIFT algorithm, interest points representing distinctive facial features can also be removed. Therefore, it is of paramount importance to properly adjust the thresholds governing the process of unstable keypoint removal, when applying the SIFT technique for the task of face recognition. In any case, the adjustment of the keypoint-removal-threshold represents a trial and error procedure that inevitably leads to suboptimal recognition performance.

Another thing to be considered, when using the SIFT technique for face recognition, are false matched keypoints. The majority of SIFT-based approaches employed for face recognition use different partitioning schemes to determine a number of subregions on the facial image and then compare the SIFT descriptors only between corresponding subregions. Due to the "local" matching, wrong matches between spatially inconsistent SIFT descriptors are partially eliminated. However, variable illumination still has significant influence on the detection of keypoints, since the keypoint detector intrinsic to the SIFT technique is not invariant to illumination.

To overcome the presented shortcomings of the original SIFT technique (for face recognition), we propose in this paper a novel SIFT-based approach to face recognition, where the SIFT descriptors are computed at fixed predefined image locations learned during the training stage. By fixing the keypoints to predefined spatial locations, we eliminate the need for threshold optimization and face image partitioning, while the developed approach gains greater illumination invariance than other SIFT adaptations found in the literature.

The proposed method, called Fixed-keypoint-SIFT (FSIFT), was compared to several other approaches found in the literature. Experimental results obtained on the Extended Yale B face database show, that, under severe illumination conditions, consistently better results can be achieved with the proposed approach than with popular face recognition methods, such as PCA and LDA or other SIFT-based approaches from the literature.

2 The Scale-Invariant Feature Transform

This section reviews the basics of the SIFT algorithm, which according to [4] consists of four computational stages: *(i)* scale-space extrema detection, *(ii)* removal of unreliable keypoints, *(iii)* orientation assignment, and *(iv)* keypoint descriptor calculation.

2.1 Scale-Space Extrema Detection

In the first stage, interest points called keypoints, are identified in the scale-space by looking for image locations that represent maxima or minima of the difference-of-Gaussian function. The scale space of an image is defined as a function $L(x, y, \sigma)$, that is produced from the convolution of a variable-scale Gaussian, $G(x, y, \sigma)$, with the input image, $I(x, y)$:

$$L(x, y, \sigma) = G(x, y, \sigma) * I(x, y), \tag{1}$$

with

$$G(x, y, \sigma) = \frac{1}{2\pi\sigma^2} e^{-(x^2+y^2)/2\sigma^2}, \tag{2}$$

where σ denotes the standard deviation of the Gaussian $G(x, y, \sigma)$.

The difference-of-Gaussian function $D(x, y, \sigma)$ can be computed from the difference of Gaussians of two scales that are separated by a factor k:

$$D(x, y, \sigma) = (G(x, y, k\sigma) - G(x, y, \sigma)) * I(x, y) = L(x, y, k\sigma) - L(x, y, \sigma) \tag{3}$$

Local maxima and minima of $D(x, y, \sigma)$ are computed based on the comparison of the sample point and its eight neighbors in the current image as well as the nine neighbors in the scale above and below. If the pixel represents a local maximum or minimum, it is selected as a candidate keypoint.

2.2 Removal of Unreliable Keypoints

The final keypoints are selected based on measures of their stability. During this stage low contrast points (sensitive to noise) and poorly localized points along edges (unstable) are discarded. Two criteria are used for the detection of unreliable keypoints. The first criterion evaluates the value of $|D(x, y, \sigma)|$ at each candidate keypoint. If the value is below some threshold, which means that the structure has low contrast, the keypoint is removed. The second criterion evaluates the ratio of principal curvatures of each candidate keypoint to search for poorly defined peaks in the Difference-of-Gaussian function. For keypoints with high edge responses, the principal curvature across the edge will be much larger than the principal curvature along it. Hence, to remove unstable edge keypoints based on the second criterion, the ratio of principal curvatures of each candidate keypoint is checked. If the ratio is below some threshold, the keypoint is kept, otherwise it is removed.

2.3 Orientation Assignment

An orientation is assigned to each keypoint by building a histogram of gradient orientations $\theta(x, y)$ weighted by the gradient magnitudes $m(x, y)$ from the keypoint's neighborhood:

$$m(x,y) = \sqrt{(L(x+1,y) - L(x-1,y))^2 + (L(x,y+1) - L(x,y-1))^2}, \quad (4)$$

$$\theta(x,y) = \tanh\left(L(x,y+1) - L(x,y-1)\right)/(L(x+1,y) - L(x-1,y)), \quad (5)$$

where L is a Gaussian smoothed image with a closest scale to that of a keypoint. By assigning a consistent orientation to each keypoint, the keypoint descriptor can be represented relative to this orientation and, therefore, invariance to image rotation is achieved.

2.4 Keypoint Descriptor Calculation

The keypoint descriptor is created by first computing the gradient magnitude and orientation at each image point of the 16×16 keypoint neighborhood (left side of Fig. 1). This neighborhood is weighted by a Gaussian window and then accumulated into orientation histograms summarizing the contents over subregions of the neighborhood of size 4×4 (see the right side of Fig. 1), with the length of each arrow in Fig. 1(right) corresponding to the sum of the gradient magnitudes near that direction within the region [4]. Each histogram contains 8 bins, therefore each keypoint descriptor features $4 \times 4 \times 8 = 128$ elements. The coordinates of the descriptor and the gradient orientations are rotated relative to the keypoint orientation to achieve orientation invariance and the descriptor is normalized to enhance invariance to changes in illumination.

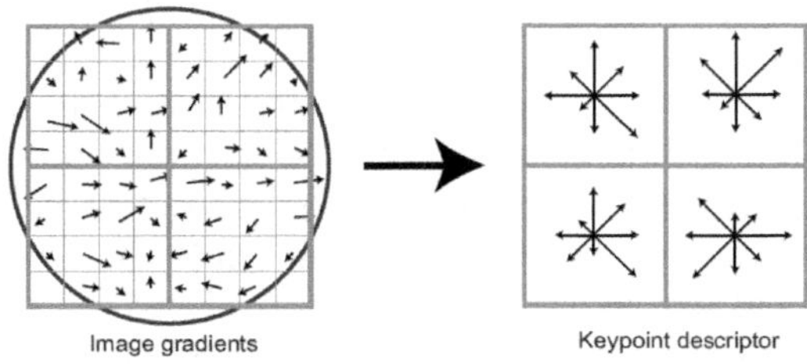

Fig. 1. In this figure the 2×2 subregions are computed from an 8×8 neighborhood, whereas in the experiments we use a 16×16 neighborhood and subregions of size 4×4

2.5 Matching

When using the SIFT algorithm for object recognition, each keypoint descriptor extracted from the query (or test) image is matched independently to the database of descriptors extracted from all training images. The best match for each descriptor is found by identifying its nearest neighbor (closest descriptor) in the database of keypoint descriptors from the training images. Generally, many features from a test image do not have any correct match in the training database, because they were either not detected in the training image or they arose from background clutter. To discard keypoints whose descriptors do not have any good match in the training database, a subsequent threshold is used, which rejects matches that are too ambiguous. If the distance ratio between the closest neighbor and the second-closest neighbor, (i.e., the closest neighbor that is known to come from a different object than the first) is below some threshold, than the match is kept, otherwise the match is rejected and the keypoint is removed. The object in the database with the largest number of matching points is considered the matched object, and is used for the classification of the object in the test image.

3 SIFT-Based Face Recognition

Over the past few years there have been some studies (from the early studies, e.g., [5], [6] to more recent ones, such as [12]) assessing the feasibility of the SIFT approach for face recognition. The progress of the SIFT technique for face recognition can be summarized as follows:

One of the first attempts to use the SIFT algorithm for face recognition was presented in [5]. The algorithm used here, differs from original SIFT algorithm in the implementation of the matching stage. Each SIFT descriptor in the test image is matched with every descriptor in each training image. Matching is done using a distance based criterion. A descriptor from the test image is said to match a descriptor from the training image, if the distance between the 2 descriptors is less than a specific fraction of the distance to the next nearest descriptor. The problem with this method is that it is very time consuming. Matching between two images has a computational complexity of $\mathcal{O}(n^2)$, where n is the average number of SIFT descriptors in each image.

In [6], the original SIFT algorithm is rendered more robust by following one of two strategies that aim at imposing local constraints on the matching procedure: the first matches only SIFT descriptors extracted from image-windows corresponding to the mouth and the two eyes, while the second relies on grid-based matching, Local matching, i.e. within a grid or a cluster, constrains the SIFT features to match features from nearby areas only. Local matching also reduces the computational complexity linearly. The computational complexity required for matching a pair of images by a local method is $\mathcal{O}(n^2/s)$, where s is the number of grids or clusters. As seen from Fig. 2, where the basic SIFT algorithm from [4] was used to match the SIFT descriptors, there are some

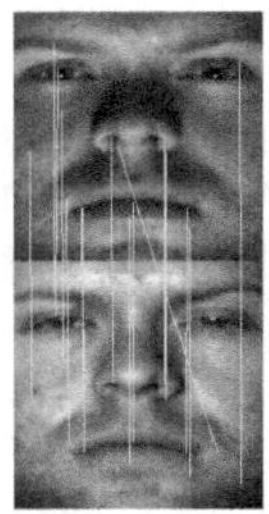 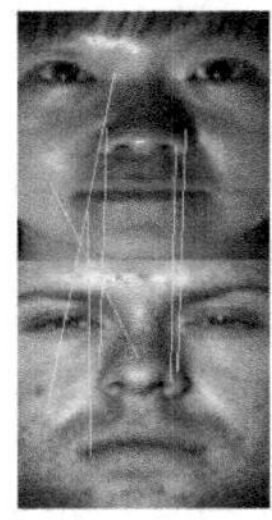 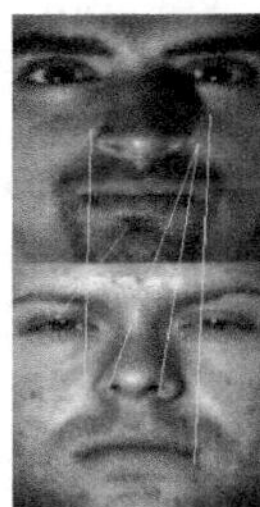

Fig. 2. Match results for one of the test images (bottom image) with a set of training faces (top) using the basic SIFT algorithm

keypoints matched, that do not represent the same characteristic of the face. Although we would expect the distance between such keypoints to be high, since they correspond to different regions of the faces, this is clearly not the case. Therefore better results are achieved, if certain subsets of SIFT keypoints are used for matching and only (spatially) corresponding subsets of SIFT descriptors are matched (as is [6] and later in [7], [9], [10] and [11]).

Both local and global information for face recognition are used in [7]. Instead of using a grid based approach, the SIFT features are clustered into 5 clusters using kmeans clustering (2 clusters for the eyes, one for the nose, and 2 clusters at the edges of the mouth). Only the SIFT descriptors between two corresponding clusters are matched. This ensures that matching is done locally. As a global matching criterion, the total number of descriptor matches (as in [4]) is used.

In [8] SIFT features are extracted from the frontal and half left and right profiles. An augmented set of SIFT features is then formed from the fusion of features from the frontal and side profiles of an individual, after removing feature redundancy. SIFT feature sets from the database and query images are matched using the Euclidean distance and Point pattern matching techniques.

In [9] a Graph Matching Technique is employed on the SIFT descriptors to to deal with false pair assignment and reduce the number of SIFT features. In [10] SIFT features are ranked according to a discriminative criterion based on Fisher's Discriminant Analysis (similar as in [2]), so that the chosen features have the minimum within-class variation and maximum variation between classes. In [11] both global and local matching strategies are used. In order to reduce the identification errors, the Dempster-Shafer decision theory is applied to fuse the two matching techniques.

In [12] an approach called Keypoints-Preserving-SIFT (KPSIFT) is proposed. The KPSIFT approach keeps all the initial keypoints for SIFT descriptor calculation. This procedure greatly differs from the basic SIFT approach, where unreliable keypoints are removed as explained in section 2. However, this removal can eliminate some keypoints and discard potentially useful discriminative information for face recognition. With the basic SIFT procedure intrinsic properties of the face images have to be considered (recall that facial images contain only a few structures with high contrast or high-edge responses, which often leads to the removal of useful keypoints), when setting the threshold values governing the process of keypoint removal. As shown in [12], recognition rates improve when adjusting thresholds on low-contrast and edge keypoints in order to accept more keypoints. Fig. 3 shows three different adjustments of the (keypoint-removal) thresholds. Here, the threshold denoted as *EdgeThreshold* controls the removal of poorly localized keypoints along edges, while the threshold denoted as *Threshold* controls the removal of low contrast keypoints (see Section 2.2 for details). The experiments in [12] show that the best recognition results are achieved with the thresholds resulting in the left image of Fig. 3.

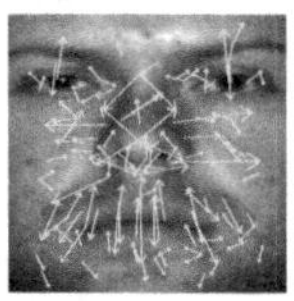

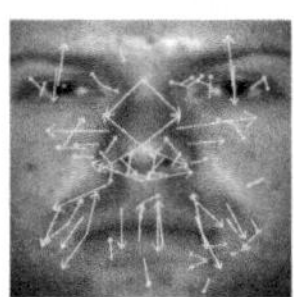

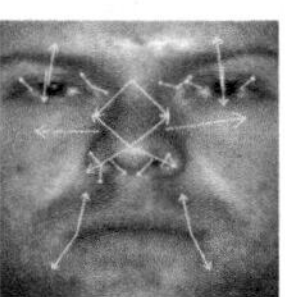

Fig. 3. Keypoints detected in a sample face image with respect to the (keypoint-removal) threshold values: Result improving values (left), common values (middle), high-elimination threshold values (right)

While the presented techniques try to compensate the imperfections of the keypoint detector by imposing local matching constraints, by relaying on subwindows of the images, by deploying graph-matching techniques, etc., we present in the remainder a simple procedure, which completely eliminates the need for the keypoint detector (in the test stage). With the proposed procedure, most shortcomings of the detector, such as susceptibility to illumination, influence of the (keypoint-removal) thresholds and false keypoint detections are solved.

4 The Fixed Keypoint SIFT Algorithm

4.1 Fixing the Keypoints

Our method, the Fixed Keypoint SIFT Algorithm or FSIFT for short, is based on the supposition that each face was preliminary localized. Thus, each image consists only of a properly registered face region of a certain person.

We assume that for the training procedure only "good" quality images are available. This assumption is reasonable, since in most operating face recognition systems the enrollment stage and with it the acquisition of training images is supervised. During training we apply the original SIFT technique and its accompanying keypoint detector (with the (keypoint-removal) threshold adjusted - Fig. 3 left) to our training images and obtain a number of candidate keypoints for each image in the set of training images (first three images of Fig. 4). Next, we apply a clustering procedure to the set of candidate keypoints to obtain $k = 100$ centroids, which serve as the *fixed* keypoints for the computation of the SIFT

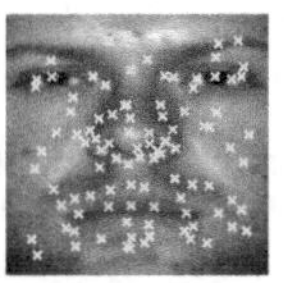 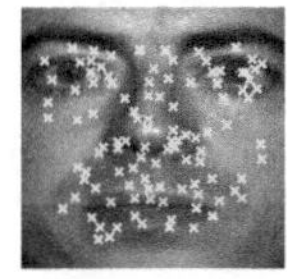 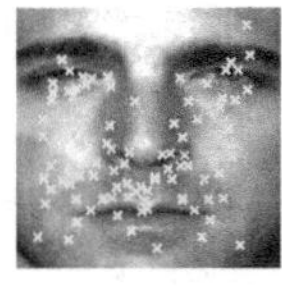 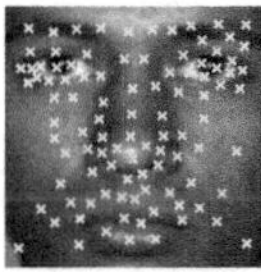

Fig. 4. Training procedure for learning the keypoint locations: sample images processed with the original keypoint detector (images one through three), the learned keypoint locations (fourth image)

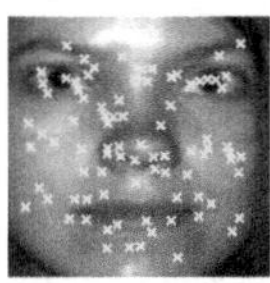 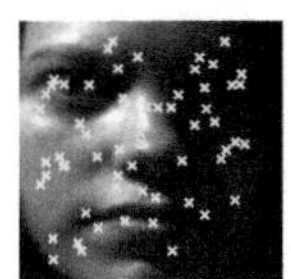 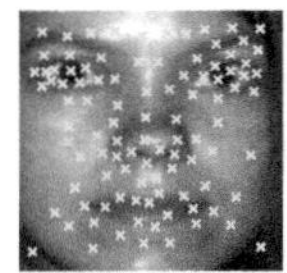 

Fig. 5. SIFT keypoints detected on the differently illuminated images of the same person: by the original keypoint detector (first two images from the left), and by the proposed method (third and fourth image)

descriptors. We can see in the fourth image of Fig. 4 that most of these centroids correspond to distinctive facial landmarks, such as the eyes, nose or the mouth.

Fig. 5 illustrates the advantages gained by the proposed approach. Here, the first image (from the left) depicts the keypoints locations found by original keypoint detector, while the second image presents the location of keypoints in the image of the same person captured in different illumination conditions. Not only the number of detected keypoints in the second image is smaller than in the first image, many of the keypoints are detected in different locations than in the first image and therefore a reduction of keypoint matches is expected. If SIFT descriptors are computed at fixed predefined locations (third and fourth image of Fig. 5) a greater robustness to illumination variations can be achieved.

4.2 Matching

As the number of descriptors for each image is the same (it equals the number of centroids k), the sum of the Euclidean distances between equally located descriptors of the two images to be compared is used as the matching criterion. By doing so, computational complexity for matching between two images is also reduced to $\mathcal{O}(2k)$. Let us denote the sets of SIFT descriptors from the training images as $\mathcal{S}_j = \{S_{i,j}(x_i, y_i); i = 1, 2, ..., k\}$, where $j = 1, 2, ..., n$ denotes the training image index, n stands for the total number of training images, i represents the descriptor index, k denotes the number of fixed keypoint locations (i.e., centroids), and (x_i, y_i) denote the image location for the i-th SIFT descriptor. Let us further assume that the n training images correspond to N different classes (i.e., subjects) with corresponding class labels $\omega_1, \omega_2, ..., \omega_N$. Then, the matching procedure can formally be written as follows:

$$\delta_{SL_2}(\mathcal{S}_g, \mathcal{S}_t) = \min_j \delta_{SL_2}(\mathcal{S}_j, \mathcal{S}_t) \rightarrow \mathcal{S}_t \in \omega_g, \tag{6}$$

where $\mathcal{S}_t$ stands for the set of SIFT descriptor extracted from the test image at the k predefined image locations, and the matching function is defined as $\delta_{SL_2}(\mathcal{S}_p, \mathcal{S}_r) = \sum_i \delta_{L_2}(S_{i,p}, S_{i,r})$.

The above expression postulates that a given test image is assigned to the class ω_g, if the sum of the Euclidian distances between spatially corresponding descriptors of the test image and one of the training images of the g-th class is the smallest among the computed distances to all n SIFT descriptor sets of the training images.

5 Experiments and Results

The experiments were done on the Extended Yale B (EYB) face database [15]. The database contains 38 subjects and each subject has approximately 64 frontal view images taken under different illuminations conditions. For the experiments the images were partitioned into five subsets. In the first image subset (S1 in the remainder), there are images captured in relatively good illumination conditions, while for the image subsets labeled S2 to S5, the lighting conditions get more extreme. S1 is used as the training set, while images in the other subsets are used as test images. It should be noted that the numbers in the brackets next to the subset label in Table 1 represent the number of images in each subset. All algorithms were implemented in Matlab relying partially on existing code available

Table 1. Rank one recognition rates (in %) obtained on the EYB database

Method	S2 (456)	S3 (455)	S4 (488)	S5 (752)
PCA	93.6	55.0	16.7	22.0
LDA	100	99.8	56.3	51. 0
SIFT	100	45.7	25.7	11.2
SIFT_CLUSTER	100	100	66.8	64.9
FSIFT	**100**	**100**	**83.2**	**82.8**

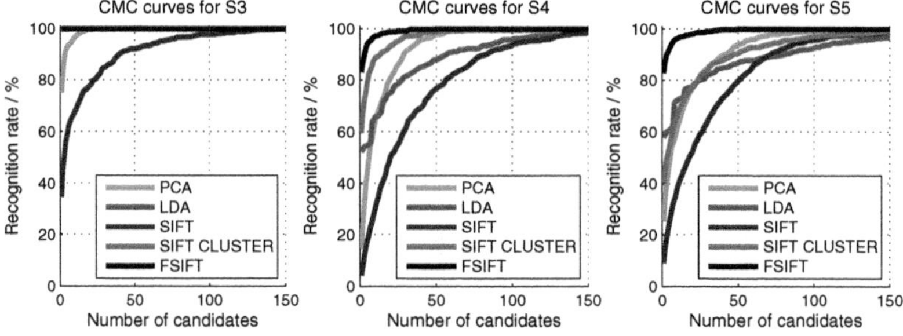

Fig. 6. Cumulative match curves for each subset

from [13] and [14]. The performance of the proposed approach was compared to the performance of some other face recognition techniques such as PCA, LDA, and to several different modifications of the SIFT algorithm. Table 1 presents the performance of the listed methods in form of rank one recognition rates for changeable illumination conditions. The recognition rates of PCA and LDA are shown in the first and second row, respectively. The original SIFT method is shown in the third row. The fourth row presents the results of the method from [7], which relies on clustering of the SIFT keypoints. With our method, denoted as FSIFT in the last row, better results are achieved in comparison with the recognition performance of the remaining techniques assessed in our experiments.

In Fig. 6, the results are presented as cumulative match curves (CMC) for subsets three through five. It should be noted that the CMCs are not shown for subset two, as all tested techniques achieve a perfect recognition rate of 100% for all ranks. From the results we can see that the FSIFT approach clearly outperformed all other techniques assessed in the comparison.

6 Conclusion and Future Work

In this paper an adaptation of the SIFT algorithm for face recognition was presented. Using the EYB database, we have shown that the performance of the proposed method is significantly better than the performance of popular techniques such as PCA or LDA and different SIFT-based recognition techniques from the literature. To be able to cope with possible pose variations, we plan to augment the proposed FSIFT technique with a pose detector and, consequently, extend it to a multi-pose version.

References

1. Turk, M., Pentland, A.: Face Recognition Using Eigenfaces. In: Proc. IEEE Conference CVPR, pp. 586–591 (1991)
2. Etemad, K., Chellappa, R.: Discriminant Analysis for Recognition of Human Face Images. J. of the Opt. Society of America A 14(8), 1724–1733 (1997)
3. Wiskott, L., Fellous, J.M., Kruger, N., von der Malsburg, C.: Face Recognition by Elastic Bunch Graph Matching. IEEE TPAMI 19(7), 775–779 (1997)
4. Lowe, D.G.: Distinctive Image Features From Scale-Invariant Keypoints. International Journal of Computer Vision 60, 91–110 (2004)
5. Aly, M.: Face Recognition using SIFT Features, CNS/Bi/EE report 186 (2006)
6. Bicego, M., Lagorio, A., Grosso, E., Tistarelli, M.: On the Use of SIFT Features for Face Authentication. In: CVPR Workshop, p. 35 (2006)
7. Luo, J., Ma, Y., Takikawa, E., Lao, S., Kawade, M., Lu, B.-L.: Person-Specific SIFT Features for Face Recognition. In: ICASSP, pp. 593–596 (2007)
8. Rattani, A., Kisku, D.R., Lagorio, A., Tistarelli, M.: Facial Template Synthesis based on SIFT Features. In: IEEE Workshop AIAT, pp. 69–73 (2007)
9. Kisku, D.R., Rattani, A., Grosso, E., Tistarelli, M.: Face Identification by SIFT-based Complete Graph Topology. In: IEEE Workshop AIAT, pp. 63–68 (2007)

10. Majumdar, A., Ward, R.K.: Discriminative SIFT Features for Face Recognition. In: Canadian Conference on Electrical and Computer Engineering, pp. 27–30 (2009)
11. Kisku, D.R., Tistarelli, M., Sing, J.K., Gupta, P.: Face Recognition by Fusion of Local and Global Matching Scores Using DS Theory: An Evaluation With Uni-Classifier and Multi-Classifier Paradigm. In: CVPR Workshop, pp. 60–65 (2009)
12. Geng, C., Jiang, X.: SIFT features for Face Recognition. In: IEEE Conference CSIT, pp. 598–602 (2009)
13. Lowe, D.G.: Software for SIFT, `http://people.cs.ubc.ca/~lowe/keypoints/`
14. Vedaldi, A.: MATLAB/C implementation of the SIFT detector and descriptor, `http://www.vlfeat.org/~vedaldi/code/sift.html`
15. Georghiades, A.S., Belhumeur, P.N., Kriegman, D.J.: From Few to Many: Illumination Cone Models for Face Recognition Under Variable Lighting and Pose. IEEE TPAMI 23(6), 643–660 (2001)

Facial Expression Recognition Using Spatiotemporal Boosted Discriminatory Classifiers

Stephen Moore, Eng Jon Ong, and Richard Bowden

Centre for Vision Speech and Signal Processing
University of Surrey, Guildford, GU2 7JW, UK
{stephen.moore,e.ong,r.bowden}@surrey.ac.uk

Abstract. This paper introduces a novel approach to facial expression recognition in video sequences. Low cost contour features are introduced to effectively describe the salient features of the face. Temporalboost is used to build classifiers which allow temporal information to be utilized for more robust recognition. Weak classifiers are formed by assembling edge fragments with chamfer scores. Detection is efficient as weak classifiers are evaluated using an efficient look up to a chamfer image. An ensemble framework is presented with all-pairs binary classifiers. An error correcting support vector machine (SVM) is utilized for final classification. The results of this research is a 6 class classifier (*joy*, *surprise*, *fear*, *sadness*, *anger* and *disgust*) with recognition results of up to 95%. Extensive experiments on the Cohn-kanade database illustrate that this approach is effective for facial exression analysis.

1 Introduction

The objective of this work is to exploit temporal information to build boosted classifiers for frontal facial expression recognition in video sequences. Facial expression recognition is a difficult task due to the natural variation in appearance of subjects. Such variation include ethnicity, age, facial hair, occlusion, pose and lighting. Many fields benefit from accurate facial expression recognition including behavioral science, security, communication and education. This paper presents an approach that relies on temporal boosted discriminatory classifiers based upon contour information. Contours are largely invariant to lighting and as will be shown, provide efficient classifiers using chamfer matching.

Cross cultural studies in Psychology signify a correlation between base emotions and facial expressions [8]. Current facial expression recognition systems highlight this observation by classifying a set of prototypical emotions such as *joy*, *surprise*, *fear*, *sadness*, *anger* and *disgust* [16,13]. Two common approaches for feature extraction for facial expression recognition are geometric based and appearance based methods [20]. Geometric features exploit shape and location information of facial components. Appearance based features capture the appearance change of the face (including wrinkles, bulges and furrows) and are

A. Campilho and M. Kamel (Eds.): ICIAR 2010, Part I, LNCS 6111, pp. 405–414, 2010.

extracted by image filters applied to the face or sub regions of the face. Geometric features are sensitive to noise and usually require reliable and accurate facial feature detection and tracking. However, appearance based features are less reliant on initialization, do not suffer from tracking errors and can encode changes in skin texture that are important for facial expression recognition. This paper investigates appearance based features based upon contour information. Humans have the ability to recognize facial expressions from a simplified line drawing or cartoon of the face. Sufficient information must therefore be present in this simplified representation for a computer to recognize facial expressions. Using only contour information provides important advantages as it offers some invariance to lighting and reduces the complexity of the problem.

Temporal information is incorporated by using a boosting framework [18] with the potential to develop weak classifiers by utilizing previous frames response in evaluating the current frame. This algorithm also incorporates temporal consistency of the data to facilitate recognition. We investigate the use of an ensemble classifier design to improve recognition. For final classification an error correcting SVM is used.

This rest of this paper is organized as follows. Related work is presented in section 2. Section 3 explains the methodology of this research. Section 4 outlines the data and experiments used to evaluate this research. Finally conclusions are presented in section 5.

2 Related Work

Facial expression recognition can be performed by using features from one image or by considering information from a image sequence. Research in psychology shows image sequences provide more accurate information than single frames. Bassili [2] conducted experiments showing that human facial expression recognition is superior when dynamic images are available. Some faces are often falsely read as expressing a particular emotion, even if their expression is neutral, because their proportions are naturally similar to those that another face would temporarily assume when emoting. Temporal information can overcome this problem by modeling the motion of the face. Utilizing temporal information can translate to more robust and accurate classification when compared with static classifiers.

Hidden Markov Models (HMM) are frequently applied to spatio temporal facial expression recognition as they model the dynamics of expressions. Oliver et al. introduce two dimensional blob features to track mouth motion and uses HMMs to classify facial expressions [14]. Cohen et al. [5] proposed a multilevel HMM that uses the state sequence of independent HMMs to segment and recognize facial expressions. However flow estimates are easily disturbed by changes in illumination and non rigid motion.

Another way to capture temporal information is to map images to low dimensional manifolds for different expressions. Chang et al. [4] created a manifold from sparse 2d points. Shan et al. [16] used local binary pattern features for the whole face to create manifolds of facial expressions and used a bayesian temporal model for facial expression recognition.

Zhao and Pietikainen introduced a novel approach for recognizing dynamic texture for classifying facial expressions [21]. Dynamic texture is an extension of texture into the temporal domain. Volume local binary patterns were proposed to capture appearance and motion. Petridis and Pantic investigated audio and visual temporal features for laughter detection [15]. Features were extracted for each frame in a temporal window. Mean, standard deviation and polynomials were calculated over the temporal window. Sheerman-Chase et al. [17] used similar temporal features for detection of non verbal facial displays. Yang et al. [19] introduces dynamic binary patterns based on harr like features to represent the dynamics of facial expressions.

Moore and Bowden [13] introduced edge and chamfer features for static facial expression recognition. Adaboost was used to learn discriminatory features and competitive results were obtained. In this research we apply the same features in a temporal framework for facial expression recognition in video sequences. Boosting techniques rarely utilize temporal information for classification. Recently Smith et al. introduced temporalboost which introduced temporal consistency in a boosting framework [18]. The algorithm averages weak classifiers from previous frames while the classification error decreases. In this research we investigate how temporal information in facial expressions can be utilized using temporalboost.

3 Methodology

3.1 Overview

The following section introduces how our facial expression classifier works, illustrated in figure 1. Images are annotated (two eyes and the tip of the nose) to allow features to be transformed to a reference co-ordinate system. The canny edge algorithm is used to create edge maps of all images. From each edge map, coherent edge fragments are extracted from the area in and around the face. A classifier bank is built containing all the edge fragments. Weak classifiers are created by combining edge fragments with a chamfer score. Temporalboost learns the optimal subset of features from the classifier bank and forms a strong classifier. Previous studies have shown a performance increase when ensemble classifiers are used [13], we adopt a similar approach resulting in 15 binary classifiers. These binary classifiers are used as input to a error correcting SVM for final classification.

3.2 Feature Extraction

Images are manually annotated to identify the two eyes and the tip of the nose, to form a 3-point basis. A 3-point basis is sufficient to align examples as only frontal faces are considered. Most approaches to frontal facial expression recognition only consider a 2 point basis (the two eyes), however head movements

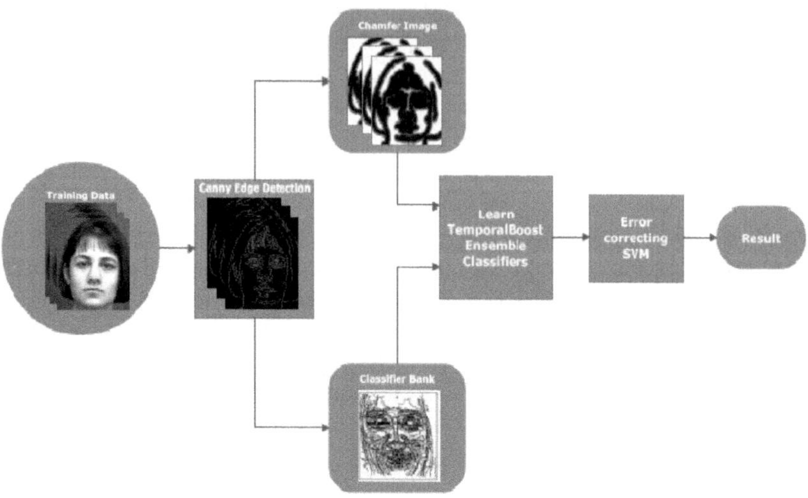

Fig. 1. Overview of Facial Expression Recognition System

are influenced by our emotions [12] so a 3 point basis as a reference co-ordinate frame is more tolerant to variations in head pose. The distance between the eyes is approximately half the width of the face and one third of the height. This identifies the region of interest (ROI) from which contours will be considered.

The canny edge detection algorithm is used to extract edges [3]. In face images, edges characterize the boundaries of salient facial features as well as facial deformation due to facial expressions. First the detector smooths the image to eliminate noise. Next the sobel operator performs spatial gradient measurement on an image. The image is then scanned along the image gradient direction and if pixels are not part of the local maxima they are set to zero. This subdues image information that is not part of local maxima and is called non maximum suppression. A threshold is then used to evaluate if the magnitudes are sufficient to be classified as an edge e. $E = \{e\}$, where E is the edge map of an image. This threshold is selected manually for the dataset so salient features of the face are visually coherent in the edge maps. Following edge detection, connected component analysis is performed. From each edge component, short edge fragments T are extracted with variable lengths (based on heuristics of the face).

3.3 Chamfer Image

To measure support for any single edge feature over a training set we need some method for measuring the edge strength along that feature in a image. This can be computed efficiently using Chamfer matching. Chamfer matching was first introduced by Barrow et al. [1]. It is a registration technique whereby a drawing consisting of a binary set of features (contour segments) is matched to an image. A distance transform converts a binary image, which consists of feature and non-feature pixels, into an image where each pixel value denotes the distance to the nearest feature pixel. Thus similarity between two shapes can be measured using their chamfer distance. The matching of a template with the chamfer image rather than the original edge image gives an advantage, as the resulting measure will be smoother as a function of the template transformation parameters [10].

All images in the training set undergo edge detection with the canny edge detector to produce an edge map E. Then a chamfer image is produced DT_E using a distance transform. Each pixel value q, is proportional to the distance to its nearest edge point in E:

$$DT_E(q) = min_{e \in E} \|q - e\|_2 \tag{1}$$

A chamfer score is evaluated for each contour fragment T, where $T = \{t\}$:

$$d_{cham}^{(T)}(DT_E) = \frac{1}{N} \sum_{t \in T} DT_E(t) \tag{2}$$

where N is the number of edge points in T. This gives the Chamfer score as a mean distance between feature T and the chamfer image DT_E. The function $d_{cham}^{(T)}(DT_E)$ is an efficient lookup to the chamfer image for all classifiers. An example of a chamfer image is shown in figure 1.

3.4 Temporalboost

Boosting is a machine learning algorithm that produces a very accurate strong classifier, by combining weak classifiers in linear combination. Adaboost was introduced by Freund and Schapire [9] and has been successfully applied to static facial expression recognition [13]. Smith et al. [18] introduced Temporalboost, a boosting algorithm that introduces temporal consistency, by averaging weak classifiers sequentially. This allows weak classifiers to utilize information from previous frame when evaluating the current frame.

Temporalboost is an extension of adaboost. Like adaboost, a distribution of weights are maintained and associated with training examples. At each iteration, a weak classifier which minimizes the weighted error rate is selected, and the distribution is updated to increase the weights of the misclassified samples and reduce the weights of correctly classified examples. However, temporalboost modifies adaboost by allowing the best weak classifier to use previous frames responses if the overall classification error is decreased. This is achieved by using an OR operation and an AND operation for the previous t responses. The OR operation will respond positively if any of the previous t frames were classified as positive. This can allow for more true positives at the cost of false positives [18]. The AND operation will respond positively if all the previous t frames were classified as positive. This operation will decrease the number of false positives at the cost of true positives [18]. For example, if a feature classifies an event correctly for the previous t frames, but missclassifies the current frame, then temporalboost allows the current frame to be classified correctly by using the previous t responses if the overall classification error is decreased. Both operations are performed for each iteration and the operation with the minimum classification error is selected. These operations allow temporal smoothing to be part of the boosting framework. The temporal window t starts at *0* and is expanded as long as the overall classification error for the current weights is decreased. The temporalboost algorithm tries to separate training examples by selecting the best weak classifier $h_j(x)$ that distinguishes between the positive and negative training examples. A weak classifier thus consists of a feature (f_j), a threshold θ_j and a parity (p_j) indicating the direction of the inequality sign.

$$h_j(x) = \begin{Bmatrix} 1 \; if \; p_j f_j < p_j \theta_j \\ 0 \; otherwise \end{Bmatrix} \tag{3}$$

Where

$$f_j = d^{(T)}_{cham}(DT_E) \tag{4}$$

θ_j is the weak classifier threshold. Setting a fixed threshold requires a priori knowledge of the feature space, an optimal θ_j is found through an exhaustive search for each weak classifier. This allows the learning algorithm to select a set of weak classifiers with low thresholds that are extremely precise allowing little deviation. Also, weak classifiers with high thresholds, which allows consistent deformation of the facial features can be selected. This increases the performance, but as will be seen, does not result in over fitting of the data. An image can have up to 500 features. Thus, over the training set, many hundreds of thousands of features are evaluated during the learning algorithm.

3.5 Ensemble Architecture

Dietterich argued that ensemble methods can often perform better than a single classifier [7]. Temporalboost is a binary classifier. There are several ways to partition the classification task into binary decisions. The simplest way is to train 1 against all. Another approach is to train all possible combinations of classes (1:1). [13] showed an all pairs ensemble (1:1) outperformed the 1 against all method. We adopt an all pairs ensemble framework, which for n classes can be broken into $\frac{n(n-1)}{2}$ binary classifiers respectively. This allows each expression to be exclusively boosted against every other expression.

An error correcting SVM is used for final classification. An SVM classifier is adopted here since it is a well understood classification technique that has been demonstrated to be effective in facial expression recognition. An SVM takes a feature vector as input in an n-dimensional space and constructs a separating hyperplane in that space, one which will maximize the margin between the positive and negative sets. The better the hyperplane, the larger the distance to the neighboring points from both classes. SVMs are usually binary classifiers, here we used a multi class SVM [6] which uses a one against all approach to solve the 6-class problem. The output from the temporalboost classifiers forms the input vector for the SVM. The SVM is trained using noisy training data by randomly perturbing the 3 point basis for each sequence.

4 Expression Classification

The Cohn-kanade facial expression database [11] was used in the following experiments. Subjects consisted of 100 university students ranging in age from 18 - 30. 65% were female, 15% were African American, and three percent were Asian or Latino. The camera was located directly in front of the subject. The expressions were captured as 640 x 480 png images. In each sequence, the subject started with a neutral expression and the sequence ends with the peak of the expression. In total 365 video sequences were chosen from the database (over 4,000 images). The only criteria was that the video sequence represented one of the prototypical expressions. This database is encoded using the Facial Action Coding System (FACS). A movement of one or more muscles of the face is called an action unit (AU) and all expressions can be described by a combination of one or more of 46 AU's. Each image has a FACS code and from this code, images are grouped into different expression categories.

Experiments were carried out using 5-fold cross validation with training and test sets divided 80-20. Due to the large number of features and training images we limited the number of boosting iterations to 500. In general about 20-30% of the weak classifiers selected have temporal paramters. Of the temporal weak classifiers selected, the majority use the OR operation. This reflects the fact that the data is not very temporally consisent and thus features using the AND operation don't minimize the classification error. Due to space restrictions the following discussion will focus on the joy ensemble classifiers (similar observation as discussed below are present for other expressions). Figure 2 shows the receiver operating characteristic (ROC) curves for all the *joy* ensemble classifiers. The more accurate classifiers are *joy* against *surprise* and *joy* against *anger*. This is as expected as the facial deformation due to the *joy* expression is very distinct from expressions *surprise* and *anger*. The worst performance is achieved with the *joy* against *disgust* classifier. This is due to the close proximity of the distinctive features (deformation around the cheek area) for these expressions.

Figure 3 visualizes the features which contribute to the classification of expressions. In general we can see that the contour around the edge of the mouth and the contour around the cheek are used to classify the *joy* expression. However as can be seen, depending on the negative expression different areas of the mouth and cheek contribute more to classification. For example in figure 3 pictures A, C and E show features from the corners of the mouth and the cheeks are prominent. Expressions *surprise*, *sad* and *anger* deform the face very differently to *joy* and thus all the deformation of the *joy* expression is captured in these classifiers. While pictures B and D show the importance of the corners of the mouth and not the deformation around the cheek. This is because the expressions *fear* and *disgust* deform the area around the cheek in a similar fashion to the *joy* expression. Another interesting observation in image B is the amount of noise. This finding can be explained by the fact that the expressions *joy* and *fear* are often difficult to disambiguate.

Table 1 shows the confusion matrix for 5 fold cross validation on the Cohn-kanade database. An overall recognition rate of 86.1% is achieved. From the

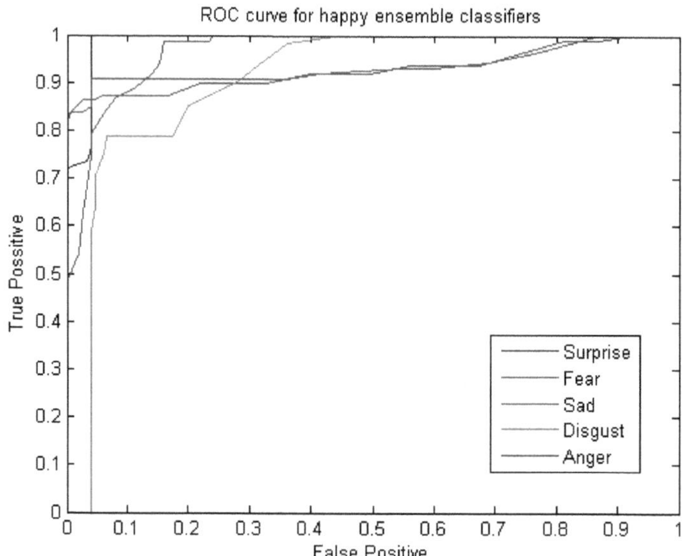

Fig. 2. Roc curves for joy ensemble classifiers

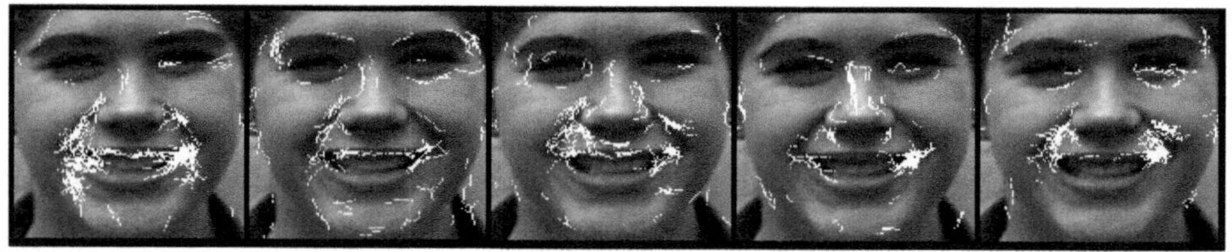

Fig. 3. Visualization of weak classifiers which contributed to classification. From left to right images represent: A) joy against surprise, B) joy against fear, C) joy against sad, D) joy against disgust and E) joy against anger.

results it is apparent that the more subtle expressions (*anger*, *fear* and *sad*) are outperformed by expressions with a large deformation (*joy*, *surprise,disgust*). Subtle changes in appearance are difficult to distinguish when using one reference co-ordinate frame due to the variability across subjects. Also it must be noted that the combination of contour and chamfer matching is varient to scale and rotation. Thus subtle expression are harder to disambiguate using these features.

The lowest recognition rate was for the *fear* classifier. Most confusion occurs between expressions *disgust* and *anger* due to similar deformation around the eyebrows. Also confusion occurs between *fear* and *sad* and between *sad* and *anger* classes respectively. In particular *sad* and *anger* expressions have little deformation when compared to expressions *surprise* or *joy*. This in itself could contribute to the confusion as a lack of distinct features makes the learning of strong classifiers more difficult. Also when posing a *sad* expression subjects can

Table 1. Confusion matrix for 5-fold cross validation on Cohn-kanade database

	Joy	Surprise	Fear	Sad	Disgust	Anger
Joy	**93.92**	0	2.94	1.67	1.47	0
Surprise	0	**95.09**	0	1.79	3.12	0
Fear	5.63	0	**75.55**	9.34	3.71	5.77
Sad	0	0	6.28	**85.36**	0	8.36
Disgust	0	0	2.78	0	**91.32**	5.9
Anger	0	0	0	9.32	15.1	**75.58**

exaggerate the expression and the mouth can have a similar appearance to the *fear* expression.

5 Conclusions

This paper presents a novel approach to frontal facial expression recognition in video sequences. Unlike other popular methods like Gabor wavelets, we present a fast efficient system that yields a recognition rate of 86.1%. Recognition is achieved on a frame by frame basis but classifiers use feature responces from previous frames to evaluate the current frame. An ensemble framework is presented which includes an all pairs architecture with an error correcting SVM for final classification. Competitive results were achieved on the Cohn-kanade database for 6 basic expressions. Expression with large deformation of the face achieved the best results with *surpise* achieving over 95% accuracy.

Acknowledgement

This work has been supported by the EPSRC project LILiR and by the FP7 project DICTASIGN (FP7/2007-2013) under grant agreement n 231135.

References

1. Barrow, H.G., Tenenbaum, J.M., Bolles, R.C., Wolf, H.C.: Parametric correspondence and chamfer matching: two new techniques for image matching. In: IJCAI 1977: Proceedings of the 5th International Joint Conference on Artificial Intelligence, pp. 659–663. Morgan Kaufmann Publishers Inc., San Francisco (1977)
2. Bassili, J.N.: Emotion recognition: the role of facial movement and the relative importance of upper and lower areas of the face. Journal of personality and social psychology 37(11), 2049–2058 (1979)
3. Canny, J.: A computational approach to edge detection. IEEE Trans. Pattern Anal. Mach. Intell. 8(6), 679–698 (1986)
4. Chang, Y., Hu, C., Turk, M.: Probabilistic expression analysis on manifolds. In: IEEE Computer Society Conference on Computer Vision and Pattern Recognition, vol. 2, pp. 520–527 (2004)
5. Cohen, I., Garg, A., Huang, T.S.: Emotion recognition from facial expressions using multilevel hmm. In: Neural Information Processing Systems (2000)

6. Crammer, K., Singer, Y., Cristianini, N., Shawe-taylor, J., Williamson, B.: On the algorithmic implementation of multiclass kernel-based vector machines. Journal of Machine Learning Research 2, 2001 (2001)
7. Dietterich, T.G.: Ensemble methods in machine learning. In: Kittler, J., Roli, F. (eds.) MCS 2000. LNCS, vol. 1857, pp. 1–15. Springer, Heidelberg (2000)
8. Ekman, P., Friesen, W.V.: Constants across cultures in the face and emotion. Journal of Personality and Social Psychology, 124–129 (1971)
9. Freund, Y., Schapire, R.E.: Experiments with a new boosting algorithm. In: International Conference on Machine Learning, pp. 148–156 (1996)
10. Gavrila, D.: Pedestrian detection from a moving vehicle. In: Vernon, D. (ed.) ECCV 2000. LNCS, vol. 1843, pp. 37–49. Springer, Heidelberg (2000)
11. Kanade, T., Tian, Y., Cohn, J.F.: Comprehensive database for facial expression analysis. In: FG 2000: Proceedings of the Fourth IEEE International Conference on Automatic Face and Gesture Recognition 2000, Washington, DC, USA, p. 46. IEEE Computer Society, Los Alamitos (2000)
12. Mignault, A., Chaudhuri, A.: The many faces of a neutral face: Head tilt and perception of dominance and emotion. Journal of Nonverbal Behavior 27(2), 111–132 (2003)
13. Moore, S., Bowden, R.: Automatic facial expression recognition using boosted discriminatory classifiers. In: Zhou, S.K., Zhao, W., Tang, X., Gong, S. (eds.) AMFG 2007. LNCS, vol. 4778, pp. 71–83. Springer, Heidelberg (2007)
14. Oliver, N., Pentland, A., Brard, F.: Lafter: Lips and face real time tracker with facial expression recognition. In: Proc. CVPR, pp. 123–129 (1997)
15. Petridis, S., Pantic, M.: Audiovisual laughter detection based on temporal features. In: IMCI 2008: Proceedings of the 10th International Conference on Multimodal Interfaces, pp. 37–44. ACM, New York (2008)
16. Shan, C.F., Gong, S.G., McOwan, P.W.: Dynamic facial expression recognition using a bayesian temporal manifold model. In: BMVC 2006, pp. 297–306 (2006)
17. Sheerman-Chase, T., Ong, E.-J., Bowden, R.: Feature selection of facial displays for detection of non verbal communication in natural conversation. In: IEEE International Workshop on Human-Computer Interaction, Kyoto (October 2009)
18. Smith, P., da Vitoria Lobo, N., Shah, M.: Temporalboost for event recognition. In: ICCV 2005: Proceedings of the Tenth IEEE International Conference on Computer Vision (ICCV 2005), Washington, DC, USA, vol. 1, pp. 733–740. IEEE Computer Society, Los Alamitos (2005)
19. Yang, P., Liu, Q.S., Cui, X.Y., Metaxas, D.N.: Facial expression recognition using encoded dynamic features, pp. 1–8 (2008)
20. Tian, Y., Kanade, T., Cohn, J.: Facial expression analysis. In: Handbook of Face Recognition, ch. 11, pp. 247–275. Springer, Heidelberg (2005)
21. Zhao, G., Pietikainen, M.: Dynamic texture recognition using local binary patterns with an application to facial expressions. IEEE Trans. Pattern Anal. Mach. Intell. 29(6), 915–928 (2007)

Recognition of Facial Expressions by Cortical Multi-scale Line and Edge Coding

R.J.R. de Sousa, J.M.F. Rodrigues, and J.M.H. du Buf

Vision Laboratory, Institute for Systems and Robotics (ISR/IST),
University of the Algarve (ISE and FCT), 8005-135 Faro, Portugal

Abstract. Face-to-face communications between humans involve emotions, which often are unconsciously conveyed by facial expressions and body gestures. Intelligent human-machine interfaces, for example in cognitive robotics, need to recognize emotions. This paper addresses facial expressions and their neural correlates on the basis of a model of the visual cortex: the multi-scale line and edge coding. The recognition model links the cortical representation with Paul Ekman's Action Units which are related to the different facial muscles. The model applies a top-down categorization with trends and magnitudes of displacements of the mouth and eyebrows based on expected displacements relative to a neutral expression. The happy vs. not-happy categorization yielded a correct recognition rate of 91%, whereas final recognition of the six expressions happy, anger, disgust, fear, sadness and surprise resulted in a rate of 78%.

1 Introduction

Currently, one of the most investigated topics of image analysis is face detection and recognition [23,11]. There are several reasons for this trend, such as the wide range of commercial vigilance and law-enforcement applications. Although state-of-the-art recognition systems have reached a certain level of maturity, their accuracy is still limited when imposed conditions are not perfect: all possible combinations of illumination changes, pose, beards, different facial expressions, etc. Solving the problem of facial expression recognition by using the same approach as used for face recognition [18] will solve part of the problem: the detected expression can be morphed to a neutral one for more robust face recognition.

Furthermore, intelligent interaction between humans and computers is an emerging research area related to interfaces and robots. Since face-to-face communications between humans involve emotions and what they convey [15], facial expressions are also important in advanced human-machine interfaces. The Facial Action Coding System or FACS [4] is probably the most well-known study about the coding of facial actions. FACS measures the behavior of the facial activity, and facial expressions are described by 44 Action Units (AUs), of which 30 are related to the contraction of muscles, 12 in the upper part of the face and 18 in the lower part.

A. Campilho and M. Kamel (Eds.): ICIAR 2010, Part I, LNCS 6111, pp. 415–424, 2010.

Pantic and Rothkrantz [16] used color images with frontal as well as profile views of faces. By detecting 10 positions of the profile views and 19 of the frontal views, describing 32 AUs, they obtained a correct recognition rate of emotions of 86%. Barlett et al. [14] created a system which detects 20 AUs in frontal views of persons in realtime video sequences. Each frame was decomposed using Gabor filters, an AdaBoost classifier was used to extract relevant AUs, and a SVM classifier yielded a recognition rate of 93%. Feitosa et al. [21] used the same database that we will use in this paper, i.e., well-framed images in order to simplify face detection. They extracted emotion features using PCA and with neural networks they achieved a recognition rate of almost 72%. Gama [7] applied the Haar transform in a cascaded classifier to segregate facial images. Using Bayesian classifiers, she achieved a recognition rate of 80% in the case of happy vs. not happy, and 55% in the case of five different expressions: anger, happy, neutral, sadness and surprise. Kumano et al. [22] proposed a method for pose-invariant expression recognition in video sequences. By using a variable-intensity template for describing different expressions, they achieved a rate of over 90% for vertical faces with a rotation range of ±40 degrees from the frontal view.

In this paper we present an approach which, like the one of Barlett et al. [14], employs Gabor filters. However, our goal is to develop more advanced models of the visual cortex. In cortical area V1 there are simple and complex cells, which are tuned to different spatial frequencies (scales) and orientations, but also disparity (depth) because of the neighboring left-right hypercolumns [9]. These cells provide input for grouping cells which code line and edge information and which probably attribute depth information to these. In V1 there also are end-stopped cells which, with complex inhibition processes, allow to extract keypoints (singularities, vertices and points of high curvature). Recently, models of simple, complex and end-stopped cells have been developed, e.g. [5], providing input for keypoint detection [5,19] and line/edge detection [8,20], including disparity extraction [6,17]. On the basis of these models and neural processing schemes, it is now possible to create a cortical architecture for figure-ground segregation, Focus-of-Attention, including object and face categorization and recognition [20].

In this paper we focus on a cortical model for the recognition of facial expressions. This model only employs the multi-scale line/edge representation based on simple and complex cells. The line and edge coding is explained in Section 2. Section 3 deals with the model devoted to facial expressions, i.e., the extraction of cortical AUs, expression classification and the cortical architecture. In Section 4 experimental results are presented and we conclude with a discussion in Section 5.

2 Multi-scale Line and Edge Coding

In order to explain the model for facial expressions in relation to the model for face recognition [18], it is necessary to explain briefly how our visual system can reconstruct, more or less, the input image. Image reconstruction can be based on one lowpass filter plus a complete set of bandpass filters, such that the frequency domain is evenly covered. This concept is the basis of many image coding schemes; it could also be used in the visual cortex because simple cells in V1 are often modeled by complex Gabor wavelets. These are bandpass filters [5], and lowpass information is available through special retinal ganglion cells with photoreceptive dendrites [2]. Activities of all cells could be combined by summing these in one cell layer that would provide a reconstruction or brightness map. But then there is a problem: it is necessary to create *yet another observer* of this map in our brain.

The solution is simple: instead of summing all cell activities, we can assume that the visual system extracts lines and edges from simple- and complex-cell responses, which is necessary for object recognition, and that responding "line cells" and "edge cells" are interpreted symbolically. For example, responding line cells along a bar signal that there is a line with a certain position, orientation, amplitude and scale, the latter being interpreted by a Gaussian cross-profile which is coupled to the scale of the underlying simple and complex cells. The same way a responding edge cell is interpreted, but with a bipolar, Gaussian-truncated, error-function profile; for more details and illustrations see [18,20].

Responses of even and odd simple cells, corresponding to the real and imaginary parts of a Gabor filter, are denoted by $R^E_{s,i}(x,y)$ and $R^O_{s,i}(x,y)$, i being the orientation (we use 8 orientations). The scale s is given by λ, the wavelength of the Gabor filters, in pixels. We use $10 \leq \lambda \leq 27$ with $\Delta\lambda = 1$. Responses of complex cells are modeled by the modulus $C_{s,i}(x,y) = [\{R^E_{s,i}(x,y)\}^2 + \{R^O_{s,i}(x,y)\}^2]^{1/2}$.

The basic scheme for line and edge detection is based on responses of simple cells: a positive (negative) line is detected where R^E shows a local maximum (minimum) and R^O shows a zero crossing. In the case of edges the even and odd responses are swapped. This gives four possibilities for positive and negative events. An improved scheme [20] consists of combining responses of simple and complex cells, i.e., simple cells serve to detect positions and event types, whereas complex cells are used to increase the confidence. Lateral and cross-orientation inhibition are used to suppress spurious cell responses beyond line and edge terminations, and assemblies of grouping cells serve to improve event continuity in the case of curved events.

Figure 1 (top row) shows one person of the JAFFE database [10] that we used in our experiments with, from left to right, anger, disgust, fear, happy, sadness and surprise. The middle row shows the neutral expression and its line and edge coding at five scales: $\lambda = \{10, 14, 18, 23, 26\}$. Different levels of grey, from white to black, show detected events: positive/negative lines and positive/negative edges, respectively. As can be seen, at fine scales many small events have been detected, whereas at coarser scales more global structures remain that convey a "sketchy" impression. The bottom row in Fig. 1 shows detected events of the non-neutral expressions (top row) at $\lambda = 16$ after applying a multi-scale stability criterion; see [20] for details. Stabilization leads to the elimination of events which are not stable over neighboring scales, and therefore to less but more reliable events.

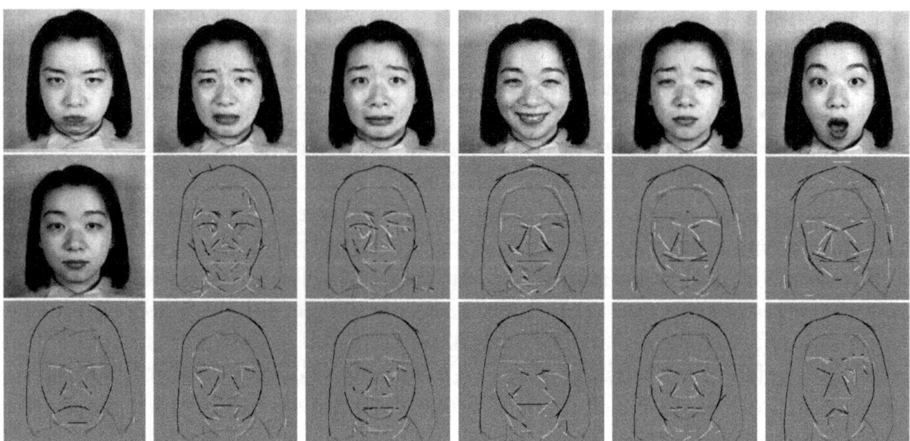

Fig. 1. Top (left to right): anger, disgust, fear, happy, sadness and surprise. Middle: neutral expression with its line/edge coding at fine (left) to coarse (right) scales; $\lambda = \{10, 14, 18, 23, 26\}$. Bottom: results at scale $\lambda = 16$, after multi-scale stabilization, for each expression on the top row.

3 Cortical Facial Expression Classification Model

Because of the multi-scale line/edge representation with deformations at coarse scales (Fig. 1) it is necessary to introduce new AUs to classify each facial expression. Therefore, three regions of interest (ROI) are defined, two covering the eyebrows and one covering the mouth. These ROIs actually correspond to the Focus-of-Attention regions as used in face recognition [18], but here we use rectangular ROIs to simplify the analysis.

Knowing the AUs involved in the different expressions [24], it is possible to estimate the positions of the line/edge events in each ROI relative to those of the neutral expression. Figure 2 (top) shows the expected movements, where + and − represent inclinations and 0 is the same as the neutral expression. The square indicates an open mouth and the arrows global trends of the events. In the bottom part of Fig. 2, the open/solid dots represent up/down trends and the number of dots the magnitudes.

3.1 Extraction of Cortical AUs

All face images in the JAFFE database are already normalized. For dealing with unnormalized face images, a cortical normalization scheme based on keypoints (end-stopped cells) can be applied [3]. The three ROIs are defined using the line/edge maps of the neutral faces. In the analysis of facial expressions, the same processing is applied as in the human visual system [13]: information at coarse scales is used for a first estimation of the expression, after which information at increasingly finer scales is added to confirm or correct the result. The basic approach is illustrated in Fig. 3: keypoints (yellow) detected at the corners of the mouth and

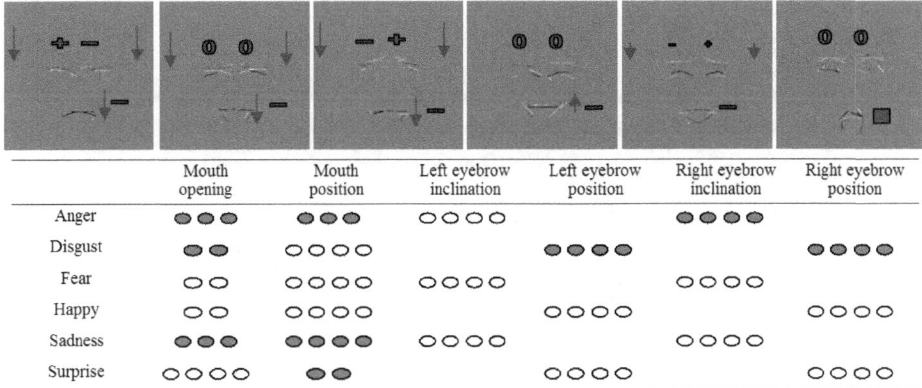

	Mouth opening	Mouth position	Left eyebrow inclination	Left eyebrow position	Right eyebrow inclination	Right eyebrow position
Anger	●●●	●●●	○○○○		●●●●	
Disgust	●●	○○○○		●●●●		●●●●
Fear	○○	○○○○	○○○○		○○○○	
Happy	○○	○○○○		○○○○		○○○○
Sadness	●●●	●●●●	○○○○		○○○○	
Surprise	○○○○	●●		○○○○		○○○○

Fig. 2. Top: movements of line/edge events relative to those of a neutral expression; left to right: anger, disgust, fear, happy, sadness and surprise. Bottom: table of movements and their magnitudes in the ROIs.

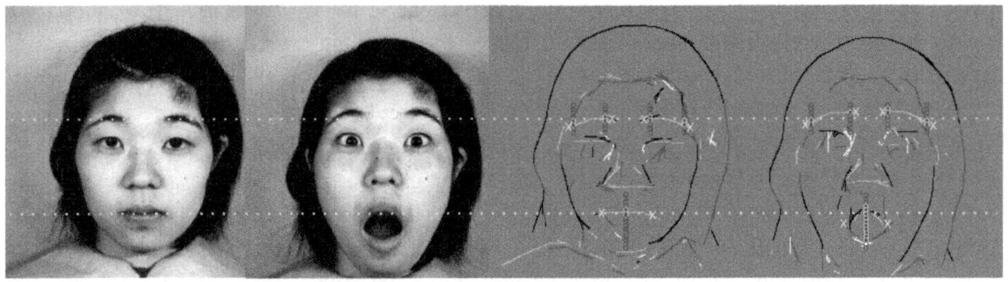

Fig. 3. Left: expressions neutral and surprise. Third image: keypoints (yellow crosses) at corners of mouth and eyebrows in the neutral face activate clusters of grouping cells (in red) which detect line or edge events in the non-neutral face (right image).

eyebrows in the neutral face (third image) activate clusters of grouping cells (red) which combine line or edge events in the non-neutral face (at right).

Positive and negative line events at any scale consist of excitated L^+ and L^- cells at positions (x, y) with output one (cell is active) or zero (cell is not active). Likewise, outputs of edge cells E^+ and E^- are also binary. Outputs of clusters of such cells are combined (summed) by grouping cells with specific dendritic fields, the outputs of which therefore correspond to the number of active cells in their fields. In the ROI of the mouth, coarse scales are screened for a negative line matching a closed mouth and for a positive line matching an open mouth; see the left two columns of Fig. 4. This is achieved by defining grouping cells S^+ and S^- with horizontal and very elongated (linear, elliptical) fields at neighboring vertical (y_i) positions. The two cells $S^+(y_1)$ and $S^-(y_2)$ with maximum output are selected using non-maximum suppression, and of these two the one with the largest response yields the state of the mouth: open (S^+) or closed (S^-). This

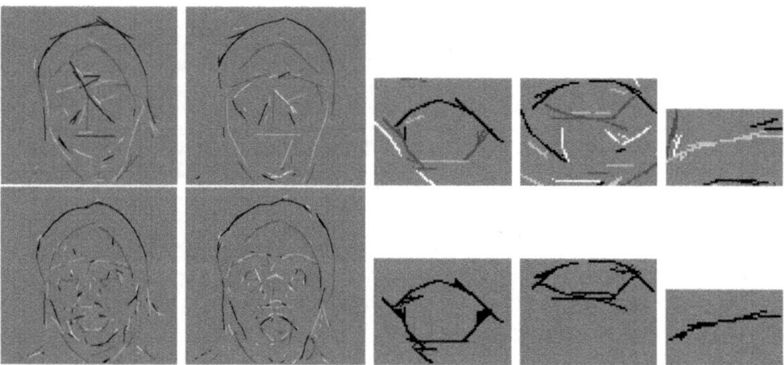

Fig. 4. First two columns: coarse scale $\lambda = 23$ (top) and fine scale $\lambda = 10$ (bottom) of a person with closed and open mouth. Third an fourth column: the ROI of the mouth with all events at scale $\lambda = 12$ (top) and only edges (bottom). Fifth column: the same as the previous two columns, but the ROI of an eyebrow and only negative lines at the bottom.

processing is applied at six coarse scales and at least four scales must yield the same result for defining the state of the mouth. If the result is not convincing, medium and fine scales are added.

At medium and fine scales, see Fig. 4 (bottom, first two columns) the edges of the lips, i.e., the transitions between lips and skin, are relevant. The figure shows in detail (3rd and 4th column) an open and a closed mouth at scale $\lambda = 10$. The analysis as described above is applied using only edge cells E^+ and E^-, and grouping cells $S^\pm$ at y_i, with horizontal elliptical dendritic fields, which combine both edge polarities. The outputs of the cells $S^\pm$ are thresholded and the two cells at the lowest and highest position y in the mouth's ROI yield the size of the mouth (magnitude of opening) as well as the vertical position of the mouth.

In each of both ROIs of the eyebrows, the processing is similar to the one for detecting the size of the mouth, but there are in each ROI two clusters of grouping cells in order to determine the inclination angle: one vertical cluster at the left and another vertical cluster at the right of the ROI. Figure 4 (last column) shows a detail image. Here we need to analyse a finer scale ($\lambda = 10$) because of the relatively thin eyebrows and only responses of L^- cells are summed. In each cluster of grouping cells the cell with maximum response is selected by non-maximum suppression, and the two selected cells on both sides of the ROI code the inclination angle as well as the eyebrow's vertical position.

In summary, the ROI of the mouth produces the state of the mouth, open or closed, and the vertical position on the mouth. The ROI of each eyebrow yields the inclination angle plus the vertical position.

3.2 Classification Schemes

Two classification schemes were tested: (1) a direct classification of the six groups anger, disgust, fear, happy, sadness and surprise [24], plus two group combinations;

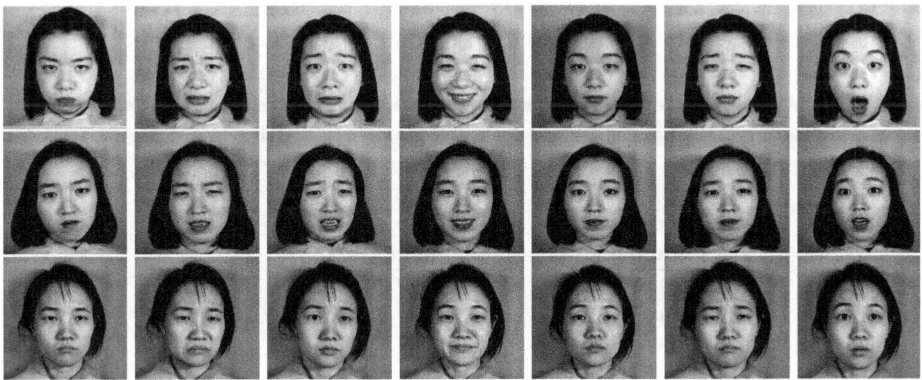

Fig. 5. Examples of images from the JAFFE [10] database

and (2) a classification with pre-categorization levels as previously applied to other, non-face objects [1,20]. The pre-categorizations consisted of the following levels. The first level discriminates happy and not-happy. At the second level, the not-happy group is further divided into anger, sadness and fear, whereas the happy group is split into happy and surprise. At the last classification level, the fear group is split into fear and surprise, and the anger group into anger and disgust. Please note that surprise is classified twice, in opposite groups, because this expression can have two contexts. The above group divisions are based on previous work by other authors. Gama [7] proposed a separation of expressions into two groups, happy and not-happy. Kumano et al. [22] proposed four groups: anger, sadness, surprise and happy. Our own scheme is slightly different because we also added the group fear which the other authors, apart from Zhang [24], left out.

4 Experimental Results

As mentioned above we used the JAFFE database [10] and selected seven expressions (neutral plus anger, disgust, fear, happy, sadness and surprise) of 10 women; see Fig. 5. The extracted facial features were processed using a Bayes minimum-distance classifier, i.e., without more advanced statistical methods like PCA, because this is a first test of the developed multi-scale method. Apart from the schemes described in the previous sub-section, we tested three feature combinations (C1–C3), taking into account that all features are relative to the features of the same woman with a neutral expression: (C1) the agreement of the *trends* of the features when compared to Fig. 2 (bottom); (C2) comparing only the *magnitudes* of the trends; and (C3) the combination of both.

Table 1 presents the results of our experiments, without pre-categorizations, i.e., direct classification of the groups of expressions, and with the three categorization levels, all with feature combinations C1, C2 and C3.

Results without and with pre-categorization into happy and not-happy are obviously equal and quite good: 91% using C1 (only trends), 84% using C2 (only

Table 1. Overview of all results. Notes: a – happy also includes surprise; b – anger also includes disgust; c – fear also includes surprise.

recognition rates (%)							
		without categorizations			with categorizations		
groups	images	C1	C2	C3	C1	C2	C3
happy	20a	80	60	60	80	60	60
not happy	50	96	94	100	96	94	100
total	70	91	84	89	91	84	89
happy	20a	80	60	60	95	95	100
anger	20b	86	55	80	85	55	80
sadness	10	60	70	50	60	70	50
fear	20c	65	80	80	75	85	80
total	70	74	66	70	81	77	81
happy	10	80	40	60	80	80	70
surprise	10	50	30	40	90	60	70
anger	10	20	60	50	60	70	80
disgust	10	70	10	30	100	20	50
sadness	10	60	70	50	60	70	50
fear	10	80	30	60	80	40	70
total	60	60	40	48	78	57	65

magnitudes), and 89% using C3 (both). This means that one can and should use generic information at coarse scales [13]. In the categorization happy-anger-sadness-fear the results are similar, which indicates that we need more detailed information at finer scales, but the use of the first pre-categorization level (happy vs. not-happy) leads to better results if compared to direct classification: 74, 66 and 70% against 81, 77 and 81%. The same trend can be seen after final recognition of all six expressions anger, disgust, fear, happy, sadness and surprise: 60, 40 and 48% against 78, 57 and 65%. Remarkable is the fact that in almost all cases the use of only binary trends (C1) yields better results than using only magnitudes (C2) and even the combination of both (C3), which requires further analysis in order to optimize the results, i.e., the relative weighting of binary and non-binary features. Comparing our first results with those of other groups who used the JAFFE database, Feitosa et al. [21] achieved a best rate of 73% for the same six facial expressions, where our method achieved 78%. Zhang and Ji [24], who used another database with the same six expressions, achieved only 72% (see Introduction). Clearly, results can and must be improved by finetuning the algorithms.

5 Discussion

The scheme for expression recognition will be part of an integrated architecture for object and face recognition with two data streams, one devoted to general objects which can be arbitrarily rotated in 3D, and the other devoted to faces which are normally seen upright and with near frontal view. In case of general objects, the multi-scale line/edge and keypoint representations of canonical views are stored as templates in memory, keypoints are used to route dynamically keypoints of an input object to a template in memory, and the same routing is applied to lines and edges for object recognition. Dynamic routing is necessary for position, size and rotation invariance, and coarse-to-fine-scale processing is applied for successive (pre-)categorizations until final recognition; see [20]. In case of faces, the same processing is applied for face detection and normalization using keypoints [3], after which faces can be recognized together with their expression using the line/edge codes. It is likely that expression is extracted before recognition, i.e., if face normalization using keypoints at eyes, nose and mouth also normalizes the expression to neutral. This is subject of ongoing research.

The framework for expressions presented in this paper is based on their neural correlates relative to the line/edge coding of neutral faces. To this purpose new AUs were defined because of the multi-scale representation with coarse-to-fine-scale processing, which allows us to apply a few categorization levels for refining the analysis. Multi-scale stabilization of the line and edge features proved to be important, because the most reliable features are used whereas other ones caused by minor variations are discarded. As expected, the positions and polarities of the lines and edges in the ROIs, combined with the AUs as proposed by Matsumoto and Ekman [12] (Fig. 2), yielded encouraging results. The use of mainly coarse-scale information in the first two-group categorization, which is very stable by definition, yielded a recognition rate of 91%. Using additional information in the subsequent four-group categorization yielded 81%, and final six-group recognition still 78%. Ongoing research addresses a detailed analysis of the data and refinement of the method, i.e., a systematic use of scales by starting with coarse scales only and then adding successively finer scales [1,20]. In addition, tests with a database of Caucasian faces are conducted, with the possibility of creating generic templates with a neutral expression, both Japanese and Caucasian.

Acknowledgements

This research is partly supported by the Foundation for Science and Technology FCT (ISR/IST pluri-annual funding) through the POS-Conhecimento Program which includes FEDER funds, and by the FCT projects PTDC/EIA/73633/2006 (SmartVision: active vision for the blind) and PTDC/PSI/67381/2006 (Neural correlates of object recognition).

References

1. Oliva, A., et al.: Top-down control of visual attention in object detection. In: IEEE Proc. Int. Conf. Image Processing, vol. 1, pp. 253–256 (2003)
2. Berson, D.: Strange vision: ganglion cells as circadian photoreceptors. TRENDS in Neurosciences 26(6), 314–320 (2003)
3. Cunha, J., Rodrigues, J., du Buf, J.M.H.: Face normalization using multi-scale cortical keypoints. In: Proc. 13rd Portuguese Conf. on Pattern Recogn., p. 2 (2007)

4. Ekman, P., Friesen, W.: Facial action coding system (FACS): Manual. Consulting Psychologists Press, Palo Alto (1978)
5. Heitger, F., et al.: Simulation of neural contour mechanisms: from simple to end-stopped cells. Vision Res. 32(5), 963–981 (1992)
6. Fleet, D.J., Jepson, A.D., Jenkin, M.R.M.: Phase-based disparity measurement. CVGIP: Image Understanding 53(2), 198–210 (1991)
7. Gama, S.: Facial emoticons: Reprodução de informação associada a expressões faciais por via do seu reconhecimento. Master Thesis, IST Lisbon, 100 p. (2009)
8. Grigorescu, C., Petkov, N., Westenberg, M.A.: Contour detection based on non-classical receptive field inhibition. IEEE Tr. IP 12(7), 729–739 (2003)
9. Hubel, D.H.: Eye, brain and vision. Scientific American Library (1995)
10. Kamachi, M., Lyons, M., Gyoba, J.: Facial expression database: Japanese female facial expression (JAFFE) database (February 2009), `http://kasrl.org/jaffe.html`
11. Massimo, T., Manuele, B., Enrico, G.: Dynamic face recognition: From human to machine vision. Image Vision Comput. 27(3), 222–232 (2009)
12. Matsumoto, D., Ekman, P.: Facial expression analysis. Scholarpedia 3(5), 4237 (2008)
13. Moshe, B.: A cortical mechanism for triggering top-down facilitation in visual object recognition. J. Cognitive Neuroscience 15(4), 600–609 (2003)
14. Bartlett, M.S., et al.: Fully automatic facial action recognition in spontaneous behavior. In: IEEE Proc. 7th Int. Conf. on Automatic Face and Gesture Recognition, pp. 223–230 (2006)
15. Neth, D., Martinez, A.M.: Emotion perception in emotionless face images suggests a norm-based representation. Journal of Vision 9(1-5), 1–11 (2006)
16. Pantic, M., Rothkrantz, L.J.M.: Facial action recognition for facial expression analysis from static face images. IEEE Tr. on Systems, Man, and Cybernetics 34(3), 1449–1461 (2004)
17. Rodrigues, J., du Buf, J.M.H.: Visual cortex frontend: integrating lines, edges, keypoints and disparity. In: Campilho, A.C., Kamel, M.S. (eds.) ICIAR 2004. LNCS, vol. 3211, pp. 664–671. Springer, Heidelberg (2004)
18. Rodrigues, J., du Buf, J.M.H.: Face recognition by cortical multi-scale line and edge representations. In: Campilho, A., Kamel, M.S. (eds.) ICIAR 2006. LNCS, vol. 4142, pp. 329–340. Springer, Heidelberg (2006)
19. Rodrigues, J., du Buf, J.M.H.: A cortical framework for invariant object categorization and recognition. Cognitive Processing 10(3), 243–261 (2009)
20. Rodrigues, J., du Buf, J.M.H.: Multi-scale lines and edges in v1 and beyond: brightness, object categorization and recognition, and consciousness. BioSystems 95, 206–226 (2009)
21. Feitosa, R.Q., et al.: Facial expression classification using RBF and back-propagation neural networks. In: Proc. 6th Int. Conf. on Information Systems Analysis and Synthesis, pp. 73–77 (2000)
22. Kumano, S., et al.: Pose-invariant facial expression recognition using variable-intensity templates. Int. J. Comput. Vision 83(2), 178–194 (2009)
23. Zhao, W., et al.: Face recognition: A literature survey. ACM Comput. Surv. 35(4), 399–458 (2003)
24. Zhang, Y., Ji, Q.: Facial expression understanding in image sequences using dynamic and active visual information fusion. In: IEEE Proc. 9th Int. Conf. on Computer Vision, 8 p. (2003)

The Analysis of Facial Beauty: An Emerging Area of Research in Pattern Analysis

Andrea Bottino and Aldo Laurentini

Dipartimento di Automatica e Informatica, Politecnico di Torino,
Corso Duca degli Abruzzi 24, 10129, Torino, Italy
{andrea.bottino,aldo.laurentini}@polito.it

Abstract. Much research presented recently supports the idea that the human perception of attractiveness is data-driven and largely irrespective of the perceiver. This suggests using pattern analysis techniques for beauty analysis. Several scientific papers on this subject are appearing in image processing, computer vision and pattern analysis contexts, or use techniques of these areas. In this paper, we will survey the recent studies on automatic analysis of facial beauty, and discuss research lines and practical applications.

Keywords: Face image analysis, facial landmarks, attractiveness.

1 Introduction

Analyzing 2D or 3D images of humans is a main area of research in pattern analysis and computer vision. The human face is by far the part of the body which conveys more information to human beings, and thus potentially to computer systems [2]. Such information span identity, intentions, emotional and health states, attractiveness, age, gender, ethnicity, attention, etc. At present, the most studied application of face image analysis is identity recognition [1], which is essentially an engineering deformable object recognition problem. Other face image analysis applications are multidisciplinary and related to human sciences and medicine. They are essentially 1) analyzing human expressions, and 2) analyzing face attractiveness.

The first is by far the most studied problem, particularly to capture human expression for animating the faces of virtual characters. A much more challenging problem is interpreting facial expressions, that is mapping expressions onto emotional states [2], [3]. The results presented are not yet convincing, since tracing backward the path from expressions (effects) to the emotions (causes), requires a shared and coherent model of the human emotions and of their effects on facial features, which psychophysiology has not yet supplied [3]. The second multidisciplinary problem, that is the analysis of human beauty and its measure, has been widely debated for centuries in human science, and, more recently, in plastic surgery and orthodontics. In the last decades, several thousands of papers on this subject have been published in these areas. The human science researchers involved in these studies are: social and developmental psychologists, cognitive psychologists and neuroscientists and evolutionary psychologists and biologists. Applying pattern analysis and computer vision techniques for analyzing beauty is a relatively new research field. The purpose of this

A. Campilho and M. Kamel (Eds.): ICIAR 2010, Part I, LNCS 6111, pp. 425–435, 2010.

paper is to survey rationale, techniques, results, applications and open problems in this emerging area.

2 Beauty in Human Sciences and Medicine

Social importance of attractiveness. What is beauty? Philosopher, scientists and artists debated the problem for centuries. A controversial long lasting question is whether beauty is objective or subjective, or if "Beauty is in the eye of the beholder", according to a sentence of the writer Margaret Wolfe Hungerford (1878). Important personages, as David Hume (1741), have supported this thesis or, as Immanuel Kant (1790), the opposite. Coming to our times, a number of recent behavioural, social and psychological studies, as well as everyday common experience, show that face and body harmony is extremely important in general social life. Looking unpleasant or different seriously affects self-esteem and can result in social isolation, depression and serious psychological disorders [35]. Thus, is not surprising that, according to a recent estimate, in the US more money is spent annually on beauty related items or services than on both education and social services [5].

Classic Beauty canons. Since ancient times, the supporters of beauty as an objective and measurable property attempted to state ideal proportions, or beauty canons, for the human body and its parts. The Greek sculptor Polykleitos was the first to define aesthetics in mathematical terms in his "*Kanon*" treatise. Marcus Vitruvius, a Roman architect, introduced the idea of facial trisection, or facial thirds, largely used in medicine and anthropometry (Fig. 1).

Fig. 1. Facial trisection, as originally described by Vitruvius

Renaissance artists, as Leonardo da Vinci, Leon Battista Alberti, Albrecht Duerer and Piero della Francesca, reformulated and documented the classic canons. Descriptions of the classic canons can be found in [6]. These canons have been used for centuries in art by sculptors, painters, and are a rough working guide for plastic surgeons.

From the classic concept of ideal proportions also stems the debate about the relevance of the golden ratio in beautiful faces. The golden ratio is an irrational number, approximately 1.618, obtained by dividing a segment into two parts, a and b, such that $a/b = (a+b)/a$. Since ancient times, the golden ratio has been used explicitly, or claimed later to have been used, by a score of sculptors, painters, architects and even composers. Today, some papers in plastic surgery and orthodontics contexts support the idea of an universal standard of beauty based on the golden ratio [31], [32]. However, several experimental studies found a little correlation between the asserted ideal proportions and the beauty scores received by human raters [33], [34].

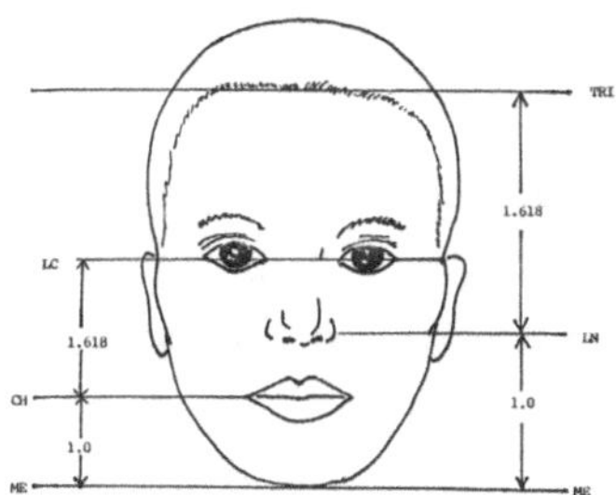

Fig. 2. The divine proportions of the face and the golden ratio

The objective nature of facial beauty. In this subsection we present the empirical results supporting the idea that facial beauty has an objective nature: *brain activity patterns*; *cross-cultural consistency of beauty ratings; infant's preference for attractive faces.*

Brain activity patterns. Psychophysiology and neuropsychology have detected the brain areas where the assessment of facial beauty is processed. Activity patterns related to explicit attractiveness judgement of 2D face images have been measured with MRI and NIRS techniques and correlated with the beauty score of the faces. Brain patterns showed greater response to highly attractive and unattractive faces [23], [47]. These results could lead to practical "ground truth" beauty assessing techniques.

Cross-cultural consistency of attractiveness ratings. Many experimental researches based on various groups of human raters have been performed. For instance, consistency of attractiveness ratings (correlation greater than 0.9) was reported in [9] for groups of Asian, Hispanic, Black and White Americans, male and female, both as subject and judges. In [10] it was reported that English, Asian, and Oriental female raters showed very close agreement in assessing the attractiveness of a selection of Greek man. Other experiments used synthetic faces [8].

Several studies compared the ratings of different professional groups, as for instance clinicians specialized in orthodontics and normal hospital clerks [11]. Attractiveness self-ratings and third-party ratings have also been compared too [41]. Moreover, most papers aimed at automatically rating beauty of previously rated images by human observers, validated the human "ground truth" ratings by checking their consistency, for instance correlating the scores of different groups.

The conclusion is that a substantial beauty rating congruence exists over ethnicity, social class, age, and sex. Rating congruence is particularly strong for very unattractive and very beautiful faces [8], which appears in agreement with the analysis of brain activity patterns.

Infant preference for attractive faces. Babies as young as three/six months were found to be able to distinguish between faces previously rated as attractive or unattractive by adult raters. This conclusion was obtained observing the time spent by the babies in looking at each face. Since very young babies should not be affected by cultural standards about beauty, these studies seem to indicate that appreciating beauty is an innate human capability [28], [39].

Concluding, even if the problem of which are the objective elements of beauty is still much debated, we can conclude that there is strong evidence that these objective elements exist, are relatively stable in time and space, and they could be measured.

3 Applications of Machine Beauty Analysis

Clearly, a fundamental application area of machine beauty analysis is supporting human sciences research. However, automatically ranking, or suggesting ways for improving attractiveness, could result in many applications in other scientific, professional and end user areas. Beauty ranking programs could be used for preparing professional carnet, screening applicants for jobs where attractiveness is a basic requirement, in social network contexts. Potentially very popular end user or professional applications could be constructed for supporting and suggesting make-up styles. A related application, automatic photo retouch, will be discussed in the following [14].

Another important application area is plastic surgery. Some computer tools have been proposed for supporting surgery planning. These tools present images of the possible effects of the surgery based on 2D images [29], or 3D scans [30], morphed with manual interfaces. How to manipulate faces, as well as the evaluation of the results, is currently left to the surgeon's judgement. Beauty scoring programs, able to evaluate the various possible surgery outcomes, or also to suggest how to enhance attractiveness would be of great help.

4 Computer-Based Beauty Analysis

In this section, we survey the papers recently appeared on the automatic analysis of beauty, and of its elements. Observe that the general approaches for most face image processing applications, including beauty analysis, are similar, and can be roughly divided into holistic, as PCA, LDA, and feature based. Holistic techniques perform an automatic extraction of significant data based on a number of face samples. The precise meaning of the data obtained, complex combination of the original 2D or 3D data, is often difficult to state. In the feature based approach the features significant for a given problem are selected *a priori.* Their meaning is clear, but elements relevant to the particular problem could have been overlooked.

Shape and texture. The relative relevance to attractiveness of face shape and colour texture has been experimentally investigated. In [24], different skin textures obtained from photographs of 170 women were applied to a common 3D face model and rendered with the same illumination. Experiments showed a high correlation between the beauty scores of the original face images and of the 3D model textured with them.

Several other results support the importance of skin colour texture for attractiveness [27], [25], especially in intersex evaluation, a thesis also supported by Darwin [26].

Fig. 3. Effects of symmetry: original face, left and right symmetries

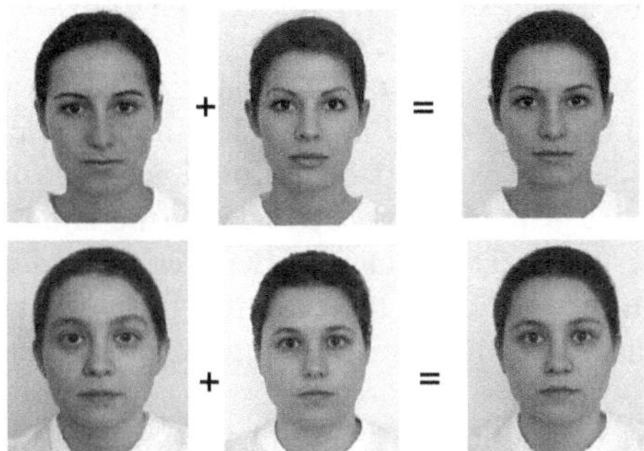

Fig. 4. Averaging faces improves (female) attractiveness

Symmetry and averageness. Several researches dealt with the role of symmetry and averageness in attractiveness. A pioneer in these studies was Sir Francis Galton, which in 1879 created photographs where the images of different faces were superimposed [40]. Today researchers use image processing techniques for finding the sagittal (symmetry) plane, locating facial landmarks, measuring asymmetry, and creating artificial symmetrical, morphed and average faces.

The effects of asymmetry on attractiveness perception have been investigated in several experiments [8], [12], [13]. Male and female images have been rated, and the ratings related with the asymmetry of the original faces and with the ratings of the faces made symmetric with left and right symmetries. The results are rather controversial. Low degrees of asymmetries do not seem to affect attractiveness. Some research even found a negative correlation between symmetry and attractiveness [13].

The effect of averageness on attractiveness perception is a related problem. According to evolutionary biology, evolutionary pressure operates against the extremes of the population, and average facial prototypes should be preferred by conspecifics [36]. Composite or average 2D face images, created by normalizing eyes and mouth position and averaging their pixel values, have been rated and their ranks compared with those of the original faces [12], [7]. Even in this case the results are controversial. For female faces, the ratings of composite faces were better than those of the original faces (see Fig. 4). However, composites are more symmetrical and rather free of facial blemishes. For male faces, composites were found less attractive than normal faces [7], possibly since attractive male faces show strong features perceived as dominance indicators and resistance to parasites [37]. According to [8], average faces are attractive, but very attractive faces are not average, as shown by the preference given to warped faces obtained by increasing the distance of facial landmarks from the average landmark position. A 3D analysis of the influence on attractiveness of averageness both of 3D shape and 2D texture is described in [19]. The 3D database included 100 young adult males and females. The 3D shapes and the 2D textures were separately averaged, and artificial face images were created in two different ways, first by morphing individual texture maps onto the average 3D head, and then the average texture onto the individual heads, using corresponding feature points. The original, texture-normalized and shape-normalized images were rated by a 36 people panel on a 5 level scale. The results show that attractiveness scores are larger for texture normalized and even more for shape-normalized images.

Enhancing beauty. An automatic system for enhancing facial attractiveness of frontal colour photographs has been presented in [14]. It is aimed at professional retouching, and requires a database of faces rated beautiful. Each face is triangulated, starting from 84 landmarks, and 234 lengths, normalized by the square root of the face area, make a representative vector. The vector of the face to beautify is compared using various techniques with the vectors of the beautiful faces. Finally, the triangulation of the original face is warped toward those of the beautiful faces more similar to the original.

A system for planning rhinoplastic surgery has been presented in [38]. In the case of rhinoplasty, the profile is the most relevant feature, and the system is based on a data-base of profiles of faces rated beautiful or at least regular. In general there is not a unique prototype of a beautiful facial feature (nose in this case), but different shapes could be more or less attractive, depending on the general harmony and integration with the rest of the face [30]. The system looks in the database for the most similar profile, excluding the nose. Then, it applies the nose profile of the selected face to the profile to improve, providing an effective suggestion for the plastic surgeon.

Assessing beauty. Several papers are aimed at automatically measuring face attractiveness. Most of these papers use the feature approach. The general idea is to look, in some particular face space, for the nearest samples of a training set of rated faces and construct a vote depending in some way on their scores.

A preliminary automatic facial beauty scoring system was described in [15]. A few face landmarks are manually determined on frontal monochromatic images, and a vector of eight ratios between landmarks distances is used to describe a face. A panel of 12 judges scored 40 training images on a four point scale. For scoring a new image, its characteristic vector is first computed. Then, the scores of the 10 nearest faces in the face space are averaged.

A similar approach is reported in [16] and [17]. Also in these cases 2D frontal images are used. 215 female images were rated on a 10 level scale from 48 human referees [16]. Standard deviation of scores for each training, showed rather compact distributions around the average vote. Smaller variances were found for very high (beautiful) and very low (unpleasant) scores. Automatically detected landmarks were used for constructing a representative vector of 13 distances ratios. Several classifiers were experimented, obtaining, on the average, score rather close to those of human referees. Investigations on the classification results in relation with age, ethnicity, gender of the referees, and with some classical beauty canons are presented in [18].

In [20], 91 color frontal images of young Caucasian female were rated on a 7 point scale by 27 raters. To validate ranks, raters were divided several times at random into 2 groups, and the ratings compared, finding 0.92 as mean correlation. For each face image, 84 feature points were automatically extracted, and a feature vector was constructed containing 3486 normalized distances between them and 3486 slopes of the distance segments. These data were reduced to 90 using PCA, and integrated with a measure of asymmetry and samples of skin colour and smoothness in selected face areas, resulting in a 98 dimension representative vector. Several rating experiments were performed with real and artificial face images, comparing human and automatic ratings, and analyzing also the relevance of the various features used. A 0.82 Pearson correlation with human ratings was found, more significant than that found in [21], owing to the larger feature vector.

A regression analysis has been used in [22] to determine the relevance to beauty of three attractiveness predictors: neoclassic canons, feature symmetry and golden ratio. The database included 420 frontal expressionless gray scale Caucasian faces selected in the FERET database, and 32 pictures of movie actors. The raters were 36, and the scores were given on a ten levels scale. Several measures were extracted from the position of 29 landmarks. The results show that several of the rules specified by these beauty predictors have actually little relation with the beauty score.
In [46], the significance to attractiveness of 17 geometric facial measures was investigated using Artificial Neural Networks. The features were classified for their relative contribution to attractiveness. Some feature dimension, as lower lip thickness, were found to be positively associated with attractiveness, other, as nose area, negatively. It has been found that the more significant are mouth width, nose width and distance between pupils, the less significant eye sizes.

Landmarks based and holistic approaches have been compared in [21]. Two data sets, each with 92 frontal images of young Caucasian American and Israeli females, were rated by 28 raters on a seven level scale, and consistency of ratings was verified. 37 normalized facial feature distances and data related to hair colour, facial symmetry and smoothness were inserted in the feature vector. PCA was used for decorrelating the geometric data. The holistic approach applied PCA on images normalized using centers of eyes and mouth. The eigenfaces most correlated with the human attractiveness ratings were selected. An interesting result is that such eigenfaces did not correspond to the highest eigenvalues, and contain clearer details of facial features like nose, eyes and lips rather than general description of hairs and face contours. For assessing attractiveness, both K-nearest neighbours and support vector machines were used, and correlation between machine and human scores was given as a function of the dimension of the feature vector. Several results are interesting. Feature based beauty prediction performed better than holistic: a top Pearson correlation of almost 0.6 versus 0.45 is reported. Probably, this is due to, the normalization of eyes and mouth position, which changes some landmark distances' ratios that are related to the general harmony of the face. A better prediction was simply obtained combining linearly the two predictions.

An automated scoring system for learning the personal preferences of individual users from example images has been presented in [42]. 70.000 web collected 2D images were used. The images were labelled for 3D pose (yaw, pitch, and roll) and for 2D positions of 6 landmarks [43]. 1000 male and 1000 female almost frontal images were selected. 8 raters were asked to state their preferences toward images of the opposite sex on a 4 points scale. For training a SVM regressor, eigenfaces, Gabor filters, edge orientation histograms, and geometric feature were used. The best average Pearson correlation with human scores (0.28) was obtained with Gabor filters, whose correlation with individual preferences was higher (up to 0.45). Some experiment was also reported for relating smile, detected through Facial Action Coding System (FACS) [4], and attractiveness.

Another large Web face database was used in [44]. From the website *hotornot.com* over 30.000 attractiveness rated images were downloaded, and the best 4000 images, almost evenly divided between the two genders, were selected and rectified with an affine transformation. Gaussian RBF kernel and a ridge regression were experimented for various textural features. The female dataset showed better prediction, and cheeks and mouth proved to be more effective predictors than eyes. A particular kernel regression technique was experimented in [45] on the same face images set.

5 Open Problems and Areas of Research

Although some interesting results are emerging, much further work is possible. In particular, the main question, which are the objective elements of facial beauty, is far from being answered. Several elements of beauty have been investigated, but not yet combined in an overall framework.

Most results have been obtained analyzing 2D images, often monochromatic, medium quality and frontal. There is little doubt that in this way much valuable information relevant to attractiveness is lost. Important applications, as supporting plastic surgery, are essentially 3D and require 3D face scans. A problem for further 3D beauty research, as well as for 3D identity recognition, is that only a few 3D face data bases exist, containing a relatively small number of face scans. In addition, selecting beautiful faces in these data bases strongly reduces the number of samples useful for attractiveness studies. Then, for carrying on further studies on attractiveness, 3D databases containing also beautiful faces should be constructed.

Another open problem concerns the density of sampling of the face and beautiful face spaces. In fact, several approaches for measuring attractiveness, or suggesting ways for improving attractiveness are based on finding the nearest face samples in some face space. To be effective this approach requires a dense sampling of the face space, or of the space of the beautiful faces only. This raises the question: how many samples are required for an adequate sampling of the face space, from the point of view of attractiveness, or of the sub-manifold of the beautiful faces?

Most papers on analyzing and assessing beauty are based on facial landmarks for constructing some representative geometric feature vector. This technique appears convenient for capturing the general harmony of face, but small details and facial texture, important elements of beauty, are essentially lost. Holistic techniques appear more suited to capture the texture. Small shape details of particularly important areas, such as mouth and eyes, are not efficiently captured neither by 2D or 3D landmarks nor by holistic techniques. A local detailed analysis could substantially improve the capture of relevant features. Then, mixed techniques could be effective.

Up to now, the matter of attractiveness research has been expressionless images. However, expressions are relevant to attractiveness: it is a common everyday experience that smiling can light a plain nondescript face. Up to now, no research has been reported aimed at extending attractiveness analysis to facial expressions.

Finally, other areas of research could concern: body attractiveness (actually limited to simple body shape indices), feasible shape or texture changes able to enhance attractiveness, and the study of *dynamic beauty,* or "grace", or "elegance", of movements.

References

[1] Zhao, W., Chellappa, R., Philips, P.J., Rosenfeld, A.: Face Recognition, a literature survey. ACM Computing Surveys 35(4), 399–458 (2003)

[2] Pantic, M., Rothkrantz, L.J.M.: Toward an affect-sensitive multimodal human-computer interaction. IEEE Proc. 91(9) (September 2003)

[3] Fasel, B., Luettin, J.: Automatic facial expression analysis: a survey. Pattern Recognition 36, 259–275 (2003)

[4] Ekman, P.: Facial Expressions. In: Handbook of Cognition and Emotion. John Wiley&Sons, New York (1999)

[5] Adamson, P., Doud Galli, S.: Modern concepts of beauty. Current Opinion in Otolaryngology& Head and Neck Surgery 11, 295–300 (2003)

[6] Bashour, M.: History and Current Concepts in the Analysis of Facial Attractiveness. Plastic and Reconstructive Surgery 118(3), 741–756 (2006)

[7] Grammer, K., Thornhill, R.: Human(*Homo hsapiens*) facial attractiveness and sexual selection: the role of symmetry and averageness. J. Comparative Psychology 108, 233–242 (1994)

[8] Perret, D., May, K., Yoshikawa, S.: Facial shape and judgement of female attractiveness. Nature 368, 239–242 (1994)

[9] Cunningham, M.R., Roberts, A.R., Barbee, A.P., Wu, C.H., Druen, P.B.: Their ideas of beauty are, on the whole, the same as ours: consistency and variability in the crosscultural perception of female physical attractiveness. J. Pers. Soc. Psychol. 68, 261–279 (1995)

[10] Thakera, J.N., Iwawaki, S.: Cross-cultural comparisons in interpersonal attraction of females towards males. Journal of Social Psychology 108, 121–122 (1979)

[11] Knight, H., Keith, O.: Ranking facial attractiveness. Europ. J. of Othodonthics 27, 340–348 (2005)

[12] Langlois, J.H., Roggman, L.A.: Attractive faces are only average. Psychological Science 1, 115–121 (1990)

[13] Swaddle, J.P., Cuthill, I.C.: Asymmetry and human facial attractiveness: simmery may not always be beautiful. Proc. R. Soc. Lond. B 261, 111–116 (1995)

[14] Leyvand, T., Cohen-Or, D., Dror, G., Lischinski, D.: Data-driven enhancement of facial attractiveness. ACM Trans. Graph. 27(3), 1–9 (2008)

[15] Aarabi, P., Hughes, D., Mohajer, K., Emami, M.: The automatic measurement of facial beauty. In: IEEE Proc. Int. Conf. on Systems, Man and Cyb., pp. 2168–2174 (2004)

[16] Gunes, H., Piccardi, M., Jan, T.: Comparative beauty classification for pre-surgery planning. In: IEEE Proc. Int. Conf. On Systems, Man and Cyb., vol. 4, pp. 2644–2647 (2001)

[17] Gunes, H., Piccardi, M., Jan, T.: Automated classification of female facial beauty by image analysis and supervised learning. In: Proc. of SPIE Symposium on Electronic Imaging 2004, Conference on Visual Communications and Image Processing, San Jose, California, USA, January 18-22, vol. 5308, Part 2, pp. 968–978 (2004)
[18] Gunes, H., Piccardi, M.: Assessing facial beauty through proportion analysis by image processing and supervised learning. Int. J. Human-Computer Studies 64, 1184–1199 (2006)
[19] O'Toole, A.J., Price, T., Vetter, T., Bartlett, J.C., Blanz, V.: 3D shape and 2D surface textures of human faces: the role of "averages" in attractiveness and age. Image and Vision Computing 18, 9–19 (1999)
[20] Kagian, A., Dror, G., Leyvand, T., Cohen-Or, D., Ruppin, E.: A humanlike predictor of facial attractiveness. Adv. Neural Info. Proc. Syst. 19, 674–683 (2008)
[21] Eisenthal, Y., Dror, G., Ruppin, E.: Facial Attractiveness: beauty and the machine. Neural Computaation 18, 119–142 (2006)
[22] Schmid, K., Marx, D., Samal, A.: Computation of face attractiveness index based on neoclassic canons, symmetry and golden ratio. Pattern Recognition 41, 2710–2717 (2008)
[23] Winston, J.S., O'Doherty, J., Kilner, J.M., Perret, D.I., Dolan, R.J.: Brain Systems for assessing facial attractiveness. Neuropsychologia 45, 195–206 (2007)
[24] Fink, B., Grammar, K., Matts, P.: Visible skin color distribution plays a role in the perception of age, attractiveness, and health in female faces. Evolution of Human Behaviour 27, 433–442 (2006)
[25] Jones, B., Little, A., Burt, D., Perret, D.: When facial attractiveness is only skin deep. Perception 33, 569–576 (2004)
[26] Darwin, C.: The descent of men and selection in relation to sex. John Murray, London (1871)
[27] Fink, B., Grammer, K., Thornhil, R.: Human (*Homo Sapiens*) facial attractiveness in relation to skin texture and color. J. of Comparative Psychology 115, 92–99 (2001)
[28] Langlois, J., Roggman, L., Casey, R., Ritter, J., Rieser-Danner, L., Jenkins, V.: Infant preference for attractive faces: rudiment of a stereotype? Dev. Psych. 23, 363–369 (1987)
[29] Ozkul, T., Ozkul, M.H.: Computer simulation tool for rhinoplasty planning. Comput. in Biol. and Med. 34, 697–718 (2004)
[30] Lee, T., Sun, Y., Lin, Y., Lin, L., Lee, C.: Three-dimensional facial model reconstruction and plastic surgery simulation. IEEE Trans. on Info. Tech. in Biomed. 3(3), 214–220 (1999)
[31] Jefferson, Y.: Facial Beauty: establishing a universal standard. Int. J. Orthod. 15, 9–22 (2004)
[32] Ricketts, R.M.: Divine proportions in facial aesthetics. Clin. Plast. Surg. 9, 401–422 (1982)
[33] Holland, E.: Marquardt's Phi Mask: pitfalls of relying on fashion models and the golden ratio to describe a beautiful face. Aesth. Plast. Surg. 32, 200–208 (2008)
[34] Baker, B.W., Woods, M.G.: The role of the divine proportions in the esthetic improvement of patients undergoing combined orthodontic/orthognathic surgical treatment. Int. J. Adult. Orthod. Orthognath. Surg. 16, 108–120 (2001)
[35] Rankin, et al.: Quality-of-life outcomes after cosmetic surgery. Discussion. Plast. and Reconstr. Surg. 12, 2139–2147 (1998)
[36] Barash, D.P.: Sociobiology and behaviour. Elsevier North Holland, New York (1982)
[37] Hausfater, G., Tornhill, R. (eds.): Parasites and sexual selection. American Zoologist (special issue) 30 (1990)

[38] Bottino, A., Laurentini, A., Rosano, L.: A New Computer-aided Technique for Planning the aesthetic Outcome of Plastic Surgery. In: Proc. WSCG 2008 (2008)
[39] Dion, K., Bertscheid, E.: Physical attractiveness and perception among children. Sociometry 37, 1–12 (1974)
[40] Galton, F.: Composite portraits, made by combining those of different persons in a single resultant figure. J. of the Anthropological Inst. 8, 132–144 (1879)
[41] Weeden, J., Sabini, J.: Subjective and objective measures of attractiveness and their relation to sexual behavior and sexual attitudes in university students. Arch. Sex Behav. 36, 79–88 (2007)
[42] Whithehill, J., Movellan, J.: Personalized facial attractiveness prediction. IEEE Proc. 8th Int. Conf. Aut. Face &Gesture Reco., 1–7 (September 17-19, 2008)
[43] Whithehill, J., Littlewort, G., Fasel, I., Bartlett, M., Movellan, J.: Developing a pratical smile detector. IEEE Trans. PAMI (2007) (submitted)
[44] White, R., Eden, A., Maire, M.: Automatic prediction of human attractiveness.UC Berkeley CS280A Project (2004)
[45] Davis, B.C., Lazebnik, S.: Analysis of human attractiveness using manifold kernel regression. In: IEEE Proc. ICIP 2008, pp. 109–112 (2008)
[46] Joy, K., Primeaux, D.: A comparison of two contributive analysis methods applied to an ANN modelling facial attractiveness. In: IEEE Proc. 4th Int. Conf. on Soft. Eng. Res. Manag. and Appl. (2006)
[47] Mitsuda, T., Yoshida, R.: Application of near-infrared spectroscopy to measuring of attractiveness of opposite-sex faces. In: Proc. IEEE 27th Conf. Engineering in Medicine and Biology, Shangai, China, September 1-4, pp. 5900–5903 (2005)

System and Analysis Used for a Dynamic Facial Speech Deformation Model

Jürgen Rurainsky

Fraunhofer Heinrich-Hertz-Institute, Einsteinufer 37, 10587 Berlin, Germany
rurainsky@hhi.fraunhofer.de
http://www.hhi.fraunhofer.de

Abstract. While facial expressions and phoneme states are analyzed and published very well, the dynamic deformation of a face is rarely described or modeled. Recently dynamic facial expressions are analyzed. We describe a capture system, processing steps and analysis results useful for modeling facial deformations while speaking. The capture system consists of a double mirror construction and a high speed camera, in order to get fluid motion. Not only major face features as well as a high accuracy of the tracked facial points, are required for such analysis. The dynamic analysis results demonstrate the potential of a reduced phoneme alphabet, because of similar 3D shape deformations. The separation of asymmetric facial motion allows to setup a personalized deformation model, besides the common symmetric deformation.

Keywords: motion, facial deformation, personalized.

1 Introduction

Different capture systems for the analysis of facial motions have been presented in recent years. Although single or multi camera approaches have been addressed with different configurations, mirrors are rarely used. One reason for this could be the resolution of the capture unit, which is shared with all virtual views.

Many different techniques for the classification of facial expressions in still images has been published. Recently also the methods for dynamic analysis of facial expressions pushing forward and described by Cohen *et al.*[1], Hu *et al.*[2] and Zhang *et al.*[3]. The dynamic deformation produced while speaking has not been analyzed so far.

Since we target for the analysis of the dynamic behavior of facial motion, the sampling rate, in which the motion states are recorded, is an important issue. Important transitions from one state to another maybe get lost if only a video frame rate of 25 fps is used and these details are not available for the natural animation of 3D models.

We present our capture system based on a high speed camera and two surface mirror. These three views are used to reconstruct a 3D model sequence of a talking person's face. This model sequence is used to compensate the rigid motion and analyzing the facial deformation to a reference model. The analysis of the dynamical performance of face while speaking phonemes is described as well.

A. Campilho and M. Kamel (Eds.): ICIAR 2010, Part I, LNCS 6111, pp. 436–444, 2010.

2 Capture Device

We have constructed a system with two mirrors which is shown in **Fig. 1**. Mirrors are widely used within capture devices and described in several publications [4,5,6], but no high speed camera was used so far as capture unit and not always a 3D reconstruction was placed in the processing steps. The four flat lights are used to illuminate the scene uniformly.

Fig. 1. High speed camera and mirror construction with two surface mirrors. Top Left Corner: Schematic scheme of the construction.

By using mirrors additional points have to be considered. The surface of the used mirror is one important parameter, because these surface properties will be incorporated into the view calibration parameter. Other parameters are the reflection properties in the sense of color, magnification and multiple reflections. Multiple reflections appear because of the reflection on the glass and coating boundary. It is very common to use not surface mirrors based on a glass body, but to use polished metal mirrors instead. The stiffness of such metal mirror is much lower than a glass body based mirror. Therefore, an additional fixation has to be considered in case of metal based mirrors. We have used surface mirrors with a glass body.

2.1 Calibration

In order to get full benefit out of such system, the system has to be calibrated. Taking one point in 3D world as reference and measuring the light ray distance

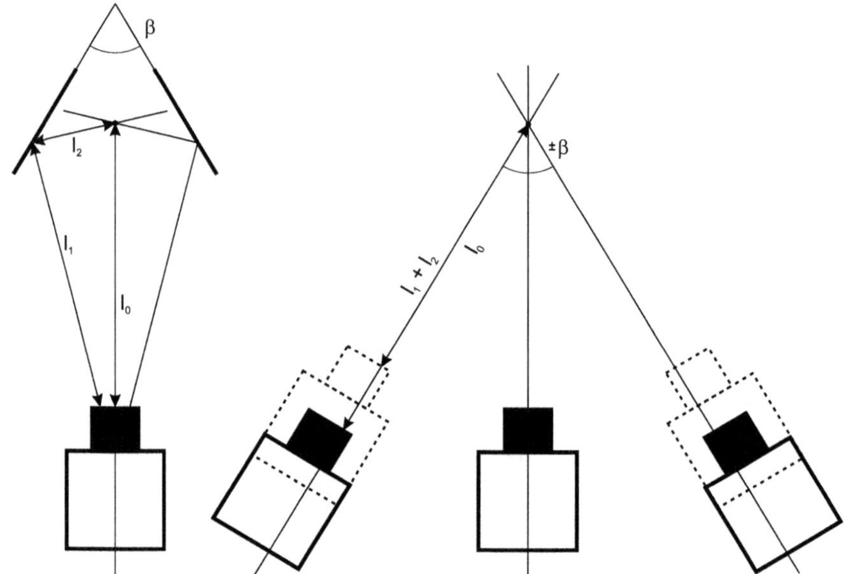

Fig. 2. (left) Capture device with two mirrors and one camera. (right) Setup interpretation used for calibration and 3D reconstruction.

from this point to each view (two mirror views and one direct view), leads us to a system with two virtual cameras and one real camera. **Fig. 2** shows this representation, where the same point viewed by the virtual cameras appears to be more far away than in reality. The distance between several 3D world points should be therefore also closer, but there are not. They have the same distance as seen by the real camera. The light ray distance can not be neglected and therefore the virtual cameras has to be adjusted, in order to magnify these views. Therefore, each view (camera) has is own position in 3D world.

Using point correspondences captured from a calibration cube and a non-linear equation solver gives as the required intrinsic and extrinsic camera parameters. The back projection of the manual selected corner points used for the calibration can be used to determine the calibration error, which is below $0.5 pixel$ and therefore sufficient for the following analysis. The accuracy of the calibration also allows the usage for depth map determination approaches, like described in [7] *et al.*for a high density of feature points.

3 Principal Analysis

The performance of a face while speaking a specific set of words was captured with 200 fps using blue tape markers and for each frame a 3D shape was reconstructed by triangulation of the tracked marker points. The total amount of

43 words has been captured with an average duration of 217 frames. Each word represents a specific British and American phoneme [8].

3.1 Motion Model

For the analysis of facial motion, we separate rigid body motion from deformations using a 3D sequence of facial points. The 3D model sequence is generated by triangulation of markers, which are placed on a human face and captured by a double mirror construction and a high speed camera. Rotation and translation for all axes (6 DOFs) of the associated 3D model describe the rigid-body motion and all other changes are regarded as deformation and noise.

4 Dynamic Comparison of Spoken Phonemes

While facial expressions and phoneme states are analyzed and published very well, the dynamic deformation of a face is rarely described or modeled. The amount of sampling points and the sampling frequency are the interesting measurements for such analysis and therefore define the value of an appropriate deformation model.

4.1 Dynamic Time Warping (DTW)

Is a very known method for the comparison and motion model description used for dynamic data. Early works of Rabiner *et al.*[9] as well as Sakoe and Chiba [10] use DTW for the comparison of audio signals of spoken words. Gestures comparison as well as recognition are well described in representative publications of Corradini [11] or Li and Greenspan [12]. The recognition of hand writing using DTW is described by Niels [13]. There are many more publications dealing with DTW in various scenarios.

The main idea behind Dynamic Time Warping (DTW) is to map two sampling sets independently from the sampling rate as well as sampling period. Ratanamahatana and Keogh analyzed the behavior of DTW with different preprocessing steps and constraints [14]. One suggestion is not equalize different datasets before mapping, which was used for our analysis.

We use DTW to analyze dynamic 3D shape deformation. Actually, we just use the weights for a specific set of eigenshapes in contrast to Angeles-Yreta and Figueroa-Nazuno, who describe a measurement method for similarities of 3D objects in [15] by using distances within the 3D shapes. The eigenshapes are the same for all phonemes and only the frame based weights are describing the difference from one phoneme data set to another phoneme data set. With other words, this could be used to compare data sets from different speakers while the sampling time do not have to be the same through all data sets. In addition facial deformations for phonemes can be correlated, in order to find an appropriate subset and therefore to reduce the amount of phonemes for the same performed deformation.

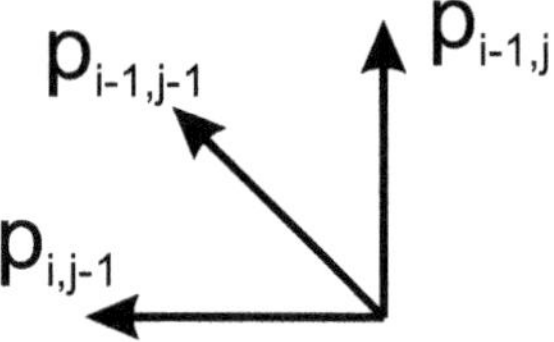

Fig. 3. Type I mapping function described by Rabiner and Juang [16]

The **Eqn. 1** shows the reconstruction of a specific shape F for a defined range of eigenshapes R after adding the average $\bar{\mathbf{B}}$ as well as the offset $\mathbf{A}^{(3N,0)}$.

$$\mathbf{A}_{recon.}^{(3N,F)} = \mathbf{A}^{(3N,0)} + \bar{\mathbf{B}} + \sum_{i=0}^{R} w_{i,F} \cdot \mathbf{Eig}_i \tag{1}$$

While the main method for DTW is well described, the used type of mapping function is based on a best performance result and therefore no specific

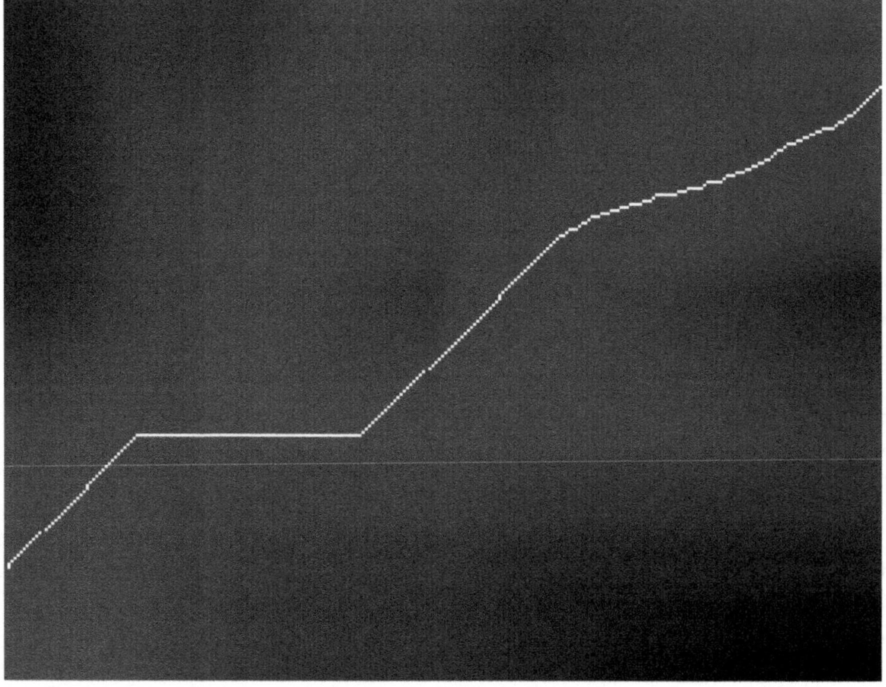

Fig. 4. Distance matrix for the DTW mapping of two phonemes represented by the words *away* and *arm*. The white colored line shows the mapping path with smallest energy. The weights for the first eight eigenshapes are incorporated into the distance matrix.

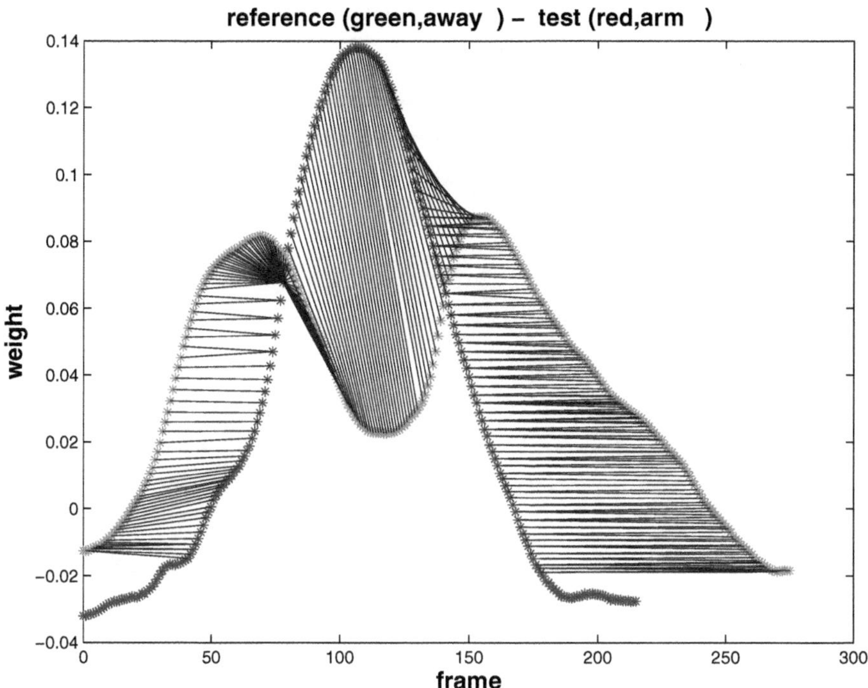

Fig. 5. Mapping between the frame based weights with the first eigenshape used to represent the words *away* and *arm*. Only the weights for the first eigenshape was incorporated into the distance matrix.

formulation is defined. In the book of Rabiner and Juang [16] 7+1 types of path specifications are described, but mostly type I is used and shown in **Fig. 3**. Besides the type of path specification the incorporation of multidimensional data can be used to change the performance of the mapping. Holt *et al.*[17] extending the DTW algorithm to the multidimensional approach MD-DTW, where different dimensions could be connected by a simple Euclidean distance for instance. For the analysis of dynamic facial deformation based on weights, the weights of several eigenshapes are included to the distance matrix by calculating the Euclidean distance. The distance matrix of two different datasets using 8 eigenshapes and therefore also 8 weights as well as path type I is shown in **Fig. 4**. The determined path is visualized with the white colored line. The mapping of these two different datasets (different length) is used to specify the correlation of the words *away* and *arm*, which represent two different visual dynamic deformations of a face during the pronunciation of these phonemes. **Fig. 5** and **Fig. 6** show the mapping results by incorporating the weights for one and eight eigenshapes to the distance matrix. Around frame 150 are difference between these both mappings can be seen, which also lead to different results. Both data

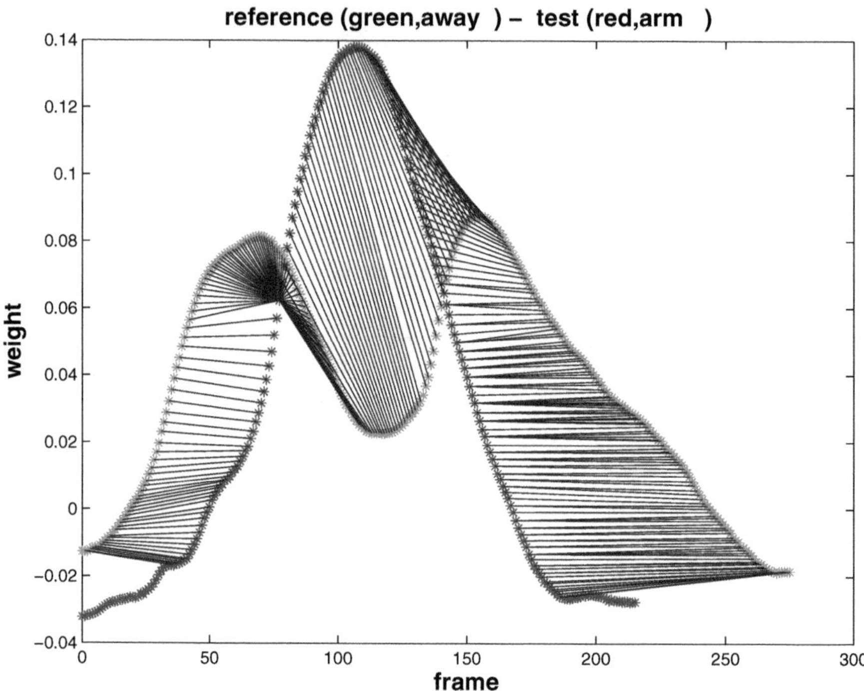

Fig. 6. Mapping between the frame based weights with the first eigenshape used to represent the words *away* and *arm*. The weights for the first eight eigenshapes were incorporated into the distance matrix.

sets have different length, but the mapping shows useful results, which supports the suggestion of Ratanamahatana and Keogh [14].

4.2 Experimental Results

The idea is to analyze the dynamic behavior of facial deformation while speaking. In order to find the right representation or subset of dynamic shape deformations, the smallest elements have to be compared. Phonemes already used to model static facial deformation and therefore we compare the dynamic shape of a phoneme alphabet. Each phoneme was compared to all other phonemes and the result will be a distance matrix showing the Euclidean distances after the DTW mapping of the to be compared data sets. We compared 52 data sets including phonemes and facial deformations, which are are done by unattended motions, like getting the lips wet. We incorporated only the first eight eigenshapes for the common facial deformation representation and left the higher order eigenshapes for asymmetric and therefore personalized deformation out. **Fig. 7** shows the result of this experiment, which leads to the awareness that the recorded phoneme examples show a high correlation and therefore this data set can be reduced.

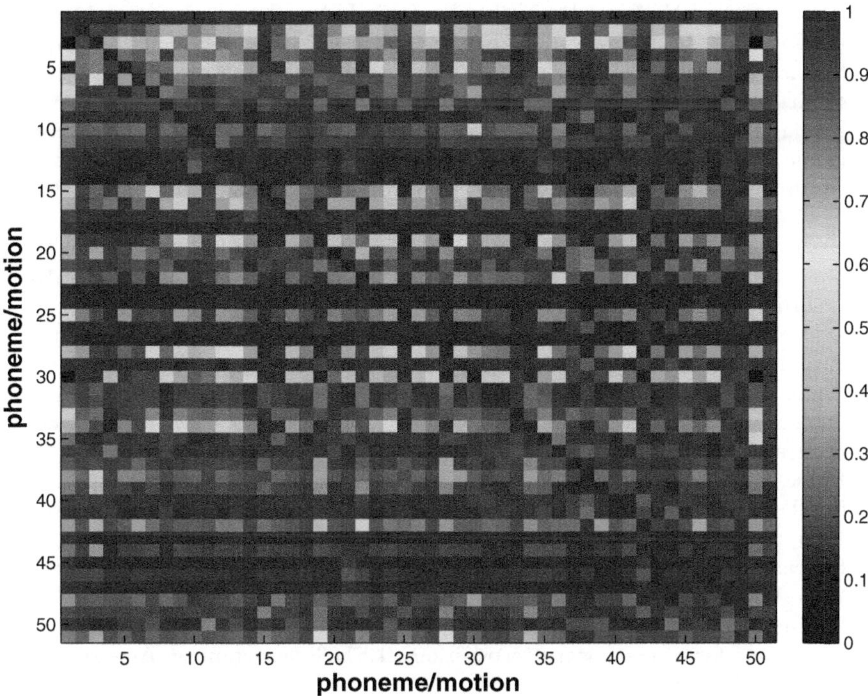

Fig. 7. Cross comparison of data sets including the dynamic facial deformation of phonemes and unattended motions. The dialog line shows the highest comparison result, because of the comparison of identical data sets.

5 Conclusion

We have shown and described a system as well as analysis methods and results for dynamic facial deformations, which can be observed during the pronunciation of words. The system allows to capture the dynamic shape motions with up to 200 fps and the double mirrors provide us with the desired 3D shapes. Extracting the rigid motion and the eigenshape representation of these observed deformations are described as well. Dynamic Time warping (DTW) is used to compare different data sets, where multidimensional data in the form of eigenshape weights are incorporated into the data set mapping. The direct comparison of a set of 52 phoneme and unattended motions is provided in the form of a distance matrix. This matrix allows the assumption, that further reduction can be applied without losing major deformations.

References

1. Cohen, I., Sebe, N., Garg, A., Lew, M.S., Huang, T.S.: Facial Expression Recognition from Video Sequences: Temporal ansd Static Modeling. In: Computer Vision and Image Understanding, September 2003, pp. 160–187 (2003)

2. Hu, C., Chang, Y., Feris, R., Turk, M.: Manifold based analysis of facial expression. In: CVPRW 2004: Proceedings of the 2004 Conference on Computer Vision and Pattern Recognition Workshop (CVPRW 2004), Washington, DC, USA, vol. 5, p. 81. IEEE Computer Society, Los Alamitos (2004)
3. Zhang, Y., Ji, Q.: Facial expression understanding in image sequences using dynamic and active visual information fusion. In: ICCV 2003: Proceedings of the Ninth IEEE International Conference on Computer Vision, Washington, DC, USA, p. 1297. IEEE Computer Society, Los Alamitos (2003)
4. Basu, S.: A Three-Dimensional Model of Human Lip Motions. Master's thesis, Massachusetts Institute of Technology Department of EECS, Cambridge, MA, USA (February 1997)
5. Odisioa, M., Baillya, G., Eliseia, F.: Tracking talking faces with shape and appearance models. Journal of Speech Communication 44, 63–82 (2004)
6. Badin, P., Bailly, G., Revéret, L., Baciu, M., Segebarth, C., Savariaux, C.: Three-dimensional articulatory modeling of tongue, lips and face, based on MRI and video images. Journal of Phonetics 30(3) (2002)
7. Dainese, G., Marcon, M., Sarti, A., Tubaro, S.: Accurate Depth-map estimation for 3D face modelling. In: 13th European Signal Processing Conference, September 2005, pp. 1883–1886 (2005)
8. International Phonetic Alphabet (IPA) for British and American phonemes. World Wide Web electronic publication
9. Rabiner, L., Rosenberg, A., Levinson, S.: Considerations in dynamic time warping algorithms for discrete word recognition. IEEE Transaction on Acoustics, Speech and Signal Processing 26, 575–582 (1978)
10. Sakoe, H., Chiba, S.: Dynamic programming algorithm optimization for spoken word recognition. IEEE Transaction on Acoustics, Speech and Signal Processing 26, 43–49 (1978)
11. Corradini, A.: Dynamic Time Warping for Off-Line Recognition of a Small Gesture Vocabulary. In: IEEE ICCV Workshop on Recognition, Analysis, and Tracking of Faces and Gestures in Real-Time Systems (RATFG-RTS 2001), Washington, DC, USA, p. 82. IEEE Computer Society, Los Alamitos (2001)
12. Li, H., Greenspan, M.: Segmentation and Recognition of Continuous Gestures. In: Proc. of the IEEE International Conference on Image Processing ICIP 2007, San Antonio, TX, USA, pp. 365–368 (September 2007)
13. Niels, R.: Dynamic Time Warping: An Intuitive Way of Handwriting Recognition? Master's thesis, Radboud University Nijmegen, Nijmegen, The Netherlands (2004)
14. Ratanamahatana, C.A., Keogh, E.: Three Myths about Dynamic Time Warping. In: Proc. of the SIAM International Conference on Data Mining (SDM 2005), Newport Beach, CA, USA, pp. 506–510 (April 2005)
15. Angeles-Yreta, A., Figueroa-Nazuno, J.: Computing Similarity Among 3D Objects Using Dynamic Time Warping. Springer, Heidelberg (2005)
16. Rabiner, L., Juang, B.H.: Fundamentals of speech recognition. Prentice-Hall, Inc., Upper Saddle River (1993)
17. ten Holt, G., Reinders, M., Hendriks, E.: Multi-dimensional dynamic time warping for gesture recognition. In: Proc. of the 13th Conference of the Advanced School of Computing and Imaging, Delft, The Netherlands, pp. 158–165 (2007)

Face Recognition from Color Images Using Sparse Projection Analysis

Vitomir Štruc and Nikola Pavešić

Faculty of Electrical Engineering,
University of Ljubljana,
Tržaška 25, SI-1000 Ljubljana, Slovenia
{vitomir.struc,nikola.pavesic}@fe.uni-lj.si.com
http://luks.fe.uni-lj.si/

Abstract. The paper presents a novel feature extraction technique for face recognition which uses sparse projection axes to compute a low-dimensional representation of face images. The proposed technique derives the sparse axes by first recasting the problem of face recognition as a regression problem and then solving the new (under-determined) regression problem by computing the solution with minimum L_1 norm. The developed technique, named Sparse Projection Analysis (SPA), is applied to color as well as grey-scale images from the XM2VTS database and compared to popular subspace projection techniques (with sparse and dense projection axes) from the literature. The results of the experimental assessment show that the proposed technique ensures promising results on un-occluded as well occluded images from the XM2VTS database.

Keywords: Image processing, biometrics, face recognition, regression problem, sparse projection axes.

1 Introduction

It is a well known fact that the existing face recognition techniques are sensitive to external factors influencing the appearance of the human face in an image. Among these factors, glasses, scarfs, hats or any other objects occluding the facial region have an immense effect on the representation of the face image in the given feature space and consequently on the performance of the face recognition technique used [1].

The reason for this sensitivity can be linked to the way the face images are usually transformed into the feature space. This transform commonly involves computing a projection of the facial image onto a low-dimensional subspace. If the image contains occlusions even on a very small part of the face, the feature (or subspace) representation may differ significantly from the true feature representation of the un-occluded face image. To derive a stable low-dimensional representation and hence to overcome the sensitivity of the existing methods to occlusions in images, robust techniques (e.g., [2]) and local approaches (e.g., [3]) have been proposed in the literature. By computing facial features in a robust manner or by considering only a small part of the image at a time (when

A. Campilho and M. Kamel (Eds.): ICIAR 2010, Part I, LNCS 6111, pp. 445–453, 2010.

calculating the low-dimensional subspace representation) these approaches are capable of improving upon the recognition performance of the holistic methods, especially when parts of images are occluded or degraded in some way.

Recent advances in sparse signal recovery and compressed sensing opened up new possibilities for the design of local techniques as well as new ways of tackling the problem of occlusion in face recognition [4]. Based on these developments we present in this paper a novel technique for facial feature extraction called Sparse Projection Analysis (SPA). SPA derives a low-dimensional face representation in the form of projection coefficients computed by projecting face images onto a set of sparse projection axes. Different from other local appearance based methods, such as, for example, independent component analysis (architecture I) [3], the non-zero elements of the projection axes are not localized but are rather distributed over the entire projection axis. Due to this property, the proposed technique should exhibit even less sensitivity to image occlusions than traditional local appearance based feature extraction techniques. It should be noted that the proposed techniques exhibits similarities with established face recognition methods such as independent component analysis [3] or non-negative matrix factorization [5] (due to the sparse nature of the projection axes), while the relation to the method presented in [4] is given only by the fact that SPA relies on the L_1 norm instead of the commonly used L_2 norm. Hence, the work presented in this paper is related more to the work presented in [3], [5] or [6], rather than [4].

The developed SPA technique is ultimately assessed in a series of face verification experiments performed on the original images of the XM2VTS database and on degraded images with an artificially added occlusion. The results of the assessment and comparative evaluations with the popular principal component analysis and independent component analysis show the effectiveness of the proposed approach for face recognition.

2 The Sparse Projection Analysis

This section commences by formulating the problem of face recognition as a regression problem and then, based on this formulation, develops the novel sparse projection analysis (SPA) for face recognition.

2.1 Face Recognition as a Regression Problem

In its most basic form the problem of face recognition can be defined as a mapping f from the given face pattern vector $\mathbf{x}$ to a class label (or identity) ω_i associated with the i-th class C_i, where $i \in \{1, 2, ..., N\}$ and N denotes the number of identities enrolled in the recognition system. Formally, this can be written as:

$$f : \mathbf{x} \mapsto \omega_i, \text{ for } \ i \in \{1, 2, ..., N\}. \tag{1}$$

While the most natural way for humans is to associate identities with names, the class labels are in general not restricted to textual descriptions, but can rather take an arbitrary form which uniquely describes the class.

To illustrate this concept let us assume that we have a set of three face pattern vectors $\mathbf{x}_1$, $\mathbf{x}_2$, $\mathbf{x}_3$ each belonging to a different class C_1, C_2, C_3. The corresponding class labels could then be written as vectors of the following form: $\omega_1 = [1\ 0\ 0]$, $\omega_2 = [0\ 1\ 0]$ and $\omega_3 = [0\ 0\ 1]$. From this example we can see that the class labels can be chosen as an orthonormal basis of a vector space, whose dimensionality is defined by the number of classes that need to be labeled.

Let us now consider a more general case and presume a set of n d-dimensional face pattern vectors $\mathcal{X} = \{\mathbf{x}_i \in \mathbb{R}^d, i = 1, 2, ..., n\}$ belonging to N classes C_1, C_2, ..., C_N with associated class labels ω_1, ω_2, ..., ω_N. If we arrange the face pattern vectors in $\mathcal{X}$ into the $n \times d$ row data matrix $\mathbf{X}$ and use the same principle as illustrated in the above example to construct our labels, then we can build the $n \times N$ label matrix $\mathbf{Y}$, whose rows represent labels of the pattern vectors in $\mathbf{X}$. In this case the label matrix $\mathbf{Y}$ takes the following form [7], [8], [9]:

$$\mathbf{Y} = \begin{bmatrix} \mathbf{1}_{n_{C_1}} & \mathbf{0}_{n_{C_1}} & \cdots & \mathbf{0}_{n_{C_1}} \\ \mathbf{0}_{n_{C_2}} & \mathbf{1}_{n_{C_2}} & \cdots & \mathbf{0}_{n_{C_2}} \\ \vdots & \vdots & \ddots & \vdots \\ \mathbf{0}_{n_{C_N}} & \mathbf{0}_{n_{C_N}} & \cdots & \mathbf{1}_{n_{C_N}} \end{bmatrix}, \tag{2}$$

where n_{C_i} represents the number of face pattern vectors corresponding to class C_i, $\mathbf{1}_{n_{C_1}}$ denotes a $n_{C_i} \times 1$ vector of all ones and $\mathbf{0}_{n_{C_i}}$ stands for a $n_{C_i} \times 1$ vector of all zeros.

Since we have encoded the class labels in the form of a (label) data matrix $\mathbf{Y}$, we can use this matrix to define the mapping f. We implement f using a simple linear transformation of the pattern vectors in $\mathbf{X}$ as follows:

$$\mathbf{Y} = \mathbf{X}\mathbf{W}, \tag{3}$$

where $\mathbf{W}$ denotes the $d \times N$ transformation matrix.

Clearly, the above expression can be thought of as being a regression problem with the goal of finding the regression matrix $\mathbf{W}$ capable of mapping the input variables in $\mathbf{X}$ to the response variables in $\mathbf{Y}$.

2.2 Computing the Projection Basis

While there are several ways to determine the regression matrix $\mathbf{W}$ (see, for example, [8]), we present in this paper a method which results in the N d-dimensional column vectors comprising the regression matrix $\mathbf{W}$ being sparse.

Let us first have a closer look at our problem defined by Eq. (3). It is easy to notice that each of the columns in $\mathbf{Y}$ represents the projection coefficients of all pattern vectors in the data matrix $\mathbf{X}$ on one of the N column vectors comprising $\mathbf{W}$. Hence, the problem of determining the regression matrix $\mathbf{W}$ can be broken down into a set of N independent sub-problems of the following form:

$$\mathbf{y}_i = \mathbf{X}\mathbf{w}_i, \text{ for } \ i = 1, 2, ..., N, \tag{4}$$

where the $n \times 1$ vector $\mathbf{y}_i$ denotes the i-th column of the response matrix $\mathbf{Y}$, and similarly the $d \times 1$ vector $\mathbf{w}_i$ stands for the i-th column of $\mathbf{W}$. It has to be noted that the vector $\mathbf{y}_i$ does not correspond to an encoded label, but is rather composed of the i-th elements of all encoded labels.

In the field of face recognition the dimensionality of the face pattern vectors in $\mathbf{X}$ is usually much larger than the available number of training pattern vectors, i.e., $d \gg n$. The equation defined by (4) is, therefore, under-determined and its solution is not unique. We can, however, overcome this problem by selecting the solution with minimum norm [4]. While usually the L_2 norm is used for this purpose, recent research in the field of compressed sensing suggests that, when adopting the L_1 norm rather than the L_2 norm, the found solution exhibits a number of desirable properties, e.g., sparseness. To solve Eq. (4) for $\mathbf{w}_i$, i.e., to find the i-th projection axis, we therefore recast the problem as follows:

$$\mathbf{w}_i = \arg\min \|\mathbf{w}\|_{L_1}, \quad \text{subject to} \quad \mathbf{y}_i = \mathbf{X}\mathbf{w}, \tag{5}$$

where $\mathbf{w}$ denotes the (non-unique) solution of the above problem. As stated in [4] the above problem can be solved in polynomial time using standard linear programming techniques (see [10],[11]).

If we examine the projection matrix $\mathbf{W}$ more closely, we can notice that the individual projection axes $\mathbf{w}_i$ are not orthogonal and, hence, do not necessary form a basis of our N-dimensional vector space. As the final processing step we, therefore, orthogonalize the projection matrix and use the result as our final mapping from $\mathbf{X}$ to $\mathbf{Y}$, i.e.:

$$\mathbf{W} = \mathbf{W}(\mathbf{W}^T\mathbf{W})^{-0.5}. \tag{6}$$

Clearly, the orthogonalization procedure also influences the label data matrix $\mathbf{Y}$, which now turns into

$$\mathbf{Y}' = \mathbf{Y}(\mathbf{W}^T\mathbf{W})^{-0.5}. \tag{7}$$

This matrix, i.e., $\mathbf{Y}'$, is ultimately employed to construct client models, which are then stored in the system's database. Note that the client model (or template) for the i-th subject is computed as the mean vector of all encoded class labels (i.e., rows of $\mathbf{Y}'$) corresponding to the subject labeled as ω_i.

When a new query image arrives at the input of the face recognition system it is simply projected into the sparse subspace using the orthogonalized version of $\mathbf{W}$.

As already indicated above, the exploited algorithm for finding the projection basis results in the computed projection axes being sparse. This fact is especially important for the feature extraction technique using this projection basis, since sparse axes insure that only a few pixels affect the value of each feature component. Such an approach should be robust to a number of image degradations including occlusion. A visual example of the sparseness of the projection basis is presented in Fig. 1, where a sample orthogonalized vector $\mathbf{w}_i$ in image form (Fig. 1 left) and its 3D surface plot (Fig. 1 right) are depicted.

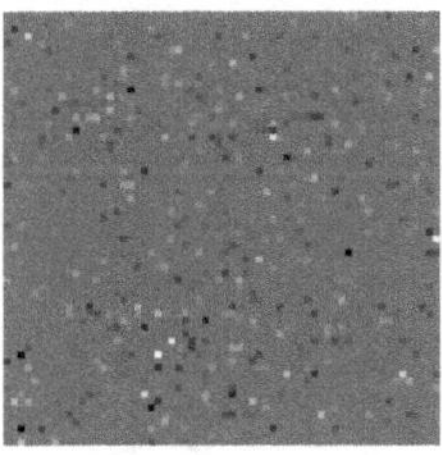
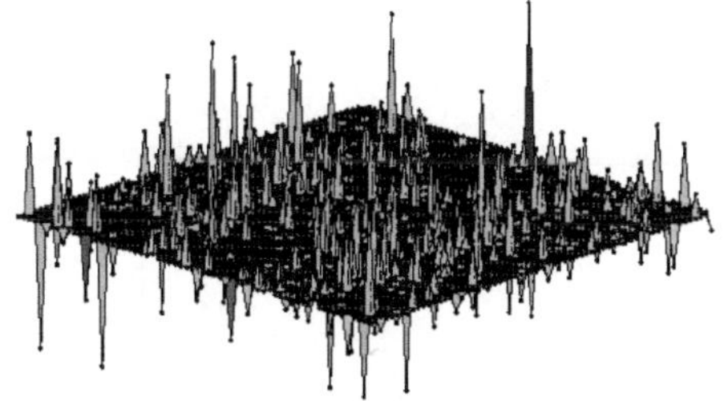

Fig. 1. A visual example of a sparse projection axis: in image form (left), as a surface (right)

3 Experiments

The feasibility of the SPA technique was empirically assessed using the XM2VTS database, which has long been the standard face-image database for evaluating novel authentication algorithms [12]. The database comprises 2360 images of 295 distinct subjects of different gender, age and race. The images were recorded in four separate sessions over a period of five months with the recording setup featuring controlled conditions, i.e., uniform background, controlled illumination, etc. Due to this recording setup, the variability in the images is induced mainly by the temporal factor. Thus, images of the same subjects differ in appearance due to changes in hairstyle, head-pose, presence or absence of make-up, glasses, etc.

For the experiments we followed the first configuration of the established experimental protocol associated with the XM2VTS database [12]. The protocol, known as the Lausanne protocol, partitions the subjects of the database into two disjoint groups of clients (200 subjects) and impostors (95 subjects) and further divides the images of these two groups into image sets used for: *(i) training and enrollment* - images which are used to train feature extractors and build client models/templates in form of mean feature vectors, *(ii) evaluation* - images which are used to determine the operating point, i.e., the decision threshold, of the face verification system and to define any potential parameters of the feature extractor (e.g., number of features, selection of features, etc.), and *(iii) testing* - images which are used to determine the verification error rates in real operating conditions.

While the first image set features only images belonging to the client group, the latter two image sets comprise images belonging to both the client and the impostor groups. These images are used in our assessment to determine the two error rates commonly exploited to quantify the performance of a given face verification system, namely, the false rejection and false acceptance error rates (FRR and FAR). These two error rates are defined as the relative frequency with which a face verification system falsely rejects a legitimate- and falsely accepts an impostor-identity-claim [7].

Unfortunately, both the FAR and the FRR depend on the value of the so-called decision threshold Δ and, hence, selecting a threshold which ensures a small value

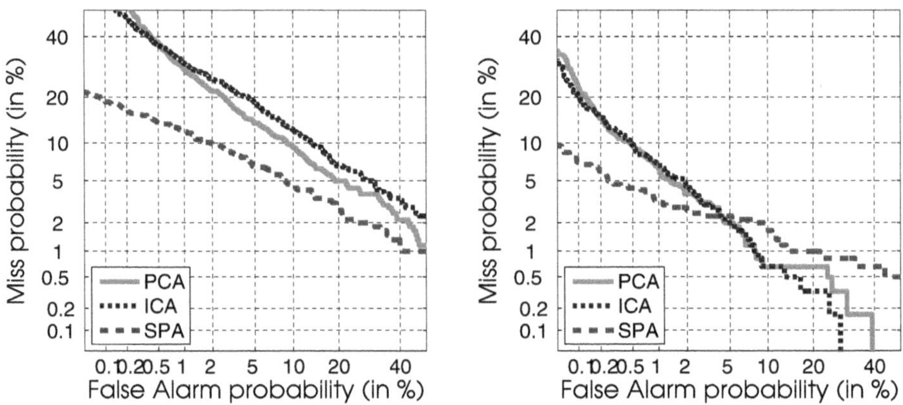

Fig. 2. DET curves generated during the experiments on the evaluation image set

of FRR inevitably results in a high value of the FAR and vice versa, selecting a threshold which ensures a small value of the FAR results in a high value of the FRR. To compare the performance of two face verification systems an operating point has to be determined or the error rates should be plotted against a number of values of the decision threshold, thus generating a performance curve. Here we choose the latter approach and present our verification results as performance curves [7].

The employed protocol results in 600 client and 40000 impostor verification attempts in the evaluation stage to determine the decision thresholds needed to construct the performance curves, and 400 client and 112000 impostor verification attempts in the test stage to determine the final performance of the assessed techniques.

In all experiments we assume that the facial images are already localized and scaled to a standard size of 64×64 pixels. To compensate for any potential illumination induced appearance changes we further normalize the images by applying histogram equalization followed by a zero mean and unit variance normalization. The presented procedure is employed on all three color components of the YIQ color space into which the images of the XM2VTS database are transformed. It should be noted that the YIQ color space [13] rather than the RGB color space is used in the experiments to reduce the correlations between the individual color components and to make additional discriminatory information not present in the commonly adopted grey-scale images available to the feature extraction techniques. As a final step, we construct the SPA feature vector (of size $1 \times 3N$) of each face image by simply concatenating the feature vectors of the three color components. In addition to the experiments on the color images, we also provide baseline comparisons with experiments performed on grey-scale images.

The goal of our first series of verification experiments is to assess the performance of the proposed SPA feature extraction technique on the evaluation image sets and to compare the obtained error rates to that of some baseline feature extraction techniques. To this end, we implement the principal component analysis (PCA)[14] and independent component analysis (ICA - architecture I)[3] face verification techniques and test them on the XM2VTS database. For all assessed techniques we employ the nearest neighbor (to the mean) classifier in conjunction with the cosine similarity measure for the matching stage. The classifier is chosen for the comparison, as it represents a non-parametric classifier and hence does not introduce any classifier-dependant bias to the results.

From the DET curves in Fig. 2[1], which plot the false acceptance error rate against the false rejection error rate at various values of the decision threshold, we can find that the proposed technique significantly outperforms the PCA and ICA methods on the grey scale images (Fig. 2 (left)), and performs better at the lower values of the FAR on the color images (Fig. 2 (right)). The reason for the good performance of the SPA technique can be found in the way the projection axes are computed. As the label data matrix encodes the class membership of the training face pattern vectors and all training samples from each class are mapped to the same encoded label, the proposed technique compresses the intra-class scatter similar to the popular linear discriminant analysis technique.

It was argued in [15] that DET curves cannot be used to effectively compare two face recognition systems, since in real operating conditions an operational threshold has to be set and the performance with this threshold on unseen data might differ from the performance achieved when setting the threshold. To this end, the so-called expected performance curves (EPC) were introduced in [15]. To construct these curves two data sets of impostor and client images are needed. The first, the evaluation data set, is used to set a decision threshold which minimizes the following weighted error rate: $\alpha FAR(\Delta)+(1-\alpha)FRR(\Delta)$, where the parameter α controls the relative importance of the two error rates FAR and FRR. This threshold is then used on the second, the test image set, to determine the value of the half total error rate (HTER) defined as HTER = (FAR+FRR)/2. When plotting the obtained HTER against a number of values of α, we obtain an example of the EPC.

In our second series of experiments we produce EPC curves for all three tested techniques. The generated curves are shown in the graphs labeled as Fig. 3(a) and Fig. 3(b) for the grey-scale and color images, respectively. We can see that as with the evaluation image set, the proposed technique performs the best, followed by a similar performance of the remaining PCA and ICA techniques. On the remaining two graphs of Fig. 3 (i.e., on the graphs labeled as Fig. 3(c) and Fig. 3(d)) we present the results of the assessment on grey-scale and color images, where the goal was to test the robustness of the two techniques to partial occlusion of the facial region. Here, the facial images were occluded by setting at most 30% of the pixels in each image to zero. The location and the size of the occlusion were chosen randomly as shown in the examples presented in Fig. 4.

From the results we can see that the proposed technique results in much better verification rates than the PCA technique, and even the ICA technique,

[1] Note that the axis labels in Fig. 2 *Miss probability* and *False Alarm Probability* correspond to the false acceptance and false rejection error rates, respectively.

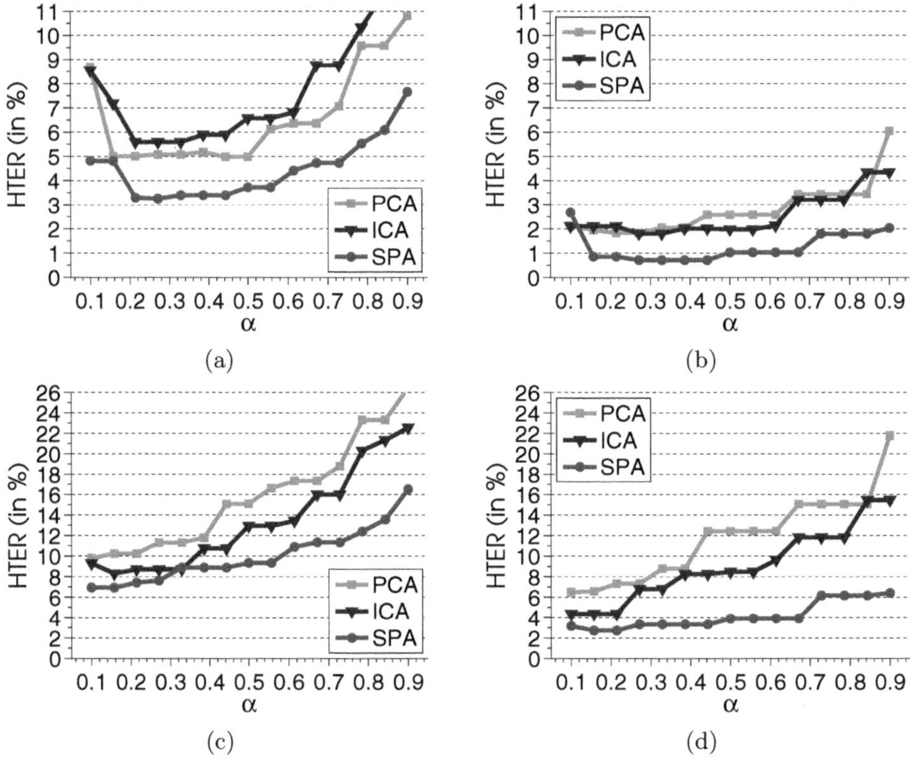

(a) (b)

(c) (d)

Fig. 3. EPC curves generated during the experiments: (a) on grey-scale images of the original test set, (b) on color images of the original test set, (c) on grey-scale images of the occluded test set, (d) on color images of the occluded test set

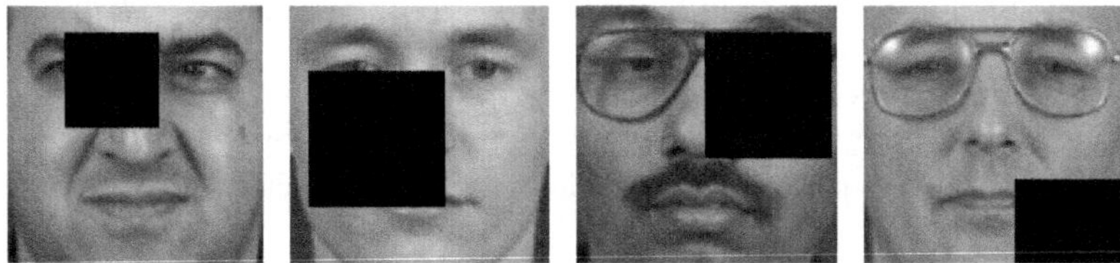

Fig. 4. Examples of occluded images

which, similarly as SPA, represents a local method. While all methods deteriorate in their performance, the deterioration is more extreme for the PCA and ICA techniques.

4 Conclusion

In the paper we have presented a novel feature extraction technique for face recognition. The technique, which uses a sparse projection basis to reduce the dimensionality of the face pattern vectors, was assessed in a series of face verification experiments performed on the XM2VTS database. In the experiments it was shown to outperform the popular PCA and ICA techniques and to perform reasonably well even if parts of the facial images are occluded.

References

1. Zhao, W., Chellappa, R., Phillips, J., Rosenfeld, A.: Face Recognition: A Literature Survey. ACM Computing Surveys 35(4), 399–458 (2003)
2. Leonardis, A., Bischof, H.: Robust Recognition Using Eigenimages. Computer Vision and Image Understanding 78(1), 99–118 (2000)
3. Bartlett, M.S., Movellan, J.R., Sejnowski, T.J.: Face Recognition by Independent Component Analysis. IEEE Transactions on Neural Networks 13(6), 1450–1464 (2002)
4. Wright, J., Yang, A.Y., Ganesh, A., Sastry, S.S., Ma, Y.: Robust Face Recognition via Sparse Representation. IEEE Transactions on Pattern Analysis and Machine Intelligence 31(2), 210–227 (2009)
5. Guillamet, D., Vitria, J.: Non-negative Matrix Factorization for Face Recognition. In: Escrig, M.T., Toledo, F.J., Golobardes, E. (eds.) CCIA 2002. LNCS (LNAI), vol. 2504, pp. 336–344. Springer, Heidelberg (2002)
6. Bociu, I., Pitas, I.: A New Sparse Image Representation Algorithm Applied to Facial Expression Recognition. In: Proc. of the IEEE Workshop on Machine Learning for Signal Processing, pp. 539–548 (2004)
7. Štruc, V., Mihelič, F., Pavešić, N.: Face Authentication Using a Hybrid Approach. Journal of Electronic Imaging 17(1), 1–11 (2008)
8. Štruc, V., Gajšek, R., Mihelič, R., Pavešić, N.: Using Regression Techniques for Coping with the One-sample-size Problem of Face Recognition. Electrotechnical Review 76(1-2), 7–12 (2009)
9. Rosipal, R.: Kernel Partial Least Squares for Nonlinear Regression and Discrimination. Neural Ntwork World 13(3), 291–300 (2003)
10. Chen, S., Donoho, D., Saunders, M.: Atomic Decomposition by Basis Pursuit. SIAM Review 43(1), 129–159 (2001)
11. Candes, E., Romberg, J., Tao, T.: Stable Signal Recovery from Incomplete and Inaccurate Measurements. Comm. Pure and Applied Math. 59(8), 1207–1223 (2006)
12. Messer, K., Matas, J., Kittler, J., Luettin, J.: XM2VTSDB: the Extended M2VTS Database. In: Proc. of AVBPA, pp. 72–77 (1999)
13. Liu, Z., Liu, C.: Fusion of the Complementary Discrete Cosine Features in the YIQ Color Space for Face Recognition. Computer Vision and Image Understanding 111(3), 249–262 (2008)
14. Turk, M., Pentland, A.: Eigenfaces for Recognition. Journal of Cognitive Neurosicence 3(1), 71–86 (1991)
15. Bengio, S., Marithoz, J.: The Expected Performance Curve: A New Assessment Measure for Person Authentication. In: Proc. of Odyssey, pp. 279–284 (2004)

Face Detection in Low-Resolution Color Images

Jun Zheng, Geovany A. Ramírez, and Olac Fuentes

Computer Science Department,
University of Texas at El Paso,
El Paso, Texas, 79968, U.S.A.

Abstract. Most face detection methods require high or medium resolution face images to attain satisfactory results. However, in many surveillance applications, where there is a need to image wide fields of view, faces cover just a few pixels, which makes their detection difficult. Despite its importance, little work has been aimed at providing reliable detection at these low resolutions. In this work, we study the relationship between resolution and the automatic face detection rate with the Modified Census Transform, one of the most successful algorithms for face detection presented to date, and propose a new Color Census Transform that provides significantly better results than the original when applied to low-resolution color images.

1 Introduction

Face detection is an important first step in several computer vision applications, including face recognition, tracking, and analysis of facial expressions, human-computer interaction, tracking, object recognition and scene reconstruction. Face detection is a difficult task, due to different factors such as varying sizes, orientations, poses, facial expressions, occlusions and lighting conditions [1]. In recent years, numerous methods for detecting faces working effectively under these various conditions have been proposed [2,3,4,5,6]. These methods usually detect face images that contain at least 20×20 pixels, but provide poor results at lower resolutions.

However, in surveillance systems, the regions of interests are often impoverished or blurred due to the large distance between the camera and the objects, or the low spatial resolution of devices. Figure 1 illustrates an image collected from a surveillance video. In this image, faces cover very small areas of the image (about 8×8 pixels), which makes face recognition and analysis difficult. A few works address face detection in lower resolution images [2,7], but the accuracies obtained are still low.

Torralba et al. [8] first studied psychologically the task of face detection in low-resolution images by humans. They investigated how face detection changes as a function of available image resolution, whether the inclusion of local context in the form of a local area surrounding the face improves face detection performance, and how contrast negation and image orientation changes affect face detection. Their results suggest that in low-resolution the internal facial features become rather indistinct and lose their effectiveness as good predictors of whether a pattern is a face or not, so that using upper-body images is better than using only face images for human beings to recognize faces images.

A. Campilho and M. Kamel (Eds.): ICIAR 2010, Part I, LNCS 6111, pp. 454–463, 2010.

Fig. 1. Faces in surveillance images

Kruppa and Schile [2] used the knowledge of Torralba's experiment and applied a local context detector, which is trained with instances that contain an entire head of one person, neck and part of the upper body, for automatic face detection in low-resolution images. They applied wavelet decomposition to capture most parts of the upper body's contours, as well as the collar of the shirt and the boundary between forehead and hair, while the facial parts such as eyes and mouth are hardly discernible in the wavelet transform of low-resolution images. In their experiments on two large data sets they found that using local context could significantly improve the detection rate, particularly in low resolution images.

Hayashi and Hasegawa [7] proposed a new face detector based on Haar features along with the a conventional AdaBoost-based classifier for low-resolution images. Their detector used four techniques to improve detection at low resolutions: using upper-body images, expansion of input image, frequency-band limitation, and combination of two detectors. This extensions allowed them to improved the face detection rate from 39% to 71% for 6×6 pixel faces of MIT+CMU frontal face test set.

In this work, we study the relationship between resolution and the automatic face detection rate with Modified Census Transform [3] and propose a new extended census transform that works better than the original one in low-resolution color images for object detection. We present experimental results showing the application of our method with the Georgia Tech color frontal face database. The experiments show that our method can attain better results than other methods using low-resolution color images as input.

2 Related Work

One of the milestones in face detection was the work of Rowley et al. who developed a frontal face detection system that scanned every possible region and scale of an image using a window of 20×20 pixels [9]. Each window is pre-processed to correct for varying lighting, then, a retinally connected neural network is used to process the pixel intensity levels of each window to determine if it contains a face. In later works, they provided invariance to rotation perpendicular to the image plane by means of another neural network that determined the rotation angle of a region, which was then be rotated by the negative of that angle and then given to the original neural network for classification [10].

Convolution neural networks, which are highly modular multilayer feedforward neural networks that are invariant to certain transformations, were originally proposed by [11] with the goal of performing handwritten character recognition. Years later Garcia and Delakis propose a novel face detection approach based on a convolutional neural architecture, to robustly detect highly variable face patterns in uncontrolled environments. [12,5].

Schneiderman and Kanade detected faces and cars from different view points using specialized detectors [13]. For faces, they used 3 specialized detectors for frontal, left profile, and right profile views. For cars, they used 8 specialized detectors. Each specialized detector is based on histograms that represent the wavelet coefficients and the position of the possible object, and then they used a statistical decision rule to eliminate false negatives.

Jesorsky et al. based their face detection system on edge images [14]. They used a coarse-to-fine detection using the Hausdorff distance between a hand-drawn model and the edge image of a possible face. In [15], the face model used by Jesorsky et al. was optimized using genetic algorithms, increasing slightly the correct detection rate.

Viola et al. [16] used Haar features in their face detection systems. They first introduced the integral image to compute Haar features rapidly. They also proposed an efficient modified version of the Adaboost algorithm that selects a small number of critical visual features in face images of 24×24 pixels, and introduced a cascade of classifiers to discard background regions of the image very quickly while spending more computation on regions of interest. Based on their work, we use a variation of the cascade of strong classifiers with the modified census transform rather than Haar features.

Sung [17] first proposed a simple lighting model followed by histogram equalization. Using a database of face window patterns and non-face window patterns, they construct a distribution-based model of face patterns in a masked 19×19 dimensional normalized image vector space. For each new window pattern to be classified, they compute a vector of distances from the new window pattern to the window pattern prototypes in the masked 19×19 pixel image feature space. Then based on the vector of distance measurements to the window pattern prototypes, they train a multi-layer perceptron (MLP) net to identify the new window as a face or non-face. Schneiderman [18] choose a functional form of the posterior probability function that captures the joint statistics of local appearance and position on the object as well as the statistics of local appearance in the visual world at large. Viola [16] applied a simpler normalization to zero mean and unit variance on the analysis window.

Wu et al. detected frontal and profile faces with arbitrary in-plane rotation and up to 90-degree out-of-plane rotation [19]. They used Haar features and a look-up table to develop strong classifiers. To create a cascade of strong classifiers they used Real AdaBoost, an extension to the conventional AdaBoost. They built a specialized detector for each of 60 different face poses. To simplify the training process, they took advantage of the fact that Haar features can be efficiently rotated by 90 degrees or reversed, thus they only needed to train 8 detectors, while the other 52 can be obtained by rotating or inverting the Haar features.

Fröba and Ernst [3] used inherently illumination-invariant local structure features for real-time face detection. They proposed the Modified Census Transform (MCT), which is a non-parametric local transform, for efficient computation. Using these local structure features and an efficient four-stage classifier, they obtained results that were comparable to the best systems presented to date.

1	0	1
0	1	1
1	0	0

1	1	1
0	1	1
1	0	0

1	0	1
0	0	0
1	0	0

1	1	1
0	1	0
0	1	0

Fig. 2. A randomly chosen subset of Local Structure Kernels

Also based on local structures, Dalal [20] introduced grids of locally normalized Histograms of Oriented Gradients (HOG)as descriptors for object detection in static images. The HOG descriptors are computed over dense and overlapping grids of spatial blocks, with image gradient orientation features extracted at fixed resolution and gathered into a high dimensional feature vector. They are designed to be robust to small changes in image contour locations and directions, and significant changes in image illumination and color, while remaining highly discriminative for overall visual form.

3 Image Features

3.1 The Modified Census Transform

The Modified Census Transform [3] is an extension of the Census Transform first introduced by Zabih and Woodfill [21].

Let (r, c) be a pixel position in image I. Let $N(r, c)$ be an ordered set containing the pixels in the 3×3 neighborhood of pixel (r, c) in I. The MCT generates a string of nine bits representing which pixels in $N(r, c)$ have an intensity that is greater than the average intensity in $N(r, c)$.

$$N(r,c) = \{(r',c')|r' \in \{r-1, r, r+1\}, c' \in \{c-1, c, c+1\}\}$$

$$\mu_I(N(r,c)) = \frac{\sum_{(r',c') \in N(r,c)} I(r',c')}{9}$$

$$MCT_I(r,c) = \bigoplus_{(r',c') \in N(r,c)} \xi(I(r',c'), \mu_I(N(r,c)))$$

where $\bigoplus$ denotes concatenation and ξ is and indicator function such that:

$$\xi(x,y) = \begin{cases} 0 \text{ if } x \leq y \\ 1 \text{ otherwise.} \end{cases}$$

3.2 The Color Census Transform

The 9-bit modified census transform is defined for gray-scale images. It has been shown to work very well when detecting faces of 24×24 pixels or above [3], but as resolution is decreased performance rapidly degrades.

We propose an extended 12-bit Color Census Transform that takes advantage of color information and yields accurate face detection even when resolution is low. We augment the MCT with three additional bits to describe the color information of each pixel's neighborhood.

Let $I_C = \langle I_R, I_G, I_B \rangle$ be a color image, where I_R, I_G, and I_B are the red, green and blue channels of I_C, and the intensity image I is given by $I = (I_R + I_G + I_B)/3$. Then:

$$\mu_{I_R}(N(r,c)) = \frac{\sum_{(r',c') \in N(r,c)} I_R(r',c')}{9}$$

$$\mu_{I_G}(N(r,c)) = \frac{\sum_{(r',c') \in N(r,c)} I_G(r',c')}{9}$$

$$\mu_{I_B}(N(r,c)) = \frac{\sum_{(r',c') \in N(r,c)} I_B(r',c')}{9}$$

The additional three bits are given by:

$$b_R(r,c) = \xi(\mu_I(N(r,c)), \mu_{I_R}(N(r,c)))$$

$$b_G(r,c) = \xi(\mu_I(N(r,c)), \mu_{I_G}(N(r,c)))$$

$$b_B(r,c) = \xi(\mu_I(N(r,c)), \mu_{I_B}(N(r,c)))$$

where ξ is the indicator function as before.

Then the 12-bit Color Census Transform is given by:

$$CCT_{I_C(r,c)} = [MCT_I(r,c), b_R(r,c), b_G(r,c), b_B(r,c)]$$

3.3 Training of Classifiers

This section describes an algorithm for constructing a cascade of classifiers. We create a cascade of strong classifiers using a variation of AdaBoost algorithm used by Viola et al. [16], where we only use three stages in low resolution face detection as displayed in Figure 3. Because of using modified census transform and less stages, our method is as powerful as but more efficient than Viola's.

Stages in the cascade are constructed by training classifiers using a version of boosting algorithms similar to Fröba's [3]. The pseudo-code of boosting is in Table 1. Boosting terminates when the minimum detection rate and the maximum false positive rate per stage in the cascade are attained. If the target false positive rate is achieved, the algorithm ends. Otherwise, all negative examples correctly classified are eliminated and the training set is balanced adding negative examples using bootstrapping. The pseudo-code of bootstrapping is in Table 2. With the updated training set, all the weak classifiers are retrained.

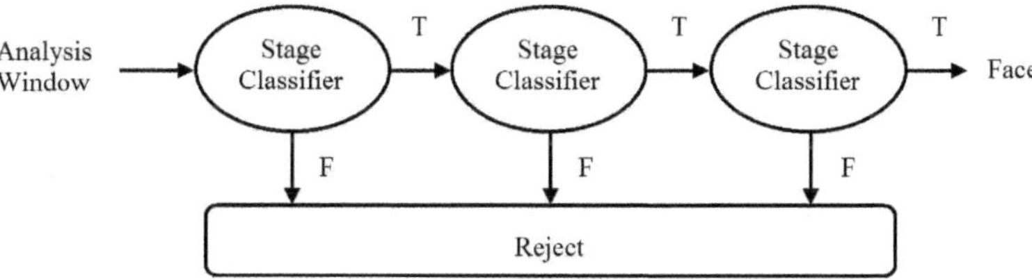

Fig. 3. The cascade has three stages of increasing complexity. Each stage has the ability to reject the current analysis window as background or pass it on to the next stage.

Table 1. Pseudo-code of boosting algorithm

- Given training examples $(\Omega_1, l_1), \ldots, (\Omega_n, l_n)$, where $l_i = 0, 1$ for negative and positive examples respectively.
- Initialize weights $\omega_{1,i} = \frac{1}{2m}, \frac{1}{2l}$ for $l_i = 0, 1$ respectively, where m and l are the number of negatives and positives respectively
- For $k = 1, \ldots, K$:
  1. Normalize the weights,
  $$\omega_{k,i} = \frac{\omega_{k,i}}{\sum_{i=1}^{n} \omega_{k,j}}$$
  2. Generate a weak classifier c_k for a single feature k, with error $\epsilon_k = \sum_i \omega_i |c_k(\Omega_i) - l_i|$
  3. Compute $\alpha_k = \frac{1}{2} \ln(\frac{1-\epsilon_k}{\epsilon_k})$
  4. Update the weights,
  $$\omega_{k+1,i} = \omega_{k,i} \times \begin{cases} e^{-\alpha_k} & c_k(\Omega_i) = l_i \\ 1 & otherwise \end{cases}$$
- The final strong classifier is:
$$C(\Omega) = \begin{cases} 1 \sum_{k=1}^{K} \alpha_k c_k(\Omega(k)) > \frac{1}{2} \sum_{k=1}^{K} \alpha_k \\ 0 \; otherwise \end{cases}$$
where K is the total number of features

A few weak classifiers are combined forming a final strong classifier. The weak classifiers consist of histograms of g_k^p and g_k^n for the feature k. Each histogram holds a weight for each feature. To build a weak classifier, we first count the kernel index statistics at each position. The resulting histograms determine whether a single feature belongs to a face or non-face. The single feature weak classifier at position k with the lowest boosting error e_t is chosen in every boosting loop. The maximal number of features on each stage is limited with regard to the resolution of analysis window. The definition of histograms is as follows:

$$\begin{cases} g_k^p(r) = \sum_i I(\Omega_i(k) = r) I(l_i = 1) \\ g_k^n(r) = \sum_i I(\Omega_i(k) = r) I(l_i = 0) \end{cases}, k = 1, \ldots, K, r = 1, \ldots, R$$

where $I(.)$ is the indicator function that takes 1 if the argument is true and 0 otherwise. The weak classifier for feature k is:

$$c_k(\Omega_i) = \begin{cases} 1 & g_k^p(\Omega_i(k)) > g_k^n(\Omega_i(k)) \\ 0 & otherwise \end{cases}$$

where $\Omega_i(k)$ is the kth feature of the ith face image, c_k is the weak classifier for the kth feature. The final stage classifier $C(\Omega)$ is the sum of all weak classifiers for the chosen features.

$$C(\Omega) = \begin{cases} 1 & \sum_{k=1}^{K} \alpha_k c_k(\Omega(k)) > \frac{1}{2}\sum_{k=1}^{K} \alpha_k \\ 0 & otherwise \end{cases}$$

where c_k is the weak classifier, $C(\Omega)$ is the strong classifier, and $\alpha_k = \frac{1}{2}\ln(\frac{1-\epsilon_k}{\epsilon_k})$.

Table 2. Pseudo-code of bootstrapping algorithm

- Set the minimum true positive rate, T_{min}, for each boosting iteration.
- Set the maximum detection error on the negative dataset, I_{neg} , for each boost-strap iteration.
- $P =$ set of positive training examples.
- $N =$ set of negative training examples.
- $K =$ the total number of features.
- While $I_{err} > I_{neg}$
  - While $T_{tpr} < T_{min}$
    - ∘ For $k = 1$ to K
      - Use P and N to train a classifier for a single feature.
      - Update the weights.
    - ∘ Test the classifier with 10-fold cross validation to determine T_{tpr}.
  - Evaluate the classifier on the negative set to determine I_{err} and put any false detections into the set N.

4 Experimental Results

The training data set consists of 6000 faces and 6000 randomly cropped non-faces. Both the faces and non-faces are down-sampled to 24 × 24, 16 × 16, 8 × 8, and 6 × 6 pixels for training detectors for different resolutions.

To test the detector, we use the Georgia Tech face database, which contains images of 50 people. All people in the database are represented by 15 color JPEG images with cluttered background taken at resolution 640 × 480 pixels. The average size of the faces in these images is 150 × 150 pixels. The pictures show frontal and tilted faces with different facial expressions, lighting conditions and scale.

We use a cascade of three strong classifiers using a variation of AdaBoost algorithm used by Viola et al. The number of Boosting iterations is not fixed. The boosting process continues until the detection rate reaches the minimum detection rate. For detecting faces of 24 × 24 pixels, the analysis window is of size 22 × 22. The maximum number of features for each stage is 20, 300, and 484 respectively. For detecting faces of 16 × 16 pixels, the analysis window is of size 14 × 14. The maximum number of features

Table 3. Face detection using Color Census Transform

Georgia tech face database			
Modified census transform		12-bit MCT	
		Detection rate	False alarms
Boosting	24×24	99.5%	1
	16×16	97.2%	10
	8×8	95.0%	136
	6×6	80.0%	149

Table 4. Face detection using 9-bit modified census transform

Georgia tech face database			
Modified census transform		9-bit MCT	
		Detection rate	False alarms
Boosting	24×24	99.2%	296
	16×16	98.4%	653
	8×8	93.5%	697
	6×6	68.8%	474

a) b)

Fig. 4. a) Sample detection results on 24×24 pixel face images; b) Detection results on 16×16 images

for each stage is 20, 150, and 196 respectively. For detecting faces of 8×8 pixels, the analysis window is of size 6×6. The maximum number of features for each stage is 20, 30, and 36 respectively. For detecting faces of 6×6 pixels, the analysis window is of size 4×4. The maximum number of features for each stage is 5, 10, and 16 respectively.

Tables 3 and 4 show the performance of the MCT and the CCT when applied to these datasets. As expected, for both types of features, decreases in resolution lead to loss of accuracy. From Table 3 and Table 4 we can conclude that, for every resolution used, the Color Census Transform yields better results than the Modified Census Transform in terms of detection rate and false detections. It can also be seen that the difference in performance between the two types of features increases as resolution decreases. This leads to the conclusion that color information is more important in situations where there is not enough information about structure to perform reliable detection, as is the case with very low resolution face images.

In Figures 4 and 5 we show some representative results from the Georgia Tech face database. The figures show results using face region sizes ranging from 6×6 to 24×24 pixels

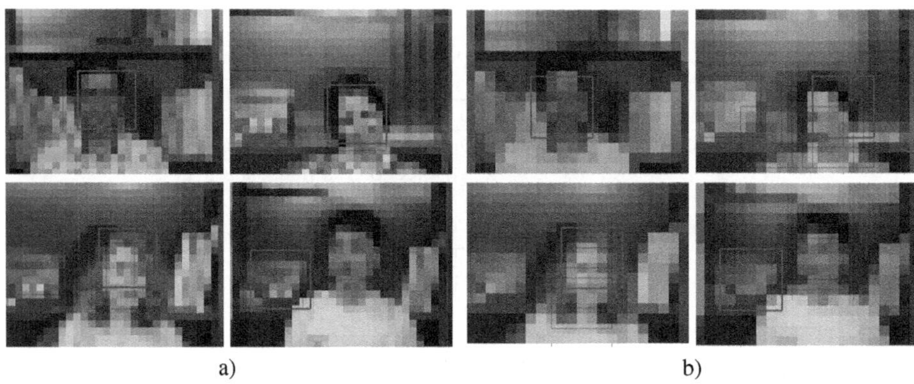

a) b)

Fig. 5. a) Sample detection results on 8×8 pixel face images; b) Detection results on 6×6 images

5 Conclusion

In this paper, we presented a 12-bit Color Census Transform that works better than the original 9-bit Modified Census Transform in low-resolution color images for object detection. According to the experiments, our method, by taking advantage of color information, can attain better results than the original 9-bit MCT detecting faces in low-resolution color images.

For future work, we will apply our system in other object detection problems such as car detection, road detection, and hand gesture detection. In addition, we plan to perform experiments using other boosting algorithms such as Float- Boost and Real AdaBoost to further improve the performance of our system in low-resolution object detection. Finally, we will take advantage of the low computational requirements of our methods and explore their hardware implementation in embedded low-power surveillance systems.

References

1. Yang, M.H., Kriegman, D.J., Ahuja, N.: Detecting faces in images: A survey. IEEE Transactions on Pattern Analysis and Machine Intelligence 24, 34–58 (2002)
2. Kruppa, H., Schiele, B.: Using local context to improve face detection. In: Proceedings of the British Machine Vision Conference, Norwich, England, pp. 3–12 (2003)

3. Fröba, B., Ernst, A.: Face detection with the modified census transform. In: Sixth IEEE International Conference on Automatic Face and Gesture Recognition, Erlangen, Germany, pp. 91–96 (2004)
4. Gevers, T., Stokman, H.: Robust histogram construction from color invariants for object recognition. IEEE Transactions on Pattern Analysis and Machine Intelligence 26, 113–118 (2004)
5. Garcia, C., Delakis, M.: Convolutional face finder: A neural architecture for fast and robust face detection. IEEE Transactions on Pattern Analysis and Machine Intelligence 26, 1408–1423 (2004)
6. Ramírez, G.A., Fuentes, O.: Face detection using combinations of classifiers. In: Proceedings of the 2nd Canadian Conference on Computer and Robot Vision, Victoria, B.C., Canada (2005)
7. Hayashi, S., Hasegawa, O.: Robust face detection for low-resolution images. Journal of Advanced Computational Intelligence and Intelligent Informatics 10, 93–101 (2006)
8. Torralba, A., Sinha, P.: Detecting faces in impoverished images. Technical Report 028, MIT AI Lab, Cambridge, MA (2001)
9. Rowley, H.A., Baluja, S., Kanade, T.: Neural network-based face detection. IEEE Transactions on Pattern Analysis and Machine Intelligence 20, 23–38 (1998)
10. Rowley, H.A., Baluja, S., Kanade, T.: Rotation invariant neural network-based face detection. In: Proceedings of 1998 IEEE Conference on Computer Vision and Pattern Recognition, Santa Barbara, CA, pp. 38–44 (1998)
11. LeCun, Y., Haffner, P., Bottou, L., Bengio, Y.: Object recognition with gradient-based learning. In: Forsyth, D. (ed.) Shape, Contour and Grouping in Computer Vision. Springer, Heidelberg (1989)
12. Garcia, C., Delakis, M.: A neural architecture for fast and robust face detection. In: IEEE IAPR International Conference on Pattern Recognition, Quebec City, pp. 40–43 (2002)
13. Schneiderman, H., Kanade, T.: A statistical model for 3-d object detection applied to faces and cars. In: IEEE Conference on Computer Vision and Pattern Recognition. IEEE, Los Alamitos (2000)
14. Jesorsky, O., Kirchberg, K., Frischholz, R.W.: Robust face detection using the hausdorff distance. In: Bigun, J., Smeraldi, F. (eds.) AVBPA 2001. LNCS, vol. 2091, pp. 90–95. Springer, Heidelberg (2001)
15. Kirchberg, K.J., Jesorsky, O., Frischholz, R.W.: Genectic model optimization for hausdorff distance-based face localization. In: Tistarelli, M., Bigun, J., Jain, A.K. (eds.) ECCV 2002. LNCS, vol. 2359, pp. 103–111. Springer, Heidelberg (2002)
16. Viola, P., Jones, M.: Rapid object detection using a boosted cascade of simple features. In: Proceedings of 2001 IEEE International Conference on Computer Vision and Pattern Recognition, pp. 511–518 (2001)
17. Sung, K.K.: Learning and Example Seletion for Object and Pattern Detection. PhD thesis, Massachusetts Institute of Technology (1996)
18. Schneiderman, H., Kanade, T.: Probabilistic modeling of local appearence and spatial relationship for object recognition. In: International Conference on Computer Vision and Pattern Recognition. IEEE, Los Alamitos (1998)
19. Wu, B., Ai, H., Huang, C., Lao, S.: Fast rotation invariant multi-view face detection based on real adaboost. In: Sixth IEEE International Conference on Automatic Face and Gesture Recognition (2004)
20. Dalal, N.: Finding People in Images and Videos. PhD thesis, Institut National Polytechnique de Grenoble (2006)
21. Zabih, R., Woodfill, J.: A non-parametric approach to visual correspondence. IEEE Transactions on Pattern Analysis and Machine Intelligence (1996)

Author Index

GPSR Compliance

The European Union's (EU) General Product Safety Regulation (GPSR) is a set of rules that requires consumer products to be safe and our obligations to ensure this.

If you have any concerns about our products, you can contact us on ProductSafety@springernature.com

In case Publisher is established outside the EU, the EU authorized representative is:

Springer Nature Customer Service Center GmbH
Europaplatz 3
69115 Heidelberg, Germany

Batch number: 09473985

Printed by Printforce, the Netherlands